Recent Advances in Seaweed Biotechnology

Nitin Trivedi • C. R. K. Reddy • Alan T. Critchley
Editors

Recent Advances in Seaweed Biotechnology

Biomass, Emerging Applications and Bioeconomy

Springer

Editors
Nitin Trivedi
Marine Biotechnology
Gujarat Biotechnology University
Gandhinagar, Gujarat, India

C. R. K. Reddy
Centre for Energy Biosciences
Institute of Chemical Technology
Mumbai, India

Alan T. Critchley
Verschuren Centre
Sydney, Nova Scotia, Canada

ISBN 978-981-96-0518-7 ISBN 978-981-96-0519-4 (eBook)
https://doi.org/10.1007/978-981-96-0519-4

This Springer imprint is published by the registered company Springer Nature Singapore Pte Ltd.
The registered company address is: 152 Beach Road, #21-01/04 Gateway East, Singapore 189721, Singapore

Preface

In the context of climate change, seaweeds have gained considerable attention worldwide. For centuries, a few Southeast Asian countries have been using seaweeds in their food preparations and some for extraction of commercially valuable hydrocolloids (agar, carrageenan, and alginates). In modern times with the advancement of scientific knowledge, there has been a significant transformation in their utilization beyond food and hydrocolloid applications, creating several new opportunities in both the biomass production and processing sectors. The use of seaweeds for making plant biostimulants for agriculture applications, biodegradable alternatives to plastic, animal feed supplements to reduce methane emissions in ruminants, and protein alternatives to animal meat are some of the new products for emerging markets with potential commercial and environmental benefits. In many developing economies in maritime countries, seaweed trade contributes to the socioeconomic development of coastal communities and promotes the bioeconomy concepts.

To advance the research and aid the researchers working in developing innovative processes and products for new and developing markets, the need was felt to bring out a consolidated document giving the latest and insightful perspectives on each topic. This book has covered a wide range of topics ranging from seaweed-based protein for human and animal nutrition, antioxidants, biomaterials, PHA from seaweed-associated bacteria, bioprospecting of secondary metabolites, polysaccharide digestion by human get microbes, natural colorants to nutraceuticals. There are also topics dedicated to seaweed farming, bioremediation, and circular economy aspects of seaweeds. Undoubtedly the information captured in this book will present you with the latest status in respective fields and point out directions for future research benefiting the young researchers working in the field of seaweeds.

We are confident that this book will be found as a valuable reference book for all seaweed stakeholders ranging from policymakers, entrepreneurs, industry representatives, investors, students, citizen scientists, and professionals.

Gandhinagar, Gujarat, India — Nitin Trivedi
Mumbai, Maharashtra, India — C. R. K. Reddy
Sydney, NS, Canada — Alan T. Critchley

Contents

About the Editors

Nitin Trivedi is an Assistant Professor in the Department of Marine Biotechnology at Gujarat Biotechnology University in Gandhinagar, Gujarat, India. He received his PhD in Biological Science from CSIR-CSMCRI in Bhavnagar, Gujarat, India, in 2015. Later, he worked as a Postdoctoral Fellow at DBT ICT CEB, Institute of Chemical Technology, Mumbai, from 2015 to 2017 and as a DST INSPIRE Faculty from 2017 to 2022. Since 2009, Dr. Trivedi has been working in the field of marine biotechnology, with a focus on seaweed biotechnology and marine microbial biotechnology. Dr. Trivedi has vast experience in seaweed biofuel, bioproducts, and biorefinery, as well as seaweed growth optimization, phycoremediation, and marine microbial enzymes. He has 35 research articles, including book chapters and reviews in SCI journals, and two patents to his credit. Dr. Trivedi has received several national and international fellowships/awards like DST INSPIRE Faculty, DSKPDF, CSIR-SRF, ARO PDF in Israel, PDF at Dalhousie University, Canada, Visiting Scientist at the University of Algarve, Portugal, and guest researcher at the University of Messina, Italy. He has 7 years of teaching experience in microbiology and biotechnology subjects in UG, PG, and PhD courses. Dr. Trivedi is affiliated to institutes of national and international importance as a research advisory committee member/examiner and also serves as a reviewer of several SCI journals/funding agencies.

C. R. K. Reddy is currently Head, Business Development (Tropical Seaweeds) at NPPL, Bangalore, and holds an Adjunct Professor position at the Institute of Chemical Technology (ICT-Mumbai), Mumbai, India. He has been exclusively focusing on marine macroalgae for more than three decades and his current research interests span from seaweed production, product development, and commercialization through the development of innovative and cutting-edge technologies complying with regulatory regimes and consumer market needs.In his research career, he had worked in various premier institutes of national and international reputation and had a long stint at CSIR-Central Salt Marine Chemicals Research Institute, Bhavnagar, Gujarat, where he worked for more than 25 years in various capacities and contributed immensely to the cellular biotechnology of seaweeds besides taking an active part in promotion and field implementation of seaweed farming technologies through active participation of coastal communities. He had a very long successful S&T career with several recognitions including awards and fellowships, and represented in several national scientific advisory committees besides serving the editorial board of scientific journals. He holds a distinguished record with established scientific credentials in S&T authoring more than 90 publications in SCI journals of high repute, 18 book chapters, 3 patents, and 5 books (some are edited). His research works have significantly contributed to the advancement of phycological research.

Alan T. Critchley grew up in Birmingham, UK. In 1981, he graduated with a PhD from Portsmouth University, in marine ecology, having been based at the university's Hayling Marine Laboratory studying the invasion of the brown seaweed *Sargassum muticum*. This was followed by a Royal Society European Postdoctoral Fellowship at the Delta Institute, Yerseke, the Netherlands. He then moved to South Africa to teach in the Botany Department of the University of KwaZulu-Natal, Pietermaritzburg (7 years), and the University of the Witwatersrand, Johannesburg (10 years). He researched sub-tropical seaweeds to the east and cold-water, upwelling-influenced species to the west and developed a much keener interest in seaweed benefits and their applications. He also collaborated with Professor M. Ohno of Kochi University, Japan, to co-edit volumes of Seaweed Cultivation and Marine Ranching and Seaweed Resources of the World. In 1998, he moved to the University of Namibia, Windhoek, to become a Research Professor at the Multidisciplinary Research and Consultancy Centre; seaweeds and their uses were at the center of a number of novel projects there. In 2001, he made the transition from academia to the commercial world. The call of carrageenan-bearing seaweeds took him to Normandy, France, and a processing facility, then operated by Degussa Texturant Systems, to be in charge of their New Raw Materials Laboratory. It was at that time that his passion for the *Kappaphycus* and *Eucheuma* spp seaweeds, their biology, production, utilization, and socio-economic importance was fostered. It was here that he was most fortunate to begin collaborative research with Dr. Anicia Q. Hurtado. In 2005, he moved to Nova Scotia, Canada, to work with a small seaweed processing company. This role involved the management of multidisciplinary, collaborative projects based on commercial seaweeds (browns and reds). Ensuring sustainable production techniques for seaweed resources and the scientific validation of biostimulant and bioactive properties of extracts was a major focus. Carrageenophyte seaweeds continued as a research focus. It was through these activities that Dr. Iain Neish and Alan were introduced. A common goal led Anne, Iain, and Alan to co-edit Tropical Sea*weed Farming* (2017) and *Topical Phyconomy Coalition Development: Focus on eucheumatoid seaweeds* (2024) both for Springer. Current collation of the broad spectrum of science associated with the *Eucheuma* seaweeds, their production, processing, applications, and most importantly socio-economic contributions to coastal communities. He helped the Global Seafood Alliance publish its first standard on seaweed cultivation in 2023, and he has been a member of the science advisors for the Safe Seaweed Coalition, now Global Seaweed Coalition.

Chapter 1
Seaweeds: A Traditional and an Alternative Source of Proteins for Human and Animal Nutrition

Joël Fleurence

Abstract Macroalgae are traditionally used in human and animal nutrition. Several species belonging to the red seaweed family are often described as containing high levels of proteins. In *Porphyra tenera*, well known under the name of Nori, the protein content is between 33% and 47% of the dry weight (DW). Other red seaweed species such as *Palmaria palmata* (Dulse) consumed in Europe as sea vegetable also has a high protein level (e.g., 25–35% of DW). Except for red seaweeds, some green algae also possess high protein content. For instance, the algae belonging to the genus *Ulva* have a protein level that can reach upto 33% of DW. The species *Ulva lactuca*, called sea lettuce and consumed frequently around the world, has this characteristic. In the past, some of these seaweed species were also used as farm animal feed.

Many red and green algae have a protein content comparable to that of soybean, which is the main protein source for animal feeding. Concerning human nutrition, some algal proteins contain a high level of methionine which is an amino acid mainly obtained by the consumption of animal protein.

In this context, the use of seaweed as a protein and amino acid source for animal feeding and for human nutrition as well as adapted to the request of vegan consumers is worth discussing.

This chapter describes the main seaweed species consumed as sea vegetables and their valorization opportunities as source proteins or amino acids. The main factors limiting the nutritional value of these proteins are also detailed. The low allergenic nature of algae and in particular of their proteins is also discussed.

Keywords Seaweeds · Proteins · Human and animal nutrition · Vegans

---

J. Fleurence (✉)
Nantes Université, ISOMER, 2 chemin de la Houssinière, Nantes Cedex 9, France
e-mail: joel.fleurence@univ-nantes.fr

N. Trivedi et al. (eds.), *Recent Advances in Seaweed Biotechnology*,
https://doi.org/10.1007/978-981-96-0519-4_1

1.1 Introduction

Marine algae were traditionally consumed as sea vegetables in Asian countries (Nisizawa et al. 1987; Fleurence 2016). In Europe, seaweeds have also been used in human nutrition (Sauvageau 1920). In Scandinavian countries, they were consumed by the local population during the dearth period. In Norway, the red seaweed, *Palmaria palmata* (Fig. 1.1), well known under the name of Dillisk was eaten by coastal communities (Delaney et al. 2016). This alga was also consumed in Ireland and Scotland. It was marketed under the name of Dulse in the streets of Edinburgh or Glasgow. In the United Kingdom, *Porphyra* sp was also used as vegetable under the name of laver in England and sloke in Scotland. More generally, in European maritime regions, the algae belonging to the genus *Fucus*, which are abundant, are frequently consumed. For example, *Fucus serratus* or *Fucus vesiculosus* (Fig. 1.2) has been used in some cooking preparations both alone or mixed with flour or milk.

In North America, there are also examples of the use of seaweed as a nutrient source for local populations. Indian tribes living on Queen Charlotte Island consume green laver (*Ulva lactuca*) and pink laver (*Porphyra* sp) (Sauvageau 1920).

Except for the use of seaweeds in human nutrition, marine algae have been often used for animal feeding. Traditionally in Norway, the algae *Alaria esculenta* is used as cow feed under the name of Kutura (cow seaweed) (Sauvageau 1920; Fleurence 2022a). Another brown seaweed *Ascophyllum nodosum* has also been used for pig nutrition under the name of Grisetang. In the island, there are also numerous examples of the use of seaweed, belonging to the genus *Palmaria*, *Chondrus*, or *Alaria*, as sheep feed.

Fig. 1.1 *Palmaria palmata* (Dulse). (Photo credit© Fleurence J, 2023)

Fig. 1.2 *Fucus vesiculosus*. (Photo credit © Fleurence J, 2023)

Algae are therefore traditionally used in human and animal nutrition. They are a source of various nutrients such as mineral elements (I, Fe, Ca, and Mg), pigments (carotenoids), or dietary fibers (alginates, agars, and carrageenans). Some red and green seaweeds contain high levels of proteins including between 26% and 47% of algal dry weight (Fujiwara-Arasaki et al. 1984; Fleurence et al. 2018). These protein levels are sometimes close to or higher than those reported from grains of leguminous plants such as soybeans, beans, or peas (20–45% of DW). Moreover, the growth of algae and their cultivation do not require freshwater. In this context, the use of seaweed as an alternative source of plant proteins deserves to be discussed. In particular, red algae are considered a sustainable and alternative source of proteins for humanity (Rawiwan et al. 2022).

1.2 Protein and Amino Acid Contents

The protein content of seaweeds can vary according to the species, the seasons, and environmental or cultivation conditions. This also applies to the amino acid composition. However, the scientific literature shows that brown seaweeds have fairly low protein levels compared to those reported for green or red algae.

1.2.1 Proteins and Amino Acids in Brown Seaweeds

Brown algae are widely used as sea vegetables in Asia. The main species consumed is *Laminaria japonica* now called *Saccharina japonica.* It is well known by the people under the name of "Kombu" or "Kelp." This seaweed remains as the first cultivated species in the world and its production has been increasing steadily for 10 years (Fig. 1.3).

In Japan, Kombu is available in several forms (Table 1.1). However, the protein levels strongly differ according to the processed product (Table 1.2) (Nisizawa et al. 1987). For example, the highest content of protein is observed with salting products.

Usually, raw brown seaweeds are often described as poor in protein content (Table 1.3). This is especially true for the species *Fucus vesiculosus* consumed in France as a sea vegetable (Munda 1977; Fleurence 2016). However, the brown seaweed *Undaria pinnatifida* or Wakame is an exception to this observation. Indeed, the protein level of Wakame can reach up to 24% of dry weight (Fleurence et al.

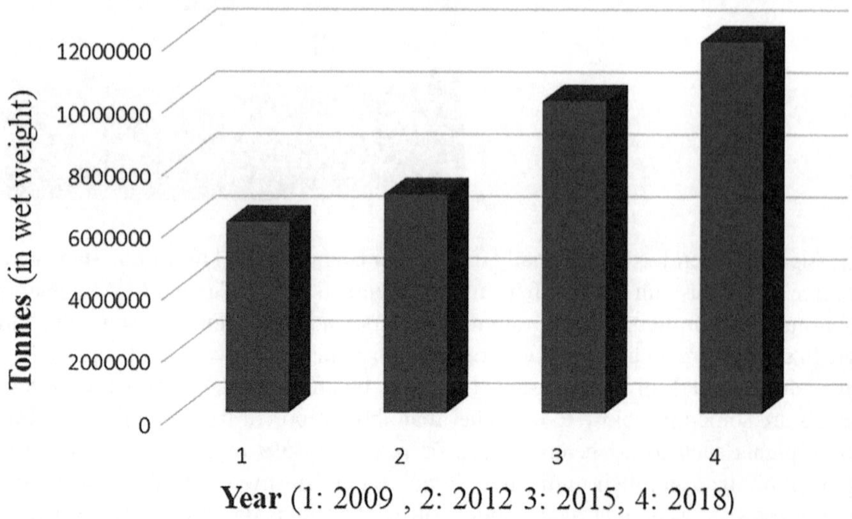

Fig. 1.3 World production of Kombu (*Laminaria japonica*) by cultivation between 2009 and 2018. (FAO 2020)

Table 1.1 Examples of different forms of Kombu proposed in Japan in human consumption

Processed forms	Boiled in soybeans	Salting	Rolling	Shredding	Slicing
Marketed product	Tsukudani-kombu	Shio-kombu	Kombu-maki	Kizami-kombu	Battera-kombu

From Nisizawa et al. (1987)

Table 1.2 Protein content of different Kombu products

Type of product	Shredded kombu	Sliced kombu	Salted kombu	Salted and boiled kombu	Rolled kombu
Protein level (in % of dry weight)	7.6	9.1	21.3	13.4	6.3

From Nisizawa et al. (1987)

2018). Like Kombu, Wakame is very widely cultivated. The main consumed part, or blades, has a protein content that can vary according to the harvest period and also according to the cultivation site (Fig. 1.4) (Zhou et al. 2015). Despite this, the protein content of *Undaria pinnatifida* generally remains higher than other brown algae.

The protein levels of leguminous plants are generally high. For example, the content of proteins in soybeans can reach up to 38% of the dry weight. No brown algae have such a high protein level even the wakame. Despite this, *Undaria pinnatifida* shows some advantages with regard to its amino acid composition. Indeed, this seaweed is characterized by a content of essential amino acids such as methionine and phenylalanine that are significantly higher than those recorded for leguminous plants (Fleurence et al. 2018). For example, the methionine level recorded for *Undaria pinnatfida* is 30% higher than those reported for leguminous plants.

In this context, brown seaweeds, and more particularly *U. pinnatifida*, can be considered a complementary source of protein to that of leguminous plants as human or animal food.

1.2.2 Proteins and Amino Acids in Green Seaweeds

Green seaweeds belonging to *Ulva* or *Caulerpa* genus are widely consumed around the world as sea vegetable. *Ulva* algae are marketed under the name of "Fisherman salad" in Europe and under the name "Sea lettuce" in Asia (Fig. 1.5).

In brown seaweeds, the protein content of green algae can differ according to species, seasons, and geographic location (Table 1.4). In some cases, the mode of production, cultivation, or harvesting in the natural environment can also influence the content of the algae. However, much data about the protein content of the main green seaweeds frequently consumed are available.

In *Ulva lactuca* and *Ulva pertusa*, the maxima levels of proteins recorded in the literature are 33.7% and 26.0% of dry weight, respectively (Table 1.4). Similar to other seaweeds, *Ulva lactuca* which has good source of proteins and the protein

Table 1.3 Examples of brown seaweed protein contents and their variations according to the season

Seaweed species	*Laminaria japonica (Kombu)* (1,2,3)	*Laminaria digitata (Breton kombu)* (3)	*Undaria pinnatifida (Wakame)* (3,4)	*Hizikia fusiforme (Hiziki)* (1,2)	*Fucus vesiculosus (Varech, Black goëmon)* (5)	*Ascophyllum nodosum* (5, 6) *(Goëmon)*
Protein level (in % of dry weight)	13.0–15.4	8.0–15.0	11.0–24.0	10.6–12.3	2.80–8.05	3.0–15.0

From (1) Arasaki and Arasaki (1983), (2) Nisizawa et al. (1987), (3) Jurković et al. (1995), (4) Fleurence et al. (2018), (5) Munda (1977), and (6) Fleurence (1999)

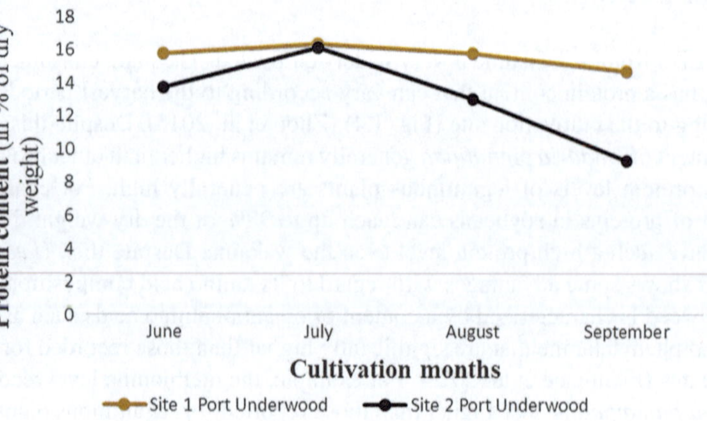

Fig. 1.4 Protein levels of *Undaria pinnatifida* cultivated in summer months and farming sites. (From Zhou et al. (2015))

content is subjected to seasonal variations (Fig. 1.6) (Abdel-Fattah and Edrees 1973). In species collected in the natural environment, the lowest level of proteins is observed at the beginning of the spring (Fig. 1.6). The use of this resource as a source of dietary protein must therefore consider this aspect into account. The cultivation of *Ulva* seaweed does not prevent the influence of seasonal variations on the protein content (Queirós et al. 2021) (Fig. 1.7). For instance, the lower protein level for *U. rigida* cultivated in the tank system is also observed during the spring season (Fig. 1.7).

Among the algae of the genus *Caulerpa*, *Caulerpa lentillifera* is the most consumed. This species is particularly popular in South East Asia (Philippines, Vietnam, and Japan). In Japan, this seaweed is called Umibudô (sea grapes) or green caviar.

Generally, the protein content of *Caulerpa* seaweeds is lower than those reported for *Ulva* sp. The values observed are often included between 10.5% and 12.5% of the dry weight of the plant (Ratana-Arporn and Chirapart 2006; Matanjun et al.

Fig. 1.5 *Ulva* sp. called "Sea lettuce" or "Fisherman salad". (Photo credit © Fleurence J, 2023)

Table 1.4 Examples of variations of protein contents for different *Ulva* species according to seasons

Seaweed species	*Ulva lactuca* (1)	*Ulva pertusa* (2, 3)	*Ulva clathrata* (4)	*Ulva rigida* (5)	*Ulva armoricana* (6)
Protein content (in % of the dry weight)	8.7–33.7	17.5–26.0	21.9–25.9	15.0–20.0	18.0–24.0

From (1) Abdel-Fattah and Edrees (1973), (2) Nisizawa et al. (1987), (3) Fleurence et al. (2018), (4) Peña-Rodríguez et al. (2011), (5) Shuuluka et al. (2013), and (6) Fleurence et al. (1999)

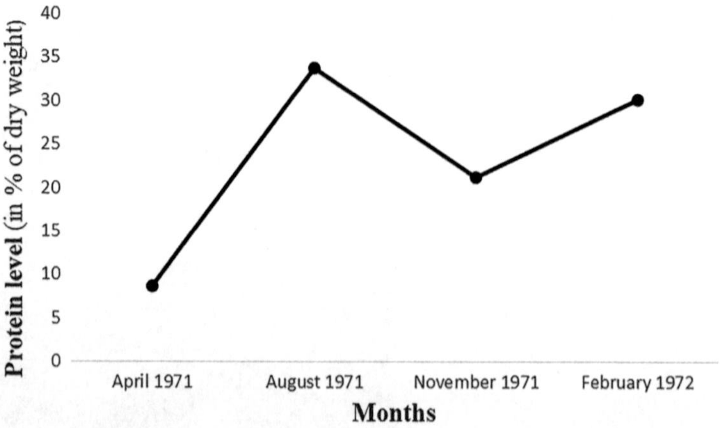

Fig. 1.6 Examples of seasonal variations of the protein levels of the green seaweed *Ulva lactuca*. (From Abdel-Fattah and Edrees (1973))

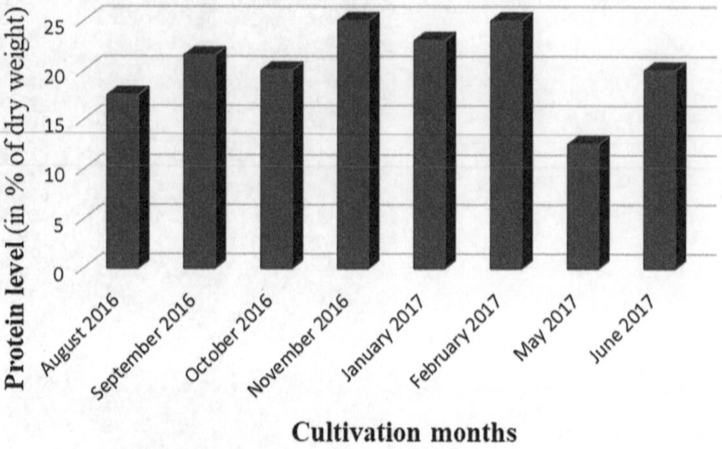

Fig. 1.7 Seasonal variation of protein content in *Ulva rigida* cultivated in ponds on land. (From Queirós et al. (2021))

2009). On the basis of this observation, *Caulerpa* species seem less interesting as a source of dietary proteins than *Ulva*.

There are several data on the amino acid composition of green algae of the *Ulva* genus. Some species appear to be an appropriate source of methionine comparable to leguminous plants such soybeans (Table 1.5). Some Ulva such as *U. rigida, U. armoricana,* and *U. lactuca* also represent an interesting source for the dietary supply of threonine (Table 1.5). *Ulva reticulata, U. armoricana U. lactuca,* and *U. pertsusa* are also a good source of isoleucine comparable to leguminous sources. The use of green algae belonging to the genus of *Ulva* is appropriate in the context of dietary diversification, especially for vegan consumers. Although less rich in

Table 1.5 Examples of essential amino acid composition of some green seaweeds belonging to the *Ulva* genus and comparison with amino acid content of leguminous plants (in g of amino acid/100 g of proteins)

	Ulva rigida (1)	*Ulva reticulata* (2)	*Ulva armoricana* (3)	*Ulva lactuca* (1)	*Ulva pertusa* (4)	*Ulva clathrata* cultivated (5)	Leguminous plants
Histidine[a]	1.4	–	1.2–2.1	1.8	4.0	1.1–1.2	3.8–4.0
Isoleucine	3.1	4.2	2.3–3.6	3.7	3.5	2.6–3.1	3.6
Leucine	5.2	7.9	4.6–6.7	6.7	6.9	1.7–1.9	7.3
Lysine	3.7	6.0	3.5–4.4	4.2	4.5	3.0–3.4	6.4–6.5
Methionine	1.5	–	1.4–2.6	1.6	1.6	1.1–1.3	1.2–1.4
Phenylalanine	3.3	5.3	5.0–7.1	4.0	3.9	3.7–4.2	2.4
Threonine	5.0	–	4.5–6.8	4.7	3.1	3.4–3.8	4.0
Tryptophan	–	–	–	–	0.3		1.6–1.9
Valine	5.6	6.3	4.0–5.2	6.2	4.9		4.5
Cysteine	1.1	–		0.4	1.2		1.1–1.3

From (1) Shuuluka et al. (2013), (2) Ratana-arporn and Chirapart (2006), (3) Fleurence et al. (1999), (4) Fujiwara-Arasaki et al. (1984), and (5) Peña-Rodríguez et al. (2011)
[a]Histidine is an essential amino acid for the children

proteins than *Ulva*, some *Caulerpa* has an amino acid composition close to *Ulva* (Ratana-Arporn and Chirapart 2006). This is the case of *Caulerpa lentillifera,* especially. The same conclusion about the opportunity to include this seaweed in the vegan diet can be made.

1.2.3 Proteins and Amino Acids in Red Seaweeds

Among the algae used in human or animal food, red seaweeds (Rhodophyta) are the richest in proteins. The protein maximum level reported in the literature is 47.5% of the dry weight in the species *Porphyra tenera* consumed as Nori in Japan (Fujiwara-Arasaki et al. 1984). This value is close to that recorded for the protein content of soybean meal (38–54% of DW) frequently used in pig feed. The high level of proteins in red seaweeds is mainly due to the presence of supplementary protein belonging to the phycobiliprotein family called phycoerythrin. In the food algae *Palmaria palmata* or Dulse, the amount of R-phycoerythrin can represent 12.2% of the total protein fraction (Galland-Irmouli et al. 2000). More generally, many red algae consumed in human food have protein levels close to or higher than that of leguminous plants (Table 1.6). The red seaweeds can in particular be used as a potential source of proteins comparable to certain leguminous plants such as peas, beans, or lentils. However, only species of *P. palmata* and *P. tenera* show protein levels close to those of soybeans (34–40% of DW).

Table 1.6 Protein contents of red seaweeds consumed as sea vegetables and comparison with other sources of dietary proteins (leguminous)

Species seaweed	*Palmaria palmata* (1,2) (Dulse)	*Porphyra tenera* (3,4) (Nori)	*Grateloupia turuturu* (3, 5)	*Chondrus crispus* (6,7) (Irish moss)	*Gracilaria verrucosa* (7) (Ogo-nori)	Leguminous (Peas, lentils, beans) (8)
Protein levels (in % of dry weight)	8.1–35.6	33.0–47.0	14.6–27.5	17. 0–21.4	24.0	20.0–25.0

From (1) Morgan et al. (1980), (2) Galland-Irmouli et al. (1999), (3) Fujiwara-Arasaki et al. (1984), (4) Fleurence et al. (2018), (5) Denis et al. (2010), (6) Carpena et al. (2021) (7) Young and Smith (1958), (7) Marrion et al. (2005), and (8) Guéguen et al. (2016)

Table 1.7 Comparison of amino acid composition of two red seaweeds and conventional and genetically modified soybeans

Amino acid (g/100 g protein)	Conventional soybeans	Genetically modified soybeans	*Palmaria palmata*	*Porphyra tenera*
Histidine[a]	1.2	1.1	0.5–1.2	1.4
Isoleucine	1.7	1.6	3.5–3.7	4.0
Leucine	2.8	2.7	5.9–5.7	8.7
Lysine	2.4	2.3	2.7–5.0	4.5
Methionine	0.0	0.6	2.7–4.5	1.1
Phenylalanine	1.9	1.8	4.4–5.3	3.9
Threonine	1.4	1.3	3.6–4.1	4.0
Valine	1.7	1.7	5.1–6.9	6.4

From Fleurence et al. (2018) and Kudełka et al. (2021)
[a]Essential amino acid for children

The red algae *Palmaria palmata* (Dulse) and *Porphyra tenera* (Nori) show an interesting essential amino acid composition compared to that of conventional or genetically modified soybeans (Table 1.7). In *P. palmata*, essential amino acids can represent 46% of the total amino acid fraction (Fleurence et al. 2018). This seaweed is also a better source of leucine, threonine, valine, and especially methionine than soybeans. This last point is important for vegan diet because the dietary intake of methionine is often by the consumption of animal proteins. For example, the maximum level of methionine recorded for *P. palmata* (4.5% of the total amino acid fraction) is higher than ovalbumin (3.1%) which is the reference food protein for the FAO. The composition of essential amino acids in *Porphyra tenera* shows an exceptional level of leucine (8.7% of the total amino acid fraction). This species, frequently consumed as Nori, is therefore an interesting food source for the supply of this amino acid.

Algae and more particularly red algae are potential sources of protein for human and animal nutrition. Already used in a traditional way, they can also satisfy vegan demand as do other protein-rich vegetable food resources (leguminous). In addition, certain species are also important as sources of sulfur amino acids such as methionine, which is generally not present in the proteins of terrestrial plants. As such, the *P. palmata* species appears as a source likely to enrich the nutritional value of a vegan diet. Regarding animal feed, algae are sources of proteins, pigments, and minerals and were traditionally used in the past (Fleurence 2022a). Their use appears today to be justified in aquaculture production. Indeed, the diet of farmed fish is often based on the use of foods based on fish meal and soybeans. Replacing the soybean ingredient with seaweed, and more specifically red algae, is a concept worth discussing. In addition, the use of red algae as a source of protein in animal feed could also have an advantage in the fight against climate change. Indeed, the contribution of red algae *Asparagopsis taxiformis* and *Asparogopsis armata* in the diet of young ruminants is described as lowering between 80 and 98% of the enteric methane production released by the digestion of these animals (Fig. 1.8) (Roque et al. 2021; Fleurence 2021; Glasson et al. 2022). The algal product involved in the inhibition of methanogenesis is bromoform ($CHBr_3$), this molecule is naturally produced by the turn over of seaweed biomass (Glasson et al. 2022).

Although the protein content of *Asparagopsis* algae (9.5–15.5% of dry weight) (Pellegrini 1969; McDermid and Stuercke 2003; Angell et al. 2012) is lower than that of other red algae, their introduction into the ruminant feed is probably a promising way in the future to improve the use of algae in animal feeding.

Most of the seaweeds are a traditional source of vegetable protein in Asia. In Western countries, they are not yet popular sources of dietary proteins.

In this particular context, what can be the future for the use of algae as food? Many plant protein sources are implicated in food allergies (e.g., rce, soybeans, and peanuts). Are algae a source of allergenic proteins?

Allergies due to seaweed consumption are a relatively rare phenomenon.

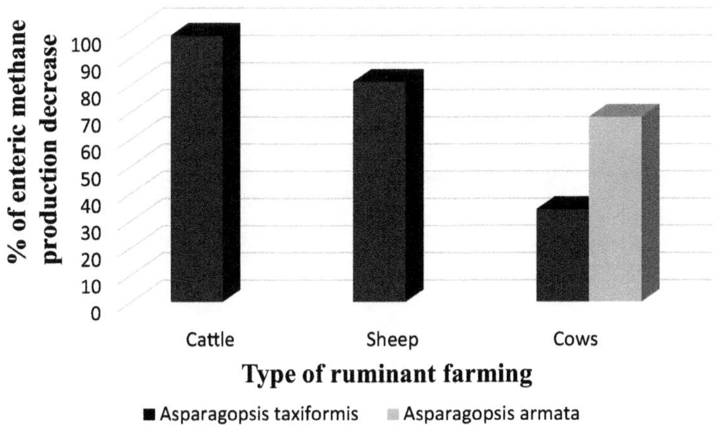

Fig. 1.8 Decrease (in %) of the enteric methane production by ruminant farming after *Asparagopsis* inclusion in their diet. (From Glasson et al. (2022))

The scientific literature reports poisoning with symptoms comparable to food allergy (diarrhea, vomiting, and hypotensive shock) following the consumption of the red algae *Gracilaria verrucosa* (Ogonori) (Noguchi et al. 1994). However, the mechanism involved in this pathologic reaction does not seem to have an allergenic origin. It was due to the excessive production of prostaglandins by an enzyme contained in Ogo-nori and active in arachidonic acid present in large quantities in the diet consumed by the people affected by this poisoning. Although prostaglandins are pro-inflammatory substances, the case described by Noguchi et al., is not based on a food allergy mechanism.

The scientific literature on the presence of allergens in seaweeds or allergenic risk appears to be very limited (Fig. 1.9) (James et al. 2023).

A first case of food allergy was reported with the red algae Nori (*Porphyra tenera* and *Porphyra yezoensis*) and more secondarily with the algae *Palmaria palmata* (Dulse) and *Chondrus crispus* (Pioca, Irish moss). The clinical symptoms observed are generalized urticaria, nose congestion, and facial angedioma (Thomas et al. 2019). The application of fresh red seaweed to patient skin shows a positive response which is stronger with Nori (Fig. 1.10) (James et al. 2023). In the patient tested, Nori, therefore, appears to be more allergenic than other species of red algae evaluated in the study by Thomas et al. (2019).

However, the biochemical nature of these allergens is still poorly understood. For some authors, sulfated polysaccharides such as carrageenans could be involved. Indeed during a study of 288 healthy people aged 18–65 years, 57 participants showed an IgE-specific immune reaction to carrageenan (Vojdani and Vojdani 2015). Other sulfated polysaccharides such as Porphyrans from *Porphyra tenera* or *Porphyra yezoensis* are, on the contrary, described for their anti-allergenic properties (Fig. 1.11) (Vo et al. 2012; Fleurence and Ar Gall 2016).

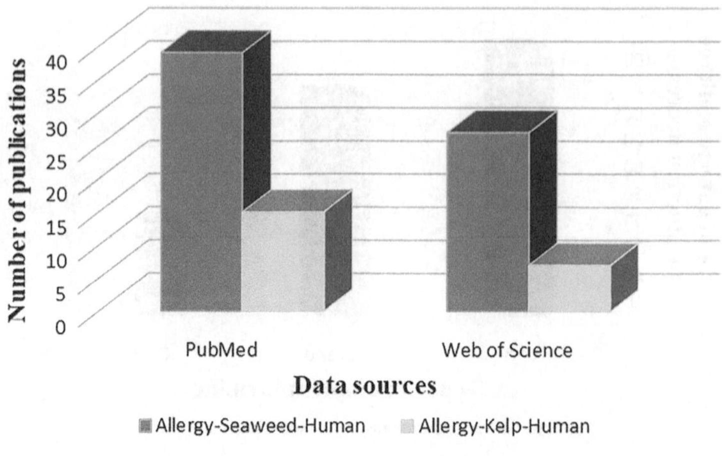

Fig. 1.9 Number of publications on allergies due to seaweeds in medical database sources. (From James et al. (2023))

Fig. 1.10 Diameter of wheal in the skin test according to the red seaweed species studied by Thomas et al. (2019). (From James et al. (2023))

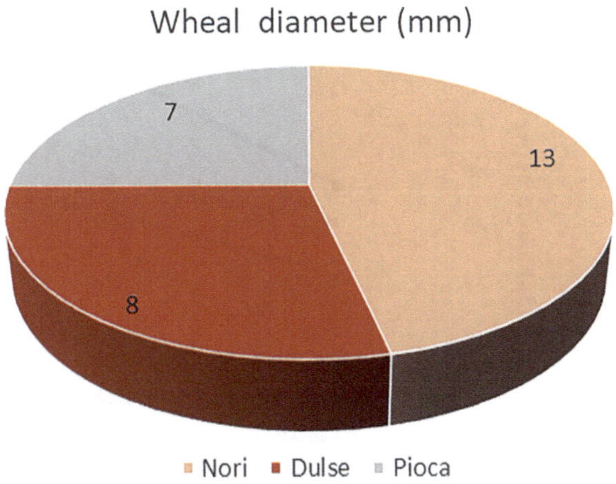

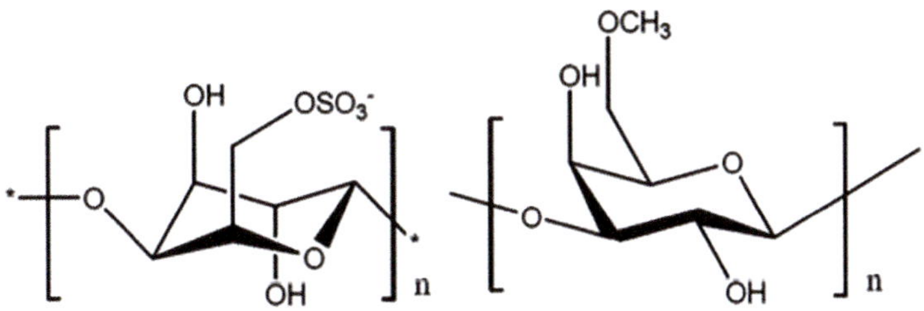

Fig. 1.11 Structure of porphyran described for its anti-allergic properties. (From Vo et al. (2012))

In a similar context, Nori sauce has been described by some authors as a healthy food devoid of the main known food allergens (Uchida et al. 2017).

Besides these aspects, algae are mainly described in the scientific literature as having anti-inflammatory or anti-allergenic properties (Fleurence and Ar Gall 2016).

1.3 Food Process for Improving Nutritional Value of Seaweed Proteins

One of the main limitations of the use of algae as a source of protein is the presence of dietary fibers acting as anti-nutritional factors. These fibers are algal polysaccharides that limit the protein digestibility by the inhibition of proteases such as pepsin, pancreatin pronase, trypsin, or chymotrypsin (Horie et al. 1995) (Galland-Irmouli et al. 1999) (Fig. 1.12)

However, these fibers do not all have the same inhibitory power. This differs according to seaweed species and the biochemical nature of the polysaccharides constituting these fibers. In raw seaweed *Palmaria palmata*, the relative digestibility

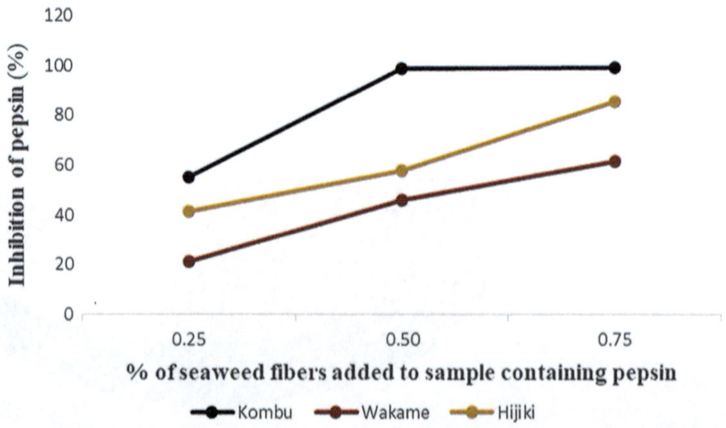

Fig. 1.12 Effect of seaweed fibers from brown algae on the inhibition of in vitro pepsin activity. (From Horie et al. (1995))

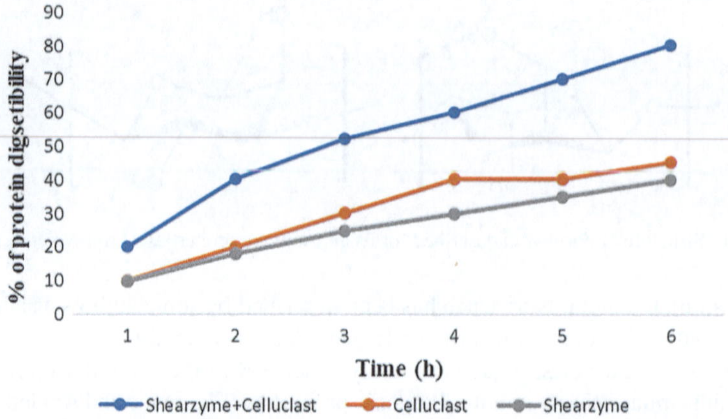

Fig. 1.13 Effect of enzymatic treatment on in vitro protein digestibility of *Palmaria palmata* (Shearzyme: xylanases, Celluclast: cellulase). (From Fleurence et al. (2018))

of protein fraction is 56% (Galland-Irmouli et al. 1999; Fleurence et al. 2018). In another red alga, *Hypnea japonica* and in the green seaweed *Ulva lactuca,* the digestibility of proteins is close to 89% and 85%, respectively (Wong and Cheung 2001). The raw extract of *Palmaria palmata* (Dulse) mainly contains polysaccharides such as xylans and cellulose. These polysaccharides strongly limit the action of protein digestion by human intestinal juice (Galland-Irmouli et al. 1999). In this condition, the use of enzymes degrading xylans and cellulose appears to be an interesting process to improve the digestibility and therefore the nutritional value of proteins contained in Dulse (Fig. 1.13) (Fleurence et al. 2018).

In *Palmaria palmata*, protein digestibility can also be improved by fermentation. Work carried out using terrestrial strains has shown that fermentation, a process

Table 1.8 Impact of fermentation process on protein digestibility

Fermentative organisms	None	*Rhizopus microscopus* var. *chinensis*	*Aspergillus oryzae*	*Trichoderma pseudokoningii*
% of protein digestibility	1.5	17	20.0	27.5

From Marrion et al. (2003)

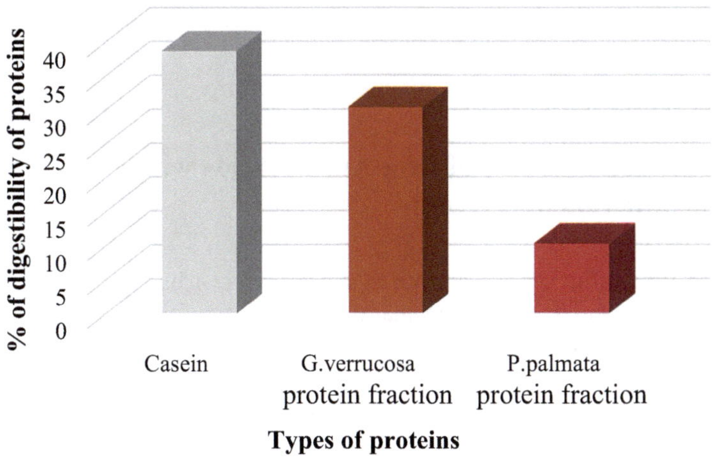

Fig. 1.14 Comparison of in vitro digestibility of red seaweed proteins in regard to the reference dietary protein (casein). (From Marrion et al. (2005))

frequently used by the food industry, is effective in increasing the digestibility of the algal protein fraction (Table 1.8) (Marrion et al. 2003).

The best result is obtained with the strain *Trichoderma pseudokoningii.* This fermentative microorganism is characterized by the production of enzymes that specifically degrade xylans and cellulose. This experimental observation confirms that these two types of polysaccharides are the cause of the low digestibility of the proteins of the red algae *P. palmata.*

Among the red algae consumed as sea vegetables, the species *Gracilaria verrucosa* or Ogo-nori can also be an interesting source of dietary protein. Indeed its protein content can reach 24% of the dry weight, which is very close to the protein content of Dulse (25%). However, the digestibility of the protein fraction of *G. verrucosa* is three times higher than that of *P. palmata* (Fig. 1.14).

G. verrucosa is characterized by a lower content of soluble fibers (30% of DW) compared to those of *P. palmata* (45% of DW) and by the chemical nature of the fibers, agar for Ogo-nori instead of xylans for Dulse.

Therefore, the use of enzymatic or fermentation treatments to improve the digestibility and nutritional value of algal proteins is not always necessary. It depends on the soluble fiber content of the algal resource and especially on the chemical nature of these polysaccharides.

Despite this, biotechnological processes (fermentation, saccharification, and enzymatic treatments) can prove interesting for producing biologically active oligosaccharides or products with improved aromas and flavors (Fleurence 2022b). This type of application is also likely to improve the consumer acceptability of protein-rich algae.

In conclusion, certain species of algae, mainly Rhodophytes or Chlorophytes, are good candidates as protein sources for human food. Their nutritional value can also be improved by technological treatments already well accomplished by the agro-food industry.

1.4 Seaweed a Source of Free Amino Acids with an Interest in Nutrition and Health

Taurine is a free sulfur amino acid with an important role in human health (Fig. 1.15). This nonprotein amino acid shows many biological activities (e.g., anti-oxidation and molecular signalization) and it is essential for cardiovascular function. In addition, taurine is implicated as a regulator of cellular homoestasis, in skeletal muscular activity, and in the functioning of the retinal tissue (Seidel et al. 2019).

The taurine level in the land plant is very low or undetectable. On the contrary, algae, and in particular red algae have higher levels of taurine in comparison to land plants and edible fungi such as "Shiitake" (Kataoka and Ohnishi 1986). The red alga *Palmaria palmata* proves to be an interesting source of taurine with a concentration of 5.5 g of taurine per kg of dry matter (Table 1.9). A 100 g food ration would provide nearly 550 mg of this free amino acid (Terriente-Palacios and Castellari 2022). This seaweed also appears as an interesting source of homotaurine (Table 1.9). Except for *Undaria pinnatifida*, brown algae are less rich sources of taurine and derivates. The same observation can be made with green algae (Table 1.9).

The dietary intake of taurine is mainly through the consumption of seafood. Scallops or Pectinidae are an important source of taurine (Fig. 1.16). In this seafood product, the taurine content can reach up to 8270 mg/kg of dry matter (El-Sayed 2014). Seaweeds such as Dulse and Nori show taurine concentrations lower than scallops but higher than other seafood such as fishes (Fig. 1.16). Therefore, some red algae are interesting resources for the supply of taurine for human and also animal nutrition. As for humans, animals can synthesize taurine. For poultry, there are two main sources of taurine, endogenous synthesis and diet. A recent study shows the protective effects of a diet enriched in taurine on the growth of poultry and their resistance to various abiotic stresses (thermal shock, chemical contamination, and

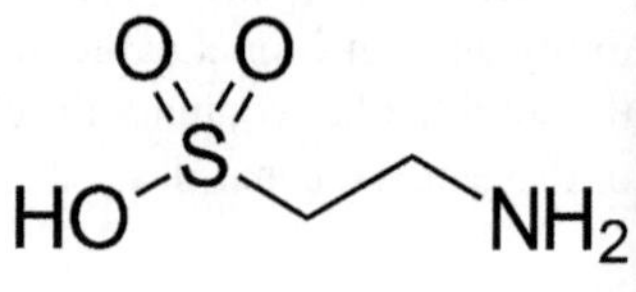

Fig. 1.15 Chemical structure of taurine

Table 1.9 Levels in taurine, homotaurine, and hypotaurine (mg/kg of DW) in some seaweeds frequently consumed as sea vegetables

Seaweed species	Taurine (mg/kg DW)	Homotaurine (mg/kg DW)	Hypotaurine (mg/kg DW)
Palmaria palmata (Dulse)	5555.0	670.0	75.6
Porphyra sp. (Nori)	5270.0	702.0	58
Chondrus crispus (Irish moss)	845.0	34.5	–
Undaria pinnatifida (Wakame)	1950.0	567.0	6.7
Hizikia fusiforme (Hijiki)	570.0	3.8	3.3
Laminaria japonica (Kombu)	650.0	20.5	16.0
Ulva lactuca (Sea lettuce)	1580.0	1590.0	221.0
Enteromorpha intestinalis (Ao-nori)	308.0	600.0	–

From Terriente-Palacios and Castellari (2022)

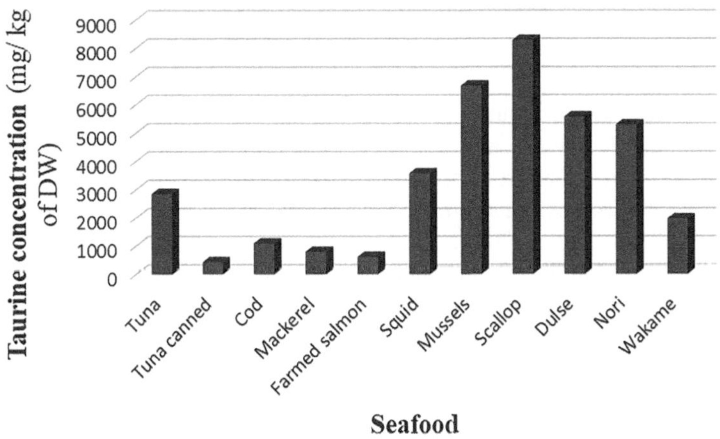

Fig. 1.16 Examples of taurine concentrations in some seafood frequently consumed. (From El-Sayed (2014))

increased breeding density) (Surai et al. 2020). In this last context, red algae by their content in taurine also in pigments (phycoerythrin) appear as sustainable sources for poultry feed.

1.5 Conclusion

Global climate change requires us to review our way of producing and consuming food. Algae are carbon consumers and their production does not require the supply of fresh water. Their production does not imply, as in the case of soya, deforestation. In this, the replacement of soy proteins by algae proteins in human food but especially animal food appears a major scientific and societal question for the future of humanity. Among algae, red seaweed is a source of proteins and sulfur amino acids that are interesting for their nutritional contribution. The digestibility of these proteins can be limited in some cases. However, it can be improved by traditional food processes such as enzymatic maceration or fermentation. This could lead to new products based on algae proteins for traditional or vegan consumers. The likely absence of major allergens in seaweed makes this resource a relatively safe seafood for allergic populations. Red seaweeds are considered a promising alternative protein source by numerous specialists (Rawiwan et al. 2022). *Porphyra tenera* and *Palmaria palmata* are often cited as the main sources of proteins for human nutrition. However, another species such as *Grateloupia turuturu*, edible seaweed consumed in Japan, is also an interesting protein source obtained from available biomass and easy to harvest. The use of seaweed as a source of protein for animals (e.g., abalone, fishes) raised by aquaculture is also an opportunity to be strengthened. Indeed, some red seaweeds, such as *P. palmata,* are known for the antibiotic activity of their extracts against the pathogenic bacterium *Vibrio harveyi* (García-Bueno et al. 2015). The use of algae in the diet of fishes is also well described. For example, diets based on using *Ulva* meal for fish nutrition significantly improve the stress response and disease resistance of farmed fishes (Satoh et al. 1987; Fleurence et al. 2012). The integration of seaweeds rich in proteins (*Palmaria* sp., *Porphyra* sp. or *Ulva* sp.) in the aquaculture process has been already experimented. This mainly includes marine species and also freshwater fishes such as Nile tilapia (*Oreochromis niloticus*) (Marinho et al. 2013). This type of approach is well known under the name of integrated multi-trophic aquaculture (IMTA). IMTA is probably the future of aquaculture and algae will probably play an essential role in the development of this activity.

Seaweeds are a traditional source of proteins for Asian countries and a future for Western countries. There are also specific and original sources of minerals, fibers, pigments, and special amino acids such as taurine. Their more frequent integration into diets could have many benefits for both human and animal health. Furthermore, algae and cyanobacteria are the only sources of vegetable proteins of aquatic origin. They are a resource that is both sustainable and renewable. Their exploitation in the natural environment or their cultivation shows less negative effects on the environment than the massive production of genetically modified soybeans or not. In such a context, the use of algae and more particularly macroalgae to produce and consume differently seems to be justified more and more.

References

Abdel-Fattah AF, Edrees M (1973) Seasonal changes in the constituents of *Ulva lactuca*. Phytochemistry 12:481–485. https://doi.org/10.1016/S0031-9422(00)84432-5

Angell AR, Pirozzi I, De Nys R, Paul NA (2012) Feeding preferences and the nutritional value of tropical algae for the abalone *Haliotis asinina*. PLoS One 7:e38857. https://doi.org/10.1371/journal.pone.0038857

Arasaki S, Arasaki T (1983) Low calories, high nutrition. Vegetables from the sea. To help you look and feel better. Japan Publications, Tokyo

Carpena M, Caleja C, Pereira E et al (2021) Red seaweeds as a source of nutrients and bioactive compounds: optimization of the extraction. Chemosensors 9:132. https://doi.org/10.3390/chemosensors9060132

Delaney A, Frangoudes K, Ii SA (2016) Society and seaweed: understanding the past and present. In: Seaweed in health and disease prevention. Academic Press, pp 7–40. https://doi.org/10.1016/B978-0-12-802772-1.00002-6

Denis C, Morançais M, Li M, Deniaud E et al (2010) Study of the chemical composition of edible red macroalgae *Grateloupia turuturu* from Brittany (France). Food Chem 119:913–917. https://doi.org/10.1016/j.foodchem.2009.07.047

El-Sayed AFM (2014) Is dietary taurine supplementation beneficial for farmed fish and shrimp? A comprehensive review. Rev Aquac 6:241–255. https://doi.org/10.1111/raq.12042

FAO (2020) The state of world fisheries and aquaculture 2020. Sustainability in action. Rome. https://doi.org/10.4060/ca9229en

Fleurence J (1999) Seaweed proteins: biochemical, nutritional aspects and potential uses. Trends Food Sci Technol 10:25–28. https://doi.org/10.1016/S0924-2244(99)00015-1

Fleurence J (2016) Seaweeds as food. In: Seaweed in health and disease prevention. Academic Press, pp 149–116. https://doi.org/10.1016/B978-0-12-802772-1.00005-1

Fleurence J (2021) Perspectives on the use of algae in agriculture and animal production. Phycology 1:79–82. https://doi.org/10.3390/phycology1020006

Fleurence J (2022a) Les algues en agrobiologie. ISTE Group

Fleurence J (2022b) Biotechnological processes applied to edible seaweeds: what perspectives? Trends Food Sci Technol. https://doi.org/10.1016/j.tifs.2022.11.007

Fleurence J, Ar Gall E (2016) Antiallergic properties. In: Seaweed in health and disease prevention. Academic Press, pp 389–406. https://doi.org/10.1016/B978-0-12-802772-1.00012-9

Fleurence J, Chenard E, Luçon M (1999) Determination of the nutritional value of proteins obtained from *Ulva armoricana*. J Appl Phycol 11:231–239. https://doi.org/10.1023/A:1008067308100

Fleurence J, Morançais M, Dumay J et al (2012) What are the prospects for using seaweed in human nutrition and for marine animals raised through aquaculture? Trends Food Sci Technol 2:57–61. https://doi.org/10.1016/j.tifs.2012.03.004

Fleurence J, Morançais M, Dumay J (2018) Seaweed proteins. In: Proteins in food processing. Woodhead Publishing, pp 245–262. https://doi.org/10.1016/B978-0-08-100722-8.00010-3

Fujiwara-Arasaki T, Mino N, Kuroda M (1984) The protein value in human nutrition of edible marine algae in Japan. In: Proceedings of the 11th International Seaweed Symposium, Qingdao, People's Republic of China, June 19–25, 1983. Springer Netherlands, pp 513–516. https://doi.org/10.1007/978-94-009-6560-7_104

Galland-Irmouli AV, Fleurence J, Lamghari R et al (1999) Nutritional value of proteins from edible seaweed *Palmaria palmata* (Dulse). J Nutr Biochem 10(6):353–359. https://doi.org/10.1016/S0955-2863(99)00014-5

Galland-Irmouli AV, Pons L, Luçon M et al (2000) One-step purification of R-phycoerythrin from the red macroalga *Palmaria palmata* using preparative polyacrylamide gel electrophoresis. J Chromatogr B Biomed Sci Appl 739:117–123. https://doi.org/10.1016/S0378-4347(99)00433-8

García-Bueno N, Dumay J, Guerin T, Turpin V et al (2015) Seasonal variation in the antivibrio activity of two organic extracts from two red seaweed: *Palmaria palmata* and the introduced *Grateloupia turuturu* against the abalone pathogen *Vibrio harveyi*. Aquat Living Resour 28:81–87. https://doi.org/10.1051/alr/2016003

Glasson CR, Kinley RD, de Nys R et al (2022) Benefits and risks of including the bromoform containing seaweed *Asparagopsis* in feed for the reduction of methane production from ruminants. Algal Res 64:102673. https://doi.org/10.1016/j.algal.2022.102673

Guéguen J, Walrand S, Bourgeois O (2016) Les protéines végétales: contexte et potentiels en alimentation humaine. Cah Nutr Diét 51:177–185. https://doi.org/10.1016/j.cnd.2016.02.001

Horie Y, Sugase K, Horie K (1995) Physiological differences of soluble and insoluble dietary fibre fractions of brown algae and mushrooms in pepsin activity in vitro and protein digestibility. Asia Pac J Clin Nutr 4:251–255

James CA, Welham S, Rose P (2023) Edible algae allergenicity–a short report. J Appl Phycol 35:339–352. https://doi.org/10.1007/s10811-022-02880-2

Jurković N, Kolb N, Colić I (1995) Nutritive value of marine algae *Laminaria japonica* and *Undaria pinnatifida*. Nahrung 39:63–66. https://doi.org/10.1002/food.19950390108

Kataoka H, Ohnishi N (1986) Occurrence of taurine in plants. Agric Biol Chem 50:1887–1888. https://doi.org/10.1271/bbb1961.50.1887

Kudełka W, Kowalska M, Popis M (2021) Quality of soybean products in terms of essential amino acids composition. Molecules 26(16):5071. https://doi.org/10.3390/molecules26165071

Marinho G, Nunes C, Sousa-Pinto I, Pereira R et al (2013) The IMTA-cultivated Chlorophyta *Ulva* spp. as a sustainable ingredient in Nile tilapia (*Oreochromis niloticus*) diets. J Appl Phycol 25:1359–1367. https://doi.org/10.1007/s10811-012-9965-3

Marrion O, Schwertz A, Fleurence J et al (2003) Improvement of the digestibility of the proteins of the red alga *Palmaria palmata* by physical processes and fermentation. Food Nahrung 47(5):339–344. https://doi.org/10.1002/food.200390078

Marrion O, Fleurence J, Schwertz A et al (2005) Evaluation of protein *in vitro* digestibility of *Palmaria palmata* and *Gracilaria verrucosa*. J Appl Phycol 17:99–102. https://doi.org/10.1007/s10811-005-5154-y

Matanjun P, Mohamed S, Mustapha N et al (2009) Nutrient content of tropical edible seaweeds, *Eucheuma cottonii, Caulerpa lentillifera* and *Sargassum polycystum*. J Appl Phycol 21:75–80. https://doi.org/10.1007/s10811-008-9326-4

McDermid KJ, Stuercke B (2003) Nutritional composition of edible Hawaiian seaweeds. J Appl Phycol 15:513–524. https://doi.org/10.1023/B:JAPH.0000004345.31686.7f

Morgan KC, Wright JL, Simpson FJ (1980) Review of chemical constituents of the red alga *Palmaria palmata* (dulse). Econ Bot 34:27–50. https://doi.org/10.1007/BF02859553

Munda IM (1977) Differences in amino acid composition of estuarine and marine fucoids. Aquat Bot 3:273–280. https://doi.org/10.1016/0304-3770(77)90029-8

Nisizawa K, Noda H, Kikuchi R, Watanabe T (1987) The main seaweed foods in Japan. Hydrobiologia 151:5–29. https://doi.org/10.1007/BF00046102

Noguchi T, Matsui T, Miyazawa K et al (1994) Poisoning by the red alga 'ogonori'(*Gracilaria verrucosa*) on the Nojima Coast, Yokohama, Kanagawa Prefecture, Japan. Toxicon 32(12):1533–1538. https://doi.org/10.1016/0041-0101(94)90311-5

Pellegrini M (1969) Contribution à l'étude chimique des Algues Méditerranéennes. Botanica Marina 12:179–184. https://doi.org/10.1515/botm.1969.12.1-4.179

Peña-Rodríguez A, Mawhinney TP, Ricque-Marie D et al (2011) Chemical composition of cultivated seaweed *Ulva clathrata* (Roth) C. Agardh. Food Chem 129:491–498. https://doi.org/10.1016/j.foodchem.2011.04.104

Queirós AS, Circuncisão AR, Pereira E et al (2021) Valuable nutrients from *Ulva rigida*: modulation by seasonal and cultivation factors. Appl Sci 11:6137. https://doi.org/10.3390/app11136137

Ratana-Arporn P, Chirapart A (2006) Nutritional evaluation of tropical green seaweeds *Caulerpa lentillifera* and *Ulva reticulata*. Agric Nat Resour 40:75–83

Rawiwan P, Peng Y, Paramayuda IGPB, Quek SY (2022) Red seaweed: a promising alternative protein source for global food sustainability. Trends Food Sci Technol. https://doi.org/10.1016/j.tifs.2022.03.003

Roque BM, Venegas M, Kinley RD et al (2021) Red seaweed (*Asparagopsis taxiformis*) supplementation reduces enteric methane by over 80 percent in beef steers. PLoS One 16:e0247820. https://doi.org/10.1371/journal.pone.0247820

Satoh K, Nakagawa H, Kasahara (1987) Effect of *Ulva* meal supplementation on disease resistance of red sea bream. Nippon Suisan Gakkaishi 53:1115–1120. https://doi.org/10.2331/suisan.53.1115

Sauvageau C (1920) Utilisation des algues marines, vol 1. Librairie Octave Doin

Seidel U, Huebbe P, Rimbach G (2019) Taurine: a regulator of cellular redox homeostasis and skeletal muscle function. Mol Nutr Food Res 63:1800569. https://doi.org/10.1002/mnfr.201800569

Shuuluka D, Bolton JJ, Anderson RJ (2013) Protein content, amino acid composition and nitrogen-to-protein conversion factors of *Ulva rigida* and *Ulva capensis* from natural populations and *Ulva lactuca* from an aquaculture system, in South Africa. J Appl Phycol 25:677–685. https://doi.org/10.1007/s10811-012-9902-5

Surai PF, Kochish II, Kidd MT (2020) Taurine in poultry nutrition. Anim Feed Sci Technol 260:114339. https://doi.org/10.1016/j.anifeedsci.2019.114339

Terriente-Palacios C, Castellari M (2022) Levels of taurine, hypotaurine and homotaurine, and amino acids profiles in selected commercial seaweeds, microalgae, and algae-enriched food products. Food Chem 368:130770. https://doi.org/10.1016/j.foodchem.2021.130770

Thomas I, Siew LQ, Watts TJ et al (2019) Seaweed allergy. J Allergy Clin Immunol Pract 7:714–715. https://doi.org/10.1016/j.jaip.2018.11.009

Uchida M, Kurushima H, Ishihara K, Murata Y et al (2017) Characterization of fermented seaweed sauce prepared from nori (*Pyropia yezoensis*). J Biosci Bioeng 123:327–332. https://doi.org/10.1016/j.jbiosc.2016.10.003

Vo TS, Ngo DH, Kim SK (2012) Marine algae as a potential pharmaceutical source for anti-allergic therapeutics. Process Biochem 47(3):386–394. https://doi.org/10.1016/j.procbio.2011.12.014

Vojdani A, Vojdani C (2015) Immune reactivities against gums. Altern Ther Health Med 21:64

Wong KH, Cheung PC (2001) Nutritional evaluation of some subtropical red and green seaweeds Part II. In vitro protein digestibility and amino acid profiles of protein concentrates. Food Chem 72:11–17. https://doi.org/10.1016/S0308-8146(00)00176-X

Young EG, Smith DG (1958) Amino acids, peptides, and proteins of Irish moss, *Chondrus crispus*. J Biol Chem 233:406–410

Zhou AY, Robertson J, Hamid N et al (2015) Changes in total nitrogen and amino acid composition of New Zealand *Undaria pinnatifida* with growth, location and plant parts. Food Chem 186:319–325. https://doi.org/10.1016/j.foodchem.2014.06.016

Chapter 2
Strategies and Major Challenges in the Manufacture, Use, and Development of Seaweed-Based Products for Agriculture

K. L. Baltrusch, M. Illera, M. E. López-Mosquera, H. Domínguez, and M. D. Torres

Abstract Obtaining seaweed-based products has been outlined as an eco-friendly alternative to the conventional use of fertilizers and pesticides, avoiding its impact on the environment and limiting the dependence on nonsustainable resources. Incorporating new eco-friendly alternatives is crucial in integrated agriculture, especially in organic farming, where only a limited number of crop management products are allowed. Seaweed-based products can enhance plant growth and health in numerous ways, which depend on their specific application strategy and pursued effect, these can be grouped into three major sections according to their properties: (i) as a fertilizer and organic amendment, (ii) as a soil quality improver, and (iii) as a plant biostimulant.

The remainder of this chapter will discuss the main advancements and challenges in this field of research focusing on (1) the agronomic interest of seaweed, (2) seaweed in agronomic soils, (3) seaweed as plant biostimulants, (4) the regulatory framework, (5) seaweed as an agricultural input and its applications, (6) nondesired effects, (7) science gaps, and (8) economic viability.

Keywords Algae · Biostimulant · Agriculture · Environmental friendly

K. L. Baltrusch · H. Domínguez · M. D. Torres (✉)
CINBIO, Universidade de Vigo (Campus Ourense), Department of Chemical Engineering, Edificio Politécnico, Ourense, Spain
e-mail: matorres@uvigo.gal

M. Illera · M. E. López-Mosquera
IBADER, Department of Plant Production, Campus Terra. University of Santiago de Compostela, Lugo, Spain

N. Trivedi et al. (eds.), *Recent Advances in Seaweed Biotechnology*,
https://doi.org/10.1007/978-981-96-0519-4_2

2.1 Agronomic Interest of Seaweed

The beneficial effects of algae on agricultural activity have been empirically verified by farmers in coastal areas since ancient times when they confirmed how they improved crop yields and harvest quality. On the other hand, anthropologists consider that the collection, transport, and application of algae on agricultural land is an agro-maritime activity. The farmers of the coast were aware of all those factors on which the supply of algae depended: the tides, the exposure of the beaches, the winds, etc. and they adapted their work tools used on land for their collection and developed new devices to collect them in the sea and on the shore of the beaches.

The benefits of algae for crops are the result of its unique composition that provides advantages on three levels: on the one hand, it acts as a fertilizer, as it is a complete long-term source of nutrients with a nitrogen content similar to most animal manures, a high potassium content, as well as a wide range of nutrients, both macro and micro. Seaweed is also an excellent soil quality improver, not only for its nutritional improvement but also for optimizing the structure, increasing the retention and availability of nutrients and water among other factors. Finally, algae also act as a plant biostimulant due to numerous bioactive compounds that generate a physiological response in plants that promote their growth and development while providing them with a greater defense against biotic and abiotic stresses.

There are numerous scientific research studies that have focused on identifying and quantifying all these benefits, finding positive results in plant growth and crop production (Popescu and Popescu 2014; Illera-Vives et al. 2015; Ali et al. 2016; Tawfeeq et al. 2016; Ghafarizadeh et al. 2017; Illera-Vives et al. 2017; Prakash et al. 2018; Taskos et al. 2019; Vijayakumar et al. 2019), increase in soil nutrition and plant nutrients absorption (Nedzarek and Rakusa-Suszczewski 2004; Ali et al. 2016), reduction of the affection of pests and diseases (Zhang et al. 2003; Raghavendra et al. 2007; Banakar et al. 2022), resistance to abiotic stresses (Mancuso et al. 2006; Shukla et al. 2019), and in improving the quality of harvests (Basak 2008; Chouliaras et al. 2009; Fan et al. 2013; Ali et al. 2016; Yuan et al. 2018).

2.1.1 Seaweed Composition Heterogeneity

The qualities mentioned so far cannot be generalized to all algae since seaweeds or macroalgae are extremely diverse multicellular marine autotrophs, comprising nearly 10,000 species (Battacharyya et al. 2015), divided into three major clades based on pigmentations: Chlorophyta (green seaweeds), Phaeophyta (brown seaweeds), and Rhodophyta (red seaweeds) (Khan et al. 2009).

Green and red seaweeds are commonly more diverse in subtropical and tropical waters, while brown seaweeds dominate cooler and more temperate waters (Santelices et al. 2009). These clades evolved in a variety of extremely dynamic habitats, with constant physical and chemical changes due to tidal cycles and environmental factors. Consequently, they adapted their physiology to survive in those habitats developing a high plasticity in terms of chemical composition, metabolite production, and response mechanisms to environmental variations (Domozych et al. 2012; Stengel et al. 2011). Therefore, we can conclude two major factors that influence the composition and effectiveness of seaweed products when considering the starting material: interspecific heterogeneity and seasonal-habitat variations.

2.1.1.1 Interspecific Heterogeneity

The different marine algae species have varying nutrient/biostimulant contents. Circuncisão et al. (2018) reviewed the mineral compositions of numerous macroalgae, showing notable differences between species and groups. Villares et al. (2007), in a comparative study carried out on algae from the coast of Galicia (NW Spain), concluded that the C and K contents were generally higher in brown algae, while green algae contained more N, Na, Mg, and Fe, and less P and C. On the other hand, the red algae showed high contents in N, P, Ca, Cu, Zn, and B and low in Na, Mg, Fe, and Mn.

A detailed list of biostimulant compounds was reviewed by Ali et al. (2021b) covering a large spectrum of bioactive molecules such as polyphenols, pholorotannins, polysaccharides, lipids, oxylipids, sterols, carotenoids, betaines, proteins, peptides, and amino acids. Cell wall matrix carboxylic and sulfated polysaccharides were pointed out as one of the major compositional differences between the three macroalgae groups (Goñi et al. 2020) and are ascribed to have important biostimulant properties (Pacheco et al. 2021). In this sense, while red seaweeds typically present carrageenans and agars; brown seaweeds are rich in alginates, fucoidan, laminarin, and mannitol; and green seaweeds are characterized by the presence of ulvans (Goñi et al. 2020). Biostimulant effectiveness can also vary depending on specific algae species-biostimulant combinations due to specific ligand-receptor mechanisms (Ali et al. 2021b).

2.1.1.2 Seasonal and Habitat Variations

The physical and chemical parameters such as wave exposure, temperature, light, salinity, pH, and mineral oxidation state are subject to seasonal variations and geographical location, which modify habitat conditions and impact seaweed's mineral and nutrient uptake mechanisms or induce new metabolic pathways which all together influence their chemical composition (Circuncisão et al. 2018). Thus, macroalgae harvested at different growth stages or seasons can have different levels and types of nutrients and biostimulant compounds. Balboa et al. (2016) showed that the mineral and protein contents correlate with seasonal variations in *Sargassum muticum*. Schiener et al. (2015) observed seasonal variations in *Laminaria digitata, Laminaria hyperborea, Saccharina latissima,* and *Alaria esculenta*, with the highest nitrogen yields in spring and lowest in autumn. The latter study also supported the general assumption that storage carbohydrates such as laminarin or mannitol are accumulated during the summer months and consumed during the winter. Brown seaweeds exposed to more hydrodynamic stress also contain higher amounts of alginate (Gillespie and Critchley 1999; McHugh 2003). The sulfated polysaccharides content increases in *Gracilaria vermiculophylla* during the reproductive stage and decreases after the vegetative stage (Sfriso et al. 2017). Similarly, fucoidan content peaked during the reproduction phase in *Sacharina japonica, Sargassum pallidum,* and *Stephanocystis crassipes* (Skriptsova 2016). Sulfate patterning of macroalgae polysaccharides might be influenced by fluctuations in salinity levels, as noted by Shepherd et al. (1999) or Skriptsova (2015). Furthermore, brown seaweeds overproduce and modify polysaccharides of the cell walls in response to heavy metal-polluted waters (Andrade et al. 2010). It is also important to note that physiological changes do not only occur between individuals of the same species but also in the organism itself. The composition of cells and tissues can vary and evolve over time (Raimundo et al. 2016; Torode et al. 2015), indicating a structural diversity which is tailored to meet the current functional demands of the specimen.

2.2 Seaweed in Agronomic Soils

Seaweed can provide numerous benefits that contribute to increasing soil quality in the three components that, according to the International Conference on Soil Quality Evaluation and Monitoring (Arshad and Martin 2002), define soil quality: a productivity component (the ability of soil to improve crop production); an environmental quality component (the ability of soil to alleviate the effects of environmental pollutants and pathogens, and to prevent offsite damage); and a health-related component (the ability of soil to improve human, animal, and plant health).

The presence of naturally occurring nitrogen, phosphorus, and particularly potassium endows seaweeds with a nutrient-rich composition, making them an interesting option as closed-loop fertilizers in agriculture (Silva et al. 2019). This is particularly beneficial for fruit crops, which have a high demand for potassium (Wang et al. 2022). Besides, seaweeds have exceptionally high mineral content when compared to land vegetables and important proportions of trace elements such as Fe, Zn, Mn, and Cu (Circuncisão et al. 2018; Rupérez 2002). The low C/N ratio of the algae (average 16, according to Illera-Vives et al. 2020) indicates that its decomposition in the soil is faster than that of other organic fertilizers. All these factors make algae an important resource as an agricultural fertilizer, especially in the present context, where climate change and an increasing global population make it necessary to increase food production. Given the economic and environmental problems of traditional chemical fertilizers, farmers increasingly demand a source of nutrients of organic origin. However, algae can provide other benefits that contribute to increasing the quality of the soil, which goes beyond the contribution of nutrients.

The seaweed polyanionic compounds can facilitate the fixation and exchange of cations. This is particularly significant as it aids in the immobilization of heavy metals, reduces soil toxicity, and enhances phytoremediation, as noted by Du Jardin (2015). Negative surface charges of cell wall polysaccharides also enable effective retention of positively charged cations or molecules from seawater (Alba and Kontogiorgos 2018). In addition, other organic molecules are present, such as organic acids, methionine, alginates, and fucoidan that can chelate some nutrients, thus increasing their bioavailability (Lynn 1972; Khan et al. 2009). The algae also have the liming capacity, due to their Ca and alginates content that complex the exchanged Al, thus neutralizing the acidity of aluminous soils (Crouch et al. 1990; Eyras et al. 1998; López-Mosquera and Pazos 1997). This complexation of Al also produces an increase in the availability of P (which precipitates with P, forming insoluble phosphates) (López-Mosquera and Pazos 1997). Seaweeds also are known to contain complex polysaccharides that function as gelling agents. These substances promote water retention and soil aeration, which are important for plant growth.

Seaweed soil applications improve soil quality through the stimulation of soil microorganisms. Recently, Wang et al. (2018) verified how algae-based fertilization induced improvements in the microbial and enzymatic activity of the soil.

2.3 Seaweed as Plant Biostimulant

The European Regulation (EU) 2019/1009 defined plant biostimulant as "Certain substances, mixtures and microorganisms, which are not as such inputs of nutrients, but nevertheless stimulate plants' natural nutrition processes. Where such products aim solely at improving the plants' nutrient use efficiency, tolerance to abiotic stress, quality traits or increasing the availability of confined nutrients in the soil or rhizosphere, they are by nature more similar to fertilizing products than to most categories of plant protection products. They act in addition to fertilizers, with the aim of optimizing the efficiency of those fertilizers and reducing the nutrient application rates."

The presence of bio-stimulating substances in algae becomes evident when they are used in extracts. These are applied in a very diluted form (1:1000 or higher); therefore, the benefits observed in crops cannot be attributed to their fertilizing effect or to the improvement of soil properties, but also to the presence of other substances that positively influence plant development. Positive effects on crops were observed when algae extracts were applied, which disappeared when the extracts were reduced to ashes. This led to the idea that the components of the algae responsible for the improvements observed in crops were due to organic compounds (Finnie and Van Staden 1985).

Initially, most of the plant responses to algae extracts were attributed to their hormone content. In fact, the effect of algae extracts is very reminiscent of the activity of phytohormones, as they promote growth at low concentrations and inhibit it at high concentrations (Battacharyya et al. 2015). Algae naturally possess phytohormones, as they, like higher plants, require these growth regulators to control morphogenic events in their life cycle (Stirk et al. 2003), although their metabolism and regulation are less complex than in higher plants due to the wide range of morphogenic events that take place in the latter (Auer 1997).

The presence of hormones such as cytokinins, auxins, and abscisic acid in algae, along with other hormone-like compounds such as sterols and polyamines, is supported by numerous authors (Stirk et al. 2003; Khan et al. 2009; Craigie 2011; Stirk et al. 2014; Battacharyya et al. 2015; du Jardin 2015). However, the hormone-like activity of seaweed extracts cannot be explained solely by their hormone content (Battacharyya et al. 2015). Evidence for this fact is that despite the large differences in phytohormone content of algae depending on the season of collection (Featonby-Smith and Van Staden 1984), the benefits of these on plants remain throughout the year (Craigie 2011). The components of algae responsible for these bio-stimulating effects, as well as their mechanism of action, are still a matter of debate (Craigie 2011). However, Stirk et al. (2020) list in their review the main biostimulant compounds that can be found in algae extracts:

- Plant hormones: cytokinins, auxins, GAs, ABA, and ethylene. Mainly involved in cell division and differentiation, germination, flowering, nutrient mobilization, senescence, apical dominance, root development, and plant growth.
- Brassinosteroids: they act at the whole-plant level, influencing plant growth and providing protection against abiotic stresses. They act on stem elongation, reproductive processes, and seed germination. They are involved in cell division, flower and fruit development, seed yield, and stimulation of the production of other metabolites involved in plant development.
- Betaines: protects the plant by accelerating repair of the photosystem, improving the elimination of reactive oxygen species, stabilizing proteins and plasmatic membranes, and acting synergistically with other protective agents against stress.
- Poliamines: improve the phytonutrient content and the quality of the fruits.
- Polymers: such as agar, alginate, carrageenans, fucans, and phlorotannins. These may act as signal molecules to regulate growth and eliciting defense mechanisms.

As mentioned, a key aspect is the synergy that occurs between these different compounds to produce a positive response in the plant, where each component can act in several metabolic networks independently or interactively, as evidenced by tests that used different fractions of the extracts, none of them separately obtaining a beneficial effect on the plant (Ali et al. 2021a, b).

In addition, these molecules do not usually have a direct effect, but instead affect the physiology of the plants, acting as inducing molecules that trigger protective responses to stresses, thus increasing the resistance of plants to pests, diseases, abiotic stresses, and improving the plant growth due to the interaction with the growth involved genes (Stirk et al. 2020; Ali et al. 2021a, b). In fact, in his review, Battacharrya (2015) indicates that there are numerous studies demonstrating that they can change the metabolome of plants. The beneficial effect of algae biostimulants on crops is varied, finding in the literature benefits such as improved germination (Silva et al. 2019; Vijayakumar et al. 2019; Tavares et al. 2020; Mendes et al. 2022), increased growth (Popescu and Popescu 2014; Rengasamy et al. 2015; Ali et al. 2016; Prakash et al. 2018; Silva et al. 2019; Vijayakumar et al. 2019), increased root system (Rengasamy et al. 2015; Wang et al. 2016), improved nutrient absorption (Popescu and Popescu 2014; Ali et al. 2016), increased chlorophyll production (Thirumaran et al. 2009; Yao et al. 2020), improved flowering (Basak 2008), high yield and harvests quality (Silva et al. 2019; Chouliaras et al. 2009; Fan et al. 2013; Ali et al. 2016; Tawfeeq et al. 2016; Yuan et al. 2018; Taskos et al. 2019; Vijayakumar et al. 2019), and increased resistance to both biotic and abiotic stresses such as pests (El-Ansary and Hamouda 2014), diseases (Vera et al. 2011), drought and salt (Xu et al. 2018), freezing (Zamani-Babgohari et al. 2019). These effects are especially notable when crops are below optimal conditions (Crouch and Van Staden 1994; Craigie 2011).

2.3.1 *The Extraction Process*

Each method has its own advantages and limitations, and their effectiveness can vary depending on the seaweed species used and the compounds being extracted. Nevertheless, there are some common technical constraints that can occur during the extraction of seaweed-based products for agriculture (Fig. 2.1).

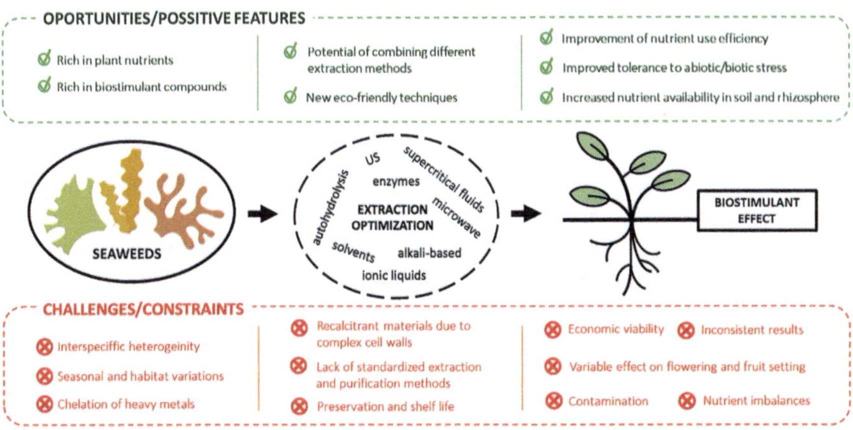

Fig. 2.1 General overview of the effect of the bioestimulant's effect of seaweeds

2.3.1.1 Obtaining Seaweed Products or Specific Compounds for Agriculture Is a Complex Process That Requires Process Optimization to Achieve High Yields and Purity

The variability of the input biomass commented on in Sect. 2.1 has to be addressed by sustainable and optimized strategies that consider the relation between the target compounds and species, seasonal growth, habitat conditions, and pre-treatments. Moreover, efficient harvest, quick processing, and adequate storing are crucial. This kind of optimization process has been developed and proven successful for some commercially important seaweeds as kelp, the main source of alginate (Peteiro 2018) or carrageenan sources as *Kappaphycus alvarezii* (Domínguez-May et al. 2017), but there is a significant requirement to optimize the harvesting of other crucial seaweeds.

Extraction protocols for seaweeds are very diverse and depend on the target compounds, the type of biomass/matrix, and the location of these compounds within the matrix. Any extraction method and optimization process aims to maximize yields of the target compounds, minimize energy consumption and solvent use, reduce waste production, ensure scalability, or reduce the amount of time taken up by the extraction process (El Boukhari et al. 2020; Kadam et al. 2013).

Seaweed-derived biostimulant products can be produced using different extraction methods such as chemical, enzymatic, or physical extraction. Minimally processed seaweed homogenates were also described as effective biostimulants (Vaghela et al. 2023). Classical extraction methods involve more solvent use or other chemical inputs and are time-consuming (Kadam et al. 2013). The prevalent technique involves extracting the active compounds from the biomass through either water or alkali-based extraction. Subsequently, the extracted material is either directly utilized or processed further to obtain a concentrated form (Craigie 2011). Unfortunately, alkali treatments lead to the production of novel compounds, which were not originally present in the seaweed biomass (Shukla et al. 2019).

More recent methods such as supercritical fluid extraction, pressurized liquid extraction, ionic liquid extraction, and enzyme-assisted extraction have been optimized to replace classical extraction techniques and improve extraction efficiencies of seaweed compounds (Kadam et al. 2015; Martins and Ventura 2020; Otero et al. 2018; Quitério et al. 2022). Nowadays, sustainable extraction technologies are still preferred in green integral biorefinery approaches (Torres et al. 2019). These techniques are based only on water as the main solvent and have been optimized in the last years for different compounds and seaweeds (Sarkar et al. 2022). Some of them include autohydrolysis (Flórez-Fernández et al. 2021), microwave-assisted extraction (Queffelec et al. 2021), and US-assisted extraction (Flórez-Fernández et al. 2019). Some new approaches also combine more than one of these novel methods, such as microwave-assisted extraction and bio-ionic liquids in different stages (André et al. 2023) or autohydrolysis with ionic liquids in the same stage (Vo Dinh et al. 2018).

2.3.1.2 The Extraction Efficiency of Biostimulant Compounds from Seaweed Is Limited Due to the Presence of Complex Cell Walls

Seaweed cell walls are composed of multiple layers of a variety of entangled sulfated and branched polysaccharides associated with proteins. The degradation of this structure is a fundamental step for the isolation of bioactive substances, but its recalcitrant complexity often limits the extraction efficiencies and isolation of the compounds of interest (Deniaud-Bouët et al. 2017; Domozych et al. 2012).

2.3.1.3 Preservation of Seaweed Extracts Is a Significant Challenge, as They Are Susceptible to Biotic Contamination and Abiotic Degradations

To increase shelf-life, it will be crucial to create stable solid formulations that are easy to use and to possess, but also to identify suitable stabilizers and additives to maintain their quality over time (Ali et al. 2021b). Especially liquid formulations tend to become unstable over time, posing challenges in terms of storage and preservation. To address these issues, drying is a recommended method as it yields a more manageable product with reduced volume and longer shelf-life. Convective air drying involves subjecting the algae to a sustained and moderately high temperature of up to 75 °C, which typically takes around 100 minutes to complete. Alternatively, lyophilization is a slow process that involves the sublimation of liquid extracts over multiple days until only solid material remains. A faster drying method is spray-drying, which involves atomizing a liquid sample into a hot gas current, yielding a fine powder as the final product in just a few hours. Recent research has shown that low-stress drying techniques such as spray-drying result in better preservation of bioactive compounds and biostimulant potential (Baltrusch et al. 2022).

2.3.1.4 Lack of Standardized Extraction and Purification Methods of Biostimulants

Standardized methods for the production and quality control of seaweed-based agricultural products would help to ensure the consistency and efficacy of biostimulant products and facilitate their regulatory approval. In agriculture, seaweed products are typically regarded as environmentally safe and organic if they consist primarily of natural ingredients. However, it is not uncommon to find products marketed as "seaweed organic biostimulants" that contain only a small percentage of seaweed-derived components and consist mainly of synthetic mineral nutrients and chemicals. These products are often marketed as biofertilizers or biostimulants, causing confusion among users. Therefore, it is necessary to establish stringent quality standards to accurately identify, certify, and label products according to the authenticity of their organic composition (Ali et al. 2021b; Caradonia et al. 2019).

Deniaud-Bouët et al. (2017) highlighted a common challenge in obtaining bioactive polysaccharides, which lies in the preparation and structural identification of homogeneous fractions. In many cases, biological response studies rely on impure mixtures that may include phenols, proteins, or other polysaccharides, which could potentially account for the observed biological activities.

The analytical techniques for the physicochemical characterization of seaweed-derived biostimulants are diverse and depend on their chemical nature. Spectroscopic methods, such as Fourier-transformed infrared (FTIR) and nuclear magnetic resonance (NMR) spectroscopy, are commonly used to identify functional groups and structural components within crude seaweed extracts (Ertani et al. 2018) or isolated compounds as carbohydrates (Pereira 2017; Pereira et al. 2009; Pereira and Van De Velde 2011). Sulfate content, protein content, total phenolic content, and antioxidant capacity of seaweed-derived extracts can be assessed through UV–Vis spectroscopy involving several colometric methods described elsewhere (Baltrusch et al. 2023; Flórez-Fernández et al. 2022). High-performance liquid chromatography (HPLC) and gas chromatography-mass spectrometry (GC-MS) are instrumental in quantifying specific compounds of seaweed metabolomics, including phytohormones or phenolic compounds, as well as for the determination of monosaccharide composition of polysaccharides, which play crucial roles in plant growth regulation (Mzibra et al. 2021; Paul et al. 2019). Additionally, elemental analysis and inductively coupled plasma mass spectrometry (ICP-MS) are commonly used to determine the mineral content of seaweed biostimulants (André et al. 2023; Flórez-Fernández et al. 2022).

2.4 Regulatory Framework: The Case of the EU

As pointed out by Ali et al. (2021a, b), labeling seaweed-based agricultural products presents certain difficulties due to their various activities and functions. Since the mid-twentieth century, the use of seaweed as a fertilizer has been shifted using different types of seaweed extracts, which are employed as biostimulants and are gaining special importance in modern agriculture (Metting et al. 1990; McHugh 2003; Battacharyya et al. 2015). However, the existing regulatory framework is intricate and inconsistent, with no uniform definition of plant biostimulant (PB) globally. The implementation of consistent regulations could simplify the establishment of a robust risk assessment framework, promoting the safe and legitimate use of these novel products worldwide. Increased harmonization would aid in the widespread acceptance and development of PBs in the market (Caradonia et al. 2019; du Jardin 2015).

Plant defense inducers (PDI) are often referred to as or described as part of PB in the literature (Ali et al. 2021a; du Jardin 2015; Rouphael and Colla 2020; Shukla et al. 2019). But, recently, article 47 of Regulation (EU) 2019/1009 excluded PDI from the PB category, describing the latter as products that enhance plant nutrition processes, regardless of their nutrient content, with the primary goal of enhancing nutrient utilization efficiency, increasing tolerance to abiotic stresses, improving crop quality characteristics, promoting the availability of nutrients in the soil and rhizosphere, as well as promoting the humification and degradation of organic compounds; but with no effect on biotic stress resistance.PDI do not yet have a specific regulatory framework in the EU. However, they may fall under the scope of the Plant Protection Products Regulation (EU) 1107/2009, which regulates the place on the market and use of plant protection products. If a seaweed-derived product contains plant defense-eliciting compounds and is marketed as a pesticide, it must comply with the regulation's requirements, including safety and efficacy assessments, labeling requirements, and restrictions on use. A more detailed disambiguation between these products is necessary and has to be accompanied by the development of a clear regulatory framework.

Organic fertilizers (OF) are regulated under the EU Organic Farming Regulation (EU) 2018/848, which sets out the rules for organic production and labeling of organic products. The regulation defines fertilizers derived from plant or animal material, raw or processed, or from mineral origin if they have been produced and processed in accordance with the regulation's requirements. OF must comply with strict rules on their composition, origin, and production process, and must be labeled as organic according to the regulation's guidelines.

Soil SI and LM are also regulated by EU legislation under Regulation (EU) 2019/1009. SI and LM must meet certain requirements in order to be placed on the market in the EU. These requirements include labeling and packaging requirements, as well as requirements for the composition and quality of the product. In addition, the regulation establishes a list of permitted substances and sets out limits for the presence of certain contaminants in these products.

Although a common regulatory basis, in the EU the legislative framework for plant biostimulants (PBs) is fragmented and varied, and may also be subject to national regulations in individual EU member states, which may impose additional requirements or restrictions (Caradonia et al. 2019).

2.5 Seaweed as an Agricultural Input

As previously mentioned, seaweed has been used for centuries directly in the cultivated soils of coastal areas. Its simplest use is its fresh application, mixing it directly with the soil, or subjecting it to simple transformations to solve some of the problems that its application entails. Thus, they can be exposed to rain to wash them and eliminate part of the salts; air-dry to facilitate handling, or compost with other residual organic materials such as straw or peat to improve its qualities as an amendment and/or fertilizer.

Another common practice is the stockpiling of algae in fields that are periodically aerated to reduce the presence of toxic sulfur compounds (Milton 1964; Craigie 2011). However, although they continue to be used fresh or sun-dried and incorporated directly into the soil, their application has become more technical and today you can find numerous (more or less complex) commercial fertilizer products made from algae. These can be presented in a wide variety of formats, from the simplest such as algae meal or compost, to other more complex forms such as biostimulant extracts. Currently, nanoparticles and nanomaterials are starting to gain attention as part of potential plant biostimulant technologies, used to increase the surface area of active ingredients, allowing them to interact more effectively with cell membranes and receptors (Juárez-Maldonado et al. 2019). In addition to their potential as plant stimulants, the use of nanomaterials in agriculture has garnered growing attention due to their ability to facilitate the controlled release of chemicals, pesticides, and plant growth regulators (Muzzalupo et al. 2020). Mohamed et al. (2023) developed seaweed-derived nanoparticles using the planetary ball milling method which showed enhanced growth parameters in date palm seedlings.

On the other hand, more massive management strategies such as composting have also gained special importance in situations where excessive algae growth causes them to become waste that needs to be managed. Numerous authors have carried out algae composting experiences to obtain fertilizers, considering it one of the most appropriate techniques from the economic and environmental point of view (Potoky et al. 1988; Mazé et al. 1993; Cuomo et al. 1995; Eyras and Sar 2003). However, the low C/N ratio of the algae means that they must be mixed with a material rich in carbon such as sawdust (Eyras et al. 1998), rice husk (Wang et al. 2009), straw (Wosnitza and Barrantes 2006), sugarcane bagasse (Cole et al. 2016), or pine bark (Illera-Vives et al. 2013) among others. The algae can also be co-composted with other nutrient-rich materials that increase the fertilizing power of the final product, such as fish remains (Illera-Vives et al. 2013) or manure (Wosnitza and Barrantes 2006).

2.5.1 Application Techniques

Seaweed products need to be applied at the appropriate dosage and time to ensure their efficacy and prevent negative effects on plant growth and development. However, the optimal dosage and application method may vary depending on the crop, the soil type, and the environmental conditions. The use of suboptimal or excessive dosages, or inappropriate application methods, may lead to reduced yield and economic losses (Table 2.1).

For liquid seaweed extract application, foliar sprays are a widely used method. It consists of the dispersion of a liquid solution directly onto the leaves allowing rapid absorption into the plant tissue leading to a rapid response. It is a suitable method for crops with a large surface area. Seaweed liquid extracts used as foliar sprays could be supplemented with other compounds such as N, P, K, chelated trace nutrients, wetting agents, or sticking agents to add properties or enhance effectiveness. However, it is important to note that the use of macroalgae extracts augmented in this way is not permitted under organic production standards for biological agriculture. To the best of the authors' knowledge no seaweed extract formulations with

Table 2.1 Examples of applications technologies

Application technology	Advantages	Constraints	Examples of seaweed-derived commercial biostimulants[a]
Foliar spray	Allows targeting specific plant tissues and organs Ensures rapid uptake and response of the plant Can be easily applied with conventional spraying equipment	Effectiveness can be affected by environmental factors (humidity, temperature, and wind) Possible phytotoxic effects if not applied with precision Repeated applications may be necessary to maintain effectiveness	Kelpak® SL (Kocira et al. 2013) Maxicrop® Whapham et al. 1993) TAM® (Ashour et al. 2021). Stimplex® (Ozbay and Demirkiran 2019) Goëmar BM 86®, (Basak 2008)
Fertigation	Efficient delivery of nutrient and active compounds to the roots	Requires a sophisticated irrigation and careful monitoring	Acadian® (Alam et al. 2014) Actiwave® (Spinelli et al. 2010)
Root dipping	Ensures direct delivery to the root	May require additional handling of the plants Not suitable for all crops	
Soil drench	Can improve soil microbiology and mitigate soilborne disease Can chelate beneficial nutrients	The effectiveness can be affected by soil type and environmental conditions Requires higher amounts of product for widespread application Can be slow to take effect	
Seed coating	Allows precise delivery to the developing plant Can improve seed germination and early plant establishment Protection from abiotic and biotic stresses at early stages	Requires specialized equipment for production Can be expensive compared to other Requires proper storage and handling to prevent damage to seed coating	Seed coating with *Sargassum muticum* (Patent no. ES 2 872 648 B2 19) Not commercialized yet
Soil amendments	Shared advantages with soil drench, but in addition: Improvement soil structure and nutrient availability Can provide long-term benefits to the plant	Shared disadvantages with soil drench, but in addition: Higher risk of heavy metal contamination	

[a]Biostimulant products can sometimes be applied both on soil and leaves, although the manufacturer may recommend a specific application method

added adjuvants were tested, but "adjuvant" properties, as surfactant effects, were identified in seaweed extracts applied as foliar sprays (Mueller and Kussow 2005).

Other methods of applying liquid biostimulants are root dripping and seed coating. Root dipping involves dipping the roots of the plant in a solution of the biostimulant product prior to transplanting. This method ensures that the biostimulant is delivered directly to the root system of the plant. Root dipping can help enhance root development, nutrient uptake, plant establishment, and reduction of the transplant shock (Khan et al. 2009). However, this method may not be suitable for all plant species and may require additional handling of the plant.

Seed coating or pelleting involves the application of seaweed products to the surface of the seed, either by coating or pelleting. Seed coating involves applying a layer of the biostimulant product on the surface of the seed, while seed pelleting involves mixing the biostimulant product with a coating material that is then applied to the seed. This method ensures that the biostimulant is delivered directly to the plant during germination and early growth stages. Seed coating and pelleting can help enhance seedling emergence, root development, and early plant growth. However, this method may not be effective for all crops, as some seeds may have an adverse reaction to the biostimulant product or the coating material.

2.6 Nondesired Effects on Plant Growth and/or Health of Seaweed-Derived Products

Although seaweed products have several potential benefits, there are also some undesirable side effects associated with seaweed-based products that require careful consideration.

Inconsistent results The effectiveness of a seaweed-based product depends on its specific composition and as commented in Sect. 2.1.1, possible seasonal and habitat variations make it difficult to have a homogeneous production, requiring, for example in the case of its use as fertilizer, a chemical analysis prior to each application. In seaweed extracts, the extraction method must also be taken into account. El Boukhari et al. (2020) pointed out the extraction method as the most important step for obtaining an efficient seaweed-derived product, with varying yields of the target compounds or extract compositions depending on the used technique.

Influence on flowering and fruit setting The use of biostimulants to improve flowering and fruit setting depend on factors such as the specific type of seaweed used, the type of plant being treated, the stage of growth, the method of application, the composition of the extract, and the duration of exposure, the outcome can be either a delay or a promotion of fruit setting (keya Tudu et al. 2022).

Contamination Seaweed-based biostimulants can potentially be contaminated with other substances, such as pollutants or toxins, which can harm plants or humans if they come into contact with them. Seaweeds grown in polluted waters can contain heavy metals and other contaminants that may affect the quality and safety of the biostimulant or biofertilizer and add legal limitations to its use (Circuncisão et al. 2018; Strain et al. 2022).

Nutrient imbalances Seaweed-based biostimulants can provide a range of nutrients to plants, but if used excessively or inappropriately, they can cause nutrient imbalances, which can lead to plant damage or reduced yields. Some seaweed-based products may contain high levels of salt, which can lead to increased soil salinity if used excessively.

2.7 Science Gaps

Despite the increasing attention and interest in seaweed-based products for agriculture, there are still many research gaps and challenges that need to be addressed.

Cell walls are a source of diverse polysaccharides exerting unique biostimulant properties, which could bring significant benefits to sustainable agriculture. However, the mechanisms of action are not well comprehended yet (Chaturvedi et al. 2022; Goñi et al. 2020; Vera et al. 2011). To the best of our knowledge, no recent review on red seaweed cell walls is available and more effort has to be made to combine agriculture and chemical engineering research. Up to now, these two research fields are mostly disconnected. Since the synthesis of polysaccharides is not as easy as in the case of nucleotides and proteins, it becomes even more important to create a detailed inventory of the variability of cell walls. This inventory would be critical to better understand the potential of the matrix carbohydrates and to ensure the obtaining of high-quality products (Deniaud-Bouët et al. 2017).

Often, the research in this field tends to concentrate on greenhouse and controlled environment trials, which may not fully capture real-world conditions. To provide growers with reliable recommendations, it is crucial to conduct rigorous experimentation in the field and collect data from diverse locations, multiple crops, and different seasons. Only through such comprehensive analysis can we ensure that our recommendations are effective and applicable (Ali et al. 2021b).

Despite the positive results obtained from the combination tests of synthetic fertilizers with seaweeds as granular organic fertilizers (Kumar et al. 2020) or as biostimulants (Ali et al. 2023), there is still a technological gap in the development of composite formulations or value-added formulations that incorporate seaweeds with adjuvants or other biological components, such as live microorganisms or plant extracts (Ali et al. 2021b).

Seaweed-based products are known for their complex nature. Added to the overlapping of a large network of physiological responses, to identify and describe the mechanism of action of the bioactive compounds is extremely challenging. In addition, the extraction methods play a crucial role in the commercial formulations of seaweed extracts. It is difficult to extract and purify potent single compounds from this mixture, making it more complex to understand the precise mechanism of action, and the significance of the synergistic action is often unclear and changes. Deeper characterization of the major active constituents of seaweeds would be beneficial for a better insight into the complex interactions between seaweed-derived compounds and plant physiology (Al-Juthery et al. 2020; Battacharyya et al. 2015).

Seaweed products contain bioactive compounds that can interact with plant receptors and signaling pathways, leading to changes in gene expression and metabolic processes. However, the specificity and intensity of these interactions can vary depending on the plant species and environmental conditions. It is important to note that different types of crops and climatic conditions respond differently to the concentration, type, and application frequency of the extracts. To address these challenges, more crop-specific research is needed to customize systematic loopholes and optimize the effectiveness of biostimulant applications (keya Tudu et al. 2022).

2.8 Economic Viability of Seaweed-Based Agriculture Products

The Food and Agriculture Organization (FAO) has observed a substantial increase in the annual production of aquatic algae worldwide. This production has grown over 60 times from 0.56 million tons in 1950 to 35.82 million tons in 2019. Seaweeds have been the primary contributor to this increase, accounting for 99.84%, with 97% of this attributed to cultivation (FAO 2021). In recent years, seaweed extracts have become increasingly popular as biostimulants, making up over 33% of the total biostimulant market globally. It is expected that the seaweed extract market will reach a value of 894 M Euro in 2022, according to Brouwers et al. (2018).

Economic viability is a significant constraint for seaweed-based agricultural products. One of the major challenges faced in the process of seaweed valorization is that it is often economically unviable to extract a single molecule or compound, unless it has an exceptionally high market value (Orozco-González et al. 2022). The high cost of seaweed extraction and processing can make these products more expensive than traditional fertilizers and other biostimulants. However, the potential benefits of increased yield and quality of crops, as well as the environmental advantages of using natural and sustainable products, may offset these costs. Additionally, the development of new and more efficient extraction and processing methods, as well as multistage cascade processes which divide seaweeds into a plethora of valuable products, may help to reduce the production costs of seaweed biostimulants, making them more economically viable by enabling the co-extraction with other bioactive ingredients, chemicals, and biofuels (El-Shenody et al. 2023; Flórez-Fernández et al. 2021; Torres et al. 2019).

Besides the costs of raw materials, processing, and required dose, the economic viability of a seaweed-based product can also be influenced by the application method. A two-year field trial conducted by Krawczuk et al. (2023) with soybean showed that the use of standard nozzles for seaweed-derived biostimulant application has been demonstrated to be more cost-effective than using air-injector nozzles.

In conclusion, to guarantee the efficiency and worth of agricultural products derived from seaweed, it is crucial to take these elements into account and establish suitable approaches for their manufacturing, usage, and marketing.

References

Alam MZ, Braun G, Norrie J, Hodges DM (2014) *Ascophyllum* extract application can promote plant growth and root yield in carrot associated with increased root-zone soil microbial activity. Can J Plant Sci 94(2):337–348. https://doi.org/10.4141/CJPS2013-135

Alba K, Kontogiorgos V (2018) Seaweed polysaccharides (agar, alginate carrageenan). In: Encyclopedia of food chemistry, pp 240–250. https://doi.org/10.1016/B978-0-08-100596-5.21587-4

Ali N, Farrell A, Ramsubhag A, Jayaraman J (2016) The effect of *Ascophyllum nodosum* extract on the growth, yield and fruit quality of tomato grown under tropical conditions. J Appl Phycol 28(2):1353–1362

Ali O, Ramsubhag A, Jayaraman J (2021a) Phytoelicitor activity of *Sargassum vulgare* and *Acanthophora spicifera* extracts and their prospects for use in vegetable crops for sustainable crop production. J Appl Phycol 33(1):639–651. https://doi.org/10.1007/s10811-020-02309-8

Ali O, Ramsubhag A, Jayaraman J (2021b) Biostimulant properties of seaweed extracts in plants: implications towards sustainable crop production. Plan Theory 10(3):1–27. https://doi.org/10.3390/plants10030531

Ali ESESES, El-Samad EEDHA, Abdelaal HA, AbouSekken MS (2023) Enhancement of Quinoa grain yield and nutritional quality by potassium fertilization combined with Foliar spraying of seaweed extract. J Ecol Eng 24(3):341–356. https://doi.org/10.12911/22998993/158384

Al-Juthery HWA, Abbas Drebee H, Al-Khafaji BMK, Hadi RF (2020) Plant biostimulants, seaweeds extract as a model (article review). IOP Conf Ser Earth Environ Sci 553(1). https://doi.org/10.1088/1755-1315/553/1/012015

Andrade LR, Leal RN, Noseda M, Duarte MER, Pereira MS, Mourão PAS, Farina M, Amado Filho GM (2010) Brown algae overproduce cell wall polysaccharides as a protection mechanism against the heavy metal toxicity. Mar Pollut Bull 60(9):1482–1488. https://doi.org/10.1016/j.marpolbul.2010.05.004

André J, Flórez-Fernández N, Domínguez H, Torres MD (2023) Microwave-assisted extraction of *Ulva* spp. including a stage of selective coagulation of ulvan stimulated by a bio-ionic liquid. Int J Biol Macromol 225(November 2022):952–963. https://doi.org/10.1016/j.ijbiomac.2022.11.158

Arshad MA, Martin S (2002) Identifying critical limits for soil quality indicators in agroecosystems. Agric Ecosyst Environ 88:153–160

Ashour M, Hassan SM, Elshobary ME, Ammar GAG, Gaber A, Alsanie WF, Mansour AT, El-shenody R (2021) Impact of commercial seaweed liquid extract (Tam®) biostimulant and its bioactive molecules on growth and antioxidant activities of hot pepper (*Capsicum annuum*). Plan Theory 10(6). https://doi.org/10.3390/plants10061045

Auer CA (1997) Cytokinin conjugation: recent advances and patterns in plant evolution. Plant Growth Regul 23(1–2):17–32

Balboa EM, Gallego-Fábrega C, Moure A, Domínguez H (2016) Study of the seasonal variation on proximate composition of oven-dried *Sargassum muticum* biomass collected in Vigo Ria, Spain. J Appl Phycol 28(3):1943–1953. https://doi.org/10.1007/s10811-015-0727-x

Baltrusch KL, Torres MD, Domínguez H, Flórez-Fernández N (2022) Spray-drying microencapsulation of tea extracts using green starch, alginate or carrageenan as carrier materials. Int J Biol Macromol 203:417–429. https://doi.org/10.1016/j.ijbiomac.2022.01.129

Baltrusch KL, Flórez-Fernández N, Illera M, Torres MD, López-Mosquera ME, Domínguez H (2023) Potential use of *Sargassum muticum* as source of plant biostimulants after three different drying methods. J Appl Phycol 35(2):921–933. https://doi.org/10.1007/s10811-023-02907-2

Banakar SN, PrasannaKumar MK, Mahesh HB et al (2022) Red-seaweed biostimulants differentially alleviate the impact of fungicidal stress in rice (*Oryza sativa* L.). Sci Rep 12:5993. https://doi.org/10.1038/s41598-022-10010-8

Basak A (2008) Effect of preharvest treatment with seaweed products, Kelpak® and Goëmar BM 86®, on fruit quality in apple. Int J Fruit Sci 8(1–2):1–14. https://doi.org/10.1080/15538360802365251

Battacharyya D, Babgohari MZ, Rathor P, Prithiviraj B (2015) Seaweed extracts as biostimulants in horticulture. Sci Hortic 196:39–48. https://doi.org/10.1016/j.scienta.2015.09.012

Brouwers E, Draisma M, Swam K van, Veen AJ, Burger L (2018) Identification of the seaweed biostimulant market (phase 1). Interreg 2 Seas project Bio4safe, WP1 Market Analysis: Deliverable document 1.1.1. phase 1, 1–64. https://www.northseafarmers.org/public/documents/Bio4safe_WP1_D112_Seaweed-biostimulant-market_North-Sea-Farm-Foundation_December-2018.pdf

Caradonia F, Battaglia V, Righi L, Pascali G, La Torre A (2019) Plant biostimulant regulatory framework: prospects in Europe and current situation at international level. J Plant Growth Regul 38(2):438–448. https://doi.org/10.1007/s00344-018-9853-4

Chaturvedi S, Kulshrestha S, Bhardwaj K (2022) Role of seaweeds in plant growth promotion and disease management. In: New and future developments in microbial biotechnology and bioengineering: sustainable agriculture: revitalization through organic products, pp 217–238. https://doi.org/10.1016/B978-0-323-85579-2.00007-1

Chouliaras V, Tasioula M, Chatzissavvidis C, Therios I, Tsabolatidou E (2009) The effects of a seaweed extract in addition to nitrogen and boron fertilization on productivity, fruit maturation, leaf nutritional status and oil quality of the olive (*Olea europaea* L.) cultivar Koroneiki. J Sci Food Agric 89:984–988

Circuncisão AR, Catarino MD, Cardoso SM, Silva AMS (2018) Minerals from macroalgae origin: health benefits and risks for consumers. Mar Drugs 16(11). https://doi.org/10.3390/md16110400

Cole AJ, Roberts DA, Garside AL, de Nys R, Paul NA (2016) Seaweed compost for agricultural crop production. J Appl Phycol 28(1):629–642

Craigie JS (2011) Seaweed extract stimuli in plant science and agriculture. J Appl Phycol 23(3):371–393. https://doi.org/10.1007/s10811-010-9560-4

Crouch IJ, Van Staden J (1994) Commercial seaweed products as biostimulants in horticulture. J Home Consum Hortic 1:19–73

Crouch IJ, Beckett RP, Van Staden J (1990) Effect of seaweed concentrate on the growth and mineral nutrition of nutrient-stressed lettuce. J Appl Phycol 2(3):269–272

Cuomo V, Perretti A, Palomba I, Verde A, Cuomo A (1995) Utilization of *Ulva rigida* biomass in the Venice Lagoon (Italy)—biotransformation in compost. J Appl Phycol 7:479–485

Deniaud-Bouët E, Hardouin K, Potin P, Kloareg B, Hervé C (2017) A review about brown algal cell walls and fucose-containing sulfated polysaccharides: cell wall context, biomedical properties and key research challenges. Carbohydr Polym 175:395–408. https://doi.org/10.1016/j.carbpol.2017.07.082

Domínguez-May R, Gasca-Leyva E, Robledo D (2017) Harvesting time optimization and risk analysis for the mariculture of *Kappaphycus alvarezii* (Rhodophyta). Rev Aquac 9(3):227–237. https://doi.org/10.1111/raq.12134

Domozych DS, Ciancia M, Fangel JU, Mikkelsen MD, Ulvskov P, Willats WGT (2012) The cell walls of green algae: a journey through evolution and diversity. Front Plant Sci 3(MAY):1–7. https://doi.org/10.3389/fpls.2012.00082

du Jardin P (2015) Plant biostimulants: definition, concept, main categories and regulation. Sci Hortic 196:3–14. https://doi.org/10.1016/j.scienta.2015.09.021

El Boukhari MEM, Barakate M, Bouhia Y, Lyamlouli K (2020) Trends in seaweed extract based biostimulants: manufacturing process and beneficial effect on soil-plant systems. Plan Theory 9(3). https://doi.org/10.3390/plants9030359

El-Ansary MSM, Hamouda RA (2014) Biocontrol of Root-Knot Nematode infected banana plants by some marine algae. Russ J Mar Biol. https://doi.org/10.1134/S1063074014020047

El-Shenody RA, Elshobary ME, Ragab GA, Huo S, Essa D (2023) Towards biorefinery: exploring the potential of seaweed-derived biodiesel and its residual biomass in improving the traits of *Eruca vesicaria* (L.) cav. S Afr J Bot 155:361–371. https://doi.org/10.1016/j.sajb.2023.02.029

Ertani A, Francioso O, Tinti A, Schiavon M, Pizzeghello D, Nardi S (2018) Evaluation of seaweed extracts from *Laminaria* and *Ascophyllum nodosum* spp. as biostimulants in *Zea mays* L. using a combination of chemical, biochemical and morphological approaches. Front Plant Sci 9(April). https://doi.org/10.3389/fpls.2018.00428

Eyras MC, Sar EA (2003) Arribazones estivales en Puerto Madryn, Argentina, como materiales para la obtención de compost. Boletín de la Sociedad Argentina de Botánica 38:105–111

Eyras MC, Rostagno CM, Defosse GE (1998) Biological evaluation of seaweed composting. Compost Sci Utiliz 6:74–81

Fan D, Hodges DM, Critchley AT, Prithiviraj B (2013) A commercial extract of brown macroalga (*Ascophyllum nodosum*) affects yield and the nutritional quality of spinach in vitro. Commun Soil Sci Plant Anal 44(12):1873–1884

FAO (2021) Global seaweed and microalgae production, 1950–2019. World Aquaculture Performance Indicators (WAPI) Factsheet

Featonby-Smith BC, Van Staden J (1984) Identification and seasonal variation of endogenous cytokinins in *Ecklonia maxima* (Osbeck) Papenf. Bot Mar 27(11):527–530

Finnie JF, Van Staden J (1985) Effect of seaweed concentrate and applied hormones on in vitro cultured tomato roots. J Plant Physiol 120:215–222

Flórez-Fernández N, Domínguez H, Torres MD (2019) A green approach for alginate extraction from *Sargassum muticum* brown seaweed using ultrasound-assisted technique. Int J Biol Macromol 124:451–459. https://doi.org/10.1016/j.ijbiomac.2018.11.232

Flórez-Fernández N, Illera M, Sánchez M, Lodeiro P, Torres MD, López-Mosquera ME, Soto M, de Vicente MS, Domínguez H (2021) Integrated valorization of *Sargassum muticum* in biorefineries. Chem Eng J 404. https://doi.org/10.1016/j.cej.2020.125635

Flórez-Fernández N, Falqué E, Domínguez H, Torres MD (2022) Green extraction of carrageenans from *Mastocarpus stellatus*. Polymers 14(3):1–15. https://doi.org/10.3390/polym14030554

Ghafarizadeh A, Seyyednejad SM, Gilani A (2017) Synergistic effect of seaweed extract and urea fertilizer on growth and biochemical parameters of *Triticum aestivum* at vegetative stage. Int J Atural Soc Sci 4:545–556

Gillespie RD, Critchley AT (1999) Phenology of *Sargassum* spp. (Sargassaceae, Phaeophyta) from Reunion Rocks, KwaZulu-Natal, South Africa. Hydrobiologia 398–399:201–210. https://doi.org/10.1007/978-94-011-4449-0_23

Goñi O, Quille P, O'Connell S (2020) Seaweed carbohydrates. In: The chemical biology of plant biostimulants, vol 56–95. https://doi.org/10.1002/9781119357254.ch3

Illera-Vives M, Seoane LS, López-Mosquera ME (2013) Production of compost from marine waste: evaluation of the product for use in ecological agriculture. J Appl Phycol 25(5):1395–1403

Illera-Vives M, Labandeira SS, Brito LM, López-Fabal A, López-Mosquera ME (2015) Evaluation of compost from seaweed and fish waste as a fertilizer for horticultural use. Sci Hortic 186:101–107

Illera-Vives M, Labandeira SS, Iglesias Loureiro L, López-Mosquera ME (2017) Agronomic assessment of a compost consisting of seaweed and fish waste as an organic fertilizer for organic potato crops. J Appl Phycol 29(3):1663–1671

Illera-Vives M, Labandeira SS, Fernández-Labrada M, López-Mosquera ME (2020) Agricultural uses of seaweed. In: Sustainable seaweed technologies. Elsevier, pp 591–612

Juárez-Maldonado A, Ortega-Ortíz H, Morales-Díaz AB, González-Morales S, Morelos-Moreno Á, Cabrera-De la Fuente M, Sandoval-Rangel A, Cadenas-Pliego G, Benavides-Mendoza A (2019) Nanoparticles and nanomaterials as plant biostimulants. Int J Mol Sci 20(1):1–19. https://doi.org/10.3390/ijms20010162

Kadam SU, Tiwari BK, O'Donnell CP (2013) Application of novel extraction technologies for bioactives from marine algae. J Agric Food Chem 61(20):4667–4675. https://doi.org/10.1021/jf400819p

Kadam SU, Álvarez C, Tiwari BK, O'Donnell CP (2015) Extraction of biomolecules from seaweeds. In: Seaweed sustainability: food and non-food applications. Elsevier Inc. https://doi.org/10.1016/B978-0-12-418697-2.00009-X

keya Tudu C, Dey A, Pandey DK, Panwar JS, Nandy S (2022) Role of plant derived extracts as biostimulants in sustainable agriculture: a detailed study on research advances, bottlenecks and future prospects. In: Singh HB, Vaishnav B (eds) New and Future Developments in Microbial Biotechnology and Bioengineering: Sustainable Agriculture: Revitalization through Organic Products. Elsevier, pp 159–179. https://doi.org/10.1016/B978-0-323-85579-2.00017-4

Khan W, Rayirath UP, Subramanian S, Jithesh MN, Rayorath P, Hodges DM, Critchley AT, Craigie JS, Norrie J, Prithiviraj B (2009) Seaweed extracts as biostimulants of plant growth and development. J Plant Growth Regul 28(4):386–399. https://doi.org/10.1007/s00344-009-9103-x

Kocira A, Kornas R, Kocira S (2013) Ocena wpływu Kelpaku sl na plonowanie fasoli zwykłej (*Phaseolus vulgaris* L.). J Cent Eur Agric 14(2):67–76. https://doi.org/10.5513/JCEA01/14.2.1234

Krawczuk A, Huyghebaert B, Rabier F, Parafiniuk S, Przywara A, Koszel M, Lorencowicz E, Kocira S (2023) The technical parameters of seaweed biostimulant spray application as a factor in the economic viability of soybean production. Appl Sci (Switzerland) 13(2). https://doi.org/10.3390/app13021051

Kumar SMS, Baradhan G, Elankavi S, Ramesh N, Ruban JS (2020) Integration of different granular organic manures with inorganic fertilizers for increasing the growth attributes of hybrid maize (*Zea mays* L.). Plant Arch 20(1):2515–2519. https://www.scopus.com/inward/record.uri?eid=2-s2.0-85065983415&partnerID=40&md5=11a21e085cca120890fa553bc6d98812

López-Mosquera ME, Pazos P (1997) Effects of seaweed on potato yields and soil chemistry. Biol Agric Hortic 14(3):199–205

Lynn LB (1972) The chelating properties of seaweed extract *Ascophyllum nodosum* vs. *Macrocystis pyrifera* on the mineral nutrition of sweet peppers, *Capsicum annuum*. Clemson University, Clemson

Mancuso S, Azzarello E, Mugnai S, Briand X (2006) Marine bioactive substances (IPA extract) improve foliar ion uptake and water stress tolerance in potted *Vitis vinifera* plants. Adv Hortic Sci 20:156–161

Martins M, Ventura SPM (2020) Emerging seaweed extraction techniques using ionic liquids. In: Sustainable seaweed technologies: cultivation, biorefinery, and applications. Elsevier Inc. https://doi.org/10.1016/B978-0-12-817943-7.00011-1

Mazé J, Morand P, Potoky P (1993) Stabilisation of 'Green tides' *Ulva* by a method of composting with a view to pollution limitation. J Appl Phycol 5(2):183

McHugh D (2003) A guide to the seaweed industry: FAO fisheries technical paper No. 441. In FAO fisheries technical paper. http://afrilib.odinafrica.org/handle/0/15549%5Cn http://www.fao.org/docrep/006/y4765e/y4765e00.htm

Mendes M, Pacheco D, Cotas J, Bahcevandziev K, Pereira L (2022) An experimental study investigating the effects on *Brassica oleracea*: Estuarine seaweeds as biostimulants in seedling development. Phycology 2(4):419–428

Metting B, Zimmerman WJ, Crouch IJ, Van Staden J (1990) Agronomic uses of seaweed and microalgae. In: Introduction to applied phycology, pp 589–627

Milton RF (1964) Liquid seaweed as a fertilizer. In: Proceedings of the international seaweed symposium, vol 4, pp 428–431

Mohamed AA, Sameeh MY, El-Beltagi HS (2023) Preparation of seaweed nanopowder particles using planetary ball milling and their effects on some secondary metabolites in date palm (*Phoenix dactylifera* L.) seedlings. Life 13(1). https://doi.org/10.3390/life13010039

Mueller SR, Kussow WR (2005) Biostimulant influences on turfgrass microbial communities and creeping bentgrass putting green quality. HortScience 40(6):1904–1910. https://doi.org/10.21273/hortsci.40.6.1904

Muzzalupo I, Badolati G, Chiappetta A, Picci N, Muzzalupo R (2020) In vitro antifungal activity of olive (*Olea europaea*) leaf extracts loaded in chitosan nanoparticles. Front Bioeng Biotechnol 8(3):1–10. https://doi.org/10.3389/fbioe.2020.00151

Mzibra A, Aasfar A, Benhima R, Khouloud M, Boulif R, Douira A, Bamouh A, Meftah Kadmiri I (2021) Biostimulants derived from Moroccan seaweeds: seed germination metabolomics and growth promotion of tomato plant. J Plant Growth Regul 40(1):353–370. https://doi.org/10.1007/s00344-020-10104-5

Nedzarek A, Rakusa-Suszczewski S (2004) Decomposition of macroalgae and the release of nutrient Admiralty Bay. King George, Antarctica

Orozco-González JG, Amador-Castro F, Gordillo-Sierra AR, García-Cayuela T, Alper HS, Carrillo-Nieves D (2022) Opportunities surrounding the use of *Sargassum* biomass as precursor of biogas, bioethanol, and biodiesel production. Front Mar Sci 8(January):1–11. https://doi.org/10.3389/fmars.2021.791054

Otero P, Quintana SE, Reglero G, Fornari T, García-Risco MR (2018) Pressurized Liquid Extraction (PLE) as an innovative green technology for the effective enrichment of Galician algae extracts with high quality fatty acids and antimicrobial and antioxidant properties. Mar Drugs 16(5). https://doi.org/10.3390/md16050156

Ozbay N, Demirkiran AR (2019) Enhancement of growth in ornamental pepper (*Capsicum annuum* L.) plants with application of a commercial seaweed product, Stimplex®. Appl Ecol Environ Res 17(2):4361–4375. https://doi.org/10.15666/aeer/1702_43614375

Pacheco D, Cotas J, Rocha CP, Araújo GS, Figueirinha A, Gonçalves AMM, Bahcevandziev K, Pereira L (2021) Seaweeds' carbohydrate polymers as plant growth promoters. Carbohydr Polym Technol Appl 2. https://doi.org/10.1016/j.carpta.2021.100097

Paul K, Sorrentino M, Lucini L, Rouphael Y, Cardarelli M, Bonini P, Reynaud H, Canaguier R, Trtílek M, Panzarová K, Colla G (2019) Understanding the biostimulant action of vegetal-derived protein hydrolysates by high-throughput plant phenotyping and metabolomics: a case study on tomato. Front Plant Sci 10:1–17. https://doi.org/10.3389/fpls.2019.00047

Pereira L (2017) Vibrational spectroscopy of seaweed polysaccharides. In: Seaweed polysaccharides: isolation, biological and biomedical applications. Elsevier Inc. https://doi.org/10.1016/B978-0-12-809816-5.00006-2

Pereira L, Van De Velde F (2011) Portuguese carrageenophytes: carrageenan composition and geographic distribution of eight species (Gigartinales, Rhodophyta). Carbohydr Polym 84(1):614–623. https://doi.org/10.1016/j.carbpol.2010.12.036

Pereira L, Amado AM, Critchley AT, van de Velde F, Ribeiro-Claro PJA (2009) Identification of selected seaweed polysaccharides (phycocolloids) by vibrational spectroscopy (FTIR-ATR and FT-Raman). Food Hydrocoll 23(7):1903–1909. https://doi.org/10.1016/j.foodhyd.2008.11.014

Peteiro C (2018) Alginate production from marine macroalgae, with emphasis on kelp farming. In: Springer series in biomaterials science and engineering, vol 11. https://doi.org/10.1007/978-981-10-6910-9_2

Popescu GC, Popescu M (2014) Effect of the brown alga *Ascophyllum nodosum* as biofertilizer on vegetative growth in grapevine (*Vitis vinifera* L.). Curr Trends Nat Sci 3(6):61–67

Potoky P, Maze J, Morand P, Schulte E (1988) Effect of seaweed inclusion in compost preparation on the quality of the composts obtained. Aquatic primary biomass (marine macroalgae): Biomass conversion, removal and use of nutrients. Proceedings of the first workshop of the COST 48 Sub Group 3, L'Homeau, France, pp 67–86

Prakash P, Mitra A, Nag R, Sunkar S (2018) Effect of seaweed liquid fertilizer and humic acid formulation on the growth and nutritional quality of *Abelmoschus esculentus*. Asian J Crop Sci 10(1):48–52

Queffelec J, Flórez-Fernández N, Domínguez H, Torres MD (2021) Microwave hydrothermal processing of *Undaria pinnatifida* for bioactive peptides. Bioresour Technol 342(September). https://doi.org/10.1016/j.biortech.2021.125882

Quitério E, Grosso C, Ferraz R, Delerue-Matos C, Soares C (2022) A critical comparison of the advanced extraction techniques applied to obtain health-promoting compounds from seaweeds. Mar Drugs 20(11):1–40. https://doi.org/10.3390/md20110677

Raghavendra VB, Lokesh S, Prakash HS (2007) Dravya, a product of seaweed extract (*Sargassum wightii*), induces resistance in cotton against *Xanthomonas campestris* pv. *malsvacearum*. Phytoparasitica 35(5):442

Raimundo SC, Avci U, Hopper C, Pattathil S, Hahn MG, Popper ZA (2016) Immunolocalization of cell wall carbohydrate epitopes in seaweeds: presence of land plant epitopes in *Fucus vesiculosus* L. (Phaeophyceae). Planta 243(2):337–354. https://doi.org/10.1007/s00425-015-2412-3

Rengasamy KR, Kulkarni MG, Stirk WA, Van Staden J (2015) Eckol-a new plant growth stimulant from the brown seaweed *Ecklonia maxima*. J Appl Phycol 27(1):581–587

Rouphael Y, Colla G (2020) Editorial: Biostimulants in agriculture. Front Plant Sci 11. https://doi.org/10.3389/fpls.2020.00040

Rupérez P (2002) Mineral content of edible marine seaweeds. Food Chem 79(1):23–26. https://doi.org/10.1016/S0308-8146(02)00171-1

Santelices B, Bolton JJ, Meneses I (2009) Chapter six marine algal communities. Marine Macroecol:153–192. http://books.google.com/books?hl=en&lr=&id=a9Qv-DgnDkoC&oi=fnd&pg=PA153&dq=kelp+forest+invertebrate+ecology+southern+california+current&ots=_6n2amSB2k&sig=EpzmR0xerYeK7XzGQcslZfuPztc

Sarkar S, Gayen K, Bhowmick TK (2022) Green extraction of biomolecules from algae using subcritical and supercritical fluids. Biomass Convers Biorefinery https://doi.org/10.1007/s13399-022-02309-3

Schiener P, Black KD, Stanley MS, Green DH (2015) The seasonal variation in the chemical composition of the kelp species *Laminaria digitata, Laminaria hyperborea, Saccharina latissima* and *Alaria esculenta*. J Appl Phycol 27(1):363–373. https://doi.org/10.1007/s10811-014-0327-1

Sfriso AA, Gallo M, Baldi F (2017) Seasonal variation and yield of sulfated polysaccharides in seaweeds from the Venice Lagoon. Bot Mar 60(3):339–349. https://doi.org/10.1515/bot-2016-0063

Shepherd VA, Beilby MJ, Heslop DJ (1999) Ecophysiology of the hypotonic response in the salt-tolerant charophyte alga *Lamprothamnium papulosum*. Plant Cell Environ 22(4):333–346. https://doi.org/10.1046/j.1365-3040.1999.00414.x

Shukla PS, Mantin EG, Adil M, Bajpai S, Critchley AT, Prithiviraj B (2019) *Ascophyllum nodosum*-based biostimulants: sustainable applications in agriculture for the stimulation of plant growth, stress tolerance, and disease management. Front Plant Sci 10(May):1–22. https://doi.org/10.3389/fpls.2019.00655

Silva LD, Bahcevandziev K, Pereira L (2019) Production of bio-fertilizer from *Ascophyllum nodosum* and *Sargassum muticum* (Phaeophyceae). J Oceanol Limnol 37(3):918–927. https://doi.org/10.1007/s00343-019-8109-x

Skriptsova AV (2015) Fucoidans of brown algae: biosynthesis, localization, and physiological role in thallus. Russ J Mar Biol 41(3):145–156. https://doi.org/10.1134/S1063074015030098

Skriptsova AV (2016) Seasonal variations in the fucoidan content of brown algae from Peter the Great Bay, Sea of Japan. Russ J Mar Biol 42(4):351–356. https://doi.org/10.1134/S1063074016040106

Spinelli F, Fiori G, Noferini M, Sprocatti M, Costa G (2010) A novel type of seaweed extract as a natural alternative to the use of iron chelates in strawberry production. Sci Hortic 125(3):263–269. https://doi.org/10.1016/j.scienta.2010.03.011

Stengel DB, Connan S, Popper ZA (2011) Algal chemodiversity and bioactivity: sources of natural variability and implications for commercial application. Biotechnol Adv 29(5):483–501. https://doi.org/10.1016/j.biotechadv.2011.05.016

Stirk WA, Novak O, Strnad M, van Staden J (2003) Cytokinins in macroalgae. Plant Growth Regul 41:13–24

Stirk WA, Tarkowska D, Turecova V, Strnad M, Van Staden J (2014) Abscisic acid, gibberellins and brassinosteroids in Kelpak®, a commercial seaweed extract made from *Ecklonia maxima*. J Appl Phycol 26:561–567

Stirk WA, Rengasamy KRR, Kulkarni MG, van Staden J (2020) Plant biostimulants from seaweed: an overview. Eds: Geelen D, Xu L. The Chemical Biology of Plant Biostimulants, Wiley and Sons Ltd. pp. 33–55.

Strain EMA, Lai RWS, White CA, Piarulli S, Leung KMY, Airoldi L, O'Brien A (2022) Editorial: marine pollution - emerging issues and challenges. Front Mar Sci 9. https://doi.org/10.3389/fmars.2022.918984

Taskos D, Stamatiadis S, Yvin JC, Jamois F (2019) Effects of an *Ascophyllum nodosum* (L.) Le Jol. extract on grapevine yield and berry composition of a Merlot vineyard. Sci Hortic 250:27–32

Tavares AR, dos Santos PLF, Zabotto AR et al (2020) Seaweed extract to enhance marigold seed germination and seedling establishment. Appl Sci 2:1792. https://doi.org/10.1007/s42452-020-03603-3

Tawfeeq A, Culham A, Davis F, Reeves M (2016) Does fertilizer type and method of application cause significant differences in essential oil yield and composition in rosemary (*Rosmarinus officinalis* L.). Indus Crop Prod 88:17–22

Thirumaran G, Arumugam M, Arumugam R, Anantharaman P (2009) Effect of seaweed liquid fertilizer on growth and pigment concentration of *Abelmoschus esculentus* (L) medikus. Am Eurasian J Agron 2(2):57–66

Torode TA, Marcus SE, Jam M, Tonon T, Blackburn RS, Hervé C, Knox JP (2015) Monoclonal antibodies directed to fucoidan preparations from brown algae. PLoS One 10(2):1–19. https://doi.org/10.1371/journal.pone.0118366

Torres MD, Kraan S, Domínguez H (2019) Seaweed biorefinery. Rev Environ Sci Biotechnol 18(2). https://doi.org/10.1007/s11157-019-09496-y

Vaghela P, Trivedi K, Anand KGV, Brahmbhatt H, Nayak J, Khandhediya K, Prasad K, Moradiya K, Kubavat D, Konwar LJ, Veeragurunathan V, Grace PG, Ghosh A (2023) Scientific basis for the use of minimally processed homogenates of *Kappaphycus alvarezii* (red) and *Sargassum wightii* (brown) seaweeds as crop biostimulants. Algal Res 70. https://doi.org/10.1016/j.algal.2023.102969

Vera J, Castro J, Gonzalez A, Moenne A (2011) Seaweed polysaccharides and derived oligosaccharides stimulate defense responses and protection against pathogens in plants. Mar Drugs 9(12):2514–2525. https://doi.org/10.3390/md9122514

Vijayakumar S, Durgadevi S, Arulmozhi P, Rajalakshmi S, Gopalakrishnan T, Parameswari N (2019) Effect of seaweed liquid fertilizer on yield and quality of *Capsicum annum* L. Acta Ecol Sin 39(5):406–410

Villares R, Carral E, Lorenzana F, Mosquera EL (2007) Drift-seaweed evaluation for fertilizer use in Galicia (Northwest Spain): tissue elemental characterization and sitesampling differences. J Sustain Agric 31(1):45–60

Vo Dinh T, Saravana PS, Woo HC, Chun BS (2018) Ionic liquid-assisted subcritical water enhances the extraction of phenolics from brown seaweed and its antioxidant activity. Sep Purif Technol 196:287–299. https://doi.org/10.1016/j.seppur.2017.06.009

Wang L, Xie L, Yang G, Yan Q, Ruan W (2009) Impact of different bulking agents and compound microbial inoculant on blue algae composting. Chinese J Environ Eng 12:2261–2265

Wang Y, Fu F, Li J, Wang G, Wu M, Zhan J, Mao Z (2016) Effects of seaweed fertilizer on the growth of *Malus hupehensis* Rehd. seedlings, soil enzyme activities and fungal communities under replant condition. Eur J Soil Biol 75:1–7

Wang M, Chen L, Li Y, Chen L, Liu Z, Wang X, Qin S (2018) Responses of soil microbial communities to a short-term application of seaweed fertilizer revealed by deep amplicon sequencing. Appl Soil Ecol 125:288–296

Wang J, Ding Z, AL-Huqail AA, Hui Y, He Y, Reichman SM, Ghoneim AM, Eissa MA, Abou-Zaid EAA (2022) Potassium source and biofertilizer influence K release and fruit yield of mango (*Mangifera indica* L.): a three-year field study in sandy soils. Sustainability (Switzerland) 14(15):1–17. https://doi.org/10.3390/su14159766

Whapham CA, Blunden G, Jenkins T, Hankins SD (1993) Significance of betaines in the increased chlorophyll content of plants treated with seaweed extract. J Appl Phycol 5(2):231–234. https://doi.org/10.1007/BF00004023

Wosnitza TMA, Barrantes JG (2006) Utilization of seaweed *Ulva* sp. in Paracas Bay (Peru): experimenting with compost. J Appl Phycol 18(1):27

Xu Y, Zheng X, Song Y, Zhu L, Yu Z, Gan L, Zhou S, Liu H, Wen F, Zhu C (2018) NtLTP4, a lipid transfer protein that enhances salt and drought stresses tolerance in *Nicotiana tabacum*. Sci Rep. https://doi.org/10.1038/s41598-018-27274-8.8

Yao Y, Wang X, Chen B, Zhang M, Ma J (2020) Seaweed extract improved yields, leaf photosynthesis, ripening time, and net returns of tomato (*Solanum lycopersicum* Mill.). ACS Omega 5(8):4242–4249

Yuan L, Ou J, Lv X, Wang J, Lin L, Deng Q (2018) The effect of seaweed fertilizer "SUNRED" on the coloring of "Red Globe" grape. In: 7th international conference on energy and environmental protection (ICEEP 2018). Advances in Engineering Research, 170, pp 856–860

Zamani-Babgohari M, Critchley AT, Norrie J, Prithiviraj B (2019) Increased freezing stress tolerance of *Nicotiana tabacum* L. Cv. bright Yellow-2 cell cultures with the medium addition of *Ascophyllum nodosum* (L.) Le Jolis extract. *In vitro* cellular and developmental biology. https://doi.org/10.1007/s11627-019-09972-

Zhang X, Ervin EH, Schmidt RE (2003) Physiological effects of liquid applications of a seaweed extract and a humic acid on creeping bentgrass. J Am Soc Hortic Sci 128(4):492–496

Chapter 3
Seaweed Polyphenols: Challenges and Future Aspects of Their Processing and Use

Weihao Meng, Hongnan Sun, Taihua Mu, and Marco Garcia-Vaquero

Abstract In recent decades, macroalgae have captured the attention of researchers as a source of multiple compounds with applications in the food, pharmaceutical, and cosmetic industries. Compared to terrestrial crops, the production of algal biomass for food-related applications has been proven to be more sustainable, as it can be done without using potable water and arable land while providing relatively high productivity when produced in large-scale seaweed farms. Despite the wide array of advantages of this biomass and its compounds, this biomass is currently underutilized. The exploitation of selected macroalgae as a source of polyphenolic compounds and/or phlorotannins (i.e. produced mainly by brown seaweeds) still remains in its infancy, although these molecules have attracted the attention of researchers and industry alike due to their wide range of health benefits and possibilities for industrial use. The exploitation of macroalgal polyphenols still faces great challenges at multiple processing stages, including extraction, purification, and characterization. This chapter briefly summarizes the type of polyphenols described in different macroalgal species and the status of the current methods available for their

W. Meng
Laboratory of Food Chemistry and Nutrition Science, Institute of Food Science and Technology, Chinese Academy of Agricultural Sciences, Beijing, P. R. China

Key Laboratory of Agro-Products Processing, Ministry of Agriculture and Rural Affairs, Haidian District, Beijing, P. R. China

School of Agriculture and Food Science, University College Dublin, Belfield, Dublin 4, Ireland

H. Sun · T. Mu
Laboratory of Food Chemistry and Nutrition Science, Institute of Food Science and Technology, Chinese Academy of Agricultural Sciences, Beijing, P. R. China

Key Laboratory of Agro-Products Processing, Ministry of Agriculture and Rural Affairs, Haidian District, Beijing, P. R. China

M. Garcia-Vaquero (✉)
School of Agriculture and Food Science, University College Dublin, Belfield, Dublin 4, Ireland
e-mail: marco.garciavaquero@ucd.ie

N. Trivedi et al. (eds.), *Recent Advances in Seaweed Biotechnology*,
https://doi.org/10.1007/978-981-96-0519-4_3

extraction, purification, and characterization. The main challenges of all these processing steps will be discussed and the most promising strategies currently being explored to overcome these bottlenecks for the industrial exploitation of polyphenols. Moreover, the possibilities of the use of these compounds and their main applications in various industries will also be discussed, providing insights into the future direction of the research, market, and prospects of the industrial exploitation of macroalgae as a source of polyphenols.

Keywords Seaweed extraction · Purification · Characterization · Innovative technologies · Nutraceutical · Biopackaging · Pickering emulsions · Tissue engineering

3.1 Introduction

Macroalgae or seaweeds have been in the traditional human diet for thousands of years, especially in Asian countries, such as China, Korea, and Japan (Agregán et al. 2017). Based on the pigmentation of the biomass, macroalgae have been classified into brown (Phaeophyta), red (Rhodophyta), and green (Chlorophyta) types (Nwosu et al. 2011). Macroalgae offer multiple advantages when used as food, as they can provide a variety of bioactive compounds (i.e. mero-terpenoids, astaxanthin, polyphenols, and alginates) to the diet while at the same time alleviating some of the major threats created by traditional agricultural production systems, such as land scarcity and climate change (Meng et al. 2022). Globally, the seaweed market is growing at an annual rate of 8.9%, and it is expected to reach a total market value of \$22.13 billion USD in 2024 (Zhu et al. 2022). In 2020, the total production of seaweed worldwide was 36 million tons, with China being the largest producer of this biomass (FAO 2022). However, some researchers believe that the world's annual production of seaweed may be about 25% overestimated (Hermans 2023). Moreover, the seaweed industry in most countries around the world is still based on wild harvesting or basic cultivation techniques, and large-scale harvesting and processing still lack automation (Zhang et al. 2022). With the growing interest of consumers in health-promoting foods and natural compounds, macroalgae are gaining increased attention for their potential as sources of health-promoting bioactive compounds, such as polyphenols (Kumar et al. 2022b). Through a literature search in ScienceDirect (www.sciencedirect.com), in the twenty-first century (2000–2022), the number of publications retrieved when using the keyword "seaweed polyphenols" showed an increasing trend year by year, and the number of publications in 2022 was 53 times higher than that in 2000 (Fig. 3.1).

Polyphenols not only affect the colour of macroalgae but also have a wide variety of biological roles in the biomass, including maintaining the physiological integrity of the cells, as well as providing resistance of the biomass to oxidative stress and

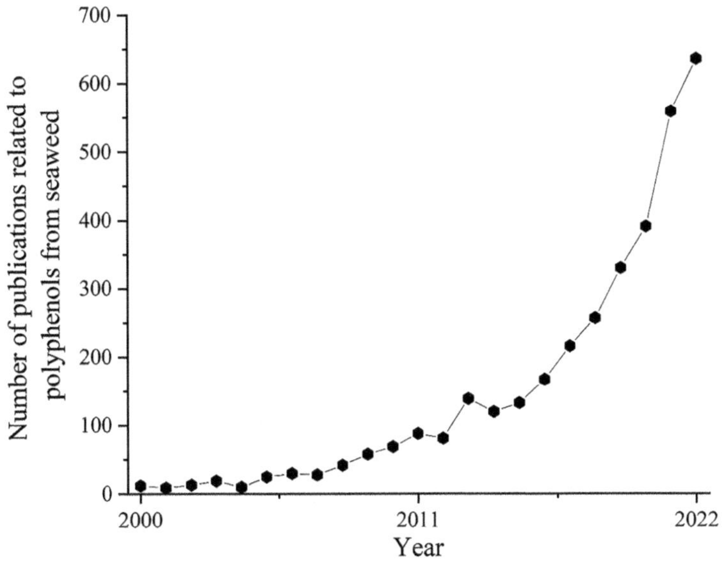

Fig. 3.1 Number of publications available when searching "seaweed polyphenols" in ScienceDirect during the period 2000–2022

chemical defence against external invaders (Martínez-López and Tuohy 2023). As shown in Fig. 3.2, recent studies explored the use of seaweed-derived polyphenols for a plethora of health-related applications, emphasizing their biological properties against various diseases (i.e. hypertension, diabetic, and cancer) (Lemesheva et al. 2023; Meng et al. 2021b; Raja et al. 2023). Despite these remarkable numbers of biological properties and potential applications, the exploitation of polyphenols is currently in its infancy due to difficulties associated with their processing, including extraction, purification, and characterization (Fig. 3.3). Therefore, this chapter summarizes the latest progress in the processing of macroalgal polyphenols, including the processes of extraction, purification, and characterization as well as further processing of these compounds needed for their final application, and the main research challenges that will need to be addressed to potentiate the use of macroalgal polyphenols in the food, pharmaceutical, and cosmeceutical industries.

3.2 Polyphenols from Macroalgae: Chemical Classification and Main Compounds

The term "phenolic compounds" refers to those compounds with one, or more, aromatic rings and hydroxyl groups. This group of compounds can be further classified into several classes depending on their chemical features, including phenolic acids, flavonoids, isoflavonoids, tannins, and phenolic polymers (Meng et al. 2021a). The

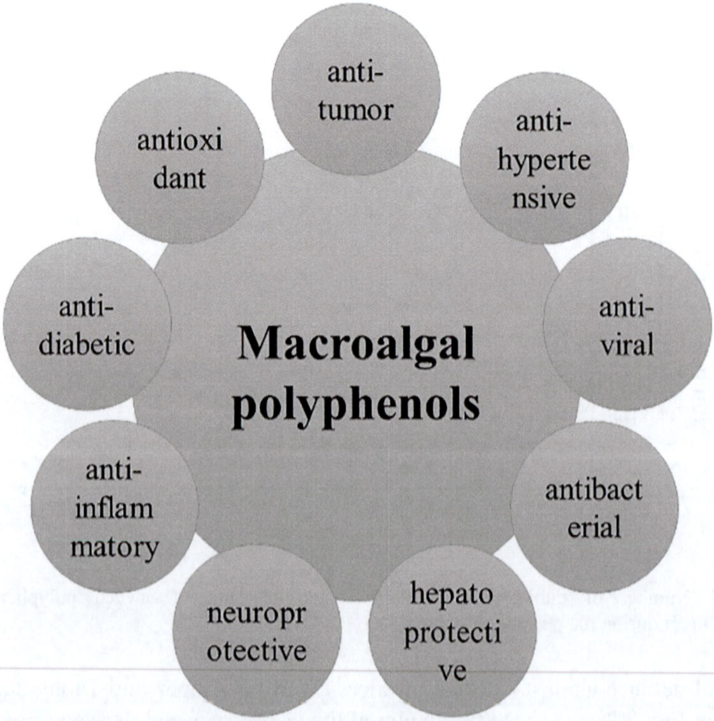

Fig. 3.2 Biological activities of macroalgal polyphenols

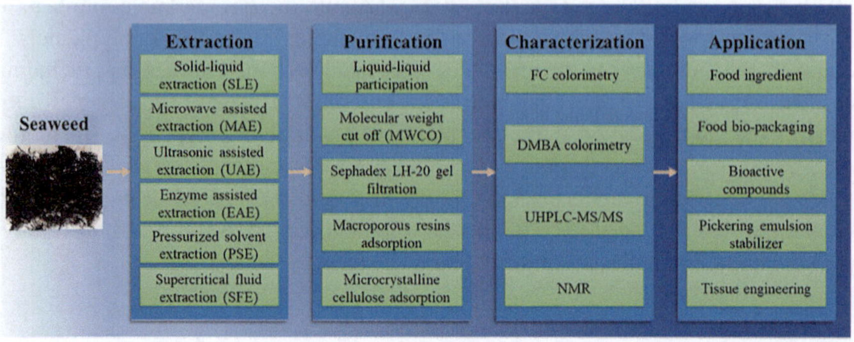

Fig. 3.3 Schematic representation of the processes for the generation of polyphenols from seaweeds

marine ecological environment, which is significantly different from the terrestrial environment, creates an environment for seaweed to produce a wide variety of chemical compounds that are unique to algae, i.e. phlorotannins, a specific type of polyphenol produced mainly by brown macroalgae (Rubiño et al. 2022). The production of phenolic compounds, similar to other products from this biomass, is influenced by multiple factors affecting seaweed biology, including the seaweed species, location, season, and year of collection as well as other environmental factors (light irradiation, evaporation, salinity, and nutrients) that will all affect the biology and thus the production of compounds and final composition of this biomass (Garcia-Vaquero et al. 2021a; Holdt and Kraan 2011). Phlorotannins are particularly abundant in brown macroalgae, while the largest proportion of polyphenols in green and red macroalgae are bromophenols, phenolic acids, and flavonoids (Besednova et al. 2020).

Phlorotannins are specific polyphenols produced by brown macroalgae consisting of polymerized phloroglucinol, with a variable molecular weight (MW), chemical structure, and degree of aggregation. The MW of phlorotannins has been described to range from 126 to 650 kDa; however, most of the phlorotannins described in the literature have a MW ranging between 10 and 100 kDa (Bogolitsyn et al. 2019; Martínez-López and Tuohy 2023; Meng et al. 2021b). Previous studies exploring polyphenols from the macroalga *Ascophyllum nodosum* revealed that 74% of these molecules had a MW > 10 kDa (Meng et al. 2023a). According to the difference in bonds between phloroglucinol units and the presence of extra hydroxyl groups, phlorotannins can also be classified into six main groups, represented in Fig. 3.4. These groups are (1) phlorethols, (2) fucols, (3) fuhalols, (4) fucophlorethols, (5) isofuhalols, and (6) eckols (Lopes et al. 2012; Meng et al. 2021b; Zhang et al. 2018). To date, approximately 150 phlorotannins have been identified, and these compounds have attracted great interest due to their low toxicity, bioactivity, and antioxidant properties (Shrestha et al. 2021). Moreover, phlorotannins are very

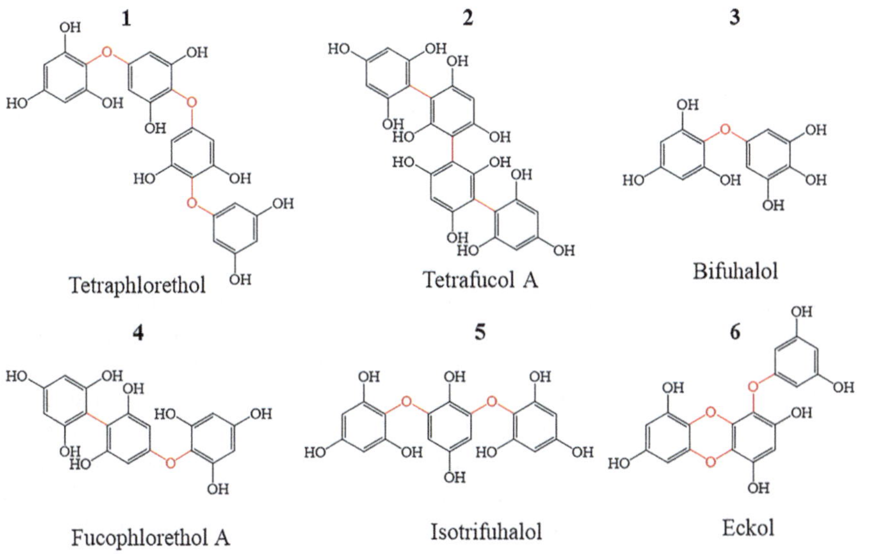

Fig. 3.4 Chemical structures and main chemical features (in red colour) used for the identification of the six main classes of phlorotannins. (Image modified from Meng et al. (2021b) and reproduced with permission from Elsevier)

complex and species specific, leading to a wide variety of compounds and huge differences in their molecular structure and degree of polymerization of polyphenols even among closely related macroalgal species (Lemesheva et al. 2023). In addition, phlorotannins can be combined through hydrogen bonding and hydrophobic interactions between the -OH group of the phloroglucinol units and the -NH or -SH group of protein side chains (Ming et al. 2023). This ability to bind to proteins has been considered an essential mechanism of these molecules when protecting proteins from oxidative damage (Ming et al. 2023).

Polyphenols derived from macroalgae have been classified as generally recognized as safe (GRAS) by the Food and Drug Administration (FDA) in the United States and are also considered safe by the European Food Safety Authority (EFSA) (Echave et al. 2022). In addition, EFSA also approved the phlorotannin-rich extract derived from *Ecklonia cava* as a novel food and considered it safe to eat (Turck et al. 2017). However, due to the complex structure of phenolic compounds in macroalgae and the high abundance of polysaccharides and proteins in macroalgal matrices, the separation and characterization of phenolic compounds is a very difficult task and has become a very challenging problem for the safety evaluation and approval of these compounds and their widespread use (Castillo et al. 2023; Meng et al. 2022).

3.3 Main Methods for the Extraction of Macroalgal Polyphenols

3.3.1 Conventional Extraction Methods

One of the first and most commonly used methods for the extraction of polyphenols from seaweed is solid–liquid extraction (SLE). Using this method, the choice of extraction solvent and its polarity play an important role in the efficiency of the extraction process. Ethanol is considered the best choice as an extraction solvent for polyphenols from seaweeds due to its high extraction efficiency, nontoxicity, low cost, and convenient processing (Kim et al. 2022a). Variations in the polarity of the solvent may influence the type of polyphenols extracted as well as co-extracted compounds. Therefore, the protein content in the aqueous extracts of *Fucus vesiculosus* was reported as 7.2% w/w, while the protein content achieved when using ethanol as solvent was approximately 1.7% (Honold et al. 2016).

Despite the widespread and ease of use of these methods, in general, the scientific community is moving towards their replacement due to the long processing times, low yields, high energy consumption, and excessive use of solvents, which limit their implementation at the industrial scale in an effective manner (Ummat et al. 2020). To address the above problems, natural deep eutectic solvents (NADES) are used in SLE to reduce the consumption of solvents and decrease the environmental risks posed by the excessive use of organic solvents in SLE. NADES are mixtures of glycerol, choline chloride, betaine, malic acid, lactic acid, and glucose in different proportions. According to their differences in polarity, different extraction efficiencies have been reported when extracting polyphenols (Obluchinskaya et al. 2019). However, due to the relatively high price of NADES, their use has not been widespread or widely accepted at the industrial level.

3.3.2 *Innovative Extraction Technologies*

Recent research dealing with the extraction of polyphenols from seaweeds has mainly focused on developing and optimizing innovative extraction technologies due to their great advantages, including increased efficiency of the processes and being more economic and environmentally friendly, as compared to conventional extraction methods. Polyphenols can be tightly bound to polysaccharides and proteins on the cell wall of algae through covalent and non-covalent bonds. Conventional extraction methods cannot effectively separate polyphenols bound to the cell wall, which greatly limits the extraction of polyphenols (Pradhan et al. 2022). Therefore, in the past few decades, a number of extraction methods have been developed and have demonstrated promising capabilities to extract target compounds from a variety of seaweeds, such as microwave-assisted extraction (MAE), ultrasonic-assisted extraction (UAE), enzyme-assisted extraction (EAE), pressurized-solvent extraction (PSE), and supercritical fluid extraction (SFE). These methods all aim to isolate polyphenols from the complex seaweed cell matrix while minimizing or even removing co-extracted compounds. Inter-molecular and intra-molecular frictions and expansion generated during MAE, cavitation during UAE, high temperature and pressure during PSE and SFE, and various enzymes used by EAE have been proven to be good at disrupting the cellular matrix of algae and allowing better release of polyphenols into the solvent (Casas et al. 2019; Ciko et al. 2018; Kadam et al. 2013; Magnusson et al. 2017; Ummat et al. 2020). The main principles of innovative and conventional technologies used for the extraction of macroalgal polyphenols, including their mechanisms of action, main advantages, and challenges for this purpose, are summarized in Table 3.1.

Compared with the traditional extraction method, the innovative extraction methods aim to further improve the extraction efficiency of polyphenols and other compounds and reduce solvent consumption, time of extraction, and the overall costs of the processes to improve their cost-effectiveness (Garcia-Vaquero et al. 2021b; Garcia-Vaquero and Rajauria 2022). These innovative extraction methods have been proven to have good application to algal polyphenols (Meng et al. 2023a; Ummat et al. 2021). However, the exploration of more efficient extraction methods has always been one of the research barriers that macroalgal polyphenols need to overcome. The optimization of parameters that are more suitable for the extraction process can improve the extraction rate of polyphenols more efficiently. Response surface methodology (RSM) can be used to evaluate the effect of different parameters in the extraction process of compounds from macroalgae (Garcia-Vaquero et al. 2017). Recently, emerging extraction strategies combining multiple extraction

Table 3.1 Evaluation of conventional and innovative technologies used for the extraction of polyphenols from macroalgae

Extraction methods	Principles for extraction	Benefits of extraction techniques	Challenges and limitations	References
Solid–liquid extraction (SLE)	Solubility differences of polyphenols and other substrates of seaweed in different solvents	Simple equipment and easy to operate	Long extraction time and unable to effectively separate polyphenols bound to seaweed matrix, such as cell walls	Pradhan et al. (2022)
Microwave-assisted extraction (MAE)	The dipole rotation of the polar solvent and the ion conduction of the dissolved ions raise the temperature and rupture the structure of the matrix	Short extraction time, less solvent, and energy consumption	Higher temperatures lead to the degradation of some phenolic compounds	Yuan et al. (2018)
Ultrasonic-assisted extraction (UAE)	Ultrasound-induced cavitation causes periodic expansion and compression cycles of media molecules, creating bubbles, bursting, and destroying the seaweed matrix	Suitable for use at an industrial-scale and easy to operate	Excessive ultrasonic time may lead to the degradation of polyphenols	Albero et al. (2019) and Meng et al. (2023a)
Enzyme-assisted extraction (EAE)	Remove matrix bound to polyphenols by enzymatic hydrolysis	Ability to use food-grade enzymes for large-scale production	Higher costs during the use of enzymes	Meng et al. (2023a)

(continued)

Table 3.1 (continued)

Extraction methods	Principles for extraction	Benefits of extraction techniques	Challenges and limitations	References
Pressurized solvent extraction (PSE)	The combination of high temperature and high pressure contributes to breakdown of seaweed matrices	The extraction time is short, the extraction rate is high, the solvent consumption is small, and allows the use of green solvents	Can lead to the isomerization and degradation of some polyphenols, and the selectivity to polyphenols is low, requiring further purification processes to separate them from co-extracted compounds	Poojary et al. (2021)
Supercritical fluid extraction (SFE)	Using solvent extraction at pressures and temperatures above the critical point	Reduced extraction time, wide range, environmental friendliness, high selectivity for polyphenols, and adjustable solvent density	The instruments used by SFE require expensive cleaning costs and long cleaning times	Ballesteros-Vivas et al. (2021) and Zhang et al. (2022)

methods have been considered to have potential applications in improving the yield of seaweed polyphenols. Recent research exploring the extraction of polyphenols from *Ascophyllum nodosum* reported higher yields of extraction of polyphenols when using ultrasonic-assisted enzyme extraction (UAEE), compared to those achieved by UAE and EAE (Meng et al. 2023a). Charoensiddhi et al. (2015) found that extracts obtained from *Ecklonia radiata* by microwave-assisted enzymatic extraction (MAEE) had higher total polyphenolic contents and antioxidant activity compared to those achieved by conventional methods (SLE) and other innovative processes using enzymes such as EAE. The synergistic effects of the combination of multiple extraction methods may contribute to the appreciated improvements in the extraction processes. However, since these combinations of technologies can be performed using multiple process parameters, further research is necessary to evaluate the influence of each individual technology as well as their combination on the yields of polyphenols, as well as to optimize all the possible combinations so as to improve the efficiency of these methods. Garcia-Vaquero et al. (2022) explored the extraction efficiencies of different powers of UAE, MAE, and ultrasound-microwave-assisted extraction for biological compounds and antioxidants in macroalgae. The synergistic use of ultrasound and microwave techniques resulted in higher yields of sulfated fucose polysaccharides and polyphenols (Garcia-Vaquero et al. 2022).

3.4 Purification and Characterization Methods of Macroalgal Polyphenols

3.4.1 Purification Methods

It is worth noting that most of the currently available algal polyphenols still exist in the form of enriched mixtures, and further separation and purification of these mixtures are needed to separate and concentrate specific compounds or separate them from undesirable co-extracted compounds. As shown in Table 3.2, many purification methods have been used to further enrich phenolic compounds in the previously described crude seaweed extracts.

3.4.1.1 Liquid–Liquid Partition

Liquid–liquid partition is based on the separation of polyphenols from other co-extracted compounds on the basis of their polarity. In this purification method, the difference in the chemical potential of polyphenols in two solvents of different polarities is the driving force for species transfer (Tong et al. 2021). Several studies applied this purification method to fractionate various seaweed-derived polyphenols. The ethyl acetate fraction of the crude extract of macroalgal polyphenol showed strong biological activity in various studies (Matsui et al. 2022; Meng et al. 2023a; Wang et al. 2012). However, liquid–liquid partition has some disadvantages, such as the use of non-food-grade solvents, large consumption of reagents, and difficulties in the separation of polyphenols from other co-extractives, i.e. polyphenols bound to proteins (Leyton et al. 2017). Although the use of ethyl acetate is permitted in food and pharmaceutical products, there are concerns about an increased risk to human health and safety from the continued exposure to solvents during the processing steps of purification of the compounds (Erpel et al. 2023). Therefore, using greener and more environmentally sustainable practices compared to liquid–liquid partition has become a major challenge that has to be faced in the continuous production of seaweed polyphenols.

3.4.1.2 Molecular Weight Cutoff (MWCO)

Due to the extremely large ranges in MW of polyphenols, MWCO may be used for the purification of polyphenols, or for the separation of polyphenols in a specific MW range. Meng et al. (2023c) obtained polyphenols from *Ascophyllum nodosum* with different MWs (i.e. <1 kDa, 1–3 kDa, 3–5 kDa, 5–10 kDa, and >10 kDa) and identified the antioxidant activity of each fraction. The results of the study showed that polyphenols with an MW of 1–3 kDa had the highest antioxidant activities measured as 2,2′-azino-bis 3-ethyl-benzothiazoline-6-sulfonic acid (ABTS) scavenging activity (1.54 mg ascorbic acid equivalent (ACE)/mg phloroglucinol

Table 3.2 Benefits and challenges of different purification methods used to purify macroalgal polyphenols

Purification methods	Principles of purification method	Benefits of purification method	Challenges and limitations	References
Liquid–liquid partition	Chemical potential of polyphenols in different polar solvent	Simple operation	High solvent consumption, environmental pollution, health risks, inability to effectively separate coextractives	Erpel et al. (2023) and Tong et al. (2021)
Molecular weight cut off	Separation according to molecular weight of compounds	The process is fast and easy to operate, no need extra solvent consumption	Unable to completely separate polyphenols from other coextracted compounds	Meng et al. (2023a)
Sephadex LH-20 gel filtration	Separation according to polarity and hydroxyl group number of compounds	Less solvent consumption, filtration materials can be recycled	Sephadex LH-20 has a separation limit of 4–5 kDa, beyond which polyphenols cannot be separated by this method	Meng et al. (2023a) and Wang et al. (2012)
Macroporous resins adsorption	Differences in the adsorption capacity of resins for polyphenols in extracts or crude extracts	Food-grade, high-efficiency, economic benefits, easy regeneration, and environmentally friendly	Optimum resins may be needed for different sources of seaweed polyphenols	Erpel et al. (2023), Leyton et al. (2017) and Mariotti-Celis et al. (2023)
Microcrystalline cellulose adsorption	Separation based on the polarity of polyphenols and other coextracted substances and usage of different elution solvents	Sustainability, biocompatibility, and suitability for large-scale industrialization of microcrystalline cellulose	Polyphenol loss, solvent usage, and residual risks need to be further reduced	Barbosa et al. (2020) and Santos et al. (2019)

equivalent (PGE)) and ferric reducing antioxidant power (FRAP) (1.65 mg Trolox equivalent (TE)/mg PGE) and that the MW ranges from 3 to 5 kDa had the highest 1,1-diphenyl-2-picrylhydrazyl (DPPH) antioxidant activity (1.57 mg ACE/mg PGE) (Meng et al. 2023a). Blanc et al. (2011) also used MWCO to obtain different MW fractions of *Ascophyllum nodosum*. The authors reported that at MW > 50 kDa, a large number of polyphenolic polymers were enriched, and the purity of the compounds was at its highest, while the phenolic content of the dialysate with MW < 2 kDa was only 10–15%. It is worth noting that only a few studies have explored the effects of different MWs on the bioactivity of seaweed-derived polyphenols. The structure–activity relationship between the MW of seaweed polyphenols and their biological activity has not been explored definitively. This is perhaps the biggest challenge for this purification method, at present, and should be further explored (Meng et al. 2021b).

3.4.1.3 Sephadex LH-20 Gel Filtration

Sephadex LH-20 is an adsorbent modified by epoxy dextran, which has been widely used for the separation and purification of polyphenols from seaweeds through suitable elution systems (Bogolitsyn et al. 2019). The separation and purification of polyphenols from algae achieved by Sephadex LH-20 gel are mainly based on the interaction between the hydroxyl groups in polyphenols and the gel due to the generation of hydrogen bonds, resulting in polyphenols being adsorbed by the gel, while the strength of this adsorption is determined by the polarity of polyphenols and the number of hydroxyl groups (Wang et al. 2012). According to the strength of the combination of polyphenols and Sephadex LH-20, the use of elution solvents of different polarities may help in the separation of polyphenols of different polarities from the gel. Nwosu et al. (2011) used Sephadex LH-20 to conduct further gel separation of polyphenols from *Ascophyllum nodosum*, resulting in the concentration of phlorotannins with potential anti-diabetic effects. Wang et al. (2012) analysed phlorotannins of different polarities separated by Sephadex LH-20 by HPLC-DAD-ESI-MS, and the authors showed that the MW of sub-fractions of phlorotannins increased with decreasing polarity of the eluent. It is worth noting that when Sephadex LH-20 is used to separate polyphenols from algae since its maximum separation limit is 4–5 kDa, many polyphenols with an MW exceeding this range cannot be adsorbed to the separation column (Meng et al. 2023a). Therefore, the biggest challenge of this separation method is to develop column materials with higher MW separation ranges to allow further separation of polyphenols due to the wide variability of these compounds in macroalgae. In addition, further clarification of the intrinsic relationship between the polarity of polyphenols from seaweeds and their biological activities is needed to improve the efficient separation of the most biologically active fractions and further improve the Sephadex LH-20 purification systems.

3.4.1.4 Macroporous Resin Adsorption

Macroporous resins have different polarities, specific surface areas, and pore sizes and have been proven to be well suited for the purification of polyphenols. Adsorbable macroporous resins have several advantages compared to other methods for the purification of these compounds, including high efficiency, low cost of operation, easy regeneration, and environmental friendliness (Erpel et al. 2023). In crude seaweed extracts containing mixtures of compounds, i.e. co-extracted substances such as sugars and proteins do not interact with the resin and can be removed with water, while the hydrophobic part of the polyphenols can be adsorbed with the resin, and the adsorbed polyphenols can be removed with a solvent (such as food-grade ethanol solution) (Mariotti-Celis et al. 2023). The optimum resin of choice will vary depending on the variety of polyphenols derived from different seaweeds. Park and Lee (2021) evaluated the ability of SP850, HP2MG, XAD2, XAD7HP, and XAD16N to purify polyphenols from *Ecklonia cava* and found that HP2MG had the best adsorption/desorption ability of polyphenols from this seaweed species. Leyton et al. 2017 proved that XAD-16 N was more efficient for the purification of polyphenols from *Macrocystis pyrifera,* as compared to HP-20, SP-850, XAD-7, XAD-4, and XAD-2. Moreover, Erpel et al. (2023) obtained food-grade *Durvillaea incurvata* and *Lessonia spicata* polyphenolic extracts by using HP-20 macroporous adsorption resin and 80% ethanol as the eluent. Compared with the crude extract, the recovery rate of this method was 66%, the purity was 29 times higher than that of the crude extract, and potentially toxic elements, such as As and Cd, in the brown seaweed extracts were significantly reduced (Erpel et al. 2023).

3.4.1.5 Microcrystalline Cellulose (MCC) Adsorption

As one of the most abundant polysaccharides in the world, cellulose and its by-products are considered to have considerable industrial potential (Nsor-Atindana et al. 2017). As one of the main derived components of cellulose, MCC is a food-grade substance and GRAS (Nsor-Atindana et al. 2017). The adsorption–desorption process MCC has been explored as a useful strategy for the purification of multiple co-extracted compounds in extracts rich in polyphenols, including pigments and polysaccharides, as well as for the purification of phlorotannins (Barbosa et al. 2017, 2020). However, during the purification process using MCC, the use of certain solvents, such as toluene, and the possible residues of these solvents in the final product, still needs to be addressed by further research improving MCC purification focused on using greener solvents. In addition, optimization of the adsorption–desorption method to reduce polyphenol loss and improve specificity is also needed when considering the use of this purification technique (Santos et al. 2019).

3.4.2 Characterization of Macroalgal Polyphenols

3.4.2.1 Quantitative Analysis to Determine the Contents of Polyphenols

The most commonly used methods for the analysis of polyphenols in seaweeds include the use of Folin-Ciocalteu (FC) and 2,4-dimethoxybenzaldehyde (DMBA) reagents to detect the colour of the solutions by spectrophotometric methods. The principle of FC colorimetry is based on the fact that electrons present in polyphenolic compounds are transferred to phosphomolybdic acid or phosphotungstic acid complexes under alkaline conditions, while DMBA can specifically react with

1,3- and 1,3,5-trihydroxybenzene of phlorotannins in the presence of polyphenols (Shrestha et al. 2021). However, the presence of other reducing derivatives in the extract, such as ascorbic acid, reducing sugars, proteins, and pigments, may lead to an overestimation of the amount of polyphenols present by FC (Castro-Alves and Cordenunsi 2015). In the case of the DMBA reagent, if the hydroxyl groups of the 2, 4, or 6 positions are substituted, there will be no colour change, resulting in an underestimation of polyphenols (Ford et al. 2019). Despite these limitations, overall, the UV-spectrophotometer-based FC and DMBA colorimetric methods can be used for efficient and rapid high-throughput quantification of polyphenols and phlorotannins, as these methods are simple, economical, and quite effective. Another limitation of this method is that there are currently three equivalent ways to express the polyphenol content of algae depending on the standards used, including pyrocatechol equivalents (PCE), gallic acid equivalents (GAE), and phloroglucinol equivalents (PGE), which may lead to complications when comparing results from different studies (Gunathilaka et al. 2020; Meng et al. 2021b).

3.4.2.2 Ultra-High-Performance Liquid Chromatography Tandem Mass Spectrometry (UHPLC–MS/MS)

UHPLC–MS/MS has been extensively used to characterize phenolic compounds from algal biomass. Kumar et al. (2022b) analysed the composition and antioxidant activity of polyphenol-rich extracts from brown and green algae by UHPLC–MS/MS. However, one of the limitations of this technique was that it could not analyse large MW polyphenols accurately and the lack of commercial standards for potential compounds (Rajauria 2018). Wang et al. (2012) proved that UHPLC–MS/MS could be effectively used to identify phlorotannins oligomers, of up to ten sub-units; however, the highly polymerized phlorotannins are not very well identified, and the high MW phlorotannins appeared as a single peak in the spectrum, which cannot be further discriminated. Kirke et al. (2017) used UHPLC–MS with tandem quadrupole mass spectrometry (TQD-MS) to analyse polyphenolic compounds from *Fucus vesiculosus*, and the authors reported that this method could be effective in analysing phlorotannins with a degree of polymerization of 3–16. Notably, related studies showed that phlorotannins could be degraded during storage, and the degradation of phlorotannins with a high degree of polymerization into phlorotannins with a low degree of polymerization can be observed by LC–MS/MS (Kirke et al. 2017). In a previous study, the authors summarized the fragmentation patterns of more than 100 phlorotannins identified by UHPLC–MS/MS (Meng et al. 2021b). However, this is only a very small part of the vast variety of phlorotannins, and the structures of many more phlorotannins remain unknown. Although the current UHPLC–MS/MS has shown good characterization of low MW polyphenols, information about the isomeric type and linkage mode of higher MW polyphenols cannot be accurately elucidated (Catarino et al. 2022). Therefore, it is urgent to conduct further research on the structure of polyphenols as well as to establish a database of currently discovered compounds. Taken together, these could help future efforts in the elucidation and discovery of new chemical structures of polyphenols. This has also been expressed in recent research, emphasizing the need for future research separating and purifying polyphenols to be able to trustworthy databases (Catarino et al. 2022).

3.4.2.3 Proton Nuclear Magnetic Resonance (^{1}H-NMR)

NMR can be used directly and non-destructively to analyse and elucidate the structural features of phlorotannins (Catarino et al. 2022). NMR is based on the principle of nuclear spins with a difference in resonance frequency, which will depend on the magnetic field strength (chemical shift) of the nuclei in the molecule (Lafarga et al. 2020). NMR has a role in distinguishing macroalgal isomers; for example, it can further determine the configuration of compounds with the same molecular weight, such as fucofuroeckol A and fucofuroeckol B (Erpel et al. 2023). Although quantitative ^{1}H NMR (qNMR) spectroscopic methods are more accurate and highly selective than colorimetric methods, sample complexity limits their general use in routine sample analysis (Martínez-López and Tuohy 2023). In general, NMR can provide accurate and reliable information on the structure of polyphenolic compounds, but it cannot currently be used widely due to the high price of equipment and difficulties in its operation and interpretation (Catarino et al. 2022).

3.5 Applications of Macroalgal Polyphenols

3.5.1 Uses as Food Ingredients

Polyphenols from the macroalga *Ecklonia cava* were approved by the FDA as dietary supplements in 2008 (U.S. Food and Drug Administration 2023). As a widely recognized natural ingredient, macroalgal polyphenols have been extensively studied to incorporate them into different food matrices. Yu et al. (2022) developed a novel bio-preservative by immobilizing phlorotannins into nanochitin (NCh). The addition of 1.5 g/kg NCh-phlorotannins to sea bass fillets was beneficial, as these products had the lowest levels of lipid oxidation and microbial loads during cold storage, extending the shelf life of sea bass by 3 days (Yu et al. 2022). Polyphenols have multiple available polar functional groups, and thus, they can selectively or non-selectively bind to various biological molecules, such as proteins (Martínez-López and Tuohy 2023). Polyphenols derived from *Sargassum fusiforme* can reduce the allergenicity of shrimp products by changing the conformational structure of tropomyosin (TM), one of the most important allergens in shrimp and shrimp-containing food products (Zhao et al. 2023). Polyphenol extracts from *Fucus vesiculosus* were also used in fish oil-enriched milk (Hermund et al. 2015), mayonnaise (Honold et al. 2016), and granola bars (Hermund et al. 2016) to inhibit the oxidation of lipids and extend the shelf life of all these food products.

Although macroalgal polyphenols have strong biological activity when they are used in the food industry, the specific, strong odour of some algae may be a limitation for their application. Therefore, encapsulation of seaweed polyphenols is an opportunity to expand the uses of polyphenols in the food industry. Savaghebi et al. (2021) encapsulated the phenolic extract of *Sargassum boveanum* in nano-liposomes and applied the product in the production process of functional mayonnaise in order to improve shelf life. The results of this study showed that free brown algae polyphenols had adverse effects on the colour, taste, and overall acceptability of mayonnaise, but the overall acceptability of the sample was significantly improved after adding encapsulated macroalgal polyphenols (Savaghebi et al. 2021).

3.5.2 Uses in Biopackaging of Food

Compared to traditional petrochemical industry derivatives, such as plastics, bio-based polymers are more environmentally friendly alternatives for the production of films for food packaging applications to extend food shelf life and provide safe transportation (Vera et al. 2023). Phlorotannins are considered to have excellent anti-bacterial activities (Besednova et al. 2020, 2021), and thus, the addition of these natural antimicrobial agents to active food packaging materials can be considered an effective strategy to effectively prevent microbial growth during storage and prolong the shelf life of food, without losing its quality (Surendhiran et al. 2019). Phlorotannins have been added to *Momordica charantia* polysaccharides (MCP) (Cui et al. 2020), carboxymethyl chitosan (Zhao et al. 2022), konjac glucomannan/cellulose nanocrystals (Wu et al. 2021), sodium alginate (SA)/poly(ethylene oxide) (PEO) (Surendhiran et al. 2019), and other materials such as matrix fibre composite membranes, and exhibit excellent anti-bacterial activity. Sodium alginate (SA)/poly(ethylene oxide) (PEO) nanofibres containing phlorotannins prepared by the electrospinning technique exhibited anti-bacterial activity against *Salmonella* Enteritidis and excellent antioxidant activities, thereby maintaining the sensory quality of marinated chicken within an acceptable range (Surendhiran et al. 2019).

Despite the fact that phlorotannins have been incorporated in packaging as antimicrobials, the large-scale application of these compounds in food packaging still requires further exploration. As previously mentioned, macroalgal polyphenols are natural compounds that can be affected by various conditions, including the environment, seaweed species, location, and time. Thus, the large-scale production of these compounds still needs to be further explored and elucidated before these compounds can be added to food packaging. Moreover, these compounds also need to be explored for their capacity to create biofilms. Some terrestrial-derived tannins have shown promising properties when used as functional additives and cross-linking agents to improve the mechanical and physicochemical properties of polymers that could be used as biofilms (Vera et al. 2023; Zhu et al. 2019).

3.5.3 Uses as Bioactive Compounds

Polyphenols from various seaweeds have been explored for their multiple bioactivities, including antioxidant (Meng et al. 2022), anti-inflammatory (Yu et al. 2019), antidiabetic (Kang et al. 2010), and anticancer (Dutot et al. 2021) properties, which could have multiple potential uses when using these molecules as functional foods or nutraceuticals. In Korea, *Ecklonia cava* has been widely used as a medicine for several centuries (Harwanto et al. 2022). Moreover, the Ministry of Food and Drug Safety (MFDS) in Korea approved the use of a polyphenol extract from this seaweed with the purpose of improving sleep quality in 2015 (Kim et al. 2022a). Nwosu et al. (2011) and Yu et al. (2019) evaluated the anti-diabetic potential of polyphenol extracts derived from *Lessonia trabeculata*, *Ecklonia cava,* and *Ascophyllum nodosum,* and these studies reported stronger α-amylase and α-glucosidase inhibitory effects of polyphenols from seaweeds compared to those of acarbose, the most commonly used anti-diabetic drug. In addition, phlorotannins are reported to exhibit some modulatory properties in multiple cancer-related pathways, including inhibition of apoptosis antigen 1 (APO-1), B-cell lymphoma 2 (Bcl-2), cysteinyl aspartate specific proteinase (CAPS), and tissue inhibitors of metalloproteinases (TIMP), as well as promotion of protein kinase B (AKT), B-cell lymphoma-extralarge (Bcl-xL), FADD-like IL-1β-converting enzyme-inhibitory protein (FLIP), matrix metalloproteinase (MMP), nuclear factor-kB (NF-kB), phosphoinositide 3-kinase (PI3K), vascular endothelial growth factor (VEGF), and X-linked inhibitor of apoptosis (XIAP) (Erpel et al. 2020; Meng et al. 2021b). Although many studies have reported that a variety of polyphenols derived from macroalgae have remarkable biological activities, the application of polyphenols as bioactive compounds has not been widely implemented to date. As we previously reported, currently, algal polyphenols are still mainly commercialized in the form of complex mixtures. Clearly more clinical trial data are required.

3.5.4 Uses as Stabilizers of Pickering Emulsions

Emulsions are oil–water mixtures widely used in functional food, nutraceutical, pharmaceutical, and cosmeceutical fields that can encapsulate and deliver a variety of high-value compounds, such as β-carotene, curcumin, and astaxanthin (Meng et al. 2023d). Within emulsions, Pickering emulsions are those that are only stabilized by solid particles. Compared with traditional emulsions stabilized by surfactants, Pickering emulsions have stronger stability and higher consumer acceptance because they do not contain surfactants (Meng et al. 2023d). Previous studies have broadened the use of water-insoluble polyphenols from *Ascophyllum nodosum* and *Laminaria japonica* as stabilizers of Pickering emulsions achieved by layer-by-layer electrostatic deposition (Meng et al. 2023b, c). Other researchers have applied various other polyphenols to Pickering emulsion stabilizers, such as chitosan-gallic acid conjugate (Yi et al. 2021) and rice bran protein with catechin (Li et al. 2020). Further research is still required to explore the application of more kinds of macroalgal polyphenols in Pickering emulsions. Moreover, the current related research focuses only on the creation of these emulsions at the laboratory scale, and their scale-up processes still need to be elucidated to broaden the application of these compounds in emulsions and their future use in functional food, nutraceutical, pharmaceutical, cosmeceutical, and other industrial fields. In addition, seaweed polyphenol-based Pickering emulsions have many potential research directions, such as customizable 3D-printing high internal emulsions and emulsion spray drying microcapsules.

3.5.5 Uses in Tissue Engineering

Tissue engineering is the field of regulating damaged organs and tissues for regeneration or repair by preparing and using biomedical scaffolds, dressings, cells, and growth factors (Im et al. 2017). To date, a large number of tissue engineering products have been designed, and some attempts have been made in the repair of tissues and organs, such as the development of artificial blood vessels, skin, cartilage, bone tissue, and ligaments (Zhang et al. 2021). However, for the human body, most tissue engineering products are detected as foreign bodies, and they may be rejected; thus, they can induce inflammation, form scars, cause immune responses, and may eventually lead to other defects in the human body (Gao et al. 2021). The application of polyphenols in tissue engineering is promising due to their antioxidant properties, as polyphenols are able to maintain the redox balance of damaged tissues and organ sites, reduce excess reactive oxygen species, and ultimately contribute to the health of the human body (Gao et al. 2021). Moreover, the physico-chemical properties of polyphenols allow their combination with other substances through covalent and non-covalent interactions, showing promising potential applications when designing material composites with added functional activities. Previous research emphasized that the addition of phlorotannins was able to provide prominent hydrophilic sites and induce high wettability without any loss of mechanical properties in composites, which is well suited to form scaffolds for bone tissue regeneration (Yeo et al. 2012). Kim et al. (2022b) generated a new type of cartilage cellular matrix (CAM) film loaded with phlorotannins derived from *Ecklonia cava*, which can be used to prevent inflammation in the process of adhesion between proliferating fibrous tissue and adjacent normal organs after surgery with fewer side effects. Im et al. (2017) also appreciated the role of phlorotannins in promoting bone tissue growth when added to bone tissue regeneration scaffolds. Their research found that compared with fish collagen/alginate, phlorotannin/fish collagen/alginate scaffolds had better cell viability, cell proliferation, calcium deposition, and in vitro biomineralization (Im et al. 2017). Although polyphenols from seaweeds have shown potential prospects in tissue engineering applications, their application still needs to be supported by further and in-depth in vitro and in vivo studies.

3.6 Future Outlook and Main Conclusions

3.6.1 Extraction Techniques

Seaweeds contain a class of polyphenols that do not exist in other organisms, namely, phlorotannins, which are compounds with abundant biological activities that have recently attracted the attention of multiple researchers (Catarino et al. 2021). However, even though the application prospects of these polyphenols from selected seaweeds are promising, there are still many challenges for the commercial development and industrial applications of these compounds. The extraction of algal-derived polyphenols is still one of the key bottlenecks during the processing and future utilization of these compounds (Kumar et al. 2022a). As we introduced in this chapter, a variety of innovative or emerging extraction technologies, applied alone or combined, provide a positive outlook for the future production of these compounds at the industrial level, as they have greatly improved the efficiency of the extraction processes compared to conventional technologies. However, further research is needed to optimize the use of these technologies and regulate the extraction process parameters aiming to achieve high yields of compounds while ideally reducing the number of co-extracted compounds and improving the sustainability, including time, cost, solvent, and energy usage of these processes (Garcia-Vaquero et al. 2021c).

3.6.2 Purification Techniques

After the extraction process, the crude extracts containing polyphenols need to be further purified to elucidate the biological properties of the compounds without the effects of compounds. The purification of polyphenols from seaweed is currently under development, and it is one of the major bottlenecks for the identification and commercialization of pure compounds. Currently, there are only commercially available polyphenols from seaweeds as crude, undefined extracts. In the current purification process, many separation methods based on the structural differences (such as MW and polarity) of polyphenols and other co-extracted substances have been widely used, such as liquid–liquid partition, MWCO, Sephadex LH-20 filtration, macroporous resins, and MCC adsorption (Mariotti-Celis et al. 2023; Meng et al. 2023a; Santos et al. 2019). These purification methods demonstrated very good efficiencies. However, due to some method limitations, they still need to be further optimized. Moreover, current trends indicate that in the future, consumers will pursue products extracted and processed following green practices, and thus, purification processes should also be similarly adapted to use food-grade solvents for the future commercialization of the products (Erpel et al. 2023). At the same time, there is still research needed to isolate and purify single polyphenolic components through complex steps, allowing the development of commercial standards.

3.6.3 Characterization Techniques

For the identification methods of seaweed polyphenols, the quantitative detection of polyphenols by spectrophotometric methods has become the standard method for the estimation of these compounds by researchers, although some compounds co-extracted may interfere with the results (Castro-Alves and Cordenunsi 2015). Thus, further exploration of extraction and purification methods is needed to reduce the presence of co-extracted substances that will also influence the levels in the characterization stage. UHPLC–MS/MS is a useful method for the characterization of the structure of most seaweed polyphenols. However, most studies have focused on the characterization of polyphenols within a maximum MW range of approximately 2 kDa, while the characterization of polyphenols with higher MW remains a field that will need further study and improvement of current characterization techniques. Moreover, there is a need to create a database for the diversity of seaweed polyphenols in order to improve and potentiate studies aiming to characterize these high-value compounds (Catarino et al. 2022). Due to the lack of commercially available standards, the variety of chemical structures of polyphenols produced by seaweeds, and the large MW of these molecules, some of these polyphenols may be undetected when using the aforementioned characterization methods (Rajauria 2018). NMR is a promising technique useful for the accurate characterization of polyphenols and their isomers (Erpel et al. 2023). Therefore, combining UHPLC–MS/MS with NMR may contribute to better characterization in future studies. Overall, the characterization of seaweed-derived compounds mainly involves two directions in the future: one is to develop further purified polyphenolic standards, and the other is to explore detection methods with higher detection limits and resolution in order to allow for the characterization of molecules of higher MW.

3.6.4 Practical Applications

Due to the wide range of biological activities of macroalgal polyphenols, these molecules, although currently under-exploited at the industrial level, have a great array and prospects for future applications as food ingredients, food bio-packaging materials, bioactive compounds such as nutraceuticals, Pickering emulsion stabilizers, and tissue engineering (Kim et al. 2022b; Meng et al. 2023d; Savaghebi et al. 2021; Surendhiran et al. 2019). However, it is worth noting that although many applications of polyphenols have been explored, and some crude extracts have been approved for use, their industrial and commercial applications are still at a very early stage. When expanding the large-scale industrial application of polyphenols from seaweeds, it is necessary to explore complex processes, including the source of macroalgae and the extraction and purification processes needed to achieve a consistent product with the ideal biological and physicochemical properties required for the aforementioned applications. Moreover, for any intended use, the bioaccessibility and degradation routes of these compounds when entering the human body by different routes depending on their intended application should all be further explored to ensure that the active compounds hit the intended organs/targets and the molecules exert their biological activities efficiently (Murray et al. 2018). Therefore, when used in the human diet, once ingested, polyphenols will undergo complex reactions and processes taking place during gastrointestinal digestion to be later absorbed and enter the body. Further research must focus on exploring in vivo and in vitro model systems and clinical trials, to further expand their functional food, nutraceutical, and pharmaceutical uses (Meng et al. 2021b).

Acknowledgements Weihao Meng is currently the recipient of a PhD Scholarship within the joint collaborative PhD programme between UCD and GSCAAS. This study is funded by the Science and Technology Innovation Project of the Chinese Academy of Agricultural Sciences (CAAS-ASTIP-202X-IFST).

References

Agregán R, Munekata PE, Domínguez R, Carballo J, Franco D, Lorenzo JM (2017) Proximate composition, phenolic content and in vitro antioxidant activity of aqueous extracts of the seaweeds *Ascophyllum nodosum, Bifurcaria bifurcata* and Fucus vesiculosus. Effect of addition of the extracts on the oxidative stability of canola oil unde. Food Res Int 99:986–994. https://doi.org/10.1016/j.foodres.2016.11.009

Albero B, Tadeo JL, Pérez RA (2019) Ultrasound-assisted extraction of organic contaminants. TrAC – Trends Anal Chem 118:739–750. https://doi.org/10.1016/j.trac.2019.07.007

Ballesteros-Vivas D, Ortega-Barbosa JP, Parada-Alfonso F, Ferreira SRS, Del Pilar Sánchez-Camargo A (2021) Supercritical fluid extraction of lipids, carotenoids, and other compounds from marine sources. In: Innovative and emerging technologies in the bio-marine food sector: applications, regulations, and prospects. https://doi.org/10.1016/B978-0-12-820096-4.00016-X

Barbosa M, Lopes G, Ferreres F, Andrade PB, Pereira DM, Gil-Izquierdo Á, Valentão P (2017) Phlorotannin extracts from Fucales: marine polyphenols as bioregulators engaged in inflammation-related mediators and enzymes. Algal Res 28:1–8. https://doi.org/10.1016/j.algal.2017.09.009

Barbosa M, Valentão P, Ferreres F, Gil-Izquierdo Á, Andrade PB (2020) In vitro multifunctionality of phlorotannin extracts from edible *Fucus* species on targets underpinning neurodegeneration. Food Chem 333:127456. https://doi.org/10.1016/j.foodchem.2020.127456

Besednova NN, Andryukov BG, Zaporozhets TS, Kryzhanovsky SP, Kuznetsova TA, Fedyanina LN, Makarenkova ID, Zvyagintseva TN (2020) Algae polyphenolic compounds and modern antibacterial strategies: current achievements and immediate prospects. Biomedicines 8(9):1–19. https://doi.org/10.3390/BIOMEDICINES8090342

Besednova NN, Andryukov BG, Zaporozhets TS, Kryzhanovsky SP, Fedyanina LN, Kuznetsova TA, Zvyagintseva TN, Shchelkanov MY (2021) Antiviral effects of polyphenols from marine algae. Biomedicines 9(2):1–23. https://doi.org/10.3390/biomedicines9020200

Blanc N, Hauchard D, Audibert L, Ar Gall E (2011) Radical-scavenging capacity of phenol fractions in the brown seaweed *Ascophyllum nodosum*: an electrochemical approach. Talanta 84(2):513–518. https://doi.org/10.1016/j.talanta.2011.01.034

Bogolitsyn K, Druzhinina A, Kaplitsin P, Ovchinnikov D, Parshina A, Kuznetsova M (2019) Relationship between radical scavenging activity and polymolecular properties of brown algae polyphenols. Chem Pap 73(10):2377–2385. https://doi.org/10.1007/s11696-019-00760-7

Casas MP, Conde E, Domínguez H, Moure A (2019) Ecofriendly extraction of bioactive fractions from *Sargassum muticum*. Process Biochem 79:166–173. https://doi.org/10.1016/j.procbio.2018.12.021

Castillo A, Celeiro M, Lores M, Grgić K, Banožić M, Jerković I, Jokić S (2023) Bioprospecting of targeted phenolic compounds of *Dictyota dichotoma, Gongolaria barbata, Ericaria amentacea, Sargassum hornschuchii* and *Ellisolandia elongata* from the Adriatic sea extracted by two green methods. Mar Drugs 21(2):97. https://doi.org/10.3390/md21020097

Castro-Alves VC, Cordenunsi BR (2015) Total soluble phenolic compounds quantification is not as simple as it seems. Food Anal Methods 8(4):873–884. https://doi.org/10.1007/s12161-014-9961-0

Catarino MD, Amarante SJ, Mateus N, Silva AMS, Cardoso SM (2021) Brown algae phlorotannins: a marine alternative to break the oxidative stress, inflammation and cancer network. Food Secur 10(7):1478. https://doi.org/10.3390/foods10071478

Catarino MD, Pires SMG, Silva S, Costa F, Braga SS, Pinto DCGA, Silva AMS, Cardoso SM (2022) Overview of phlorotannins' constituents in Fucales. Mar Drugs 20(12):1–27. https://doi.org/10.3390/md20120754

Charoensiddhi S, Franco C, Su P, Zhang W (2015) Improved antioxidant activities of brown seaweed *Ecklonia radiata* extracts prepared by microwave-assisted enzymatic extraction. J Appl Phycol 27(5):2049–2058. https://doi.org/10.1007/s10811-014-0476-2

Ciko AM, Jokić S, Šubarić D, Jerković I (2018) Overview on the application of modern methods for the extraction of bioactive compounds from marine macroalgae. Mar Drugs 16(10):348. https://doi.org/10.3390/md16100348

Cui H, Yang X, Abdel-Samie MA, Lin L (2020) Cold plasma treated phlorotannin/*Momordica charantia* polysaccharide nanofiber for active food packaging. Carbohydr Polym 239:116214. https://doi.org/10.1016/j.carbpol.2020.116214

Dutot M, Olivier E, Fouyet S, Magny R, Hammad K, Roulland E, Rat P, Fagon R (2021) In vitro chemopreventive potential of phlorotannins-rich extract from brown algae by inhibition of benzo[a]pyrene-induced P2X7 activation and toxic effects. Mar Drugs 19(1):1–12. https://doi.org/10.3390/MD19010034

Echave J, Lourenço-Lopes C, Cassani L, Fraga-Corral M, Garcia-Perez P, Otero P, Carreira-Casais A, Perez-Gregorio R, Baamonde S, Saa FF, Simal-Gandara J, Prieto MA (2022) Evidence and perspectives on the use of phlorotannins as novel antibiotics and therapeutic natural molecules. Med Sci Forum 12:43. https://doi.org/10.3390/eca2022-12728

Erpel F, Mateos R, Pérez-Jiménez J, Pérez-Correa JR (2020) Phlorotannins: from isolation and structural characterization, to the evaluation of their antidiabetic and anticancer potential. Food Res Int 137:109589. https://doi.org/10.1016/j.foodres.2020.109589

Erpel F, Camilo C, Mateos R, Ricardo Pérez-Correa J (2023) A macroporous resin purification process to obtain food-grade phlorotannin-rich extracts with α-glucosidase inhibitory activity from Chilean brown seaweeds: an UHPLC-MSn profiling. Food Chem 402. https://doi.org/10.1016/j.foodchem.2022.134472

FAO (2022) FAO global fishery and aquaculture production statistics. FishStatJ. www.fao.org/fishery/statistics/software/fishstatj/en

Ford L, Theodoridou K, Sheldrake GN, Walsh PJ (2019) A critical review of analytical methods used for the chemical characterisation and quantification of phlorotannin compounds in brown seaweeds. Phytochem Anal 30(6):587–599. https://doi.org/10.1002/pca.2851

Gao X, Xu Z, Liu G, Wu J (2021) Polyphenols as a versatile component in tissue engineering. Acta Biomater 119:57–74. https://doi.org/10.1016/j.actbio.2020.11.004

Garcia-Vaquero M, Rajauria G (2022) Chapter 1 – Overview of the application of innovative and emerging technologies in the bio-marine food sector. In: Innovative and emerging technologies in the bio-marine food sector. Elsevier Inc. https://doi.org/10.1016/B978-0-12-820096-4.00021-3

Garcia-Vaquero M, Rajauria G, O'Doherty JV, Sweeney T (2017) Polysaccharides from macroalgae: recent advances, innovative technologies and challenges in extraction and purification. Food Res Int 99:1011–1020. https://doi.org/10.1016/j.foodres.2016.11.016

Garcia-Vaquero M, Rajauria G, Miranda M, Sweeney T, Lopez-Alonso M, O'doherty J (2021a) Seasonal variation of the proximate composition, mineral content, fatty acid profiles and other phytochemical constituents of selected brown macroalgae. Mar Drugs 19(4):204. https://doi.org/10.3390/MD19040204

Garcia-Vaquero M, Ravindran R, Walsh O, O'doherty J, Jaiswal AK, Tiwari BK, Rajauria G (2021b) Evaluation of ultrasound, microwave, ultrasound–microwave, hydrothermal and high pressure assisted extraction technologies for the recovery of phytochemicals and antioxidants from brown macroalgae. Mar Drugs 19(6):309. https://doi.org/10.3390/md19060309

Garcia-Vaquero M, Sweeney T, O'Doherty J, Rajauria G (2021c) Recent advances in the use of greener extraction technologies for the recovery of valuable bioactive compounds from algae. In: Recent advances in micro and macroalgal processing, pp 96–122. https://doi.org/10.1002/9781119542650.ch4

Garcia-Vaquero M, Ummat V, Tiwari B, Rajauria G (2022) Exploring ultrasound, microwave and ultrasound – microwave assisted extraction technologies to increase the extraction of bioactive compounds and antioxidants from. Mar Drugs 18:172. https://doi.org/10.3390/md18030172

Gunathilaka TL, Samarakoon K, Ranasinghe P, Peiris LDC (2020) Antidiabetic potential of marine brown algae – a mini review. J Diabetes Res 2020. https://doi.org/10.1155/2020/1230218

Harwanto D, Negara BFSP, Tirtawijaya G, Meinita MDN, Choi JS (2022) Evaluation of toxicity of crude phlorotannins and phloroglucinol using different model organisms. Toxins 14(5). https://doi.org/10.3390/toxins14050312

Hermans S (2023) 2023 Seaweed state of the industry. Phyconomy. https://phyconomy.net/articles/2022-seaweed-review/

Hermund DB, Yeşiltaş B, Honold P, Jónsdóttir R, Kristinsson HG, Jacobsen C (2015) Characterisation and antioxidant evaluation of Icelandic *F. vesiculosus* extracts in vitro and in fish-oil-enriched milk and mayonnaise. J Funct Foods 19:828–841. https://doi.org/10.1016/j.jff.2015.02.020

Hermund DB, Karadağ A, Andersen U, Jónsdóttir R, Kristinsson HG, Alasalvar C, Jacobsen C (2016) Oxidative stability of granola bars enriched with multilayered fish oil emulsion in the presence of novel brown seaweed based antioxidants. J Agric Food Chem 64(44):8359–8368. https://doi.org/10.1021/acs.jafc.6b03454

Holdt SL, Kraan S (2011) Bioactive compounds in seaweed: functional food applications and legislation. J Appl Phycol 23(3):543–597. https://doi.org/10.1007/s10811-010-9632-5

Honold PJ, Jacobsen C, Jónsdóttir R, Kristinsson HG, Hermund DB (2016) Potential seaweed-based food ingredients to inhibit lipid oxidation in fish-oil-enriched mayonnaise. Eur Food Res Technol 242(4):571–584. https://doi.org/10.1007/s00217-015-2567-y

Im JH, Choi CH, Mun F, Lee JH, Kim H, Jung WK, Jang CH, Kim GH (2017) A polycaprolactone/fish collagen/alginate biocomposite supplemented with phlorotannin for hard tissue regeneration. RSC Adv 7(4):2009–2018. https://doi.org/10.1039/c6ra25182j

Kadam SU, Tiwari BK, O'Donnell CP (2013) Application of novel extraction technologies for bioactives from marine algae. J Agric Food Chem 61(20):4667–4675. https://doi.org/10.1021/jf400819p

Kang C, Jin YB, Lee H, Cha M, Sohn E-t, Moon J, Park C, Chun S, Jung ES, Hong JS, Kim SB, Kim JS, Kim E (2010) Brown alga *Ecklonia cava* attenuates type 1 diabetes by activating AMPK and Akt signaling pathways. Food Chem Toxicol 48(2):509–516. https://doi.org/10.1016/j.fct.2009.11.004

Kim S, Kim D, Um MY, Yoon M, Choi JS, Choi YH, Cho S (2022a) Marine polyphenol phlorotannins as a natural sleep aid for treatment of insomnia: a review of sedative–hypnotic effects and mechanism of action. Mar Drugs 20(12). https://doi.org/10.3390/md20120774

Kim TH, Heo SY, Oh GW, Park WS, Choi IW, Kang HW, Kim HW, Kim YM, Jo SH, Park SH, Jung WK (2022b) A phlorotannins-loaded homogeneous acellular matrix film modulates post-implantation inflammatory responses. J Tissue Eng Regen Med 16(1):51–62. https://doi.org/10.1002/term.3258

Kirke DA, Smyth TJ, Rai DK, Kenny O, Stengel DB (2017) The chemical and antioxidant stability of isolated low molecular weight phlorotannins. Food Chem 221:1104–1112. https://doi.org/10.1016/j.foodchem.2016.11.050

Kumar LRG, Paul PT, Anas KK, Tejpal CS, Chatterjee NS, Anupama TK, Mathew S, Ravishankar CN (2022a) Phlorotannins–bioactivity and extraction perspectives. J Appl Phycol 34(4):2173–2185. https://doi.org/10.1007/s10811-022-02749-4

Kumar Y, Tarafdar A, Kumar D, Saravanan C, Badgujar PC, Pharande A, Pareek S, Fawole OA (2022b) Polyphenols of edible macroalgae: estimation of in vitro bio-accessibility and cytotoxicity, quantification by LC-MS/MS and potential utilization as an antimicrobial and functional food ingredient. Antioxidants 11(5). https://doi.org/10.3390/antiox11050993

Lafarga T, Acién-Fernández FG, Garcia-Vaquero M (2020) Bioactive peptides and carbohydrates from seaweed for food applications: natural occurrence, isolation, purification, and identification. Algal Res 48. https://doi.org/10.1016/j.algal.2020.101909

Lemesheva V, Islamova R, Stepchenkova E, Shenfeld A, Birkemeyer C, Tarakhovskaya E (2023) Antibacterial, antifungal and algicidal activity of phlorotannins, as principal biologically active components of ten species of brown algae. Plan Theory 12(4):821. https://doi.org/10.3390/plants12040821

Leyton A, Vergara-Salinas JR, Pérez-Correa JR, Lienqueo ME (2017) Purification of phlorotannins from *Macrocystis pyrifera* using macroporous resins. Food Chem 237:312–319. https://doi.org/10.1016/j.foodchem.2017.05.114

Li D, Zhao Y, Wang X, Tang H, Wu N, Wu F, Yu D, Elfalleh W (2020) Effects of (+)-catechin on a rice bran protein oil-in-water emulsion: droplet size, zeta-potential, emulsifying properties, and rheological behavior. Food Hydrocoll 98:105306. https://doi.org/10.1016/j.foodhyd.2019.105306

Lopes G, Sousa C, Silva LR, Pinto E, Andrade PB, Bernardo J, Mouga T, Valentão P (2012) Can phlorotannins purified extracts constitute a novel pharmacological alternative for microbial infections with associated inflammatory conditions? PLoS One 7(2):e31145. https://doi.org/10.1371/journal.pone.0031145

Magnusson M, Yuen AKL, Zhang R, Wright JT, Taylor RB, Maschmeyer T, de Nys R (2017) A comparative assessment of microwave assisted (MAE) and conventional solid-liquid (SLE) techniques for the extraction of phloroglucinol from brown seaweed. Algal Res 23:28–36. https://doi.org/10.1016/j.algal.2017.01.002

Mariotti-Celis MS, Rivera-Tovar PR, Huamán-Castilla NL, Pérez-Correa JR (2023) Purification and fractionation of crude seaweed extracts by adsorption-desorption processes. In: Marine phenolic compounds, pp 187–215. https://doi.org/10.1016/b978-0-12-823589-8.00009-1

Martínez-López R, Tuohy MG (2023) Rapid and cost-efficient microplate assay for the accurate quantification of total phenolics in seaweeds. Food Chem: Mol Sci 6:1–8. https://doi.org/10.1016/j.fochms.2023.100166

Matsui T, Ito C, Itoigawa M, Shibata T (2022) Three phlorotannins from *Sargassum carpophyllum* are effective against the secretion of allergic mediators from antigen-stimulated rat basophilic leukemia cells. Food Chem 377:131992. https://doi.org/10.1016/j.foodchem.2021.131992

Meng W, Mu T, Marco GV (2021a) Seaweeds and microalgal biomass: the future of food and nutraceuticals. In: Future foods: global trends, opportunities, and sustainability challenges. Elsevier, pp 183–201. https://doi.org/10.1016/B978-0-323-91001-9.00014-1

Meng W, Mu T, Sun H, Garcia-Vaquero M (2021b) Phlorotannins: a review of extraction methods, structural characteristics, bioactivities, bioavailability, and future trends. Algal Res 60:102484. https://doi.org/10.1016/j.algal.2021.102484

Meng W, Mu T, Sun H, Garcia-Vaquero M (2022) Evaluation of the chemical composition and nutritional potential of brown macroalgae commercialised in China. Algal Res 64:102683. https://doi.org/10.1016/j.algal.2022.102683

Meng W, Sun H, Mu T, Garcia-Vaquero M (2023a) Extraction, purification, chemical characterization and antioxidant properties in vitro of polyphenols from the brown macroalga *Ascophyllum nodosum*. Algal Res 70:102989. https://doi.org/10.1016/j.algal.2023.102989

Meng W, Sun H, Mu T, Garcia-Vaquero M (2023b) Pickering emulsions with chitosan and macroalgal polyphenols stabilized by layer-by-layer electrostatic deposition. Carbohydr Polym 300:120256. https://doi.org/10.1016/j.carbpol.2022.120256

Meng W, Sun H, Mu T, Garcia-Vaquero M (2023c) Effects of environmental stimuli on the physicochemical and rheological properties of chitosan-macroalgal polyphenol stabilized Pickering emulsion. Int J Biol Macromol 227:1245–1257. https://doi.org/10.1016/j.ijbiomac.2022.11.314

Meng W, Sun H, Mu T, Garcia-Vaquero M (2023d) Chitosan-based Pickering emulsion: a comprehensive review on their stabilizers, bioavailability, applications and regulations. Carbohydr Polym 304:120491. https://doi.org/10.1016/j.carbpol.2022.120491

Ming Y, Wang Y, Xie Y, Dong X, Nakamura Y, Chen X, Qi H (2023) Polyphenol extracts from *Ascophyllum nodosum* protected sea cucumber (*Apostichopus japonicas*) body wall against thermal degradation during tenderization. Food Res Int 164:112419. https://doi.org/10.1016/j.foodres.2022.112419

Murray M, Dordevic AL, Ryan L, Bonham MP (2018) The impact of a single dose of a polyphenol-rich seaweed extract on postprandial glycaemic control in healthy adults: a randomised cross-over trial. Nutrients 10(3). https://doi.org/10.3390/nu10030270

Nsor-Atindana J, Chen M, Goff HD, Zhong F, Sharif HR, Li Y (2017) Functionality and nutritional aspects of microcrystalline cellulose in food. Carbohydr Polym 172:159–174. https://doi.org/10.1016/j.carbpol.2017.04.021

Nwosu F, Morris J, Lund VA, Stewart D, Ross HA, McDougall GJ (2011) Anti-proliferative and potential anti-diabetic effects of phenolic-rich extracts from edible marine algae. Food Chem 126(3):1006–1012. https://doi.org/10.1016/j.foodchem.2010.11.111

Obluchinskaya ED, Daurtseva AV, Pozharitskaya ON, Flisyuk EV, Shikov AN (2019) Natural deep eutectic solvents as alternatives for extracting phlorotannins from brown algae. Pharm Chem J 53(3):243–247. https://doi.org/10.1007/s11094-019-01987-0

Park JJ, Lee WY (2021) Adsorption and desorption characteristics of a phenolic compound from *Ecklonia cava* on macroporous resin. Food Chem 338:128150. https://doi.org/10.1016/j.foodchem.2020.128150

Poojary MM, Laurora A, Lund MN, Tiwari BK (2021) Application of pressurized liquids to extract high-value compounds from marine biomass. In: Innovative and emerging technologies in the bio-marine food sector: applications, regulations, and prospects. Elsevier Inc. https://doi.org/10.1016/B978-0-12-820096-4.00005-5

Pradhan B, Nayak R, Bhuyan PP, Patra S, Behera C, Sahoo S, Ki JS, Quarta A, Ragusa A, Jena M (2022) Algal phlorotannins as novel antibacterial agents with reference to the antioxidant modulation: current advances and future directions. Mar Drugs 20(6). https://doi.org/10.3390/md20060403

Raja R, Hemaiswarya S, Arunkumar K, Mathiyazhagan N, Kandasamy S, Arun A, Ramasamy P (2023) Efficacy of *Eisenia bicyclis* phlorotannins in the treatment of diabetes and reducing inflammation. Food Biosci 52:102381. https://doi.org/10.1016/j.fbio.2023.102381

Rajauria G (2018) Optimization and validation of reverse phase HPLC method for qualitative and quantitative assessment of polyphenols in seaweed. J Pharm Biomed Anal 148:230–237. https://doi.org/10.1016/j.jpba.2017.10.002

Rubiño S, Peteiro C, Aymerich T, Hortós M (2022) Brown macroalgae (Phaeophyceae): a valuable reservoir of antimicrobial compounds on northern coast of Spain. Mar Drugs 20(12):1–21. https://doi.org/10.3390/md20120775

Santos SAO, Félix R, Pais ACS, Rocha SM, Silvestre AJD (2019) The quest for phenolic compounds from macroalgae: a review of extraction and identification methodologies. Biomolecules 9(12):1–56. https://doi.org/10.3390/biom9120847

Savaghebi D, Ghaderi-Ghahfarokhi M, Barzegar M (2021) Encapsulation of *Sargassum boveanum* algae extract in nano-liposomes: application in functional mayonnaise production. Food Bioprocess Technol 14(7):1311–1325. https://doi.org/10.1007/s11947-021-02638-7

Shrestha S, Zhang W, Smid SD (2021) Phlorotannins: a review on biosynthesis, chemistry and bioactivity. Food Biosci 39:100832. https://doi.org/10.1016/j.fbio.2020.100832

Surendhiran D, Cui H, Lin L (2019) Encapsulation of phlorotannin in alginate/PEO blended nanofibers to preserve chicken meat from Salmonella contaminations. Food Packag Shelf Life 21:100346. https://doi.org/10.1016/j.fpsl.2019.100346

Tong T, Liu X, Yu C (2021) Extraction and nano-sized delivery systems for phlorotannins to improve its bioavailability and bioactivity. Mar Drugs 19(11):625. https://doi.org/10.3390/md19110625

Turck D, Bresson J, Burlingame B, Dean T, Fairweather-Tait S, Heinonen M, Hirsch-Ernst KI, Mangelsdorf I, McArdle HJ, Naska A, Neuhäuser-Berthold M, Nowicka G, Pentieva K, Sanz Y, Siani A, Sjödin A, Stern M, Tomé D, Vinceti M et al (2017) Safety of Ecklonia cava phlorotannins as a novel food pursuant to Regulation (EC) No 258/97. EFSA J 15(10):5003. https://doi.org/10.2903/j.efsa.2017.5003

U.S. Food & Drug Administration (2023) Submitted 75-day premarket notifications for new dietary ingredients. New Dietary Ingredient Notifications (NDINs). https://www.fda.gov/food/new-dietary-ingredient-ndi-notification-process/submitted-75-day-premarket-notifications-new-dietary-ingredients

Ummat V, Tiwari BK, Jaiswal AK, Condon K, Garcia-Vaquero M, O'Doherty J, O'Donnell C, Rajauria G (2020) Optimisation of ultrasound frequency, extraction time and solvent for the recovery of polyphenols, phlorotannins and associated antioxidant activity from brown seaweeds. Mar Drugs 18(5). https://doi.org/10.3390/md18050250

Ummat V, Sivagnanam SP, Rajauria G, O'Donnell C, Tiwari BK (2021) Advances in pre-treatment techniques and green extraction technologies for bioactives from seaweeds. Trends Food Sci Technol 110:90–106. https://doi.org/10.1016/j.tifs.2021.01.018

Vera M, Mella C, García Y, Jiménez VA, Urbano BF (2023) Recent advances in tannin-containing food biopackaging. Trends Food Sci Technol 133:28–36. https://doi.org/10.1016/j.tifs.2023.01.014

Wang T, Jónsdóttir R, Liu H, Gu L, Kristinsson HG, Raghavan S, Ólafsdóttir G (2012) Antioxidant capacities of phlorotannins extracted from the brown algae *Fucus vesiculosus*. J Agric Food Chem 60(23):5874–5883. https://doi.org/10.1021/jf3003653

Wu Z, Tong C, Zhang J, Sun J, Jiang H, Duan M, Wen C, Wu C, Pang J (2021) Investigation of the structural and physical properties, antioxidant and antimicrobial activity of konjac glucomannan/cellulose nanocrystal bionanocomposite films incorporated with phlorotannin from *Sargassum*. Int J Biol Macromol 192:323–330. https://doi.org/10.1016/j.ijbiomac.2021.09.200

Yeo M, Jung WK, Kim G (2012) Fabrication, characterisation and biological activity of phlorotannin-conjugated PCL/β-TCP composite scaffolds for bone tissue regeneration. J Mater Chem 22(8):3568–3577. https://doi.org/10.1039/c2jm14725d

Yi J, Huang H, Wen Z, Fan Y (2021) Fabrication of chitosan-gallic acid conjugate for improvement of physicochemical stability of β-carotene nanoemulsion: impact of Mw of chitosan. Food Chem 362:130218. https://doi.org/10.1016/j.foodchem.2021.130218

Yu Y, Wang L, Fu X, Wang L, Fu X, Yang M, Han Z, Mou H, Jeon YJ (2019) Anti-oxidant and anti-inflammatory activities of ultrasonic-assistant extracted polyphenol-rich compounds from *Sargassum muticum*. J Oceanol Limnol 37(3):836–847. https://doi.org/10.1007/s00343-019-8138-5

Yu S, Duan M, Sun J, Jiang H, Zhao J, Tong C, Pang J, Wu C (2022) Immobilization of phlorotannins on nanochitin: a novel biopreservative for refrigerated sea bass (*Lateolabrax japonicus*) fillets. Int J Biol Macromol 200:626–634. https://doi.org/10.1016/j.ijbiomac.2022.01.089

Yuan Y, Zhang J, Fan J, Clark J, Shen P, Li Y, Zhang C (2018) Microwave assisted extraction of phenolic compounds from four economic brown macroalgae species and evaluation of their antioxidant activities and inhibitory effects on α-amylase, α-glucosidase, pancreatic lipase and tyrosinase. Food Res Int 113:288–297. https://doi.org/10.1016/j.foodres.2018.07.021

Zhang R, Yuen AKL, Magnusson M, Wright JT, de Nys R, Masters AF, Maschmeyer T (2018) A comparative assessment of the activity and structure of phlorotannins from the brown seaweed *Carpophyllum flexuosum*. Algal Res 29:130–141. https://doi.org/10.1016/j.algal.2017.11.027

Zhang X, Li Z, Yang P, Duan G, Liu X, Gu Z, Li Y (2021) Polyphenol scaffolds in tissue engineering. Mater Horiz 8(1):145–167. https://doi.org/10.1039/d0mh01317j

Zhang L, Liao W, Huang Y, Wen Y, Chu Y, Zhao C (2022) Global seaweed farming and processing in the past 20 years. Food Prod Process Nutr 4(1). https://doi.org/10.1186/s43014-022-00103-2

Zhao J, Jiang H, Huang Q, Xu J, Duan M, Yu S, Zhi Z, Pang J, Wu C (2022) Carboxymethyl chitosan incorporated with gliadin/phlorotannin nanoparticles enables the formation of new active packaging films. Int J Biol Macromol 203:40–48. https://doi.org/10.1016/j.ijbiomac.2022.01.128

Zhao Y, Zhang Z, Li Z, Li XM, Wang H, Lin H (2023) Insight into the conformational and allergenicity alterations of shrimp tropomyosin induced by *Sargassum fusiforme* polyphenol. Food Res Int 165:112521. https://doi.org/10.1016/j.foodres.2023.112521

Zhu X, Hou X, Ma B, Xu H, Yang Y (2019) Chitosan/gallnut tannins composite fiber with improved tensile, antibacterial and fluorescence properties. Carbohydr Polym 226:115311. https://doi.org/10.1016/j.carbpol.2019.115311

Zhu X, Soro AB, Tiwari BK, Garcia-Vaquero M (2022) Seaweeds in Ireland: main components, applications, and industrial prospects. In: Sustainable global resources of seaweeds, vol 1, pp 163–183. https://doi.org/10.1007/978-3-030-91955-9_10

Chapter 4
Seaweed-Based Biomaterials for Emerging Biotechnological Applications

Wilson Thau Lym Yong, Aliyah Madihah Asran, Syafiqah Syazwani Jaffar, Ming She See, Vun Yee Thien, Mailin Misson, Suryani Saallah, Nyuk Ling Ma, and Kenneth Francis Rodrigues

Abstract Marine organisms have produced an increasing number of biomaterials over the last decade, ranging from coralline bone grafts to polysaccharide-based biomaterials. Polysaccharides derived from seaweed have been studied for their antiviral, antitumor, anticoagulant, antidiabetic, antimicrobial, and anti-inflammatory properties. Researchers have been looking for and developing novel seaweed-based biomaterials that are fully biodegradable to replace traditional petroleum-based materials to promote sustainability and avoid environmental issues for decades. This chapter aims to better understand the recent uses of seaweed-based biomaterials in a wide range of emerging biotechnological applications, including medical applications, bioactive agents, drug delivery agents, disease rehabilitation, tissue engineering, and polysaccharide-based nanomaterials. More

W. T. L. Yong (✉)
Biotechnology Research Institute, Universiti Malaysia Sabah, Jalan UMS, Kota Kinabalu, Sabah, Malaysia

Seaweed Research Unit, Faculty of Science and Natural Resources, Universiti Malaysia Sabah, Jalan UMS, Kota Kinabalu, Sabah, Malaysia
e-mail: wilsonyg@ums.edu.my

A. M. Asran · S. S. Jaffar · M. Misson · S. Saallah · K. F. Rodrigues
Biotechnology Research Institute, Universiti Malaysia Sabah, Jalan UMS, Kota Kinabalu, Sabah, Malaysia

M. S. See · N. L. Ma
Biological Security and Sustainability (BIOSES) Research Interest Group, Faculty of Science and Marine Environment, Universiti Malaysia Terengganu, Kuala Nerus, Terengganu Darul Iman, Malaysia

V. Y. Thien
Innovation and Entrepreneurship Centre, Xiamen University Malaysia, Jalan Sunsuria, Bandar Sunsuria, Sepang, Selangor Darul Ehsan, Malaysia

Kelip-kelip! Center of Excellence for Light Enabling Technologies, School of Energy and Chemical Engineering, Xiamen University Malaysia, Jalan Sunsuria, Bandar Sunsuria, Sepang, Selangor Darul Ehsan, Malaysia

N. Trivedi et al. (eds.), *Recent Advances in Seaweed Biotechnology*,
https://doi.org/10.1007/978-981-96-0519-4_4

research is needed to decipher the mechanisms of action of seaweed biomaterials to promote sustainable and healthier products for various industries, create a sustainable society, and achieve the Sustainable Development Goals for a better future.

Keywords Seaweed-based biomaterials · Biotechnological applications · Marine biopolymers · Sustainable development · Emerging technologies

4.1 Introduction

4.1.1 Background and Significance of Seaweed-Based Biomaterials

Seaweeds, also known as marine macroalgae, have piqued human interest for centuries due to their astounding diversity and abundance in marine ecosystems. These intriguing creatures have long been valued for their applications in food, fertilizers, and cosmetics (Biris-Dorhoi et al. 2020; Sultana et al. 2023). However, in recent years, the discovery of their inherent biomaterial properties has unlocked a new realm of possibilities, resulting in the emergence of seaweed-based biomaterials as a fascinating area of biotechnology research and development. Seaweed has a rich composition of polysaccharides, proteins, lipids, and bioactive compounds that can generate various biomaterials (Ghosh et al. 2022; Hentati et al. 2020; Lomartire and Gonçalves 2022; Thiviya et al. 2022). These components can be extracted and used to form materials with diverse properties, including gels, films, fibers, and scaffolds, making seaweed an appealing resource for various applications.

One of the key advantages of seaweed-based biomaterials is their long-term availability, as seaweed is abundant and fast-growing organisms that thrive in marine environments. They do not require arable land or freshwater, and they can even grow in places where other crops cannot, such as rocky coastlines, making seaweed production an environmentally friendly alternative to terrestrial-based biomaterial sources (Baghel et al. 2015; Farghali et al. 2023; Gomez-Zavaglia et al. 2019; Tabassum et al. 2017). Furthermore, using seaweed for biomaterial production can reduce the demand for land-based resources while also contributing to the preservation of terrestrial ecosystems. Another aspect that has sparked significant interest in the scientific and industrial communities is the versatility of seaweed-based biomaterials, as the unique composition of seaweeds allows for the development of biomaterials with desirable properties such as biocompatibility, biodegradability, and antimicrobial activity (Chudasama et al. 2021; Iliou et al. 2022; Sood et al. 2021; Wan et al. 2021). These properties make them suitable for a wide range of uses, including medicine, agriculture, environmental remediation, and packaging.

Seaweed-based biomaterials show potential in tissue engineering and regenerative medicine due to their biocompatibility, cell adhesion, and proliferation (Farshidfar et al. 2023), and seaweed bioactive chemicals have anti-inflammatory and antioxidant characteristics that could be used to treat several ailments (Fu et al. 2016; Ganesan et al. 2019). Furthermore, seaweed polysaccharides and growth-promoting chemicals promote plant growth and crop output, stimulate plant defense mechanisms, and reduce the demand for chemical fertilizers and pesticides, promoting sustainable farming and reducing the environmental impact of conventional agriculture (Ali et al. 2021; Ammar et al. 2022; Mamede et al. 2023). Because they are biodegradable and low-carbon, seaweed-based biomaterials may potentially help to alleviate environmental challenges such as plastic pollution as a sustainable alternative to fossil fuel-based polymers, reducing plastic waste and resulting in a greener world with a lower environmental impact (Atiwesh et al. 2021; Thiruchelvi et al. 2021). Exploration of the vast and diverse resource of seaweed-based biomaterials has the potential to pave the way for a more sustainable and environmentally friendly future in a wide range of industries, opens endless opportunities for innovation, and contributes to the advancement of biotechnological applications.

4.1.2 Overview of the Chapter

This chapter provides a comprehensive overview of seaweed-based biomaterials, focusing on their types, properties, extraction methods, and emerging biotechnological applications (Fig. 4.1). It aims to shed light on the immense potential of seaweed-based biomaterials and their transformative impact across diverse industries. Section 4.2 delves into the use of seaweed as a source of biomaterials, diving into the types of seaweed commonly used in biotechnology and discussing their unique properties. It also addresses the numerous extraction methods used to obtain seaweed-based biomaterials, demonstrating their versatility and adaptability. Section 4.3 examines the emerging biotechnological applications of seaweed-based biomaterials in diverse areas, including medical and pharmaceutical, food and

Fig. 4.1 Emerging biotechnological applications of seaweed-based biomaterials

beverage, agricultural and environmental, and industrial and technological, highlighting their advantages and potential benefits. Section 4.4 focuses on seaweed-based biomaterials for medical and pharmaceutical applications, investigating their biocompatibility and biodegradability, and their roles in wound healing, drug delivery systems, and tissue engineering, revealing promising advances in these fields. Section 4.5 studies the use of seaweed-based biomaterials in the food and beverage industries, discussing their application as food additives, edible films and coatings, and nutraceuticals, and demonstrating the innovative ways in which seaweed can enhance the quality and functionality of food products. Section 4.6 explores the potential of seaweed-based biomaterials for agricultural and environmental applications, highlighting their use as biofertilizers, biostimulants, and in bioremediation to promote sustainable agriculture and environmental preservation. Section 4.7 focuses on seaweed-based biomaterials for industrial and technological applications, notably their potential use in bioplastics and packaging materials, energy production, and the textile and paper industries, emphasizing their versatility and relevance in these sectors. Section 4.8 addresses the challenges and future directions of seaweed-based biomaterials, outlining the limitations and obstacles that need to be overcome for wider adoption and commercialization. It also indicates potential future research and development directions, paving the way for further innovation in this discipline. Through this chapter, we embark on a journey into the realm of seaweed-based biomaterials, revealing their immense potential and providing a foundation for further exploration and advancement in biotechnology.

4.2 Seaweed as a Source of Biomaterials

4.2.1 Types of Seaweed Used in Biotechnology

Seaweed has increasingly attracted the biotechnology industry in recent years, owing to its inherent renewable capacity and promising potential for sustainable biomaterials (Laurens et al. 2020; Yong et al. 2022). With their remarkable diversity and unique chemical compositions, seaweed has integrated into the industry and become a vital component. These unique organisms can be divided into three primary groups based on their pigmentation: brown, red, and green seaweeds. Each group has diverse properties, ranging from their appearance and habitat to their bioactive compounds and applications in various biotechnological processes (El-Beltagi et al. 2022; Lomartire and Gonçalves 2022). From pharmaceuticals and biofuels to food additives and fertilizers, the various species of seaweed provide a treasure mine of possibilities for sustainable and innovative biotechnology solutions (Hong et al. 2014).

Brown seaweed, scientifically known as Phaeophyta, is a prominent type of seaweed renowned for its abundant presence and wide-ranging properties (Rathour et al. 2021). It is distinguishable by its brown or dark green coloration, which is primarily due to the pigment fucoxanthin (Wijesinghe and Jeon 2011). Brown seaweed thrives in diverse marine habitats, including coastal areas and subtidal zones. It is the most plentiful type of seaweed and contains a significant concentration of alginates, natural polysaccharides that can be extracted and employed as effective gelling and thickening agents across diverse industries (Jumaidin et al. 2018). Moreover, brown seaweed harbors fucoidans, which have been recognized for their anticoagulant, antioxidant, and anticancer properties, making them a valuable resource (Remya et al. 2022).

Red seaweed, also known as Rhodophyta, exhibits its distinctive red coloration due to the presence of pigments called phycoerythrins. These phycoerythrins are protein-pigment complexes that absorb blue light and reflect red light, contributing to the vibrant red color characteristic of red seaweed (Ganesan and Shanmugam 2020). Red seaweed is renowned for its high content of carrageenans, which are widely used as thickening agents, stabilizers, and gelling agents in the food and cosmetic industries (Shafie et al. 2022). The unique properties of carrageenans contribute to the texture, stability, and sensory attributes of various products, including dairy desserts and sauces (Peñalver et al. 2020). Additionally, red seaweed is a valuable source of agar, which is used as a culture medium in microbiology and cell biology research (Fu and Kim 2010). Agar provides a solid substrate for the growth and cultivation of microorganisms, making it an essential component in laboratory settings.

Green seaweed, scientifically known as Chlorophyta, is a diverse group of seaweeds that are characterized by their green coloration. Green seaweed is green in color due to the presence of chlorophyll pigments. According to Sanjivkumar et al. (2020), the notable class of compounds found in green seaweed is ulvans. They have shown that ulvans possess antioxidant properties, meaning they can scavenge free radicals and protect cells from oxidative damage. Additionally, ulvans exhibit anti-inflammatory effects, potentially modulating inflammatory responses in the body. Moreover, green seaweed serves as a potential source of bioactive peptides and polysaccharides with diverse biological activities. These compounds have been investigated for their antimicrobial properties, inhibiting the growth of pathogenic bacteria and fungi, thus showing promise as natural alternatives to conventional antimicrobial agents (Pradhan et al. 2023). Furthermore, certain bioactive compounds derived from green seaweed have demonstrated antitumor effects, exhibiting potential in cancer treatment and prevention (Saadaoui et al. 2020).

Seaweed has emerged as a highly promising and desirable resource, with great properties and a wide range of applications, presenting it as a sustainable and renewable supply of biomaterials for a variety of biotechnological endeavors. From its rapid growth and effective resource utilization to its low environmental impact, seaweed possesses intrinsic benefits that make it an appealing choice to produce eco-friendly biomaterials. As research and technological advances continue, seaweed-based biomaterials are poised to play an increasingly important role in propelling biotechnology forward, paving the way for a greener and more sustainable future.

4.2.2 Properties of Seaweed-Based Biomaterials

Seaweed presents significant potential as a valuable source of biomaterials, including hydrogels, films, fibers, and nanoparticles. Compared to synthetic materials, seaweed-based biomaterials offer numerous advantages, such as biocompatibility, biodegradability, and renewability. These properties establish them as a sustainable and environmentally friendly alternative to conventional biomaterials (Pacheco-Quito et al. 2020). One notable attribute of seaweed-based biomaterials is their exceptional water-holding capacity, making them ideal natural hydrocolloids for use in food and pharmaceutical formulations (Fernández-Segovia et al. 2018). Furthermore, seaweed serves as a natural reservoir of bioactive compounds, including antioxidants and anti-inflammatory agents, which can have various applications in the development of nutraceuticals and functional foods (Khalid et al. 2018). Beyond these applications, seaweed-based biomaterials show promise in areas such as wound dressings, drug delivery systems, and tissue engineering scaffolds (Bilal and Iqbal 2020). This is primarily due to their remarkable biocompatibility and biodegradability. By leveraging these properties, seaweed-based biomaterials can contribute to the advancement of medical treatments and regenerative medicine.

A study conducted by Ngo et al. (2011) revealed that brown seaweed is rich in alginates, laminarin, and fucoidan. Conversely, red seaweed, like *Chondrus crispus*, contains significant amounts of carrageenan and agar, as indicated by Sekkal et al. (1993). Green seaweed, such as *Ulva lactuca*, is characterized by the presence of ulvans and sulfated polysaccharides, as highlighted by Yaich et al. (2017). Furthermore, the protein content of seaweed-based biomaterials can significantly vary depending on the specific species and growth conditions, as noted by Kostas et al. (2021). Seaweed also contains minerals like calcium, magnesium, and potassium, albeit in varying quantities (Circuncisão et al. 2018). In order to enhance the properties of seaweed-based biomaterials, various processing methods, including hydrolysis, oxidation, and sulfation, can be employed, as discussed by Lakshmi et al. (2020). These methods allow for modifications in the chemical composition of the biomaterials, thereby optimizing their characteristics.

Khaliq et al. (2022) conducted a study on carrageenan-based hydrogels, examining their mechanical properties. They discovered that the tensile strength and elastic modulus of the hydrogels varied based on the degree of crosslinking, which was regulated by the concentration of crosslinking agents. Similarly, Matyash et al. (2014) investigated the mechanical properties of alginate-based hydrogels and found that the stiffness and strength were influenced by the degree of crosslinking and the presence of divalent cations. Other than that, the impact of processing methods on the mechanical properties of seaweed-based biomaterials has also been explored. Kumar et al. (2022) explored the effect of freeze-drying on agar-based films and observed that reducing the water content increased the mechanical strength of the films. Conversely, thermal treatment was found to affect the mechanical properties of carrageenan-based hydrogels, with the elastic modulus decreasing as the temperature increased, as reported by Kozlowska et al. (2018). Moreover, environmental conditions, including temperature and humidity, have been found to influence the mechanical properties of seaweed-based biomaterials. Liu et al. (2019) reported that increasing temperature and humidity led to a decrease in the tensile strength and elongation at the break of carrageenan-based films, indicating that these properties are highly sensitive to environmental factors.

In a study conducted by Bagheri et al. (2019), the thermal properties of agarose-based hydrogels were examined. The researchers discovered that as the concentration of agarose increased, the hydrogels exhibited a higher glass transition temperature. Another study by Guidara et al. (2019) focused on the thermal properties of various seaweed species, revealing that the melting temperature of the samples varied based on the composition of polysaccharides. Furthermore, the thermal properties of seaweed-based biomaterials can be influenced by the processing methods employed during their preparation. Hezaveh and Muhamad (2013) conducted research on carrageenan-based hydrogels using different techniques, and they found that the melting temperature of the hydrogels increased with both the heating rate and the concentration of carrageenan.

The viscosity of seaweed extracts plays a significant role in food formulations, impacting the texture and mouthfeel of the final product, as highlighted by Hotchkiss et al. (2016). In the cosmetics industry, the rheological properties of seaweed-based biomaterials are crucial for their effectiveness as thickeners, emulsifiers, and stabilizers. For instance, a study conducted by Shafie et al. (2022) examined the rheological properties of agar and carrageenan as natural thickeners for cosmetics. They reported that by adjusting the concentration and type of seaweed extract used, it was possible to control the viscosity of the formulations. Moreover, the gelation behavior of seaweed-based biomaterials holds significance in the development of hydrogels for wound healing and drug delivery applications, as indicated by Pacheco-Quito et al. (2020).

Research has provided evidence that seaweed-based biomaterials can undergo degradation by enzymes present in the human body, such as hyaluronidase and lysozyme. However, the rates of degradation can be controlled through chemical modification, as observed by D'Ayala et al. (2008). Furthermore, the biocompatibility of seaweed-based biomaterials has been extensively investigated, revealing their capacity to support cell growth and proliferation, as demonstrated by Otari and Jadhav (2021). For instance, Ditta et al. (2020) conducted a study assessing the biocompatibility of carrageenan-based hydrogels. The findings indicated that the hydrogels were effective at keeping the area around the wound moist while also acting as a barrier against bacteria, highlighting their favorable biocompatibility. In addition, the potential of seaweed-based biomaterials in the fields of tissue engineering and drug delivery has been explored, as noted by Jafari et al. (2022).

4.2.3 Extraction Methods for Obtaining Seaweed-Based Biomaterials

The extraction methods employed for obtaining seaweed-based biomaterials can vary based on the desired properties and intended applications. Each type of seaweed offers unique characteristics. For instance, brown seaweed is notable for its abundance of alginates, which possess remarkable gelling and film-forming properties. On the other hand, red seaweeds contain carrageenans, highly sulfated compounds known for their exceptional thickening and emulsifying capabilities. The extraction of seaweed-based biomaterials typically involves a range of methods, including physical, chemical, and biological approaches.

4.2.3.1 Physical Extraction Method

The extraction of seaweed can involve three primary physical methods: mechanical pressing, milling, and centrifugation. Each method operates on distinct principles and presents its own set of advantages and disadvantages. Mechanical pressing, which involves the application of pressure to the seaweed biomass to extract both intracellular and extracellular biomolecules, is well-suited for extracting oils and lipids from seaweeds, as it eliminates the need for solvents and can yield high quantities of desired compounds with minimal processing steps. However, it is important to control the applied pressure carefully to prevent thermal degradation and oxidation of the extracted biomolecules.

Another physical extraction method is milling, also known as grinding or homogenization. This approach involves reducing the seaweed biomass into smaller particles to increase the surface area, facilitating the release of biomolecules. Milling is particularly effective for extracting polysaccharides and proteins from seaweed, as it can disrupt cell walls and release intracellular contents (Abdul Khalil et al. 2018). Nonetheless, excessive milling can lead to physical and chemical modifications of the extracted biomolecules, potentially impacting their functionality.

On the other hand, centrifugation is commonly used for solid—liquid separation, concentration, and purification in seaweed extraction processes. After initial extraction steps, such as acid or alkaline treatment, centrifugation is employed to separate the solid residue from the liquid extract. The centrifugal force causes the solid particles to sediment, allowing for the collection of the supernatant containing the desired compounds (Livshits et al. 2015). Centrifugation is also used for concentrating seaweed extracts by removing water or solvents, thereby increasing the concentration of target compounds (Anaëlle et al. 2013). Centrifugation aids in the fractionation of compounds with different densities or molecular weights, allowing for the separation of specific fractions and therefore it is an effective technique for purifying seaweed extracts by removing impurities or unwanted components.

4.2.3.2 Chemical Extraction Method

Chemical extraction methods, including acid/base extraction and solvent extraction each possess unique principles underlying their respective techniques, along with distinct advantages and disadvantages. The acid/base extraction techniques use acidic or basic solutions to dissolve the seaweed biomass and extract specific biomolecules in which well-suited for obtaining sulfated polysaccharides like carrageenans and fucoidans. For instance, alkaline extraction involves treating seaweed biomass with alkaline solutions, typically using sodium hydroxide (NaOH) or potassium hydroxide (KOH) (Nogueira et al. 2022). This process aims to solubilize and extract specific components from seaweed, such as polysaccharides (e.g., alginates) or proteins. Alkaline extraction is commonly used in the production of alginate, a widely utilized polysaccharide derived from seaweed. However, the use of harsh chemical conditions in this method can lead to biomolecule degradation and necessitates proper management of waste streams.

Meanwhile, the solvent extraction methods use organic solvents to extract hydrophobic biomolecules such as lipids, pigments, and other compounds from seaweeds. Solvent extraction is particularly effective in obtaining valuable substances like carotenoids and lutein. Nonetheless, the use of organic solvents entails potential hazards and requires special handling. Furthermore, the presence of solvent residues in the extracted biomolecules can impact their purity and safety (Warkoyo and Saati 2011).

4.2.3.3 Biological Extraction Method

Biological extraction methods, such as fermentation and microbial degradation, are employed to extract valuable compounds from seaweed. Enzymatic hydrolysis relies on enzymes to break down the seaweed biomass into smaller molecules and release the target biomolecules. This method is well-suited for extracting proteins and peptides, as it selectively cleaves peptide bonds and enables the release of bioactive peptides. However, it is crucial to carefully optimize the selection of enzymes and their optimal conditions to ensure the high yield and purity of the extracted biomolecules (Teixeira-Guedes et al. 2023). Via fermentation process, microorganisms such as bacteria, yeast, or fungi are used to convert organic substrates into desired products. During fermentation, microorganisms metabolize the substrate and produce enzymes or other bioactive compounds as byproducts. These products can be harvested and utilized for various purposes. The advantages of fermentation include high specificity, as specific microorganisms can be selected to produce desired biomolecules and the potential for high yields. However, fermentation can be a time-consuming process and requires optimal conditions for microorganism growth and product formation (Norakma et al. 2021).

On the other hand, microbial degradation involves the use of microorganisms to break down complex biomaterials into simpler compounds. Microbes possess enzymes capable of degrading various organic substrates, and by manipulating the conditions, specific biomolecules can be targeted for extraction. This method can be advantageous in terms of its ability to degrade complex materials that may be challenging to extract using other techniques. However, microbial degradation may require careful control of environmental conditions and specific strains of microorganisms to achieve optimal results (Leadbeater et al. 2022).

4.2.3.4 Combination Extraction Method

Sequential extraction and ultrasound-assisted extraction are the two types of combination methods usually used in seaweed extraction. Sequential extraction involves the use of multiple extraction steps with different solvents or conditions to extract different biomolecules from seaweed biomass. This method can improve the selectivity and purity of the extracted biomaterials by targeting specific compounds. The advantages of sequential extraction include the ability to extract a wide range of biomolecules and the potential for high yield and purity. However, this method can be time-consuming and requires careful optimization of the extraction conditions (Birgersson et al. 2023). Ultrasound-assisted extraction (UAE) involves the use of high-frequency ultrasound waves to enhance the extraction efficiency and selectivity of seaweed biomaterials. This method can improve the mass transfer and cell wall disruption of the seaweed biomass, resulting in higher extraction yields. The advantages of the UAE include high extraction efficiency, reduced extraction time, and mild operating conditions. However, this method can be energy-intensive and may require careful optimization of the ultrasound parameters to prevent the degradation of the extracted biomaterials (Gordalina et al. 2021).

4.3 Emerging Biotechnological Applications of Seaweed-Based Biomaterials

4.3.1 Medical and Pharmaceutical Applications

Seaweed bioactive compounds, which include polysaccharides, proteins, polyphenols, and pigments, have a wide range of biological functions and have shown tremendous promise for medical and pharmaceutical applications (Ghosh et al. 2022; Lomartire and Gonçalves 2022). Polysaccharides extracted from seaweeds, such as alginate, carrageenan, and fucoidan, have excellent gelling properties, biocompatibility, and biodegradability, making them appropriate for encapsulating and delivering therapeutic substances (Bilal and Iqbal 2020; Chudasama et al. 2021). Alginate-based hydrogels, for example, have been used for the controlled release of drugs, growth factors, and cells, with promising results in wound healing and tissue regeneration (Abasalizadeh et al. 2020). Carrageenan-based nanoparticles have also been investigated for targeted drug delivery for cancer therapy due to their ability to encapsulate and protect anticancer drugs while enhancing bioavailability and lowering negative effects (Pacheco-Quito et al. 2020).

Seaweed also contains bioactive compounds with antibacterial, antiviral, and anti-inflammatory effects, making them ideal candidates for creating novel pharmaceuticals (Farghali et al. 2023; Lomartire and Gonçalves 2022; Maliki et al. 2022). Fucoidan, a sulfated polysaccharide found in brown seaweeds, has demonstrated significant antiviral action against multiple viruses, including herpes simplex virus (HSV), human immunodeficiency virus (HIV), and coronaviruses (Krylova et al. 2020; Pradhan et al. 2022). Furthermore, seaweed-derived polyphenols, such as phlorotannins, have antimicrobial activity against bacteria and fungi (Khan et al. 2022), and their antioxidant properties have piqued interest for potential applications in the prevention and treatment of chronic diseases, such as cancer, cardiovascular disease, and neurodegenerative disorders (El-Beltagi et al. 2022; Gómez-Guzmán et al. 2018).

Furthermore, seaweed-based biomaterials have shown potential in tissue engineering and regenerative medicine (Bilal and Iqbal 2020; Jafari et al. 2022). Seaweed extracellular matrix components promote cell proliferation, differentiation, and tissue regeneration (Bar-Shai et al. 2021). Seaweed-derived scaffolds mixed with cells and growth factors have been used to regenerate diverse tissues such as bone, cartilage, and skin (Bilal and Iqbal 2020). The unique bioactive components and features of seaweed-based biomaterials have opened interesting potential for their application in the medical and pharmaceutical areas, as discussed more in Sect. 4.4. Their potential in drug delivery systems, antibacterial and antiviral medicines, and tissue engineering holds promise for the development of novel therapeutics. More research and exploration of the great diversity of seaweed species and their bioactive chemicals is required to fully realize their potential.

4.3.2 Food and Beverage Industry Applications

Seaweed has long been consumed as part of traditional diets in coastal regions, and their nutritional richness, unique flavors, and potential health advantages are gaining popularity globally (Bizzaro et al. 2022; Rioux et al. 2017). In recent years, there has been a surge of interest in using seaweed-based biomaterials in the food and beverage industries for a variety of applications ranging from natural additives and functional ingredients to unique culinary products (Peñalver et al. 2020). Seaweed is used extensively in the food industry as a natural source of hydrocolloids such as carrageenan, agar, and alginates (Liao et al. 2021; Rhein-Knudsen and Meyer 2021). These hydrocolloids have excellent gelling, thickening, and stabilizing qualities, making them useful ingredients in a variety of food products such as dairy, confectionery, meat products, and beverages (Saha and Bhattacharya 2010). Carrageenan, for example, is often used as a stabilizer in dairy products to prevent protein aggregation and improve texture (Biglarian et al. 2022). Alginate is used in the preparation of fruit-based sweets, jellies, and sauces as a thickening and gelling agent (de Avelar and Efraim 2020; Saha and Bhattacharya 2010).

Seaweed includes a high concentration of dietary fiber, minerals (including iodine, calcium, and magnesium), vitamins, and bioactive substances (Roohinejad et al. 2017). Seaweed-based ingredients also provide functional benefits to food items by enhancing nutritional profiles, increasing fiber content, and improving sensory features (Peñalver et al. 2020). Seaweed powder or flakes, for example, can be used in bakery products, snacks, or ready-to-eat meals to enhance nutritional content while also introducing new flavors and sensations (Quitral et al. 2022). Furthermore, seaweed extracts and bioactive compounds are being investigated for potential health advantages. Seaweed contains a variety of bioactive compounds, including antioxidants, polyphenols, and phycobiliproteins, which have been proven to have anti-inflammatory, anticancer, and immunomodulatory properties (Ganesan et al. 2019; Lau et al. 2014; Lomartire and Gonçalves 2022). To enhance health and well-being, these bioactive compounds can be isolated and combined into functional meals, nutraceuticals, and dietary supplements (Choudhary et al. 2021; Shannon and Abu-Ghannam 2019; Thiviya et al. 2022).

Seaweed production and utilization in the food and beverage industry also provide environmental benefits, as using seaweed as a food source decreases reliance on traditional land-based agriculture, which can have negative environmental consequences such as deforestation and water pollution (Koesling et al. 2021; Ullmann and Grimm 2021). As discussed further in Sect. 4.5, seaweed-based biomaterials have enormous potential in the food and beverage industry, offering opportunities for innovation and sustainability ranging from natural hydrocolloids and functional ingredients to novel food products with enhanced nutritional profiles and bioactive compounds. Further research and development are required to investigate novel uses and optimize extraction and processing procedures to assure the quality and safety of seaweed-based food items.

4.3.3 Agricultural and Environmental Applications

The application of seaweed-based biomaterials in agriculture and the environment has gained popularity as a green approach to promote plant growth, improve soil health, and reduce environmental hazards (Ammar et al. 2022; Kholssi et al. 2022; Mamede et al. 2023; Nabti et al. 2017). Seaweed extracts, which include a variety of plant growth-promoting substances such as phytohormones (e.g., auxins, cytokinins, and gibberellins), vitamins, amino acids, and trace elements, have shown substantial potential as biostimulants in agriculture. When applied to crops, seaweed extracts or bioactive compounds have been found to improve seed germination, root development, nutrient uptake, and overall plant growth (Ali et al. 2021). They also promote plant tolerance to biotic and abiotic stressors such as drought, salinity, and pest infestations (Lau et al. 2022; Shukla et al. 2019). Seaweed extracts can be used as foliar sprays, seed treatments, or soil amendments, providing a sustainable and organic alternative to synthetic fertilizers and pesticides (Ali et al. 2021; El Boukhari et al. 2020).

In addition to their biostimulant properties, seaweeds rich in organic matter improve soil structure, moisture retention, and nutrient-holding capacity when introduced into the soil (Espinosa-Antón et al. 2023; Ma et al. 2022). The organic matter generated from seaweed also serves as a substrate for beneficial soil microorganisms, enhancing microbial diversity and activity (Chen et al. 2020; Costa et al. 2023). These microorganisms, such as plant growth-promoting bacteria and mycorrhizal fungi, create symbiotic interactions with plants, improving nutrient acquisition and disease resistance (González-González et al. 2020; Lau et al. 2022). Seaweed-based soil amendments have been proven to promote soil microbial populations and enzymatic activities, improving nutrient cycling and reducing environmental consequences (Garbowski et al. 2023; Wadduwage et al. 2023; Zhou et al. 2019).

Additionally, seaweed has the potential to reduce environmental challenges, particularly in aquaculture and wastewater treatment (Farghali et al. 2023; Tanaka et al. 2020). Seaweed can be grown in integrated multitrophic aquaculture (IMTA) systems, where they absorb and use excess nutrients, minimizing water pollution and adding economic value (Chopin et al. 2012; Hossain et al. 2022; Ingle et al. 2022; Rosa et al. 2020). Seaweeds have also been studied for their potential to bioaccumulate heavy metals and pollutants from contaminated waters, providing a low-cost and ecologically benign method to wastewater treatment and phytoremediation (Deniz and Ersanli 2018; Saldarriaga-Hernandez et al. 2020). The use of seaweed-based biomaterials in agriculture and environmental applications is consistent with the principles of sustainable and regenerative practices, as it provides a natural and renewable resource that promotes soil health, reduces reliance on synthetic inputs, and contributes to ecosystem resilience, as outlined further in Sect. 4.6. More research is needed, however, to optimize the composition, application methods, and doses of seaweed-based products and to comprehend their long-term effects on crop productivity, soil fertility, and environmental sustainability.

4.3.4 Industrial and Technological Applications

Seaweed-based biomaterials have emerged as a useful resource for a wide range of industrial and technological applications, particularly in the development of sustainable alternatives for bioplastics and packaging materials, energy production, and the textile and paper sectors (Baghel 2023; Baghel et al. 2021; Lim et al. 2021; Lomartire et al. 2022). Agar, carrageenan, and alginate are seaweed-derived polysaccharides with excellent film-forming properties, biodegradability, and mechanical strength, making them appealing as alternatives to synthetic plastics in producing packaging materials, coatings, and other industrial applications (Jayakody et al. 2022; Tennakoon et al. 2023). Seaweed-based bioplastics have the advantage of being renewable and biodegradable, decreasing the environmental impact of traditional plastics, which frequently end up in landfills or contribute to marine pollution (Atiwesh et al. 2021; Coppola et al. 2021).

Another promising application for seaweed-based biomaterials is in energy production (Fasahati et al. 2022; Zhao et al. 2022). Seaweed is excellent at absorbing solar energy and transforming it into biomass via photosynthesis (Yantovski 2008). They grow quickly, can be grown in huge quantities, and do not compete for arable land or freshwater resources. Seaweed biomass can be transformed into biofuels such as bioethanol and biodiesel through various procedures, including fermentation and thermochemical conversion (Cassani et al. 2022; Michalak 2018). Furthermore, seaweed can be used in anaerobic digestion to generate biogas, a renewable energy source (Deng et al. 2020; Thakur et al. 2022). Using seaweed-based biofuels and biogas provides a sustainable alternative to fossil fuels while lowering greenhouse gas emissions and reliance on finite resources.

Seaweed-based biomaterials are also used in the textile and paper sectors. Seaweed fibers have unique properties such as strength, flexibility, and resistance to microbial degradation, making them suitable for textile production such as clothing, upholstery, and home furnishings, providing a sustainable alternative to synthetic materials and contributing to the reduction of environmental impact associated with traditional textile production (Han et al. 2021; Ju et al. 2023). Furthermore, seaweed extracts and derivatives can be used in the production of paper, giving desired qualities such as increased strength, water resistance, and biodegradability (Saleh et al. 2021). The use of seaweed-based biomaterials in the bioplastics and packaging materials, energy production, and textile and paper industries has been discussed further in Sect. 4.7, highlighting their unique properties, such as film-forming abilities, renewable energy potential, and textile fiber characteristics, that make them appealing alternatives in these sectors. Continued research and development efforts are required to optimize production processes, investigate new uses, and encourage the wider acceptance of seaweed-based biomaterials in a variety of industrial and technical applications.

4.4 Seaweed-Based Biomaterials for Medical and Pharmaceutical Applications

4.4.1 Biocompatibility and Biodegradability

Applications of seaweed began in the 1990s and have since grown, with extensive research conducted in pre-clinical and clinical studies. The beneficial functions of seaweed extracts are attributed to their antioxidation, anti-inflammation, antiviral, antitumor, and anticancer properties (Table 4.1). These properties make seaweed extracts a promising candidate for developing medical appliances and pharmaceutical products. Seaweed's polysaccharides, including alginate, carrageenan, ulvan, and laminarin, have diverse applications in biomedical utilization due to their

Table 4.1 Biomaterials isolated from various sources of seaweed, their potential medical and pharmaceutical applications, and associated mechanisms

Bioactive composition	Sources	Applications/Mechanism	References
Polysaccharides			
Agar	–	Applicable in controlled drug release	Wakhet et al. (2015)
(1–4)-linked 3,6-anhydro-α-L-galactose alternating with (1–3)-linked β-D-galactopyranose	*Gelidium amansii*	Antioxidant activity	Xu et al. (2018)
Alginate	*Cissus quadrangularis*	Osteoinductive herbal scaffold for bone engineering	Soumya et al. (2012)
Sodium alginate	–	Removal of heavy metal from body	Kim et al. (2005)
	Macrocystis pyrifera	Wound dressing material	Balakrishinan et al. (2005)
Alginic acid	*Sargassum vulgare*	Antitumor	de Souza et al. (2007)
Ascophyllan	*Ascophyllum nodosum*	Anticancer-inhibit MMP expression	Abu et al. (2015)
Carrageenan	–	Functionized graphene oxide for bone regeneration and implantation	Liu et al. (2014)
	Hypnea musciformis	Anticancer	Souza et al. (2018)
	Champia feldmannii	Antitumor	Lins et al. (2009)
Galactose	*Tribonem minus*	Anticancer activity	Chen et al. (2019b)
κ-carrageenan	*Kappaphycus striatum*	Antitumoral activity against human nasopharynx carcinoma, human gastric carcinoma, and cervical cancer cell lines	Yuan and Song (2005)
Sulfated carrageenans	*Stenogramme interrupta*	Antiviral	Cáceres et al. (2000)
Sulfated κ-carrageenan	*Eucheuma spinosum*	Anticoagulant, antithrombotic	Campo et al. (2009)
Chitosan	–	Carriers of fibroblast growth factor-2	Nakamura et al. (2008)
O-carboxymethyl chitosan	*Cissus quadrangularis*	Osteoinductive herbal scaffold for bone engineering	Soumya et al. (2012)
Fucan	*Fucus vesiculosus*	Antithrombin	Mourao (2004)

(continued)

Table 4.1 (continued)

Bioactive composition	Sources	Applications/Mechanism	References
Sulfated fucan	*Laminaria cichorioides*	In vitro anticoagulant activity	Yoon et al. (2007)
	Ascophyllum nodosum	Immunomodulatory, anti-inflammatory, anticoagulant, antithrombotic	Cumashi et al. (2007)
	Fucus spp.	Immunostimulant, antiviral, antitumor, antiproliferative, antiadhesive	Costa et al. (2010)
Fucoidan	*Chnoospora minima*	Anti-inflammation	Fernando et al. (2017)
	Fucus evanescens	Anticancer-inhibit cell proliferation	Anastyuk et al. (2012)
	Fucus evanescens	In vitro and in vivo anticoagulant activity	Drozd et al. (2006)
Galactofucan	*Spatoglossum schröederi*	Antithrombotic	Costa et al. (2010)
	Adenocystis utricularis	Inhibitory against HSV 1 and 2	Ponce et al. (2003)
Fucose and Xylose	*Fucus vesiculosus*	Antioxidant activity	Obluchinskaya et al. (2022)
l-fucose	*Cladisiphon okamuranus*	Anticancer activity	Nagamine et al. (2009)
Laminarin	*Laminaria japonica*	Antiapoptotic	Kim et al. (2006)
β-(1,3)-glucan	*Laminaria digitata*	Antioxidant protection	Ozanne et al. (2020)
Sulfated fucoidan	*Sargassum horneri*	Antitumor	Ale et al. (2011)
Ulvan	*Ulva rigida*	Antigenotoxic activity in human lymphocytes; hypoglycaemic effect in vivo experiment	Celikler et al. (2008, 2009)
	Ulva fasciata	Antioxidant and good mechanical properties; antiviral property	Ganesan et al. (2018) and Mendes et al. (2010)
Rhamnose, and xylose	*Ulva pertusa*	Antioxidant and antihyperlipidemic activity	Jiang et al. (2020)
	Ulva pertusa	Antioxidant activity	Li et al. (2020)
Sulfated arabinogalactans	*Codium* spp.	Anticoagulant, antithrombotic, antiviral	Takano et al. (1998) and Lee et al. (2004a)
Sulfated rhamnans	*Monostroma latissimum*	Antiviral, anticoagulant	Lee et al. (1999), Zhang et al. (2008), and Mao et al. (2009)

(continued)

Table 4.1 (continued)

Bioactive composition	Sources	Applications/Mechanism	References
Polymeric nanoparticle			
Chitosan-carrageenan nanoparticles	–	Carriers of therapeutic macromolecule in blood	Grenha et al. (2009)
Montmorillonite-Alginate nanocomposite	–	Intestinal delivery system of Vitamin B1 and B6	Kevadiya et al. (2009)
Ulvan/chitosan polyelectrolyte nanofibrous	–	Substrate material for the cultivation of osteoblast	Toskas et al. (2012)
Proteins			
Deoxygadusol, palythene, and usujirene	*Rhodymenia pseudopalmata*	Antioxidant	Pliego-Cortés et al. (2019)
Griffith (Protein)	*Griffithsia* sp.	Antiviral activity against MERS-CoV-2 virus and SARS-CoV-2 glycoprotein	Millet et al. (2016) and Zumla et al. (2016)
Lactate and progerin	*Alaria esculenta*	Reduce synthesis antielastase and anticollagenase	Verdy et al. (2011)
Palythine, palythinol, porphyra-334, asterina-330, shinorine, or usujirene	*Palmaria palmata, Mastocarpus stellatus, Chondrus crispus*	Antioxidant, antiproliferative	Athukorala et al. (2016)
Peptide PPY1	*Pyropia yezoensis*	Antiaging	Lee et al. (2015)
Peptides PYP1–5 and porphyra 334	*Porphyra yezoensis f. coreana Ueda*	Boost synthesis of gelatine	Ryu et al. (2014)
Phycobiliproteins	*Gracilaria gracilis*	Antioxidant	Pereira et al. (2020)
Porphyra-334, shinorine, palythine and asterina-330	*Gracilaria vermiculophylla*	Antioxidant, UV-protective effect	Barceló-Villalobos et al. (2017)
Lipid			
Arachidonic acid (ARA)	*Pelvetia purpureum, Porphyridium cruentum*	Improves growth and development of neonates	Udayan et al. (2017)
Docosahexaenoic acid (DHA)	*Crypthecodinium cohnii, Schizochytrium* sp., *Ulkenis* sp.	Brain and eye health, cardiovascular benefits, nervous system development	Udayan et al. (2017) and Weill et al. (2020)
Eicosapentaenoic acid (EPA)	*Nannochloropsis, Phaeodactylum tricornutum, Porphyridium cruentum*	Cognition, heart health, protection against arthrosclerosis, anti-inflammatory	Weill et al. (2020)
Fucosterol	*Hizikia fusiformis*	Anti-inflammatory, antiphotodamage	Lee et al. (2004b)
	Hizikia fusiformis	Antiaging, inhibit MMP expression	Shin et al. (2018)

(continued)

Table 4.1 (continued)

Bioactive composition	Sources	Applications/Mechanism	References
Omega 3 fatty acids	*Brown algae*	Antioxidant	Neto et al. (2018)
Other			
Ethanolic extract	*Ulva rigida*	In vivo antihyperglycaemic, antioxidative, and genotoxic/antigenotoxic activities	Celikler et al. (2009)
Phenolic compounds and flavonoids	*Caulerpa racemosa*	Antioxidant, scavenging, antiproliferative activities of cancer line cells	Tanna et al. (2018) and Paul et al. (2014)
Phenolic, flavonoid, and alkaloid compounds	*Gracilaria edulis*	Hypoglycaemic activity	Gunathilaka et al. (2019)
Fucoxanthin	*Cladosiphon okamuranus*	Antioxidant-DPPH inhibition	Mise et al. (2011)
	Gayralia oxysperma	Antioxidant-high FRAP value	Kelman et al. (2012)
	Ulva fasciata	Antioxidant-DPPH inhibition	Premalatha et al. (2011)
	Colpomenia sinuosa, Sargassum prismaticum	Antitumoral activity on MCF-7, HepG-2, HCT-116 cells	Atya et al. (2021)
	Undaria pinnatifida	Antitumoral activity on SiHa, Malme-3 M cells	Wang et al. (2014)
	Fucus evanescens	Antiviral activity against ECHO-1, HIV-1, HSV-1, HSV-2	Krylova et al. (2020)
	Laminaria japonica	Antitumoral activity on lung cancer cell	Mei et al. (2017)
Phlorotannin	*Eisenia arborea*	Anti-inflammation-inhibit release of histamine	Sugiura et al. (2006b)
	Ecklonia cava	Antiviral activities	Karadeniz et al. (2014)
	Eisenia arborea	Antiallergic effects	Sugiura et al. (2006a)
Eckol	*Ecklonia stolonifera, Eisenia bicyclis*	Anti-inflammatory	Eom et al. (2017) and Manandhar et al. (2019)
Eckol, 8,8′-bieckol, 6,6′-bieckol, phlorofucofuroeckol-A, and dieckol	*Ecklonia stolonifera, Ecklonia cava*	Antiallergic	Sugiura et al. (2021)
8,80-bieckol; 8,40 0 0dieckol, 6,60-bieckol	*Ecklonia cava*	Inhibitor of HIV-1 RT	Ahn et al. (2004) and Artan et al. (2008)
Phlorofucofuroeckol A	*Eisenia bicyclis*	Antioxidant and hepatoprotective effect	Kim et al. (2011)

(continued)

Table 4.1 (continued)

Bioactive composition	Sources	Applications/Mechanism	References
Phlorofucofuroeckol B	*Eisenia arborea*	Antiallergy	Sugiura et al. (2007)
Dioxinodehydroeckol	*Ecklonia cava*	Anticancer	Kong et al. (2009)
Fucosterol	*Pelvetia siliquosa*	Antidiabetic	Lee et al. (2004c)
Phloroglucinol	*Ecklonia cava*	Anti-inflammatory activity	Kim and Kim (2010)
Diterpene and sesquiterpene	*Laurencia obtuse*	Actions against different cancer cell lines (KB, HepG2, and MCF-7)	Tannoury et al. (2017)
Diterpene-8,10,18-trihydroxy-2,6-dolabelladiene	*Dictyota pfaffii*	Inhibitory against HSV-1	Abrantes et al. (2010)
Halogenated metabolites	*Laurencia catarinensis*	Antitumoral activity	Tannoury et al. (2017)
Phlorotannins, grasshopper ketone, fucoidan, and polysaccharides	*Sargassum fulvellum*	Antioxidant, anticancer, anti-inflammatory, antibacterial, and anticoagulant activities	Kang et al. (2008)

biocompatibility and biodegradability (Bilal and Iqbal 2020; Fertah 2017; Venkatesan et al. 2015, 2016; Zia et al. 2017). Biocompatibility refers to the low risk of adverse reactions or toxicity when these materials come into contact with living tissue. Guzman-Puyol et al. (2017) demonstrated the biocompatibility of seaweed's polysaccharides, specifically alginate, using an in vitro model with human foreskin fibroblast (HFF-1) cells. They found that seaweed's polysaccharides were nontoxic at a ratio of 0.5 cm^2/ml of film surface area to extraction medium and even had additional anti-inflammatory properties, indicating their potential in wound healing mechanisms.

4.4.2 Wound Healing Applications

Wound healing is a complex process that requires extra care to prevent secondary injury, infection, and sequelae. The implementation of wound dressing can provide a suitable environment that encourages rapid healing while improving patient compliance and comfort. Preliminary studies on poly(N-vinyl-2-pyrrolidone)-k-carrageenan hydrogels demonstrated their suitability as wound dressings due to good mechanical strength, biocompatibility, prevention of bacterial infection, gas permeability, and moisture control on the wound surface (Şen and Avcı 2005). Additionally, the wound healing-promoting properties of alginate were described in a study conducted by Balakrishinan et al. (2005). The study highlighted that maintaining a moist environment in the wound bed using a hydrogel dressing could enhance the migration of epithelial cells during the healing process. In their experiment, a hydrogel containing periodate oxidized alginate, gelatin, and borax was applied to full-thickness wounds on laboratory mice. The researchers observed rapid re-epithelialization without any significant adverse reactions within a two-week period. Thus, replacing traditional polymer dressings with seaweed's polysaccharides could reduce medical waste and improve patient comfort, as these materials are biodegradable and pose no harm to humans or the environment (Guzman-Puyol et al. 2017; Lalzawmliana et al. 2019).

4.4.3 Drug Delivery Systems

In drug delivery, the encapsulation and biodegradation properties of alginate ensure controlled drug release and targeted delivery to specific tissues (Klöck et al. 1994; Lee and Mooney 2012; Janarthanan and Senthil Kumar 2018). Advancements in nanotechnology have led to the development of alternative approaches for cancer treatment utilizing polysaccharides derived from seaweed. Alginate nanoparticles (NPs) have emerged as promising candidates for cancer drug delivery, including drugs like doxorubicin and 5-fluorouracil. Compared to conventional chemotherapy, the delivery of drugs using alginate nanoparticles offers a less aggressive treatment option by enabling the controlled release of the drug specifically to the target tissue, thus reducing damage to normal healthy cells (Venkatesan et al. 2016). In the preclinical of liver cancer treatment, Zhang et al. (2009, 2012) discovered that doxorubicin-loaded glycyrrhetinic acid-alginate NPs exhibited the ability to selectively target and kill hepatocellular carcinoma cells. Furthermore, these NPs demonstrated a sustained release rate of the drug for up to 20 days after treatment. Fucoidan, another polysaccharide derived from seaweed, is increasingly recognized as an excellent candidate material for cancer drug delivery due to its biocompatibility, biodegradability, and anticancer properties. Fucoidan nanoparticles have been developed with encapsulation capabilities and specific cell/environmental release functions. For instance, chitosan-fucoidan nanoparticles have been utilized to encapsulate curcumin, aiming to minimize the loss of bioavailability during digestion. The release of curcumin from these nanoparticles was optimized between pH 6.0 and 7.0 to ensure optimal drug efficiency for cancer treatment (Anitha et al. 2014; Huang and Kuo 2016; Huang and Lam 2011).

4.4.4 Tissue Engineering

In scaffold engineering, seaweed's polysaccharide-based compounds possess properties such as biocompatibility, nontoxicity, and the ability to stimulate cell adhesion and proliferation. They also have a structural similarity to macromolecular-based human tissue (Beata Łabowska et al. 2019). Seaweed's polysaccharide-based scaffolds can be used to reconstruct lost bone and treat various organ defects (Barralet et al. 2005; Detsch et al. 2015; Sun and Tan 2013). Combining alginate, O-carboxymethyl chitosan (O-CMC), and *Cissus quadrangularis* extract (CQ-E), a novel "herbal scaffold" (Alg/O-CMC/CQ-E scaffold) was developed as a potential candidate for bone tissue engineering therapeutics (Soumya et al. 2012). This scaffold has demonstrated a reduction in the recovery time for bone damage in preclinical studies. The presence of alginate and O-carboxymethyl chitosan in the scaffold facilitates initial cell attachment, proliferation, and an increase in calcified mineral deposition within a two-week incubation period. Additionally, carrageenan-functionalized graphene oxide (GO-Car) has been shown to promote the mineralization of hydroxyapatite through in vitro studies on MC3T3-E1 cells (Liu et al. 2014). The GO-Car nanocomposites facilitate cell attachment, cell proliferation, and calcium deposition in bone tissue, offering promising effects in bone tissue recovery. Moreover, these alginate scaffolds gradually degrade over time, reducing the need for additional surgical procedures for implant removal.

4.5 Seaweed-Based Biomaterials for Food and Beverage Industry Applications

4.5.1 Food Additives

Seaweed-based biomaterials have gained popularity in the food industry as a sustainable alternative to conventional food additives. According to Abdul Khalil et al. (2017), seaweed contains various polysaccharides, including alginate, carrageenan, agar, ulvan, and fucoidan, which possess unique functional properties suitable for food applications. For instance, alginate exhibits excellent gelling properties and can serve as a thickening and stabilizing agent in a wide range of food products. Carrageenan, on the other hand, enhances texture and stability in dairy products. Seaweed also contains bioactive compounds like polyphenols, carotenoids, and phlorotannins, which act as natural antioxidants and offer health benefits to consumers. Seaweed has been utilized as a natural flavor enhancer in various food products such as soups, sauces, and dressings. A study by Jensen et al. (2022) explored the potential of seaweed extracts from different species as natural flavor enhancers. The findings led to the development of umami and salty flavorings that have the potential to replace sodium in some food products and produce new flavors. Seaweed-based biomaterials also act as emulsifiers and stabilizers in food and beverage formulations, preventing phase separation and improving product appearance and shelf-life (Song et al. 2009). Furthermore, seaweed extracts such as carrageenan, alginate, and agar serve as thickening and gelling agents in various food products, including dairy, bakery, and meat products (Rupert et al. 2022). These biomaterials enhance texture, stability, and mouthfeel. Carrageenan and alginate, for instance, are commonly used as gelling agents in dairy products like ice cream and yogurt to achieve a smooth texture and enhance overall quality, as shown in a study by Qin et al. (2018). Additionally, agar and carrageenan can effectively function as thickening agents in meat products such as sausages and burgers, improving their texture and reducing cooking loss, as demonstrated by Amini Sarteshnizi et al. (2015).

4.5.2 Edible Films and Coatings

In recent years, seaweed-based films and coatings have emerged as a promising solution for packaging food products, offering a biodegradable and compostable alternative to traditional plastic packaging that poses significant environmental challenges. Seaweed-based packaging materials have the added advantage of being derived from a renewable resource abundantly available in the oceans. For instance, a prior study explored the potential of seaweed-based films for food packaging applications, revealing their favorable mechanical and barrier properties, which make them suitable for food packaging purposes (De-Oliveira et al. 2019). Further research demonstrated that seaweed-based coatings could effectively extend the shelf life of fresh strawberries while also maintaining the firmness of raspberries and blueberries, prolonging their shelf life under refrigeration, and improving their appearance, highlighting their potential as a natural and sustainable alternative to synthetic coatings (Falcó et al. 2019). Similarly, a study done by Sharma et al. (2020) found that seaweed-based films and coatings could successfully prolong the shelf life of chicken breasts. The seaweed-based coating exhibited the ability to delay lipid oxidation and bacterial growth on chicken breasts during 25 days of refrigerated storage, demonstrating its potential as a natural and sustainable alternative to synthetic coatings (Perera et al. 2021).

4.5.3 Nutraceuticals

Seaweeds are renowned for their rich nutrient content, encompassing vitamins, minerals, dietary fiber, and antioxidants, which positions them as promising ingredients for fortifying food products (Shannon and Abu-Ghannam 2019). Their nutritional value has been leveraged in various food items, including bread, pasta, and snacks, to enhance their nutritional composition (Palmieri and Forleo 2020). Seaweed-based ingredients have demonstrated the ability to enhance the nutritional profile of meat products such as burgers and sausages, augmenting their protein and fiber content (Debbarma et al. 2022). Additionally, seaweed has recently been harnessed to produce low-calorie noodles as an alternative to traditional wheat-based noodles (Kumoro et al. 2016). These noodles offer a distinct texture, flavor profile, and are rich in fiber and protein. Moreover, seaweed contains a diverse array of bioactive compounds, including antioxidants and polysaccharides, which have been associated with various health benefits, such as anti-inflammatory and antitumor effects (Shannon and Abu-Ghannam 2019). Considering these factors, seaweed-derived biomaterials present a significant opportunity as a source of bioactive compounds for integration into developing and formulating nutraceutical food and beverage products (Lomartire and Gonçalves 2022). The development of a novel functional kombucha beverage using laver (*Porphyra dentata*) with competitive physicochemical properties, antioxidant effects, and nutraceutical properties, by fermentation through kombucha consortia of yeast and bacteria, suggests that fermentation of seaweed may be a viable strategy for further enhancing its bioactive components and health benefits (Aung and Eun 2021). The incorporation of seaweed-based ingredients in the food industry aligns with the increasing demand for natural and sustainable components, fostering the creation of healthier and more environmentally friendly food and beverage products.

4.6 Seaweed-Based Biomaterials for Agricultural and Environmental Applications

4.6.1 Biofertilizers

Seaweed-derived fertilizers present a highly promising substitute for conventional synthetic fertilizers. According to Hamouda et al. (2022), *Ulva linza* and *Corallina officinalis* are rich in macronutrients including nitrogen (N), phosphorus (P), potassium (K), calcium (Ca), magnesium (Mg), and several trace elements. Additionally, these macroalgae possess growth regulators, amino acids, and antioxidants, making them valuable sources of natural fertilizers. Seaweed extracts have gained popularity as commercial foliar fertilizers for various fruit and vegetable crops. Examples of such products are Maxicrop Soluble Seaweed Powder (US), Alga Mix (Italy), Takamin Alga (Spain), and SeaTop (China), which are readily accessible in the market. Farmers can apply seaweed-based fertilizers through many methods, including foliar sprays, soil soaking, or seed treatments. According to Mishra et al. (2020), the extracts possess inherent biodegradability and eco-friendliness, hence exhibiting no adverse effects on both human health and the environment. Several studies have documented that the utilization of fertilizers derived from seaweed has resulted in an augmentation of soil organic matter content and microbial biomass (de Sosa et al. 2023; Prasedya et al. 2022; Wang et al. 2016). This, in turn, has contributed to the promotion of a more sustainable and resilient soil ecosystem.

The study conducted by Ammar et al. (2022) provides a comprehensive examination of the utilization of seaweed-derived biomaterials in enhancing soil health and promoting plant growth. The authors emphasize the significance of seaweed extracts and biofertilizers in augmenting plant development, promoting nutrient absorption, and maximizing agricultural productivity. Numerous studies have provided evidence of the effective incorporation of seaweed-derived fertilizers into standard agricultural methodologies, leading to notable enhancements in crop productivity and the overall quality of fruits across diverse crop species, including tomatoes (Khan et al. 2009). Chrysargyris et al. (2018) demonstrated enhanced growth and improved quality of harvested lettuce with the use of foliar spray containing *Ascophyllum nodosum*. In a study conducted by Yang et al. (2023), it was shown that the application of foliar fertilizer derived from seaweed extract has positive effects on the accumulation of nutrients and the development of flavor in "Fuji" apple (*Malus robusta*). The study conducted by Rajan et al. (2020) provided evidence that the application of aqueous extracts derived from *Padina gymnospora*, *Gracilaria edulis*, and *Ulva fasciata* as a seaweed biofertilizer in *Capsicum annuum* had a positive impact on seed germination. Specifically, this treatment resulted in increased root and shoot length, as well as a decrease in mean germination time. Studies conducted on the utilization of red seaweed fertilizers have yielded encouraging outcomes across a range of crops. In a study conducted by Agarwal et al. (2013), it was demonstrated that the utilization of a 5% solution of *Kappaphycus alvarezii* sap resulted in increased absorption of nitrogen (N), phosphorus (P), potassium (K), sulfur (S), calcium (Ca), and magnesium (Mg) in the leaves of rice plants. Basavaraja et al. (2018) conducted a study to explore the impact of foliar spraying with *Kappaphycus alvarezii* and *Gracilaria edulis* on grain yield in hybrid maize in India. The findings of the research indicated that the utilization of seaweeds at varying concentrations (2.5%, 5.0%, 7.5%, and 15.0% v/v) resulted in a notable increase of 18.5% and 26.0% in the production of grains. In a recent study conducted by Ghosh et al. (2020), it was observed that the inclusion of seaweed, specifically *Kappaphycus alvarezii*, at a concentration of 15% in black gram crops grown in sandy loam soil within the red and lateritic belt of West Bengal, led to an enhanced uptake and accessibility of several inorganic elements such as calcium (Ca), sodium (Na), potassium (K), magnesium (Mg), nitrogen (N), zinc (Zn), and copper (Cu).

4.6.2 Biostimulants

Seaweed-derived biomaterials exhibit biostimulant properties by harnessing their bioactive chemicals, growth regulators, and nutrient content. The plant growth regulator (PGR) found in seaweed extracts is considered a significant bioactive chemical due to its ability to influence various physiological processes in plants (Ali et al. 2021; Deolu-Ajayi et al. 2022; Sani and Yong 2022). These processes encompass cell division, elongation, differentiation, and flowering. Seaweed extracts are known to include many phytohormones, including cytokinins, auxins, gibberellins, abscisic acid, and polyamines (Agarwal et al. 2021; Benítez García et al. 2020; De Saeger et al. 2020). These compounds have been found to have a significant impact on plant growth and development, exerting their effects through multiple mechanisms such as hormone regulation, stress tolerance enhancement, and interactions with microorganisms. Diverse types of seaweed display variations in the composition and quantity of growth hormones. For instance, the *Sargassum heterophyllum* seaweed extract is recognized for its cytokinin content (Stirk et al. 2003), whereas the *Ascophyllum nodosum* seaweed extract is noted for containing indole-3-acetic acid (Shukla et al. 2019). According to Pramanick et al. (2013), the use of seaweed biostimulant results in the improvement of various physiological processes in plants, including meristematic growth, translocation of photosynthate, activation of enzymes, elongation of cells, and enhancement of cellular stability. According to Mohy El-Din (2015), the use of seaweed extracts that are abundant in magnesium and minerals has been found to enhance the concentrations of total chlorophyll and carotenoids in plant leaves. The increased protein and sugar levels observed in seaweed priming can be ascribed to the uptake of key minerals, specifically magnesium, which has the capacity to stimulate chlorophyll synthesis and thus enhance rates of photosynthesis (Castellanos-Barriga et al. 2017). Table 4.2 presents the effects of utilizing diverse seaweed extracts as biostimulants on distinct growth phases of numerous economically significant crops.

4.6.3 *Bioremediation*

Seaweed-derived biomaterials play a pivotal role in the field of environmental management as they effectively contribute to pollution mitigation and ecosystem restoration. For example, the utilization of dried seaweed biomass, seaweed biochar, and seaweed cell wall polymers has been found to be effective in the remediation of heavy metals. The primary constituents of cell walls are sulfated polysaccharides such as carrageenan, alginates, and fucoidans. These polymers possess a diverse range of functional groups, including COO–, OSO_3–, and –NH_2, which are located on the cell wall's surface. These functional groups serve as binding sites for heavy metals, enabling their removal from the surrounding environment (Zhao et al. 2023). The process of metal biosorption encompasses various mechanisms, such as ion exchange, electrostatic interactions, and the establishment of both weak Van der Waals forces and strong covalent bonds. The study conducted by Znad et al. (2022) examines the application of seaweed-based adsorbents in the removal of heavy metals. The authors provide a comprehensive analysis of the performance of these adsorbents, considering various significant factors such as the toxicity and sources of heavy metals, adsorption mechanisms, and the factors that influence the adsorption process. Table 4.3 presents the application of adsorbents derived from algae or seaweed for the purpose of removing heavy metals.

Seaweed biochar is a highly promising biomaterial derived from seaweed, with considerable promise for many uses in the field of environmental remediation. Bird et al. (2011) reported that seaweed-derived biochar has a lower carbon content, surface area, and cation exchange capacity than lignocellulosic feedstock, but it contains significant concentrations of nitrogen (N), phosphorus (P), calcium (Ca), and magnesium (Mg). The biochars derived from *Gracilaria*, *Saccharina*, and *Undaria* exhibit a carbon-to-nitrogen (C/N) ratio of less than 20. This suggests that

Table 4.2 Effects of seaweed extract biostimulants on agricultural productivity of prominent crop species

Crop	Seaweed extract	Effects on agricultural productivity	References
Tomato (*Solanum lycopersicum*)	*Ascophyllum nodosum* *Sargassum* spp. *Cystoseira myriophylloides* *Gelidium serrulatum* *Ulva lactuca* *Laminaria digitata* *Fucus spiralis* *Acanthophora spicifera* *Ecklonia maxima* *Euglena gracilis* *Sargassum horneri*	Enhanced germination rate and seedling vigor Heightened shoot and root growth Elevated chlorophyll content Increased mineral content, specifically iron (Fe) and zinc (Zn) Augmented presence of antioxidants and soluble carbohydrates in fruits Improved fruit quality Strengthened resistance to pathogens Heightened tolerance to salinity, drought, and cold stress Increased levels of endogenous phytohormones Enhanced gene transcription of protective heat shock proteins Expression of six flowering genes, resulting in amplified flower bud and fruit production	Demir et al. (2006), Khan et al. (2009), Parađiković et al. (2011), Jayaraman and Ali (2015), Vinoth et al. (2017), Esserti et al. (2017), Ali et al. (2018), Murtic et al. (2018), Al-Ghamdi and Elansary (2018), Digruber et al. (2018), Casadesús et al. (2019), Barsanti et al. (2019), and Yao et al. (2020)
Sweet pepper (*Capsicum annuum*)	*A. nodosum* *Sargassum* spp. *A. spicifera*	Increased shoot and root growth Heightened chlorophyll content Enhanced flowering Increased fruit yield Improved fruit quality Strengthened resistance to pathogens Heightened tolerance to salinity and drought stress	Khan et al. (2009), Jayaraman and Ali (2015), Vinoth et al. (2017), Dalal et al. (2019), and Di Mola et al. (2019)

(continued)

Table 4.2 (continued)

Crop	Seaweed extract	Effects on agricultural productivity	References
Lettuce (*Lactuca sativa*)	*A. nodosum* *Durvillaea potatorum* *Durvillaea antarctica* *E. maxima*	Heightened root and shoot growth Increased chlorophyll content Enhanced photochemical efficiency and greater activity of photosystem II Increased marketable yield	Yusuf et al. (2019) and Di Mola et al. (2019)
Cauliflower (*Brassica oleracea*)	*A. nodosum*	Enlarged cauliflower heart size Expanded curd diameter	Abetz and Young (1983)
Soybean (*Glycine max*)	*A. nodosum* *Kappaphycus alvarezii*	Enhanced nutrient uptake Improved yield parameters Increased drought tolerance Heightened protein content Augmented activity of antioxidant enzymes, stomatal conductance, and higher energy efficiency	Martynenko et al. (2016), Shukla et al. (2018), Kocira et al. (2018), and do Rosário Rosa et al. (2021)
Strawberry (*Fragaria × ananassa*)	*A. nodosum* *Sargassum* sp. *Laminaria* sp. *D. potatorum*	Heightened vegetative growth Increased levels of carbohydrates in the crown, phosphorus and potassium in leaves Augmented yield Improved fruit quality and taste Strengthened resistance to pathogens	Washington et al. (1999), Meszka and Bielenin (2009), El-Miniawy et al. (2014), Mattner et al. (2018), and Kapur et al. (2018)
Cucumber (*Cucumis sativus*)	*A. nodosum* *Macrocystis pyrifera* *Ulva armoricana*	Increased fruit yield Elevated mineral content (Fe, Mn, Zn) in the fruit Improved nutritional value of the fruit Decreased incidence of fungal infections and powdery mildew	Jayaraman et al. (2011), Jaulneau et al. (2011), Ahmed and Shalaby (2012), and Hidangmayum and Sharma (2017)

(continued)

Table 4.2 (continued)

Crop	Seaweed extract	Effects on agricultural productivity	References
Onion (*Allium cepa*)	*A. nodosum*	Enhanced germination rate and seedling vigor Augmented bulb diameter and weight Increased mineral content Elevated levels of ascorbic acid Reduction in disease incidence caused by downy mildew (Peronospora destructor) Improved resistance to water stress and increased uptake of nitrogen (N), phosphorus (P), and potassium (K)	Hidangmayum and Sharma (2017), Abbas et al. (2020), Almaroai and Eissa (2020), and Haider et al. (2012)
Potato (*Solanum tuberosum*)	*A. nodosum* *K. alvarezii* *Gracilaria edulis* *E. maxima*	Enhanced growth Increased yield and tuber quality Strengthened resistance to drought stress	Blunden and Wildgoose (1977), Dwelle and Hurley (1984), Mattner et al. (2013), Prajapati et al. (2016), and Dziugieł and Wadas (2020)
Broccoli (*Brassica oleracea* var. *italica*	*A. nodosum* *Laminaria ochroleuca*	Enhanced biomass production Increased nitrogen (N), phosphorus (P), potassium (K), magnesium (Mg), copper (Cu), and manganese (Mn) levels Heightened protein content Increased presence of glucosinolates and phenolic compounds	Gajc-Wolska et al. (2013), Mattner et al. (2013), Rengasamy et al. (2016), and Flores et al. (2021)
Spinach (*Spinacea oleracea*)	*A. nodosum* *E. maxima* *Codium liyengarii*	Increased fresh yield, dry biomass, and leaf area Increased chlorophyll content Improved micro/macronutrient profile Strengthened resistance to drought stress Elevated levels of phytohormones and phenolic acids	Zahid (1999), Xu and Leskovar (2015), Rouphael et al. (2018), and Pobereżny et al. (2020)

(continued)

Table 4.2 (continued)

Crop	Seaweed extract	Effects on agricultural productivity	References
Carrot (*Daucus carota*)	*A. nodosum* *E. maxima*	Heightened harvest index (HI) Enhanced nutritional content Decreased severity of fungal diseases	Jayaraj et al. (2008), Eggen et al. (2011), Alam et al. (2014), Kasim et al. (2015), and Kiraci (2018)
Wheat (*Triticum aestivum*)	*A. nodosum* *E. maxima* *K. alvarezii* *G. edulis* *Gracilaria dura* *Sargassum latifolium* *U. lactuca* *Sargassum* sp. *Ulva linza* *Corallina officinalis*	Increased chlorophyll content Increased yield Increased micro/macronutrients in roots, leaves, and grains Heightened protein content Enhanced tolerance to drought and salinity Increased abscisic acid content and upregulation of stress-protective genes	Beckett and van Staden (1989), Shah et al. (2013), Amaral de Carvalho et al. (2014), Salim (2016), Begum et al. (2018), Sharma et al. (2019), Feitosa de Vasconcelos (2020), Hamouda et al. (2022), and Ali et al. (2022)
Rice (*Oryza sativa*)	*A. nodosum* *Kappaphycus* sp. *Gracilaria* sp. *Hydroclathrus* sp. *Sargassum* sp. *S. horneri*	Higher percentage of germination and enhanced seedling vigor Increased yield Improved uptake of nutrients Enhanced protein content Beneficial modulation of the bacterial community during the tillering and heading stages	Spinelli et al. (2009), Sunarpi et al. (2010), Pramanick et al. (2014), Layek et al. (2018), and Chen et al. (2022)
Apple (*Malus domestica*)	*A. nodosum* *Codium tomentosum*	Reduced alternate bearing phenomenon Increased chlorophyll content Augmented rates of photosynthesis and respiration Increased fruit set and fruit yield Elevated anthocyanin content Enhanced intensity of red coloration Minimized fruit browning after harvesting	Jeannin et al. (1991), Malaguti et al. (2002), Soppelsa et al. (2018), Valencia et al. (2018), and de Sousa et al. (2019)

(continued)

Table 4.2 (continued)

Crop	Seaweed extract	Effects on agricultural productivity	References
Maize (*Zea mays*)	*A. nodosum* *Laminaria* sp. *G. edulis* *K. alvarezii*	Enhanced germination percentage and rate Improved seedling vigor Heightened shoot and root growth Increased net carbon assimilation Improved overall grain yield	Crouch et al. (1990), Pal et al. (2015), Singh et al. (2015), Layek et al. (2016), Basavaraja et al. (2018), and Al-Musawi (2018)
Orange (*Citrus* spp.)	*A. nodosum* *E. maxima* *Sargassum horridum* *Laurencia johnstonii* *Caulerpa sertularioides*	Higher maturity index (MI) Reduced fruit drop Enhanced yield Greater vitamin C content Increased total soluble solids (TSS) Effective control of the Asian citrus psyllid (*Diaphorina citri*)	Fornes et al. (1995), Fornes et al. (2002), Asaraja and Sahayaraj (2013), Arioli et al. (2015), and Gomathi et al. (2017)
Cowpea	*Sargassum swartzii*	Heightened levels of phenolic and flavonoid compounds	Vasantharaja et al. (2019)
Sugarcane (*Saccharum officinarum*)	*A. nodosum*	Increased biomass in plantlets Enhanced plant growth parameters, including chlorophyll content, height, and leaf area index Increased yield Improved sugar content Heightened water retention capacity and water content Reduced risk of wilting Effective control of borers, aphids, and thrips	da Silveira et al. (2015), Chen et al. (2019a), and Gomathi et al. (2017)
Common bean	*E. maxima*	Heightened yield and anthocyanin content in the seeds Increased synthesis of phenolics, flavonoids, anthocyanins, and antioxidant activities	Kocira et al. (2018) and Kocira et al. (2020)

(continued)

Table 4.2 (continued)

Crop	Seaweed extract	Effects on agricultural productivity	References
Black eyed pea (*Vigna sinensis*)	*Caulerpa chemnitzia* *S. wightii*	Increased protein content	Sivasankari et al. (2006)
Mung bean (*Vigna radiata*)	*U. lactuca*	Improved seed germination rates Increased yield and biomass Enhanced protein content Increased chlorophyll content	Castellanos-Barriga et al. (2017)
Okra (*Abelmoschus esculentus*)	*S. wightii*	Increased protein content	Jothinayagi and Anbazhagan (2009)
Banana (*Musa* spp.)	*K. alvarezii*	Increased protein content	Karthikeyan and Shanmugam (2016)
Eggplant (*Solanum melongene* L.)	*Caulerpa* sp. *Sargassum* sp. *K. alvarezii* *Ulva* sp.	Improved growth and yield	Yusuf et al. (2021)

Adopted and modified from Ali et al. (2021), Deolu-Ajayi et al. (2022), and Sani and Yong (2022)

these biochars have the potential to directly supply bioavailable nitrogen (N) and phosphorus (P) to soil. Furthermore, they can also enhance the retention of additional nutrients that are applied as fertilizers (Roberts et al. 2015). In addition, Michalak et al. (2019) conducted an experimental study on the utilization of *Cladophora glomerata*, a type of freshwater macroalgae, for the manufacture of biochar. The findings of their research demonstrated that *Cladophora* biochar exhibits a notable capacity for the concurrent removal of metal ions from synthetic wastewater. Seaweed biochar has also been identified as a promising means of carbon sequestration, as highlighted by Sun et al. (2022). The long-term storage capability of seaweed biochar in soil, estimated to span thousands of years, enables the effective immobilization of solid carbon, hence facilitating the removal and retention of atmospheric carbon. Several studies have indicated that the introduction of seaweed biochar into soil has the potential to enhance the population of methane-oxidizing microbes, hence leading to a reduction in methane emissions from agricultural fields (Wu et al. 2019a, b). The potential of seaweed biochar to achieve long-term carbon sequestration is evident from a comparative analysis of the carbon content derived from biomass (9931.5 tons CO_2 yr^{-1}) and the carbon emissions (7976 tons CO_2 yr^{-1}) associated with seaweed in Ailian Bay, China (Lian et al. 2023). Nevertheless, the authors emphasized the importance of addressing the issue of excessive carbon emissions that arise from seaweed biochar processing. In their study, Wen et al. (2022) conducted an evaluation of the life cycle of beach-cast

Table 4.3 Biosorption of various metals by different seaweed species

Metal ions	Adsorbent material	Removal efficiency (%)/ adsorption capacity (mg/g, mmol/g)	References
Lead	*Turbinaria ornata*	99.8%	Abdullah Al-Dhabi and Arasu (2022)
	Gelidium amansii	100%	El-Naggar et al. (2018)
	Fucus vesiculous	1.04 mmol/g	Brinza et al. (2007)
	Ascophylhum nodosum	204.1 mg/g	Kumar and Oommen (2012)
	Hizikia fusiformis	10.9%	Pham et al. (2021)
Chromium	*Cystoseira barbata* and *Cystoseira crinite*	Cr(III): 70.7% (*C. barbata*), 73.34% (*C. crinite*) Cr(VI): 35% (*C. barbata*), 28% (*C. crinite*)	Yalçın and Özyürek (2018)
	Sargassum dentifolium	Cr(VI): 99.68%	Husien et al. (2019)
	Sargassum filipendula	0.964 mmol/g	Costa et al. (2020)
Nickel	*Sargassum filipendula*	45%	Moino et al. (2017)
	Fucus vesiculous	0.80 mmol/g	Brinza et al. (2007)
	Ascophylhum nodosum	50 mg/g	Kumar and Oommen (2012)
	Sargassum filipendula	0.347 mmol/g	Costa et al. (2020)
	Hizikia fusiformis	10.9%	Pham et al. (2021)
Cobalt	*Gracilaria corticata*	87.8%	Raju et al. (2021)
	Hypnea valentiae	Raw adsorbent 10.98 mg/g Modified adsorbent 16.66%	Vafajoo et al. (2018)
Copper	*Fucus vesiculous*	0.97 mmol/g	Brinza et al. (2007)
	Ascophylhum nodosum	70.9 mg/g	Kumar and Oommen (2012)
	Hizikia fusiformis	31.6%	Pham et al. (2021)
Cadmium	*Sargassum filipendula*	0.43 mmol/g	Nishikawa et al. (2018)
	Bifurcaria bifurcata	95 mg/g	Gin et al. (2002)
	Sargassum tenerrimmum	0.4 mmol/g	Hajar (2009)
	Fucus vesiculous	1.12 mmol/g	Brinza et al. (2007)
	Sargassum glauceseens	84.7 mg/g	Romera et al. (2007)
	Ascophylhum nodosum	114.9 mg/g	Kumar and Oommen (2012)
	Hizikia fusiformis	14.9%	Pham et al. (2021)
	Sargassum polycystum	86.2%	Jayakumar et al. (2022)
Arsenic	*Sargassum muticum*	≈100%	Vieira et al. (2017)

(continued)

Table 4.3 (continued)

Metal ions	Adsorbent material	Removal efficiency (%)/ adsorption capacity (mg/g, mmol/g)	References
Mercury	*Ulva intestinalis*	95%	Fabre et al. (2020)
	Ulva lactuca	90%	
	Fucus spiralis	85%	
	Fucus vesiculosus	80%	
	Gracilaria sp.	90%	
	Osmundea pinnatifida	80%	
Zinc	*Ulva lactuca*	128 mg/g	Senthilkumar et al. (2019)
	Caulerpa scalpelliformis	83.3 mg/g	Jayakumar et al. (2021)
	Ascophylhum nodosum	53.2 mg/g	Kumar and Oommen (2012)
	Sargassum polycystum	92.9%	Jayakumar et al. (2022)
	Sargassum filipendula	0.302 mmol/g	Costa et al. (2020)

seaweed using pyrolysis as the analytical method. Their findings indicate that subjecting washed seaweed to pyrolysis at a temperature of 600 °C leads to the release of approximately 790.89 kg of carbon dioxide equivalent emissions.

4.7 Seaweed-Based Biomaterials for Industrial and Technological Applications

4.7.1 Bioplastics and Packaging Materials

The issue of plastic pollution has become a major global concern, largely driven by the widespread use of petroleum-based plastics in various applications such as packaging, electronics, and household appliances, among others. While these plastics offer versatility, they can take thousands of years to degrade and end up accumulating in the environment or landfills. To mitigate this, biodegradable plastics made from renewable resources are recommended due to their faster degradation rates, nontoxicity, and reduced energy consumption during manufacturing (Dang et al. 2022). Biomaterials such as polysaccharides, proteins, and lipids are commonly used to produce different types of bioplastics that can be applied in various sectors. The most widely used bioplastics today include starch-based, polylactic acid (PLA), poly-3-hydroxybutyrate (PHB), polyamide 11 (PA 11), and organic polyethylene (PE) (Lim et al. 2021).

Seaweeds are a promising new source of bioplastics due to their ability to form films and their benefits as a raw material, including reduced greenhouse gas (GHG) emissions, high yields, and the potential for use in food packaging industries (Dang et al. 2022; Lomartire et al. 2022). According to Future Market Insights, the seaweed packaging market is expected to increase rapidly, with an estimated worth of US$ 682.1 million by 2023 and a CAGR of 6.0% over the forecast period, reaching US$ 1224.5 million by 2033. This growth is primarily due to the increasing demand for sustainable packaging solutions in response to the plastic waste problem. One significant advantage of seaweed-based packaging over traditional plastic packaging is its natural antimicrobial properties, which help products last longer. Although seaweed-based packaging is currently more expensive than conventional plastic packaging, costs are likely to fall as the industry expands and new manufacturing methods are developed (Future Market Insights 2023).

The assessment of bioplastics and packaging materials involves evaluating their mechanical properties, including tensile strength, elongation at break, thermal resistance, and water vapor permeability. Plant fibers with higher cellulose content typically exhibit enhanced tensile strength and Young's modulus (Lomartire et al. 2022). Seaweed-based bioplastics are currently under investigation as potential biopackaging materials. In many cases, the addition of plasticizers like glycerol and PEG is necessary to enhance the mechanical properties and water barrier capabilities of the seaweed-based bioplastics (Hidayati et al. 2021; Santana et al. 2022). Additionally, the incorporation of nanoparticles such as zinc oxide (ZnONPs), cupric oxide (CuONPs), and silicon dioxide (SiO_2NPs) can provide bioplastics with additional properties such as antibacterial activity (Sudhakar et al. 2022). Ongoing research on the utilization of seaweed from various species for the development of bioplastics and packaging materials is summarized in Table 4.4.

In addition to the extensive research and development activities surrounding seaweed-based bioplastics, several companies have acknowledged the immense potential of seaweed as a valuable feedstock for bioplastic production. These companies have actively pushed the boundaries in this field, achieving significant progress and innovation. Table 4.5 presents some notable companies and startups that have been actively engaged in the production of bioplastics derived from seaweed.

4.7.2 Energy Production

Seaweed has gained significant attention as a renewable resource with great potential for energy production. Its application in this field encompasses various avenues, including biofuel production, biogas generation, and hydrogen fuel extraction. These applications demonstrate the potential of seaweed as a valuable and sustainable resource for meeting energy needs while reducing carbon emissions (Milledge et al. 2014). Seaweed biomass can be converted into biofuels such as ethanol and biodiesel through various processes, including fermentation and anaerobic digestion. During fermentation, microorganisms break down the carbohydrates present in seaweed into sugars, which are subsequently fermented to yield ethanol.

Table 4.4 Ongoing research on the utilization of seaweed from various sources for the development of bioplastics

Type of seaweed	Bioplastic formulation	Properties	References
Kappaphycus alvarezii	Carrageenan from the seaweed was blended with metal oxide nanoparticles such as zinc oxide (ZnONPs), cupric oxide (CuONPs), and silicon dioxide (SiO_2NPs)	Improved mechanical strength, optical, water barrier, and antibacterial properties. Degradation time: 10–15 days	Sudhakar et al. (2022)
Sargassum sp.	Alginate blended with 0–15% inverted sugar (IS)	Bioplastic that comprised of alginate (Alg) 6% with 5% IS exhibited the highest tensile strength Degradation time: 4 days	Kanagesan et al. (2022)
Sargassum natans	6 wt% alginate, 0.263 wt% starch, 0.35 wt% carboxy methyl cellulose (CMC), 0.065 g/g sorbitol, and 0.025 g/g PEG 200	No migration of additives into a simulated aqueous food system in 10 days Improved optical and water barrier properties Comparable material properties to high-density polyethylene (HDPE), polyethylene terephthalate (PET), and polylactic acid (PLA) Degradation time: 14 days	Mohammed et al. (2023)
Rugulopteryx okamurae	The raw seaweed *Rugulopteryx okamurae* blended with glycerol (ratios: 50/50, 60/40, and 70/30)	Film with higher seaweed content exhibited better viscoelastic properties, rigidity, and resistance	Santana et al. (2022)
Eucheuma cottonii	*Eucheuma cottonii* waste blended with glycerol CMC	Tensile strength comparable to poly tetra fluoro ethyne (PTFE) and polypropylene (PP) Degradation time: 14 days	Hidayati et al. (2021)

Table 4.5 Companies and start-ups producing seaweed-based bioplastics

Company	Country of origin	Product
Kelpi	United Kingdom	Seaweed-based composite film
Evoware	Indonesia	Ello Jello seaweed-based disposable cup
Loliware	United States	Hypercompostable straw, utensils, SEA Tech™ Resins
B'ZEOS	Norway	Edible drinking straws and food films
Sway	United States	Seaweed-based packaging

Fermentation of seaweed offers advantages such as high ethanol yields and the utilization of diverse seaweed species. On the other hand, anaerobic digestion primarily facilitates biogas production. In this process, microorganisms degrade the organic matter in seaweed in an oxygen-deprived environment, resulting in the production of biogas, predominantly composed of methane. Seaweed-derived biogas finds application in heat and power generation. Furthermore, anaerobic digestion offers the supplementary benefit of generating digestate, a nutrient-rich byproduct suitable for use as fertilizer (Godvin Sharmila et al. 2021; Zhao et al. 2022).

The rich carbohydrate content in seaweed can be harnessed to extract and produce hydrogen through processes such as gasification or fermentation. Gasification involves subjecting seaweed to high temperatures in the presence of a gasifying agent, leading to the production of a gas mixture containing hydrogen, while fermentation releases hydrogen as a byproduct. Seaweed-based hydrogen fuel offers several advantages, including clean energy source that solely produces water vapor when utilized in fuel cells. These characteristics make seaweed-derived hydrogen a compelling alternative to conventional fuels, contributing to the ongoing transition towards a low-carbon economy (Ahmed et al. 2021; Zhang et al. 2020). Current research efforts focus on enhancing the efficiency and scalability of seaweed-based hydrogen production methods. This entails optimizing gasification and fermentation processes to maximize hydrogen yield and exploring innovative techniques like photocatalysis and biophotolysis. Photocatalysis exploits light energy to drive the splitting of water molecules into hydrogen and oxygen, while biophotolysis involves utilizing microorganisms or enzymes to catalyze the same reaction (Ahmed et al. 2021; Davis et al. 2021). These endeavors contribute to advancing hydrogen as a renewable energy source and its integration across diverse sectors, including transportation, industry, and power generation.

4.7.3 Textile and Paper Industries

Studies have shown that seaweed fibers possess desirable characteristics for application in textile and paper industries such as high strength, natural antimicrobial properties, and biodegradability. Research done by Yu et al. (2013) have shown that alginate extracted from brown seaweed could be used to manufacture nonwoven textiles, which are typically made from materials such as polyester or polypropylene. In recent times, certain types of seaweed have been employed in the production of fibers for the manufacturing of technical textiles. These textiles utilize the functional properties of seaweed components to cater to the health and body care sector. Martinez and Becherucci (2022) successfully produced a soft bilaminate textile that exhibited a high crease recovery and efficient release of mucilage by incorporating fine dust of *Undaria* particles, measuring less than 0.26 mm, between two layers of nonwoven material. The extract of *Undaria* contained minerals and vitamins that had positive effects on human skin. Additionally, it proved to be nonirritating when applied to both healthy and abraded skin. Apart from that, the high cellulose content of seaweed makes it a suitable candidate for papermaking. Research conducted by Moral et al. (2019) explored the feasibility of using *Ulva* sp. seaweed for producing paper pulp. The findings indicated that the seaweed-based paper exhibited favorable qualities, even outperforming softwood fibers in tear resistance, suggesting that seaweed might be exploited as a new alternative source of fibers for papermaking.

4.8 Challenges and Future Directions

4.8.1 Challenges and Limitations of Seaweed-Based Biomaterials

Seaweed-based biomaterials possess immense potential as a sustainable and environmentally friendly alternative to conventional materials. Their distinctive properties, widespread availability, and rapid growth rate make them highly appealing for diverse industrial applications. Nonetheless, the broad adoption of seaweed-based biomaterials faces significant challenges that necessitate effective resolution to fully exploit their capabilities and ensure seamless integration across various domains.

One critical challenge pertains to the processing and extraction techniques involved in obtaining usable polymers, such as alginate and carrageenan from seaweed. Seaweed comprises numerous constituents, and the isolation of targeted polymers from other compounds presents considerable technical complexity. Developing scalable and cost-effective extraction methods that efficiently separate these polymers is essential for enabling large-scale production, enhancing efficiency, and reducing costs, thereby facilitating the commercial viability of seaweed-based biomaterials.

Furthermore, seaweed-derived biomaterials often exhibit distinct material characteristics compared to their conventional counterparts. These properties, encompassing mechanical strength, flexibility, and durability, significantly influence their suitability for diverse applications. Achieving consistent performance and desired material properties is a multifaceted endeavor requiring continuous research and optimization. By refining the composition and processing conditions of seaweed-based biomaterials, it is possible to enhance their mechanical and thermal properties, rendering them more comparable and competitive with traditional materials. This optimization process promotes broader acceptance and utilization of seaweed-based biomaterials in various industries.

Stability and shelf-life present additional challenges for seaweed-based biomaterials. Moisture absorption can lead to dimensional alterations, diminished mechanical properties, and potential degradation, thereby limiting their long-term performance and viability. Surmounting these stability concerns is crucial to ensuring the reliability and durability of seaweed-based biomaterials in diverse applications. Innovative solutions, such as incorporating additives or modifying the material composition, are necessary to enhance their moisture resistance, stability, and shelf life. By addressing these challenges, the longevity and dependability of seaweed-based biomaterials can be significantly enhanced.

The supply and cultivation of seaweed are fundamental considerations for the successful adoption of seaweed-based biomaterials. Meeting the escalating demand necessitates large-scale expansion of seaweed cultivation. To accomplish this, sustainable and efficient cultivation protocols must be established, optimizing factors such as nutrient supply, water quality, and environmental implications. Additionally, ensuring a consistent and dependable supply of high-quality seaweed biomass for large-scale production is of paramount importance. Cultivation practices need to be optimized to maximize yield, minimize environmental impact, and preserve the ecological balance of marine ecosystems. Factors such as meticulous location selection, advanced farming techniques, and refined harvesting methods are under careful evaluation to ensure sustainable and responsible seaweed cultivation practices.

Economic considerations and market competitiveness significantly influence the widespread adoption of seaweed-based biomaterials. The costs associated with seaweed cultivation, processing, and extraction techniques, coupled with limited economies of scale, can render these biomaterials relatively expensive compared to conventional alternatives. However, substantial potential for cost reduction exists through improved production efficiency, technological advancements, and heightened demand. Continued research and innovation in these areas can drive down costs and augment the market competitiveness of seaweed-based biomaterials, rendering them economically viable and appealing to businesses and consumers alike.

4.8.2 Future Directions

To tackle existing challenges and explore new opportunities in the application of seaweed-based biomaterials, future directions should focus on several key areas. First, there is a need to optimize cultivation and processing techniques. Research efforts should be dedicated to refining seaweed cultivation methods by optimizing factors such as nutrient supply, water quality, and cultivation systems. Furthermore, it is imperative to explore innovative techniques for extracting and processing seaweed biomass into biomaterials to enhance overall production efficiency and scalability. Additionally, diversifying the range of seaweed species used is crucial. By identifying species with desirable characteristics such as high polymer content and rapid growth rates, the availability of feedstocks for biomaterial production can be expanded. This expansion would lead to increased versatility and improved performance of seaweed-based biomaterials.

Another promising avenue is the development of seaweed-based biomaterials with value-added properties. Incorporating nanoparticles, plasticizers, or other additives into seaweed-based biomaterials can augment their mechanical strength, thermal stability, and barrier properties, and introduce functionalities like antibacterial activity. By optimizing the composition and processing conditions, seaweed-based biomaterials can be tailored to meet specific application requirements. Furthermore, it is essential to establish sustainable end-of-life options for seaweed-based biomaterials. This involves promoting recycling, composting, or biodegradation strategies to minimize environmental impact and foster a circular economy approach. Research efforts should focus on developing efficient and environmentally friendly methods for the disposal and repurposing of seaweed-based biomaterials.

Collaboration among researchers, industry stakeholders, and policymakers is vital in driving innovation and facilitating the adoption of seaweed-based biomaterials. Establishing standards, sharing knowledge, and implementing supportive policies and regulations that incentivize the use of sustainable materials and promote circular economy principles will expedite the development and adoption of seaweed-based biomaterials.

References

Abasalizadeh F, Moghaddam SV, Alizadeh E, Akbari E, Kashani E, Fazljou SMB, Torbati M, Akbarzadeh A (2020) Alginate-based hydrogels as drug delivery vehicles in cancer treatment and their applications in wound dressing and 3D bioprinting. J Biol Eng 14(1):8. https://doi.org/10.1186/s13036-020-0227-7

Abbas M, Anwar J, Zafar-ul-Hye M, Khan RI, Saleem M, Rahi AA, Danish S, Datta R (2020) Effect of seaweed extract on productivity and quality attributes of four onion cultivars. Horticulturae 6(2):28. https://doi.org/10.3390/horticulturae6020028

Abdul Khalil HPS, Saurabh CK, Tye YY, Lai TK, Easa AM, Rosamah E, Fazita MRN, Syakir MI, Adnan AS, Fizree HM, Aprilia NAS, Banerjee A (2017) Seaweed based sustainable films and composites for food and pharmaceutical applications: a review. Renew Sustain Energ Rev 77:353–362. https://doi.org/10.1016/j.rser.2017.04.025

Abdul Khalil HPS, Lai TK, Tye YY, Rizal S, Chong EWN, Yap SW, Hamzah AA, Nurul Fazita MR, Paridah MT (2018) A review of extractions of seaweed hydrocolloids: properties and applications. Express Polym Lett 12(4):296–317. https://doi.org/10.3144/expresspolymlett.2018.27

Abdullah Al-Dhabi N, Arasu MV (2022) Biosorption of hazardous waste from the municipal wastewater by marine algal biomass. Environ Res 204(B):112115. https://doi.org/10.1016/j.envres.2021.112115

Abetz P, Young CL (1983) The effect of seaweed extract sprays derived from *Ascophyllum nodosum* on lettuce and cauliflower crops. Bot Mar 26(10):487–492. https://doi.org/10.1515/botm.1983.26.10.487

Abrantes JL, Barbosa J, Cavalcanti D, Pereira RC, Fontes CLF, Teixeira VL, Moreno Souza TL, Paixão ICP (2010) The effects of the diterpenes isolated from the Brazilian brown algae *Dictyota pfaffii* and *Dictyota menstrualis* against the Herpes Simplex Type-1 replicative cycle. Planta Med 76(4):339–344. https://doi.org/10.1055/s-0029-1186144

Abu R, Jiang Z, Ueno M, Isaka S, Nakazono S, Okimura T, Oda T, Yamaguchi K, Kim D, Oda T (2015) Anti-metastatic effects of the sulfated polysaccharide ascophyllan isolated from *Ascophyllum nodosum* on B16 melanoma. Biochem Biophys Res Commun 458(4):727–732. https://doi.org/10.1016/j.bbrc.2015.01.061

Agarwal PK, Shukla PS, Gupta K, Jha B (2013) Bioengineering for salinity tolerance in plants: state of the art. Mol Biotechnol 54:102–123. https://doi.org/10.1007/s12033-012-9538-3

Agarwal PK, Dangariya M, Agarwal P (2021) Seaweed extracts: potential biodegradable, environmentally friendly resources for regulating plant defence. Algal Res 58:102363. https://doi.org/10.1016/j.algal.2021.102363

Ahmed YM, Shalaby EA (2012) Effect of different seaweed extracts and compost on vegetative growth, yield and fruit quality of cucumber. J Hortic Sci Ornam Plants 4(3):235–240. https://doi.org/10.5829/idosi.jhsop.2012.4.3.252

Ahmed SF, Rafa N, Mofijur M, Badruddin IA, Inayat A, Ali MS, Farrok O, Yunus Khan TM (2021) Biohydrogen production from biomass sources: metabolic pathways and economic analysis. Front Energy Res 9:753878. https://doi.org/10.3389/fenrg.2021.753878

Ahn M-J, Yoon K-D, Min S-Y, Lee JS, Kim JH, Kim TG, Kim SH, Kim N-G, Huh H, Kim J (2004) Inhibition of HIV-1 reverse transcriptase and protease by phlorotannins from the brown alga *Ecklonia cava*. Biol Pharm Bull 27(4):544–547. https://doi.org/10.1248/bpb.27.544

Alam MZ, Braun G, Norrie J, Hodges DM (2014) *Ascophyllum* extract application can promote plant growth and root yield in carrot associated with increased root-zone soil microbial activity. Can J Plant Sci 94:337–348. https://doi.org/10.4141/CJPS2013-135

Ale MT, Maruyama H, Tamauchi H, Mikkelsen JD, Meyer AS (2011) Fucoidan from *Sargassum* sp. and *Fucus vesiculosus* reduces cell viability of lung carcinoma and melanoma cells in vitro and activates natural killer cells in mice in vivo. Int J Biol Macromol 49(3):331–336. https://doi.org/10.1016/j.ijbiomac.2011.05.009

Al-Ghamdi AA, Elansary HO (2018) Synergetic effects of 5-aminolevulinic acid and *Ascophyllum nodosum* seaweed extracts on Asparagus phenolics and stress related genes under saline irrigation. Plant Physiol Biochem 129:273–284. https://doi.org/10.1016/j.plaphy.2018.06.008

Ali O, Ramsubhag A, Jayaraman J (2018) *Ascophyllum nodosum* (Linnaeus) Le Jolis seaweed extract improves seed germination in tomato and sweet pepper under NaCl-induced salt stress. Trop Agric 95(2):141–148

Ali O, Ramsubhag A, Jayaraman J (2021) Biostimulant properties of seaweed extracts in plants: implications towards sustainable crop production. Plan Theory 10(3):531. https://doi.org/10.3390/plants10030531

Ali AH, Said EM, Abdelgawad ZA (2022) The role of seaweed extract on improvement drought tolerance of wheat revealed by osmoprotectants and DNA (cpDNA) markers. Rev Bras Bot 45(3):857–867. https://doi.org/10.1007/s40415-022-00820-5

Almaroai YA, Eissa MA (2020) Role of marine algae extracts in water stress resistance of onion under semiarid conditions. J Soil Sci Plant Nutr 20(3):1092–1101. https://doi.org/10.1007/s42729-020-00195-0

Al-Musawi MAHM (2018) Effect of foliar application with algae extracts on fruit quality of sour orange, *Citrus aurantium* L. J Environ Sci Pollut Res 4(1):250–252. https://doi.org/10.30799/jespr.122.18040104

Amaral de Carvalho ME, de Camargo e Castro PR, Gallo LA, Castro Ferraz Junior MV (2014) Seaweed extract provides development and production of wheat. Agrarian 7(23):166–170. https://ojs.ufgd.edu.br/index.php/agrarian/article/view/2459

Amini Sarteshnizi R, Hosseini H, Mousavi Khaneghah A, Karimi N (2015) A review on application of hydrocolloids in meat and poultry products. Int Food Res J 22(3):872–887

Ammar EE, Aioub AAA, Elesawy AE, Karkour AM, Mouhamed MS, Amer AA, El-Shershaby NA (2022) Algae as bio-fertilizers: between current situation and future prospective. Saudi J Biol Sci 29(5):3083–3096. https://doi.org/10.1016/j.sjbs.2022.03.020

Anaëlle T, Serrano Leon E, Laurent V, Elena I, Mendiola JA, Stéphane C, Nelly K, Stéphane LB, Luc M, Valérie S-P (2013) Green improved processes to extract bioactive phenolic compounds from brown macroalgae using *Sargassum muticum* as model. Talanta 104:44–52. https://doi.org/10.1016/j.talanta.2012.10.088

Anastyuk SD, Shevchenko NM, Ermakova SP, Vishchuk OS, Nazarenko EL, Dmitrenok PS, Zvyagintseva TN (2012) Anticancer activity in vitro of a fucoidan from the brown alga *Fucus evanescens* and its low-molecular fragments, structurally characterized by tandem mass-spectrometry. Carbohydr Polym 87(1):186–194. https://doi.org/10.1016/j.carbpol.2011.07.036

Anitha A, Sreeranganathan M, Chennazhi KP, Lakshmanan V-K, Jayakumar R (2014) In vitro combinatorial anticancer effects of 5-fluorouracil and curcumin loaded N,O-carboxymethyl chitosan nanoparticles toward colon cancer and in vivo pharmacokinetic studies. Eur J Pharm Biopharm 88(1):238–251. https://doi.org/10.1016/j.ejpb.2014.04.017

Arioli T, Mattner SW, Winberg PC (2015) Applications of seaweed extracts in Australian agriculture: past, present and future. J Appl Phycol 27(5):2007–2015. https://doi.org/10.1007/s10811-015-0574-9

Artan M, Li Y, Karadeniz F, Lee S-H, Kim M-M, Kim S-K (2008) Anti-HIV-1 activity of phloroglucinol derivative, 6,6′-bieckol, from *Ecklonia cava*. Bioorg Med Chem 16(17):7921–7926. https://doi.org/10.1016/j.bmc.2008.07.078

Asaraja A, Sahayaraj K (2013) Screening of insecticidal activity of brown macroalgal extracts against *Dysdercus cingulatus* (Fab.) (Hemiptera: Pyrrhocoridae). J Biopest 6(2):193–203

Athukorala Y, Trang S, Kwok C, Yuan YV (2016) Antiproliferative and antioxidant activities and mycosporine-like amino acid profiles of wild-harvested and cultivated edible Canadian marine red macroalgae. Molecules 21(1):119. https://doi.org/10.3390/molecules21010119

Atiwesh G, Mikhael A, Parrish CC, Banoub J, Le T-AT (2021) Environmental impact of bioplastic use: a review. Heliyon 7(9):e07918. https://doi.org/10.1016/j.heliyon.2021.e07918

Atya ME, El-Hawiet A, Alyeldeen MA, Ghareeb DA, Abdel-Daim MM, El-Sadek MM (2021) In vitro biological activities and in vivo hepatoprotective role of brown algae-isolated fucoidans. Environ Sci Pollut Res 28(16):19664–19676. https://doi.org/10.1007/s11356-020-11892-9

Aung T, Eun J-B (2021) Production and characterization of a novel beverage from laver (*Porphyra dentata*) through fermentation with kombucha consortium. Food Chem 350:129274. https://doi.org/10.1016/j.foodchem.2021.129274

Baghel RS (2023) Developments in seaweed biorefinery research: a comprehensive review. Chem Eng J 454:140177. https://doi.org/10.1016/j.cej.2022.140177

Baghel RS, Trivedi N, Gupta V, Neori A, Reddy CRK, Lali A, Jha B (2015) Biorefining of marine macroalgal biomass for production of biofuel and commodity chemicals. Green Chem 17(4):2436–2443. https://doi.org/10.1039/C4GC02532F

Baghel RS, Reddy CRK, Singh RP (2021) Seaweed-based cellulose: applications, and future perspectives. Carbohydr Polym 267:118241. https://doi.org/10.1016/j.carbpol.2021.118241

Bagheri B, Zarrintaj P, Surwase SS, Baheiraei N, Saeb MR, Mozafari M, Kim Y-C, Park OO (2019) Self-gelling electroactive hydrogels based on chitosan–aniline oligomers/agarose for neural tissue engineering with on-demand drug release. Colloids Surf B: Biointerfaces 184:110549. https://doi.org/10.1016/j.colsurfb.2019.110549

Balakrishinan B, Mohanty M, Umashankar P, Jayakrishnan A (2005) Evaluation of an in situ forming hydrogel wound dressing based on oxidized alginate and gelatin. Biomaterials 26(32):6335–6342. https://doi.org/10.1016/j.biomaterials.2005.04.012

Barceló-Villalobos M, Figueroa FL, Korbee N, Álvarez-Gómez F, Abreu MH (2017) Production of mycosporine-like amino acids from *Gracilaria vermiculophylla* (Rhodophyta) cultured through one year in an integrated multi-trophic aquaculture (IMTA) system. Mar Biotechnol 19:246–254. https://doi.org/10.1007/s10126-017-9746-8

Barralet JE, Wang L, Lawson M, Triffitt JT, Cooper PR, Shelton RM (2005) Comparison of bone marrow cell growth on 2D and 3D alginate hydrogels. J Mater Sci Mater Med 16(6):515–519. https://doi.org/10.1007/s10856-005-0526-z

Barsanti L, Coltelli P, Gualtieri P (2019) Paramylon treatment improves quality profile and drought resistance in *Solanum lycopersicum* L. cv. Micro-Tom. Agronomy 9(7):394. https://doi.org/10.3390/agronomy9070394

Bar-Shai N, Sharabani-Yosef O, Zollmann M, Lesman A, Golberg A (2021) Seaweed cellulose scaffolds derived from green macroalgae for tissue engineering. Sci Rep 11(1):11843. https://doi.org/10.1038/s41598-021-90903-2

Basavaraja PK, Yogendra ND, Zodape ST, Prakash R, Ghosh A (2018) Effect of seaweed sap as foliar spray on growth and yield of hybrid maize. J Plant Nutr 41(14):1851–1861. https://doi.org/10.1080/01904167.2018.1463381

Beata Łabowska M, Michalak I, Detyna J (2019) Methods of extraction, physicochemical properties of alginates and their applications in biomedical field – a review. Open Chem 17(1):738–762. https://doi.org/10.1515/chem-2019-0077

Beckett RP, van Staden J (1989) The effect of seaweed concentrate on the growth and yield of potassium stressed wheat. Plant Soil 116:29–36. https://doi.org/10.1007/BF02327254

Begum M, Bordoloi BC, Singha DD, Ojha NJ (2018) Role of seaweed extract on growth, yield and quality of some agricultural crops: a review. Agric Rev 39(4):321–326. https://doi.org/10.18805/ag.R-1838

Benítez García I, Dueñas Ledezma AK, Martínez Montaño E, Salazar Leyva JA, Carrera E, Osuna Ruiz I (2020) Identification and quantification of plant growth regulators and antioxidant compounds in aqueous extracts of *Padina durvillaei* and *Ulva lactuca*. Agronomy 10(6):866. https://doi.org/10.3390/agronomy10060866

Biglarian N, Rafe A, Shahidi S-A, Lorenzo JM (2022) Rheological, textural and structural properties of dairy cream as affected by some natural stabilizers. Chem Biol Technol Agric 9(1):96. https://doi.org/10.1186/s40538-022-00371-7

Bilal M, Iqbal HMN (2020) Marine seaweed polysaccharides-based engineered cues for the modern biomedical sector. Mar Drugs 18(1):7. https://doi.org/10.3390/md18010007

Bird MI, Wurster CM, de Paula Silva PH, Bass AM, de Nys R (2011) Algal biochar – production and properties. Bioresour Technol 102(2):1886–1891. https://doi.org/10.1016/j.biortech.2010.07.106

Birgersson PS, Oftebro M, Strand WI, Aarstad OA, Sætrom GI, Sletta H, Arlov Ø, Aachmann FL (2023) Sequential extraction and fractionation of four polysaccharides from cultivated brown algae *Saccharina latissima* and *Alaria esculenta*. Algal Res 69:102928. https://doi.org/10.1016/j.algal.2022.102928

Biris-Dorhoi ES, Michiu D, Pop CR, Rotar AM, Tofana M, Pop OL, Socaci SA, Farcas AC (2020) Macroalgae – a sustainable source of chemical compounds with biological activities. Nutrients 12(10). https://doi.org/10.3390/nu12103085

Bizzaro G, Vatland AK, Pampanin DM (2022) The One-Health approach in seaweed food production. Environ Int 158:106948. https://doi.org/10.1016/j.envint.2021.106948

Blunden G, Wildgoose PB (1977) The effects of aqueous seaweed extract and kinetin on potato yields. J Sci Food Agric 28(2):121–125. https://doi.org/10.1002/jsfa.2740280203

Brinza L, Dring MJ, Gavrilescu M (2007) Marine micro- and macro-algal species as biosorbents for heavy metals. Environ Eng Manag J 6(3):237–251

Cáceres PJ, Carlucci M J, Damonte EB, Matsuhiro B, Zúñiga EA (2000) Carrageenans from chilean samples of *Stenogramme interrupta* (Phyllophoraceae): structural analysis and biological activity. Phytochemistry 53(1):81–86. https://doi.org/10.1016/s0031-9422(99)00461-6

Campo VL, Kawano DF, da Silva DB, Carvalho I (2009) Carrageenans: biological properties, chemical modifications and structural analysis – a review. Carbohydr Polym 77(2):167–180. https://doi.org/10.1016/j.carbpol.2009.01.020

Casadesús A, Polo J, Munné-Bosch S (2019) Hormonal effects of an enzymatically hydrolyzed animal protein-based biostimulant (Pepton) in water-stressed tomato plants. Front Plant Sci 10:758. https://doi.org/10.3389/fpls.2019.00758

Cassani L, Lourenço-Lopes C, Barral-Martinez M, Chamorro F, Garcia-Perez P, Simal-Gandara J, Prieto MA (2022) Thermochemical characterization of eight seaweed species and evaluation of their potential use as an alternative for biofuel production and source of bioactive compounds. Int J Mol Sci 23(4):2355. https://doi.org/10.3390/ijms23042355

Castellanos-Barriga LG, Santacruz-Ruvalcaba F, Hernández-Carmona G, Ramírez-Briones E, Hernández-Herrera RM (2017) Effect of seaweed liquid extracts from *Ulva lactuca* on seedling growth of mung bean (*Vigna radiata*). J Appl Phycol 29(5):2479–2488. https://doi.org/10.1007/s10811-017-1082-x

Celikler S, Yildiz G, Vatan O, Bilaloglu R (2008) In vitro antigenotoxicity of *Ulva rigida* C. Agardh (Chlorophyceae) extract against induction of chromosome aberration, sister chromatid exchange and micronuclei by mutagenic agent MMC. Biomed Environ Sci 21(6):492–498. https://doi.org/10.1016/s0895-3988(09)60008-8

Celikler S, Tas S, Vatan O, Ziyanok-Ayvalik S, Yildiz G, Bilaloglu R (2009) Anti-hyperglycemic and antigenotoxic potential of *Ulva rigida* ethanolic extract in the experimental diabetes mellitus. Food Chem Toxicol 47(8):1837–1840. https://doi.org/10.1016/j.fct.2009.04.039

Chen D, Huang Y, Shen D, Zhou W, Ao J, Jiang Y, Huang Z, Hassan M, Fahd R (2019a) Effects of seaweed extracts on promoting growth and improving stress resistance in sugarcane. Asian Agric Res 11(4):69–76. https://doi.org/10.22004/ag.econ.292416

Chen X, Song L, Wang H, Liu S, Yu H, Wang X, Li R, Liu T, Li P (2019b) Partial characterization, the immune modulation and anticancer activities of sulfated polysaccharides from filamentous microalgae *Tribonema* sp. Molecules 24(2):322. https://doi.org/10.3390/molecules24020322

Chen Y, Li J, Huang Z, Su G, Li X, Sun Z, Qin Y (2020) Impact of short-term application of seaweed fertilizer on bacterial diversity and community structure, soil nitrogen contents, and plant growth in maize rhizosphere soil. Folia Microbiol 65(3):591–603. https://doi.org/10.1007/s12223-019-00766-4

Chen C-L, Song W-L, Sun L, Qin S, Ren C-G, Yang J-C, Feng D-W, Liu N, Yan J, Cui B-B, Zhong Z-H, Li Q-Q, Liu Z-D, Liu Z-Y (2022) Effect of seaweed extract supplement on rice rhizosphere bacterial community in tillering and heading stages. Agronomy 12(2):342. https://doi.org/10.3390/agronomy12020342

Chopin T, Cooper JA, Reid G, Cross S, Moore C (2012) Open-water integrated multi-trophic aquaculture: environmental biomitigation and economic diversification of fed aquaculture by extractive aquaculture. Rev Aquac 4(4):209–220. https://doi.org/10.1111/j.1753-5131.2012.01074.x

Choudhary B, Chauhan OP, Mishra A (2021) Edible seaweeds: a potential novel source of bioactive metabolites and nutraceuticals with human health benefits. Front Mar Sci 8:740054. https://doi.org/10.3389/fmars.2021.740054

Chrysargyris A, Xylia P, Anastasiou M, Pantelides I, Tzortzakis N (2018) Effects of *Ascophyllum nodosum* seaweed extracts on lettuce growth, physiology and fresh-cut salad storage under potassium deficiency. J Sci Food Agric 98(15):5861–5872. https://doi.org/10.1002/jsfa.9139

Chudasama NA, Sequeira RA, Moradiya K, Prasad K (2021) Seaweed polysaccharide based products and materials: an assessment on their production from a sustainability point of view. Molecules 26(9):2608. https://doi.org/10.3390/molecules26092608

Circuncisão AR, Catarino MD, Cardoso SM, Silva AMS (2018) Minerals from macroalgae origin: health benefits and risks for consumers. Mar Drugs 16(11):400. https://doi.org/10.3390/md16110400

Coppola G, Gaudio MT, Lopresto CG, Calabro V, Curcio S, Chakraborty S (2021) Bioplastic from renewable biomass: a facile solution for a greener environment. Earth Syst Environ 5(2):231–251. https://doi.org/10.1007/s41748-021-00208-7

Costa LS, Fidelis GP, Cordeiro SL, Oliveira RM, Sabry DA, Câmara RBG, Nobre LTDB, Costa MSSP, Almeida-Lima J, Farias EHC, Leite EL, Rocha HAO (2010) Biological activities of sulfated polysaccharides from tropical seaweeds. Biomed Pharmacother 64(1):21–28. https://doi.org/10.1016/j.biopha.2009.03.005

Costa CSD, Bertagnolli C, Boos A, da Silva MGC, Vieira MGA (2020) Application of a dealginated seaweed derivative for the simultaneous metal ions removal from real and synthetic effluents. J Water Process Eng 37:101546. https://doi.org/10.1016/j.jwpe.2020.101546

Costa OYA, Pijl A, Houbraken J, van Lith W, Kuramae EE (2023) Soil substrate source drives the microbes involved in the degradation of gelatin used as a biostimulant. Appl Soil Ecol 189:104906. https://doi.org/10.1016/j.apsoil.2023.104906

Crouch IJ, Beckett RP, van Staden J (1990) Effect of seaweed concentrate on the growth and mineral nutrition of nutrient-stressed lettuce. J Appl Phycol 2(3):269–272. https://doi.org/10.1007/BF02179784

Cumashi A, Ushakova NA, Preobrazhenskaya ME, D'Incecco A, Piccoli A, Totani L, Tinari N, Morozevich GE, Berman AE, Bilan MI, Usov AI, Ustyuzhanina NE, Grachev AA, Sanderson CJ, Kelly M, Rabinovich GA, Iacobelli S, Nifantiev NE (2007) A comparative study of the anti-inflammatory, anticoagulant, antiangiogenic, and antiadhesive activities of nine different fucoidans from brown seaweeds. Glycobiology 17(5):541–552. https://doi.org/10.1093/glycob/cwm014

D'Ayala G, Malinconico M, Laurienzo P (2008) Marine derived polysaccharides for biomedical applications: chemical modification approaches. Molecules 13(9):2069–2106. https://doi.org/10.3390/molecules13092069

da Silveira LCI, Mattos P, Mógor ÁF, Daros E, Bettini MO, Norrie J (2015) Effect of kelp extract on sugarcane plantlets biomass accumulation. Idesia 33(3):31–33. https://doi.org/10.4067/S0718-34292015000300005

Dalal A, Bourstein R, Haish N, Shenhar I, Wallach R, Moshelion M (2019) Dynamic physiological phenotyping of drought-stressed pepper plants treated with "productivity-enhancing" and "survivability-enhancing" biostimulants. Front Plant Sci 10:905. https://doi.org/10.3389/fpls.2019.00905

Dang B-T, Bui X-T, Tran DPH, Hao Ngo H, Nghiem LD, Hoang T-K-D, Nguyen P-T, Nguyen HH, Vo T-K-Q, Lin C, Andrew Lin KY, Varjani S (2022) Current application of algae derivatives for bioplastic production: a review. Bioresour Technol 347:126698. https://doi.org/10.1016/j.biortech.2022.126698

Davis KA, Yoo S, Shuler EW, Sherman BD, Lee S, Leem G (2021) Photocatalytic hydrogen evolution from biomass conversion. Nano Convergence 8:6. https://doi.org/10.1186/s40580-021-00256-9

de Avelar MHM, Efraim P (2020) Alginate/pectin cold-set gelation as a potential sustainable method for jelly candy production. LWT 123:109119. https://doi.org/10.1016/j.lwt.2020.109119

Debbarma J, Viji P, Rao BD, Ravishankar CN (2022) Seaweeds: potential applications of the aquatic vegetables to augment nutritional composition, texture, and health benefits of food and food products. In: Rao AR, Ravishankar GA (eds) Sustainable global resources of seaweeds, vol 2. Springer, Cham, pp 3–54. https://doi.org/10.1007/978-3-030-92174-3_1

De Saeger J, Van Praet S, Vereecke D, Park J, Jacques S, Han T, Depuydt S (2020) Toward the molecular understanding of the action mechanism of *Ascophyllum nodosum* extracts on plants. J Appl Phycol 32(1):573–597. https://doi.org/10.1007/s10811-019-01903-9

de Sosa LL, Navarro-Fernández CM, Panettieri M, Madejón P, Pérez-de-Mora A, Madejón E (2023) Application of seaweed and pruning residue as organic fertilizer to increase soil fertility and vine productivity. Soil Use Manag 39(2):794–804. https://doi.org/10.1111/sum.12882

de Sousa AM, Ayub RA, Viencz T, Botelho RV (2019) Fruit set and yield of apple trees cv. Gala treated with seaweed extract of *Ascophyllum nodosum* and thidiazuron. Rev Bras Frutic 41(1):e-072. https://doi.org/10.1590/0100-29452019072

de Souza MCR, Marques CT, Dore CMG, da Silva FRF, Rocha HAO, Leite EL (2007) Antioxidant activities of sulfated polysaccharides from brown and red seaweeds. J Appl Phycol 19(2):153–160. https://doi.org/10.1007/s10811-006-9121-z

Demir N, Dural B, Yildirim K (2006) Effect of seaweed suspensions on seed germination of tomato, pepper and aubergine. J Biol Sci 6(6):1130–1133. https://doi.org/10.3923/jbs.2006.1130.1133

Deng C, Lin R, Kang X, Wu B, O'Shea R, Murphy JD (2020) Improving gaseous biofuel yield from seaweed through a cascading circular bioenergy system integrating anaerobic digestion and pyrolysis. Renew Sustain Energ Rev 128:109895. https://doi.org/10.1016/j.rser.2020.109895

Deniz F, Ersanli ET (2018) An ecofriendly approach for bioremediation of contaminated water environment: potential contribution of a coastal seaweed community to environmental improvement. Int J Phytoremediation 20(3):256–263. https://doi.org/10.1080/15226514.2017.1374335

De-Oliveira JP, Bruni GP, Fabra MJ, da Rosa Zavareze E, López-Rubio A, Martínez-Sanz M (2019) Development of food packaging bioactive aerogels through the valorization of *Gelidium sesquipedale* seaweed. Food Hydrocoll 89:337–350. https://doi.org/10.1016/j.foodhyd.2018.10.047

Deolu-Ajayi AO, van der Meer IM, van der Werf A, Karlova R (2022) The power of seaweeds as plant biostimulants to boost crop production under abiotic stress. Plant Cell Environ 45(9):2537–2553. https://doi.org/10.1111/pce.14391

Detsch R, Sarker B, Zehnder T, Frank G, Boccaccini AR (2015) Advanced alginate-based hydrogels. Mater Today 18(10):590–591. https://doi.org/10.1016/j.mattod.2015.10.013

Di Mola I, Cozzolino E, Ottaiano L, Giordano M, Rouphael Y, Colla G, Mori M (2019) Effect of vegetal- and seaweed extract-based biostimulants on agronomical and leaf quality traits of plastic tunnel-grown baby lettuce under four regimes of nitrogen fertilization. Agronomy 9(10):571. https://doi.org/10.3390/agronomy9100571

Digruber T, Sass L, Cseri A, Paul K, Nagy AV, Remenyik J, Molnár I, Vass I, Toldi O, Gyuricza C, Dudits D (2018) Stimulation of energy willow biomass with triacontanol and seaweed extract. Ind Crop Prod 120:104–112. https://doi.org/10.1016/j.indcrop.2018.04.047

Ditta LA, Rao E, Provenzano F, Sánchez JL, Santonocito R, Passantino R, Costa MA, Sabatino MA, Dispenza C, Giacomazza D, San Biagio PL, Lapasin R (2020) Agarose/κ-carrageenan-based hydrogel film enriched with natural plant extracts for the treatment of cutaneous wounds. Int J Biol Macromol 164:2818–2830. https://doi.org/10.1016/j.ijbiomac.2020.08.170

do Rosário Rosa V, Farias dos Santos AL, Alves da Silva A, Vicentini Sab MP, Germino GH, Cardoso FB, de Almeida Silva M (2021) Increased soybean tolerance to water deficiency through biostimulant based on fulvic acids and *Ascophyllum nodosum* (L.) seaweed extract. Plant Physiol Biochem 158:228–243. https://doi.org/10.1016/j.plaphy.2020.11.008

Drozd NN, Tolstenkov AS, Makarov VA, Kuznetsova T, Zaporozhets TS, Shevchenko NN, Zvyagintseva TN (2006) Pharmacodynamic parameters of anticoagulants based on sulfated polysaccharides from marine algae. Bull Exp Biol Med 142(5):591–593. https://doi.org/10.1007/s10517-006-0426-3

Dwelle RB, Hurley PJ (1984) The effects of foliar application of cytokinins on potato yields in southeastern Idaho. Am Potato J 61(5):293–299. https://doi.org/10.1007/BF02854141

Dziugieł T, Wadas W (2020) Possibility of increasing early crop potato yield with foliar application of seaweed extracts and humic acids. J Cent Eur Agric 21(2):300–310. https://doi.org/10.5513/JCEA01/21.2.2576

Eggen T, Asp TN, Grave K, Hormazabal V (2011) Uptake and translocation of metformin, ciprofloxacin and narasin in forage- and crop plants. Chemosphere 85(1):26–33. https://doi.org/10.1016/j.chemosphere.2011.06.041

El Boukhari MEM, Barakate M, Bouhia Y, Lyamlouli K (2020) Trends in seaweed extract based biostimulants: manufacturing process and beneficial effect on soil-plant systems. Plants 9(3):359. https://doi.org/10.3390/plants9030359

El-Beltagi HS, Mohamed AA, Mohamed HI, Ramadan KMA, Barqawi AA, Mansour AT (2022) Phytochemical and potential properties of seaweeds and their recent applications: a review. Mar Drugs 20(6):342. https://doi.org/10.3390/md20060342

El-Miniawy SM, Ragab ME, Youssef SM, Metwally AA (2014) Influence of foliar spraying of seaweed extract on growth, yield and quality of strawberry plants. J Appl Sci Res 10(2):88–94

El-Naggar NE-A, Hamouda RA, Mousa IE, Abdel-Hamid MS, Rabei NH (2018) Biosorption optimization, characterization, immobilization and application of *Gelidium amansii* biomass for complete Pb^{2+} removal from aqueous solutions. Sci Rep 8:13456. https://doi.org/10.1038/s41598-018-31660-7

Eom S-H, Lee E-H, Park K, Kwon J-Y, Kim P-H, Jung W-K, Kim Y-M (2017) Eckol from *Eisenia bicyclis* inhibits inflammation through the Akt/NF-κB signaling in *Propionibacterium acnes*-induced human keratinocyte HaCaT cells. J Food Biochem 41(2):e12312. https://doi.org/10.1111/jfbc.12312

Espinosa-Antón AA, Zamora-Natera JF, Zarazúa-Villaseñor P, Santacruz-Ruvalcaba F, Sánchez-Hernández CV, Águila Alcántara E, Torres-Morán MI, Velasco-Ramírez AP, Hernández-Herrera RM (2023) Application of seaweed generates changes in the substrate and stimulates the growth of tomato plants. Plants 12(7):1520. https://doi.org/10.3390/plants12071520

Esserti S, Smaili A, Rifai LA, Koussa T, Makroum K, Belfaiza M, Kabil EM, Faize L, Burgos L, Alburquerque N, Faize M (2017) Protective effect of three brown seaweed extracts against fungal and bacterial diseases of tomato. J Appl Phycol 29(2):1081–1093. https://doi.org/10.1007/s10811-016-0996-z

Fabre E, Dias M, Costa M, Henriques B, Vale C, Lopes CB, Pinheiro-Torres J, Silva CM, Pereira E (2020) Negligible effect of potentially toxic elements and rare earth elements on mercury removal from contaminated waters by green, brown and red living marine macroalgae. Sci Total Environ 724:138133. https://doi.org/10.1016/j.scitotenv.2020.138133

Falcó I, Randazzo W, Sánchez G, López-Rubio A, Fabra MJ (2019) On the use of carrageenan matrices for the development of antiviral edible coatings of interest in berries. Food Hydrocoll 92:74–85. https://doi.org/10.1016/j.foodhyd.2019.01.039

Farghali M, Mohamed IMA, Osman AI, Rooney DW (2023) Seaweed for climate mitigation, wastewater treatment, bioenergy, bioplastic, biochar, food, pharmaceuticals, and cosmetics: a review. Environ Chem Lett 21(1):97–152. https://doi.org/10.1007/s10311-022-01520-y

Farshidfar N, Iravani S, Varma RS (2023) Alginate-based biomaterials in tissue engineering and regenerative medicine. Mar Drugs 21(3):189. https://doi.org/10.3390/md21030189

Fasahati P, Dickson R, Saffron CM, Woo HC, Liu JJ (2022) Seaweeds as a sustainable source of bioenergy: techno-economic and life cycle analyses of its biochemical conversion pathways. Renew Sustain Energ Rev 157:112011. https://doi.org/10.1016/j.rser.2021.112011

Feitosa de Vasconcelos AC (2020) Amelioration of drought stress on plants under biostimulant sources. In: Hossain A (ed) Plant stress physiology. InTech, Rijeka. https://doi.org/10.5772/intechopen.91975

Fernández-Segovia I, Lerma-García MJ, Fuentes A, Barat JM (2018) Characterization of Spanish powdered seaweeds: composition, antioxidant capacity and technological properties. Food Res Int 111:212–219. https://doi.org/10.1016/j.foodres.2018.05.037

Fernando IPS, Sanjeewa KKA, Samarakoon KW, Lee WW, Kim H-S, Kang N, Ranasinghe P, Lee H-S, Jeon Y-J (2017) A fucoidan fraction purified from *Chnoospora minima*; a potential inhibitor of LPS-induced inflammatory responses. Int J Biol Macromol 104:1185–1193. https://doi.org/10.1016/j.ijbiomac.2017.07.031

Fertah M (2017) Isolation and characterization of alginate from seaweed. In: Venkatesan J, Anil S, Kim S-K (eds) Seaweed polysaccharides. Elsevier, Amsterdam, pp 11–26. https://doi.org/10.1016/B978-0-12-809816-5.00002-5

Flores P, Pedreño MA, Almagro L, Hernández V, Fenoll J, Hellín P (2021) Increasing nutritional value of broccoli with seaweed extract and trilinolein. J Food Compos Anal 98:103834. https://doi.org/10.1016/j.jfca.2021.103834

Fornes F, Sánchez-Perales M, Guardiola JL (1995) Effect of a seaweed extract on citrus fruit maturation. Acta Hortic 379:75–82. https://doi.org/10.17660/ActaHortic.1995.379.6

Fornes F, Sánchez-Perales M, Guardiola JL (2002) Effect of a seaweed extract on the productivity of 'de Nules' clementine mandarin and Navelina orange. Bot Mar 45(5):486–489. https://doi.org/10.1515/BOT.2002.051

Fu XT, Kim SM (2010) Agarase: review of major sources, categories, purification method, enzyme characteristics and applications. Mar Drugs 8(1):200–218. https://doi.org/10.3390/md8010200

Fu CWF, Ho CW, Yong WTL, Abas F, Tan TB, Tan CP (2016) Extraction of phenolic antioxidants from four selected seaweeds obtained from Sabah. Int Food Res J 23(6):2363–2369

Future Market Insights (2023) Seaweed packaging market outlook. Delaware: Future Market Insights. https://www.futuremarketinsights.com/reports/seaweed-packaging-market. Assessed 31 July 2023

Gajc-Wolska J, Spiżewski T, Grabowska A (2013) The effect of seaweed extracts on the yield and quality parameters of broccoli (*Brassica oleracea* var. *cymosa* L.) in open field production. Acta Hortic 1009:83–90. https://doi.org/10.17660/ActaHortic.2013.1009.9

Ganesan AR, Shanmugam M (2020) Isolation of phycoerythrin from *Kappaphycus alvarezii*: a potential natural colourant in ice cream. J Appl Phycol 32(6):4221–4233. https://doi.org/10.1007/s10811-020-02214-0

Ganesan AR, Munisamy S, Bhat R (2018) Producing novel edible films from semi refined carrageenan (SRC) and ulvan polysaccharides for potential food applications. Int J Biol Macromol 112:1164–1170. https://doi.org/10.1016/j.ijbiomac.2018.02.089

Ganesan AR, Tiwari U, Rajauria G (2019) Seaweed nutraceuticals and their therapeutic role in disease prevention. Food Sci Human Wellness 8(3):252–263. https://doi.org/10.1016/j.fshw.2019.08.001

Garbowski T, Bar-Michalczyk D, Charazińska S, Grabowska-Polanowska B, Kowalczyk A, Lochyński P (2023) An overview of natural soil amendments in agriculture. Soil Tillage Res 225:105462. https://doi.org/10.1016/j.still.2022.105462

Ghosh A, Shankar T, Malik GC, Banerjee M, Ghosh A (2020) Effect of seaweed extracts on the growth, yield and nutrient uptake of black gram (*Vigna mungo* L.) in the red and lateritic belt of West Bengal. Int J Chem Stud 8(3):799–802. https://doi.org/10.22271/chemi.2020.v8.i3j.9300

Ghosh S, Sarkar T, Pati S, Kari ZA, Edinur HA, Chakraborty R (2022) Novel bioactive compounds from marine sources as a tool for functional food development. Front Mar Sci 9:832957. https://doi.org/10.3389/fmars.2022.832957

Gin KY-H, Tang Y-Z, Aziz MA (2002) Derivation and application of a new model for heavy metal biosorption by algae. Water Res 36(5):1313–1323. https://doi.org/10.1016/S0043-1354(01)00332-3

Godvin Sharmila V, Dinesh Kumar M, Pugazhendi A, Bajhaiya AK, Gugulothu P, Rajesh Banu J (2021) Biofuel production from macroalgae: present scenario and future scope. Bioengineered 12(2):9216–9238. https://doi.org/10.1080/21655979.2021.1996019

Gomathi R, Kohila S, Ramachandiran K (2017) Evaluating the effect of seaweed formulations on the quality and yield of sugarcane. Madras Agric J 104(4–6):161–165. https://doi.org/10.29321/MAJ2019.000213

Gómez-Guzmán M, Rodríguez-Nogales A, Algieri F, Gálvez J (2018) Potential role of seaweed polyphenols in cardiovascular-associated disorders. Mar Drugs 16(8):250. https://doi.org/10.3390/md16080250

Gomez-Zavaglia A, Prieto Lage MA, Jimenez-Lopez C, Mejuto JC, Simal-Gandara J (2019) The potential of seaweeds as a source of functional ingredients of prebiotic and antioxidant value. Antioxidants 8(9):406. https://doi.org/10.3390/antiox8090406

González-González MF, Ocampo-Alvarez H, Santacruz-Ruvalcaba F, Sánchez-Hernández CV, Casarrubias-Castillo K, Becerril-Espinosa A, Castañeda-Nava JJ, Hernández-Herrera RM (2020) Physiological, ecological, and biochemical implications in tomato plants of two plant biostimulants: arbuscular mycorrhizal fungi and seaweed extract. Front Plant Sci 11:999. https://doi.org/10.3389/fpls.2020.00999

Gordalina M, Pinheiro HM, Mateus M, da Fonseca MMR, Cesário MT (2021) Macroalgae as protein sources — a review on protein bioactivity, extraction, purification and characterization. Appl Sci 11(17):7969. https://doi.org/10.3390/app11177969

Grenha A, Gomes ME, Rodrigues M, Santo VE, Mano JF, Neves NM, Reis RL (2009) Development of new chitosan/carrageenan nanoparticles for drug delivery applications. J Biomed Mater Res A 92A(4):1265–1272. https://doi.org/10.1002/jbm.a.32466

Guidara M, Yaich H, Richel A, Blecker C, Boufi S, Attia H, Garna H (2019) Effects of extraction procedures and plasticizer concentration on the optical, thermal, structural and antioxidant properties of novel ulvan films. Int J Biol Macromol 135:647–658. https://doi.org/10.1016/j.ijbiomac.2019.05.196

Gunathilaka TL, Samarakoon KW, Ranasinghe P, Peiris LCD (2019) In-vitro antioxidant, hypoglycemic activity, and identification of bioactive compounds in phenol-rich extract from the marine red algae *Gracilaria edulis* (Gmelin) Silva. Molecules 24(20):3708. https://doi.org/10.3390/molecules24203708

Guzman-Puyol S, Russo D, Penna I, Ceseracciu L, Palazon F, Scarpellini A, Cingolani R, Bertorelli R, Bayer IS, Heredia-Guerrero JA, Athanassiou A (2017) Facile production of seaweed-based biomaterials with antioxidant and anti-inflammatory activities. Algal Res 27:1–11. https://doi.org/10.1016/j.algal.2017.08.015

Haider MW, Ayyub CM, Pervez MA, Asad HU, Manan A, Raza SA, Ashraf I (2012) Impact of foliar application of seaweed extract on growth, yield and quality of potato (*Solanum tuberosum* L.). Soil Environ 31(2):157–162

Hajar M (2009) Biosorption of cadmium from aqueous solution using dead biomass of brown alga *Sargassum* sp. Chem Eng Trans 17:1173–1178. https://doi.org/10.3303/CET0917196

Hamouda MM, Saad-Allah KM, Gad D (2022) Potential of seaweed extract on growth, physiological, cytological and biochemical parameters of wheat (*Triticum aestivum* L.) seedlings. J Soil Sci Plant Nutr 22(2):1818–1831. https://doi.org/10.1007/s42729-022-00774-3

Han J, Han X, Xue Z, Wang Q, Xia Y, Zhao Z (2021) An eco-friendly procedure for achieving high-yield carrageenan from *Hypnea cervicornis* suitable for wet spinning. J Appl Polym Sci 138(33):50833. https://doi.org/10.1002/app.50833

Hentati F, Tounsi L, Djomdi D, Pierre G, Delattre C, Ursu AV, Fendri I, Abdelkafi S, Michaud P (2020) Bioactive polysaccharides from seaweeds. Molecules 25(14):3152. https://doi.org/10.3390/molecules25143152

Hezaveh H, Muhamad II (2013) Modification and swelling kinetic study of kappa-carrageenan-based hydrogel for controlled release study. J Taiwan Inst Chem Eng 44(2):182–191. https://doi.org/10.1016/j.jtice.2012.10.011

Hidangmayum A, Sharma R (2017) Effect of different concentrations of commercial seaweed liquid extract of *Ascophyllum nodosum* as a plant bio stimulant on growth, yield and biochemical constituents of onion (*Allium cepa* L.). J Pharmacogn Phytochem 6(4):658–663

Hidayati S, Zulferiyenni N, Maulidia U, Satyajaya W, Hadi S (2021) Effect of glycerol concentration and carboxy methyl cellulose on biodegradable film characteristics of seaweed waste. Heliyon 7(8):E07799. https://doi.org/10.1016/j.heliyon.2021.e07799

Hong IK, Jeon H, Lee SB (2014) Comparison of red, brown and green seaweeds on enzymatic saccharification process. J Ind Eng Chem 20(5):2687–2691. https://doi.org/10.1016/j.jiec.2013.10.056

Hossain A, Senff P, Glaser M (2022) Lessons for coastal applications of IMTA as a way towards sustainable development: a review. Appl Sci 12(23):11920. https://doi.org/10.3390/app122311920

Hotchkiss S, Brooks M, Campbell R, Philp K, Trius A (2016) The use of carrageenan in food. In: Pereira L (ed) Carrageenans: sources and extraction methods, molecular structure, bioactive properties and health effects. Nova Science Publishers, New York, pp 229–243

Huang Y-C, Kuo T-H (2016) O-carboxymethyl chitosan/fucoidan nanoparticles increase cellular curcumin uptake. Food Hydrocoll 53:261–269. https://doi.org/10.1016/j.foodhyd.2015.02.006

Huang Y-C, Lam U-I (2011) Chitosan/Fucoidan pH sensitive nanoparticles for oral delivery system. J Chin Chem Soc 58(6):779–785. https://doi.org/10.1002/jccs.201190121

Husien S, Labena A, El-Belely EF, Mahmoud HM, Hamouda AS (2019) Adsorption studies of hexavalent chromium [Cr (VI)] on micro-scale biomass of *Sargassum dentifolium*, Seaweed. J Environ Chem Eng 7(6):103444. https://doi.org/10.1016/j.jece.2019.103444

Iliou K, Kikionis S, Ioannou E, Roussis V (2022) Marine biopolymers as bioactive functional ingredients of electrospun nanofibrous scaffolds for biomedical applications. Mar Drugs 20(5):314. https://doi.org/10.3390/md20050314

Ingle KN, Polikovsky M, Fenta MC, Ingle AS, Golberg A (2022) Integration of multitrophic aquaculture approach with marine energy projects for management and restoration of coastal ecosystems of India. Ecol Eng 176:106525. https://doi.org/10.1016/j.ecoleng.2021.106525

Jafari A, Farahani M, Sedighi M, Rabiee N, Savoji H (2022) Carrageenans for tissue engineering and regenerative medicine applications: a review. Carbohydr Polym 281:119045. https://doi.org/10.1016/j.carbpol.2021.119045

Janarthanan M, Senthil Kumar M (2018) Extraction of alginate from brown seaweeds and evolution of bioactive alginate film coated textile fabrics for wound healing application. J Ind Text 49(3):328–351. https://doi.org/10.1177/1528083718783331

Jaulneau V, Lafitte C, Corio-Costet M-F, Stadnik MJ, Salamagne S, Briand X, Esquerré-Tugayé M-T, Dumas B (2011) An *Ulva armoricana* extract protects plants against three powdery mildew pathogens. Eur J Plant Pathol 131(3):393–401. https://doi.org/10.1007/s10658-011-9816-0

Jayakody MM, Vanniarachchy MPG, Wijesekara I (2022) Seaweed derived alginate, agar, and carrageenan based edible coatings and films for the food industry: a review. J Food Meas Charact 16(2):1195–1227. https://doi.org/10.1007/s11694-021-01277-y

Jayakumar V, Govindaradjane S, Rajasimman M (2021) Efficient adsorptive removal of zinc by green marine macro alga *Caulerpa scalpelliformis* – characterization, optimization, modeling, isotherm, kinetic, thermodynamic, desorption and regeneration studies. Surf Interfaces 22:100798. https://doi.org/10.1016/j.surfin.2020.100798

Jayakumar V, Govindaradjane S, Rajamohan N, Rajasimman M (2022) Biosorption potential of brown algae, *Sargassum polycystum*, for the removal of toxic metals, cadmium and zinc. Environ Sci Pollut Res 29:41909–41922. https://doi.org/10.1007/s11356-021-15185-7

Jayaraj J, Wan A, Rahman M, Punja ZK (2008) Seaweed extract reduces foliar fungal diseases on carrot. Crop Prot 27(10):1360–1366. https://doi.org/10.1016/j.cropro.2008.05.005

Jayaraman J, Ali N (2015) Use of seaweed extracts for disease management of vegetable crops. In: Ganesan S, Vadivel K, Jayaraman J (eds) Sustainable crop disease management using natural products. CABI, Wallingford, pp 160–183. https://doi.org/10.1079/9781780643236.0160

Jayaraman J, Norrie J, Punja ZK (2011) Commercial extract from the brown seaweed *Ascophyllum nodosum* reduces fungal diseases in greenhouse cucumber. J Appl Phycol 23(3):353–361. https://doi.org/10.1007/s10811-010-9547-1

Jeannin I, Lescure J-C, Morot-Gaudry J-F (1991) The effects of aqueous seaweed sprays on the growth of maize. Bot Mar 34:469–473. https://doi.org/10.1515/botm.1991.34.6.469

Jensen S, Ólafsdóttir A, Einarsdóttir B, Hreggviðsson GÓ, Guðmundsson H, Jónsdóttir LB, Friðjónsson ÓH, Jónsdóttir R (2022) New wave of flavours – on new ways of developing and processing seaweed flavours. Int J Gastron Food Sci 29:100566. https://doi.org/10.1016/j.ijgfs.2022.100566

Jiang N, Li B, Wang X, Xu X, Liu X, Li W, Chang X, Li H, Qi H (2020) The antioxidant and antihyperlipidemic activities of phosphorylated polysaccharide from *Ulva pertusa*. Int J Biol Macromol 145:1059–1065. https://doi.org/10.1016/j.ijbiomac.2019.09.198

Jothinayagi N, Anbazhagan C (2009) Effect of seaweed liquid fertilizer of *Sargassum wightii* on the growth and biochemical characteristics of *Abelmoschus esculentus* (L.) Medikus. Recent Res Sci Technol 1(4):155–158

Ju J, Yang J, Zhang W, Wei Y, Yuan H, Tan Y (2023) Seaweed polysaccharide fibers: solution properties, processing and applications. J Mater Sci Technol 140:1–18. https://doi.org/10.1016/j.jmst.2022.07.053

Jumaidin R, Sapuan SM, Jawaid M, Ishak MR, Sahari J (2018) Seaweeds as renewable sources for biopolymers and its composites: a review. Curr Anal Chem 14(3):249–267. https://doi.org/10.2174/1573411013666171009164355

Kanagesan K, Abdulla R, Derman E, Sabullah MK, Govindan N, Gansau JA (2022) A sustainable approach to green algal bioplastics production from brown seaweeds of Sabah, Malaysia. J King Saud Univ Sci 34(7):102268. https://doi.org/10.1016/j.jksus.2022.102268

Kang JY, Khan MNA, Park NH, Cho JY, Lee MC, Fujii H, Hong YK (2008) Antipyretic, analgesic, and anti-inflammatory activities of the seaweed *Sargassum fulvellum* and *Sargassum thunbergii* in mice. J Ethnopharmacol 116(1):187–190. https://doi.org/10.1016/j.jep.2007.10.032

Kapur B, Sarıdaş MA, Çeliktopuz E, Kafkas E, Paydaş Kargı S (2018) Health and taste related compounds in strawberries under various irrigation regimes and bio-stimulant application. Food Chem 263:67–73. https://doi.org/10.1016/j.foodchem.2018.04.108

Karadeniz F, Kang K-H, Park JW, Park S-J, Kim S-K (2014) Anti-HIV-1 activity of phlorotannin derivative 8,4‴-dieckol from Korean brown alga *Ecklonia cava*. Biosci Biotechnol Biochem 78(7):1151–1158. https://doi.org/10.1080/09168451.2014.923282

Karthikeyan K, Shanmugam M (2016) Grain yield and functional properties of red gram applied with seaweed extract powder manufactured from *Kappaphycus alvarezii*. Int J Recent Adv Multidiscip Res 3(3):1353–1359

Kasim WA, Hamada EAM, Shams El-Din NG, Eskander SK (2015) Influence of seaweed extracts on the growth, some metabolic activities and yield of wheat grown under drought stress. Int J Agron Agric Res 7(2):173–189

Kelman D, Posner EK, McDermid KJ, Tabandera NK, Wright PR, Wright AD (2012) Antioxidant activity of Hawaiian marine algae. Mar Drugs 10(12):403–416. https://doi.org/10.3390/md10020403

Kevadiya BD, Joshi GV, Patel HS, Ingole PP, Mody HM, Bajaj HC (2009) Montmorillonite-alginate nanocomposites as a drug delivery system: intercalation and in vitro release of vitamin B_1 and vitamin B_6. J Biomater Appl 25(2):161–177. https://doi.org/10.1177/0885328209344003

Khalid S, Abbas M, Saeed F, Bader-Ul-Ain H, Suleria HAR (2018) Therapeutic potential of seaweed bioactive compounds. In: Maiti S (ed) Seaweed biomaterials. InTech, Rijeka. https://doi.org/10.5772/intechopen.74060

Khaliq T, Sohail M, Minhas MU, Ahmed Shah S, Jabeen N, Khan S, Hussain Z, Mahmood A, Kousar M, Rashid H (2022) Self-crosslinked chitosan/κ-carrageenan-based biomimetic membranes to combat diabetic burn wound infections. Int J Biol Macromol 197:157–168. https://doi.org/10.1016/j.ijbiomac.2021.12.100

Khan W, Rayirath UP, Subramanian S, Jithesh MN, Rayorath P, Mark Hodges D, Critchley AT, Craigie JS, Norrie J, Prithiviraj B (2009) Seaweed extracts as biostimulants of plant growth and development. J Plant Growth Regul 28(4):386–399. https://doi.org/10.1007/s00344-009-9103-x

Khan F, Jeong G-J, Khan MSA, Tabassum N, Kim Y-M (2022) Seaweed-derived phlorotannins: a review of multiple biological roles and action mechanisms. Mar Drugs 20(6):384. https://doi.org/10.3390/md20060384

Kholssi R, Lougraimzi H, Grina F, Lorentz JF, Silva I, Castaño-Sánchez O, Marks EAN (2022) Green agriculture: a review of the application of micro- and macroalgae and their impact on crop production on soil quality. J Soil Sci Plant Nutr 22(4):4627–4641. https://doi.org/10.1007/s42729-022-00944-3

Kim M-M, Kim S-K (2010) Effect of phloroglucinol on oxidative stress and inflammation. Food Chem Toxicol 48(10):2925–2933. https://doi.org/10.1016/j.fct.2010.07.029

Kim YJ, Park HG, Yang YL, Yoon Y, Kim S, Oh E (2005) Multifunctional drug delivery system using starch-alginate beads for controlled release. Biol Pharm Bull 28(2):394–397. https://doi.org/10.1248/bpb.28.394

Kim K-H, Kim Y-W, Kim HB, Lee BJ, Lee DS (2006) Anti-apoptotic activity of laminarin polysaccharides and their enzymatically hydrolyzed oligosaccharides from *Laminaria japonica*. Biotechnol Lett 28(6):439–446. https://doi.org/10.1007/s10529-005-6177-9

Kim SM, Kang K, Jeon J-S, Jho EH, Kim CY, Nho CW, Um B-H (2011) Isolation of phlorotannins from *Eisenia bicyclis* and their hepatoprotective effect against oxidative stress induced by *tert*-butyl hyperoxide. Appl Biochem Biotechnol 165:1296–1307. https://doi.org/10.1007/s12010-011-9347-3

Kiraci S (2018) Effects of seaweed and different farm manures on growth and yield of organic carrots. J Plant Nutr 41(6):716–721. https://doi.org/10.1080/01904167.2018.1425435

Klöck G, Frank H, Houben R, Zekorn T, Horcher A, Siebers U, Wöhrle M, Federlin K, Zimmermann U (1994) Production of purified alginates suitable for use in immunoisolated transplantation. Appl Microbiol Biotechnol 40(5):638–643. https://doi.org/10.1007/bf00173321

Kocira A, Świeca M, Kocira S, Złotek U, Jakubczyk A (2018) Enhancement of yield, nutritional and nutraceutical properties of two common bean cultivars following the application of seaweed extract (*Ecklonia maxima*). Saudi J Biol Sci 25(3):563–571. https://doi.org/10.1016/j.sjbs.2016.01.039

Kocira S, Szparaga A, Hara P, Treder K, Findura P, Bartoš P, Filip M (2020) Biochemical and economical effect of application biostimulants containing seaweed extracts and amino acids as an element of agroecological management of bean cultivation. Sci Rep 10:17759. https://doi.org/10.1038/s41598-020-74959-0

Koesling M, Kvadsheim NP, Halfdanarson J, Emblemsvåg J, Rebours C (2021) Environmental impacts of protein-production from farmed seaweed: comparison of possible scenarios in Norway. J Clean Prod 307:127301. https://doi.org/10.1016/j.jclepro.2021.127301

Kong C-S, Kim J-A, Yoon N-Y, Kim S-K (2009) Induction of apoptosis by phloroglucinol derivative from *Ecklonia cava* in MCF-7 human breast cancer cells. Food Chem Toxicol 47(7):1653–1658. https://doi.org/10.1016/j.fct.2009.04.013

Kostas ET, Adams JMM, Ruiz HA, Durán-Jiménez G, Lye GJ (2021) Macroalgal biorefinery concepts for the circular bioeconomy: a review on biotechnological developments and future perspectives. Renew Sustain Energ Rev 151:111553. https://doi.org/10.1016/j.rser.2021.111553

Kozlowska J, Pauter K, Sionkowska A (2018) Carrageenan-based hydrogels: effect of sorbitol and glycerin on the stability, swelling and mechanical properties. Polym Test 67:7–11. https://doi.org/10.1016/j.polymertesting.2018.02.016

Krylova NV, Ermakova SP, Lavrov VF, Leneva IA, Kompanets GG, Iunikhina OV, Nosik MN, Ebralidze LK, Falynskova IN, Silchenko AS, Zaporozhets TS (2020) The comparative analysis of antiviral activity of native and modified fucoidans from brown algae *Fucus evanescens* in vitro and in vivo. Mar Drugs 18(4):224. https://doi.org/10.3390/md18040224

Kumar JIN, Oommen C (2012) Removal of heavy metals by biosorption using freshwater alga *Spirogyra hyaline*. J Environ Biol 33:27–31

Kumar S, Jahan K, Verma A, Agarwal M, Chandraprakash C (2022) Agar-based composite films as effective biodegradable sound absorbers. ACS Sustain Chem Eng 10:26. https://doi.org/10.1021/acssuschemeng.2c00168

Kumoro AC, Johnny D, Alfilovita D (2016) Incorporation of microalgae and seaweed in instant fried wheat noodles manufacturing: nutrition and culinary properties study. Int Food Res J 23(2):715–722

Lakshmi DS, Sankaranarayanan S, Gajaria TK, Li G, Kujawski W, Kujawa J, Navia R (2020) A short review on the valorization of green seaweeds and ulvan: FEEDSTOCK for chemicals and biomaterials. Biomol Ther 10(7):991. https://doi.org/10.3390/biom10070991

Lalzawmliana V, Anand A, Mukherjee P, Chaudhuri S, Kundu B, Nandi SK, Thakur NL (2019) Marine organisms as a source of natural matrix for bone tissue engineering. Ceram Int 45(2):1469–1481. https://doi.org/10.1016/j.ceramint.2018.10.108

Lau TY, Vittal DF, Chew CSY, Yong WTL (2014) Antiproliferative potential of extracts from *Kappaphycus* seaweeds on HeLa cancer cell lines. Sains Malays 43(12):1895–1900

Lau S-E, Teo WFA, Teoh EY, Tan BC (2022) Microbiome engineering and plant biostimulants for sustainable crop improvement and mitigation of biotic and abiotic stresses. Discov Food 2(1):9. https://doi.org/10.1007/s44187-022-00009-5

Laurens LML, Lane M, Nelson RS (2020) Sustainable seaweed biotechnology solutions for carbon capture, composition, and deconstruction. Trends Biotechnol 38(11):1232–1244. https://doi.org/10.1016/j.tibtech.2020.03.015

Layek J, Das A, Ramkrushna GI, Ghosh A, Panwar AS, Krishnappa R, Ngachan SV (2016) Effect of seaweed sap on germination, growth and productivity of maize (*Zea mays*) in North Eastern Himalayas. Indian J Agron 61(3):354–359

Layek J, Das A, Idapuganti RG, Sarkar D, Ghosh A, Zodape ST, Lal R, Yadav GS, Panwar AS, Ngachan S, Meena RS (2018) Seaweed extract as organic bio-stimulant improves productivity and quality of rice in eastern Himalayas. J Appl Phycol 30(1):547–558. https://doi.org/10.1007/s10811-017-1225-0

Leadbeater DR, Bruce NC, Tonon T (2022) In silico identification of bacterial seaweed-degrading bioplastic producers. Microb Genomics 8(9):1–15. https://doi.org/10.1099/mgen.0.000866

Lee KY, Mooney DJ (2012) Alginate: properties and biomedical applications. Prog Polym Sci 37(1):106–126. https://doi.org/10.1016/j.progpolymsci.2011.06.003

Lee J-B, Hayashi K, Hayashi T, Sankawa U, Maeda M (1999) Antiviral activities against HSV-1, HCMV, and HIV-1 of rhamnan sulfate from *Monostroma latissimum*. Planta Med 65(5):439–441. https://doi.org/10.1055/s-2006-960804

Lee J-B, Hayashi K, Maeda M, Hayashi T (2004a) Antiherpetic activities of sulfated polysaccharides from green algae. Planta Med 70(9):813–817. https://doi.org/10.1055/s-2004-827228

Lee Y-S, Jung S-H, Lee S-H, Shin K-H (2004b) Effects of the extracts from the marine algae *Pelvetia siliquosa* on hyperlipidemia in rats. Korean J Pharmacogn 35(2):143–146

Lee YS, Shin KH, Kim B-K, Lee S (2004c) Anti-diabetic activities of fucosterol from *Pelvetia siliquosa*. Arch Pharm Res 27(11):1120–1122. https://doi.org/10.1007/BF02975115

Lee H-A, Kim I-H, Nam T-J (2015) Bioactive peptide from *Pyropia yezoensis* and its anti-inflammatory activities. Int J Mol Med 36(6):1701–1706. https://doi.org/10.3892/ijmm.2015.2386

Li B, Xu H, Wang X, Wan Y, Jiang N, Qi H, Liu X (2020) Antioxidant and antihyperlipidemic activities of high sulfate content purified polysaccharide from *Ulva pertusa*. Int J Biol Macromol 146:756–762. https://doi.org/10.1016/j.ijbiomac.2019.11.061

Liao YC, Chang CC, Nagarajan D, Chen CY, Chang JS (2021) Algae-derived hydrocolloids in foods: applications and health-related issues. Bioengineered 12(1):3787–3801. https://doi.org/10.1080/21655979.2021.1946359

Lian Y, Wang R, Zheng J, Chen W, Chang L, Li C, Yim SC (2023) Carbon sequestration assessment and analysis in the whole life cycle of seaweed. Environ Res Lett 18(7):074013. https://doi.org/10.1088/1748-9326/acdae9

Lim C, Yusoff S, Ng CG, Lim PE, Ching YC (2021) Bioplastic made from seaweed polysaccharides with green production methods. J Environ Chem Eng 9(5):105895. https://doi.org/10.1016/j.jece.2021.105895

Lins KOAL, Bezerra DP, Paula A, Alencar NMN, Douglas M, Torres VM, Lobo R, Pessoa C, de Moraes MO, Costa-Lotufo LV (2009) Antitumor properties of a sulfated polysaccharide from the red seaweed *Champia feldmannii* (Diaz-Pifferer). J Appl Toxicol 29(1):20–26. https://doi.org/10.1002/jat.1374

Liu H, Cheng J, Chen F, Hou F, Bai D, Xi P, Zeng Z (2014) Biomimetic and cell-mediated mineralization of hydroxyapatite by carrageenan functionalized graphene oxide. ACS Appl Mater Interfaces 6(5):3132–3140. https://doi.org/10.1021/am4057826

Liu Y, Qin Y, Bai R, Zhang X, Yuan L, Liu J (2019) Preparation of pH-sensitive and antioxidant packaging films based on κ-carrageenan and mulberry polyphenolic extract. Int J Biol Macromol 134:993–1001. https://doi.org/10.1016/j.ijbiomac.2019.05.175

Livshits MA, Khomyakova E, Evtushenko EG, Lazarev VN, Kulemin NA, Semina SE, Generozov EV, Govorun VM (2015) Isolation of exosomes by differential centrifugation: theoretical analysis of a commonly used protocol. Sci Rep 5(1):17319. https://doi.org/10.1038/srep17319

Lomartire S, Gonçalves AMM (2022) An overview of potential seaweed-derived bioactive compounds for pharmaceutical applications. Mar Drugs 20(2):141. https://doi.org/10.3390/md20020141

Lomartire S, Marques JC, Gonçalves AMM (2022) An overview of the alternative use of seaweeds to produce safe and sustainable bio-packaging. Appl Sci 12(6):3123. https://doi.org/10.3390/app12063123

Ma C, Song W, Yang J, Ren C, Du H, Tang T, Qin S, Liu Z, Cui H (2022) The role and mechanism of commercial macroalgae for soil conditioner and nutrient uptake catalyzer. Plant Growth Regul 97(3):455–476. https://doi.org/10.1007/s10725-022-00819-8

Malaguti D, Rombolà AD, Gerin M, Simoni G, Tagliavini M, Marangoni B (2002) Effect of seaweed extracts-based leaf sprays on the mineral status, yield and fruit quality of apple. Acta Hortic 594:357–359. https://doi.org/10.17660/ActaHortic.2002.594.44

Maliki IM, Misson M, Teoh PL, Rodrigues KF, Yong WTL (2022) Production of lectins from marine algae: current status, challenges, and opportunities for non-destructive extraction. Mar Drugs 20(2):102. https://doi.org/10.3390/md20020102

Mamede M, Cotas J, Bahcevandziev K, Pereira L (2023) Seaweed polysaccharides in agriculture: a next step towards sustainability. Appl Sci 13(11):6594. https://doi.org/10.3390/app13116594

Manandhar B, Wagle A, Seong SH, Paudel P, Kim HR, Jung HA, Choi JS (2019) Phlorotannins with potential anti-tyrosinase and antioxidant activity isolated from the marine seaweed *Ecklonia stolonifera*. Antioxidants 8(8):240. https://doi.org/10.3390/antiox8080240

Mao W, Li H, Li Y, Zhang H, Qi X, Sun H, Chen Y, Guo S (2009) Chemical characteristic and anticoagulant activity of the sulfated polysaccharide isolated from *Monostroma latissimum* (Chlorophyta). Int J Biol Macromol 44(1):70–74. https://doi.org/10.1016/j.ijbiomac.2008.10.003

Martinez MA, Becherucci ME (2022) Study of the potential use of the invasive marine algae *Undaria pinnatifida* in the preliminary development of a functional textile. J Ind Text 51(5_suppl):8127S–8141S. https://doi.org/10.1177/1528083720967067

Martynenko A, Shotton K, Astatkie T, Petrash G, Fowler C, Neily W, Critchley AT (2016) Thermal imaging of soybean response to drought stress: the effect of *Ascophyllum nodosum* seaweed extract. SpringerPlus 5:1393. https://doi.org/10.1186/s40064-016-3019-2

Mattner SW, Wite D, Riches DA, Porter IJ, Arioli T (2013) The effect of kelp extract on seedling establishment of broccoli on contrasting soil types in southern Victoria, Australia. Biol Agric Hortic 29(4):258–270. https://doi.org/10.1080/01448765.2013.830276

Mattner SW, Milinkovic M, Arioli T (2018) Increased growth response of strawberry roots to a commercial extract from *Durvillaea potatorum* and *Ascophyllum nodosum*. J Appl Phycol 30(5):2943–2951. https://doi.org/10.1007/s10811-017-1387-9

Matyash M, Despang F, Ikonomidou C, Gelinsky M (2014) Swelling and mechanical properties of alginate hydrogels with respect to promotion of neural growth. Tissue Eng Part C Methods 20(5):401–411. https://doi.org/10.1089/ten.tec.2013.0252

Mei C, Zhou S, Zhu L, Ming J, Zeng F, Xu R (2017) Antitumor effects of *Laminaria* extract fucoxanthin on lung cancer. Mar Drugs 15(2):39. https://doi.org/10.3390/md15020039

Mendes GS, Soares AR, Martins FO, de Albuquerque MCM, Costa SS, Yoneshigue-Valentin Y, Gestinari LMS, Santos N, Romanos MTV (2010) Antiviral activity of the green marine alga *Ulva fasciata* on the replication of human metapneumovirus. Rev Inst Med Trop Sao Paulo 52(1):3–10. https://doi.org/10.1590/S0036-46652010000100001

Meszka B, Bielenin A (2009) Bioproducts in control of strawberry *Verticillium* wilt. Phytopathologia 52:21–27

Michalak I (2018) Experimental processing of seaweeds for biofuels. WIREs Energy Environ 7(3):e288. https://doi.org/10.1002/wene.288

Michalak I, Baśladyńska S, Mokrzycki J, Rutkowski P (2019) Biochar from a freshwater macroalga as a potential biosorbent for wastewater treatment. Water 11(7):1390. https://doi.org/10.3390/w11071390

Milledge JJ, Smith B, Dyer PW, Harvey P (2014) Macroalgae-derived biofuel: a review of methods of energy extraction from seaweed biomass. Energies 7(11):7194–7222. https://doi.org/10.3390/en7117194

Millet JK, Séron K, Labitt RN, Danneels A, Palmer KE, Whittaker GR, Dubuisson J, Belouzard S (2016) Middle East respiratory syndrome coronavirus infection is inhibited by griffithsin. Antivir Res 133:1–8. https://doi.org/10.1016/j.antiviral.2016.07.011

Mise T, Ueda M, Yasumoto T (2011) Production of fucoxanthin-rich powder from *Cladosiphon okamuranus*. Adv J Food Sci Technol 3(1):73–76

Mishra A, Sahni S, Kumar S, Prasad BD (2020) Seaweed – an eco-friendly alternative of agrochemicals in sustainable agriculture. Curr J Appl Sci Technolol 39(27):71–78. https://doi.org/10.9734/cjast/2020/v39i2730921

Mohammed A, Gaduan A, Chaitram P, Pooran A, Lee KY, Ward K (2023) *Sargassum* inspired, optimized calcium alginate bioplastic composites for food packaging. Food Hydrocoll 135:108192. https://doi.org/10.1016/j.foodhyd.2022.108192

Mohy El-Din SM (2015) Utilization of seaweed extracts as bio-fertilizers to stimulate the growth of wheat seedlings. Egypt J Exp Biol (Bot) 11(1):31–39

Moino BP, Costa CSD, da Silva MGC, Vieira MGA (2017) Removal of nickel ions on residue of alginate extraction from *Sargassum filipendula* seaweed in packed bed. Can J Chem Eng 95(11):2120–2128. https://doi.org/10.1002/cjce.22859

Moral A, Aguado R, Castelló R, Tijero A, Ballesteros M (2019) Potential use of green alga *Ulva* sp. for papermaking. BioResources 14(3):6851–6862. https://doi.org/10.15376/biores.14.3.6851-6862

Mourao PAS (2004) Use of sulfated fucans as anticoagulant and antithrombotic agents: future perspectives. Curr Pharm Des 10(9):967–981. https://doi.org/10.2174/1381612043452730

Murtic S, Oljaca R, Smajic Murtic M, Vranac A, Akagic A, Civic H (2018) Cherry tomato productivity as influenced by liquid organic fertilizer under different growth conditions. J Cent Eur Agric 19(3):503–516. https://doi.org/10.5513/JCEA01/19.3.2168

Nabti E, Jha B, Hartmann A (2017) Impact of seaweeds on agricultural crop production as biofertilizer. Int J Environ Sci Technol 14(5):1119–1134. https://doi.org/10.1007/s13762-016-1202-1

Nagamine T, Hayakawa K, Kusakabe T, Hisashi T, Nakazato K, Hisanaga E, Iha M (2009) Inhibitory effect of fucoidan on Huh7 hepatoma cells through downregulation of CXCL12. Nutr Cancer 61(3):340–347. https://doi.org/10.1080/01635580802567133

Nakamura S, Nambu M, Ishizuka T, Hattori H, Kanatani Y, Takase B, Kishimoto S, Amano Y, Aoki H, Kiyosawa T, Ishihara M, Maehara T (2008) Effect of controlled release of fibroblast growth factor-2 from chitosan/fucoidan micro complex-hydrogel on in vitro and in vivo vascularization. J Biomed Mater Res A 85A(3):619–627. https://doi.org/10.1002/jbm.a.31563

Neto R, Marçal C, Queirós A, Abreu H, Silva A, Cardoso S (2018) Screening of *Ulva rigida*, *Gracilaria* sp., *Fucus vesiculosus* and *Saccharina latissima* as functional ingredients. Int J Mol Sci 19(10):2987. https://doi.org/10.3390/ijms19102987

Ngo D-H, Wijesekara I, Vo T-S, Van Ta Q, Kim S-K (2011) Marine food-derived functional ingredients as potential antioxidants in the food industry: an overview. Food Res Int 44(2):523–529. https://doi.org/10.1016/j.foodres.2010.12.030

Nishikawa E, da Silva MGC, Vieira MGA (2018) Cadmium biosorption by alginate extraction waste and process overview in Life Cycle Assessment context. J Clean Prod 178:166–175. https://doi.org/10.1016/j.jclepro.2018.01.025

Nogueira MT, Chica LR, Yamashita C, Nunes NSS, Moraes ICF, Branco CCZ, Branco IG (2022) Optimal conditions for alkaline treatment of alginate extraction from the brown seaweed *Sargassum cymosum* C. Agardh by response surface methodology. Appl Food Res 2(2):100141. https://doi.org/10.1016/j.afres.2022.100141

Norakma MN, Zaibunnisa AH, Wan Razarinah WAR (2021) The changes of phenolics profiles, amino acids and volatile compounds of fermented seaweed extracts obtained through microbial fermentation. Mater Today Proc 48(4):815–821. https://doi.org/10.1016/j.matpr.2021.02.366

Obluchinskaya ED, Pozharitskaya ON, Zakharov DV, Flisyuk EV, Terninko II, Generalova YE, Smekhova IE, Shikov AN (2022) The biochemical composition and antioxidant properties of *Fucus vesiculosus* from the Arctic region. Mar Drugs 20(3):193. https://doi.org/10.3390/md20030193

Otari SV, Jadhav JP (2021) Seaweed-based biodegradable biopolymers, composite, and blends with applications. In: Pant D, Bhatia SK, Patel AK, Giri A (eds) Bioremediation using weeds. Springer, Singapore, pp 121–149. https://doi.org/10.1007/978-981-33-6552-0_6

Ozanne H, Toumi H, Roubinet B, Landemarre L, Lespessailles E, Daniellou R, Cesaro A (2020) Laminarin effects, a β-(1,3)-glucan, on skin cell inflammation and oxidation. Cosmetics 7(3):66. https://doi.org/10.3390/cosmetics7030066

Pacheco-Quito E-M, Ruiz-Caro R, Veiga M-D (2020) Carrageenan: drug delivery systems and other biomedical applications. Mar Drugs 18(11):583. https://doi.org/10.3390/md18110583

Pal A, Dwivedi SK, Maurya PK, Kanwar P (2015) Effect of seaweed saps on growth, yield, nutrient uptake and economic improvement of maize (sweet corn). J Appl Nat Sci 7(2):970–975. https://doi.org/10.31018/jans.v7i2.716

Palmieri N, Forleo MB (2020) The potential of edible seaweed within the western diet. A segmentation of Italian consumers. Int J Gastron Food Sci 20:100202. https://doi.org/10.1016/j.ijgfs.2020.100202

Parađiković N, Vinković T, Vinković Vrček I, Žuntar I, Bojić M, Medić-Šarić M (2011) Effect of natural biostimulants on yield and nutritional quality: an example of sweet yellow pepper (*Capsicum annuum* L.) plants. J Sci Food Agric 91(12):2146–2152. https://doi.org/10.1002/jsfa.4431

Paul NA, Neveux N, Magnusson M, de Nys R (2014) Comparative production and nutritional value of "sea grapes" — the tropical green seaweeds *Caulerpa lentillifera* and *C. racemosa*. J Appl Phycol 26(4):1833–1844. https://doi.org/10.1007/s10811-013-0227-9

Peñalver R, Lorenzo JM, Ros G, Amarowicz R, Pateiro M, Nieto G (2020) Seaweeds as a functional ingredient for a healthy diet. Mar Drugs 18(6). https://doi.org/10.3390/md18060301

Perera KY, Sharma S, Pradhan D, Jaiswal AK, Jaiswal S (2021) Seaweed polysaccharide in food contact materials (active packaging, intelligent packaging, edible films, and coatings). Foods 10(9):2088. https://doi.org/10.3390/foods10092088

Pereira T, Barroso S, Mendes S, Amaral RA, Dias JR, Baptista T, Saraiva JA, Alves NM, Gil MM (2020) Optimization of phycobiliprotein pigments extraction from red algae *Gracilaria gracilis* for substitution of synthetic food colorants. Food Chem 321:126688. https://doi.org/10.1016/j.foodchem.2020.126688

Pham BN, Kang J-K, Lee C-G, Park S-J (2021) Removal of heavy metals (Cd^{2+}, Cu^{2+}, Ni^{2+}, Pb^{2+}) from aqueous solution using *Hizikia fusiformis* as an algae-based bioadsorbent. Appl Sci 11(18):8604. https://doi.org/10.3390/app11188604

Pliego-Cortés H, Bedoux G, Boulho R, Taupin L, Freile-Pelegrín Y, Bourgougnon N, Robledo D (2019) Stress tolerance and photoadaptation to solar radiation in *Rhodymenia pseudopalmata* (Rhodophyta) through mycosporine-like amino acids, phenolic compounds, and pigments in an Integrated Multi-Trophic Aquaculture system. Algal Res 41:101542. https://doi.org/10.1016/j.algal.2019.101542

Pobereżny J, Szczepanek M, Wszelaczyńska E, Prus P (2020) The quality of carrot after field biostimulant application and after storage. Sustainability 12(4):1386. https://doi.org/10.3390/su12041386

Ponce NMA, Pujol CA, Damonte EB, Flores ML, Stortz CA (2003) Fucoidans from the brown seaweed *Adenocystis utricularis*: extraction methods, antiviral activity and structural studies. Carbohydr Res 338(2):153–165. https://doi.org/10.1016/s0008-6215(02)00403-2

Pradhan B, Nayak R, Patra S, Bhuyan PP, Behera PK, Mandal AK, Behera C, Ki JS, Adhikary SP, MubarakAli D, Jena M (2022) A state-of-the-art review on fucoidan as an antiviral agent to combat viral infections. Carbohydr Polym 291:119551. https://doi.org/10.1016/j.carbpol.2022.119551

Pradhan B, Bhuyan PP, Ki JS (2023) Immunomodulatory, antioxidant, anticancer, and pharmacokinetic activity of ulvan, a seaweed-derived sulfated polysaccharide: an updated comprehensive review. Mar Drugs 21(5):300. https://doi.org/10.3390/md21050300

Prajapati A, Patel CK, Singh N, Jain SK, Chongtham SK, Maheshwari MN, Patel CR, Patel RN (2016) Evaluation of seaweed extract on growth and yield of potato. Environ Ecol 34(2):605–608

Pramanick B, Brahmachari K, Ghosh A (2013) Effect of seaweed saps on growth and yield improvement of green gram. Afr J Agric Res 8(13):1180–1186. https://doi.org/10.5897/AJAR12.1894

Pramanick B, Brahmachari K, Ghosh A, Zodape ST (2014) Effect of seaweed saps on growth and yield improvement of transplanted rice in old alluvial soil of West Bengal. Bangladesh J Bot 43(1):53–58. https://doi.org/10.3329/bjb.v43i1.19746

Prasedya ES, Kurniawan NSH, Kirana IAP, Ardiana N, Abidin AS, Ilhami BTK, Jupri A, Widyastuti S, Sunarpi H, Nikmatullah A (2022) Seaweed fertilizer prepared by EM-fermentation increases abundance of beneficial soil microbiome in paddy (*Oryzasativa* L.) during vegetative stage. Fermentation 8(2):46. https://doi.org/10.3390/fermentation8020046

Premalatha M, Dhasarathan P, Theriappan P (2011) Phytochemical characterization and antimicrobial efficiency of seaweed samples, *Ulva fasciata* and *Chaetomorpha antennina*. Int J Pharm Bio Sci 2(1):288–293

Qin Y, Jiang J, Zhao L, Zhang J, Wang F (2018) Applications of alginate as a functional food ingredient. In: Grumezescu AM, Holban AM (eds) Biopolymers for food design. Academic Press, London, pp 409–429. https://doi.org/10.1016/b978-0-12-811449-0.00013-x

Quitral V, Sepúlveda M, Gamero-Vega G, Jiménez P (2022) Seaweeds in bakery and farinaceous foods: a mini-review. Int J Gastron Food Sci 28:100403. https://doi.org/10.1016/j.ijgfs.2021.100403

Rajan MSA, Thriunavukkarasu R, Joseph J, Aruni W (2020) Effect of seaweed on seed germination and biochemical constituents of *Capsicum annuum*. Biocatal Agric Biotechnol 29:101761. https://doi.org/10.1016/j.bcab.2020.101761

Raju CAI, Anitha J, Kalyani RM, Satyanandam K, Jagadeesh P (2021) Sorption of cobalt using marine macro seaweed *Graciliaria corticata* red algae powder. Mater Today Proc 44(1):1816–1827. https://doi.org/10.1016/j.matpr.2020.12.009

Rathour RK, Mehta S, Sharma P, Bhatia RK, Bhatt AK (2021) Seaweed cultivation and its biobusiness status around the world. In: Pant D, Bhatia SK, Patel AK, Giri A (eds) Bioremediation using weeds. Springer, Singapore, pp 151–182. https://doi.org/10.1007/978-981-33-6552-0_7

Remya RR, Samrot AV, Kumar SS, Mohanavel V, Karthick A, Chinnaiyan VK, Umapathy D, Muhibbullah M (2022) Bioactive potential of brown algae. Adsorpt Sci Technol 2022:e9104835. https://doi.org/10.1155/2022/9104835

Rengasamy KRR, Kulkarni MG, Pendota SC, Van Staden J (2016) Enhancing growth, phytochemical constituents and aphid resistance capacity in cabbage with foliar application of eckol – a biologically active phenolic molecule from brown seaweed. New Biotechnol 33(2):273–279. https://doi.org/10.1016/j.nbt.2015.11.002

Rhein-Knudsen N, Meyer AS (2021) Chemistry, gelation, and enzymatic modification of seaweed food hydrocolloids. Trends Food Sci Technol 109:608–621. https://doi.org/10.1016/j.tifs.2021.01.052

Rioux L-E, Beaulieu L, Turgeon SL (2017) Seaweeds: a traditional ingredients for new gastronomic sensation. Food Hydrocoll 68:255–265. https://doi.org/10.1016/j.foodhyd.2017.02.005

Roberts DA, Paul NA, Dworjanyn SA, Bird MI, de Nys R (2015) Biochar from commercially cultivated seaweed for soil amelioration. Sci Rep 5:9665. https://doi.org/10.1038/srep09665

Romera E, González F, Ballester A, Blázquez ML, Muñoz JA (2007) Comparative study of biosorption of heavy metals using different types of algae. Bioresour Technol 98(17):3344–3353. https://doi.org/10.1016/j.biortech.2006.09.026

Roohinejad S, Koubaa M, Barba FJ, Saljoughian S, Amid M, Greiner R (2017) Application of seaweeds to develop new food products with enhanced shelf-life, quality and health-related beneficial properties. Food Res Int 99(3):1066–1083. https://doi.org/10.1016/j.foodres.2016.08.016

Rosa J, Lemos MFL, Crespo D, Nunes M, Freitas A, Ramos F, Pardal MÂ, Leston S (2020) Integrated multitrophic aquaculture systems – potential risks for food safety. Trends Food Sci Technol 96:79–90. https://doi.org/10.1016/j.tifs.2019.12.008

Rouphael Y, Giordano M, Cardarelli M, Cozzolino E, Mori M, Kyriacou MC, Bonini P, Colla G (2018) Plant- and seaweed-based extracts increase yield but differentially modulate nutritional quality of greenhouse spinach through biostimulant action. Agronomy 8(7):126. https://doi.org/10.3390/agronomy8070126

Rupert R, Rodrigues KF, Thien VY, Yong WTL (2022) Carrageenan from *Kappaphycus alvarezii* (Rhodophyta, Solieriaceae): metabolism, structure, production, and application. Front Plant Sci 13:859635. https://doi.org/10.3389/fpls.2022.859635

Ryu J, Park S-J, Kim I-H, Choi YH, Nam T-J (2014) Protective effect of porphyra-334 on UVA-induced photoaging in human skin fibroblasts. Int J Mol Med 34(3):796–803. https://doi.org/10.3892/ijmm.2014.1815

Saadaoui I, Rasheed R, Abdulrahman N, Bounnit T, Cherif M, Al Jabri H, Mraiche F (2020) Algae-derived bioactive compounds with anti-lung cancer potential. Mar Drugs 18(4):197. https://doi.org/10.3390/md18040197

Saha D, Bhattacharya S (2010) Hydrocolloids as thickening and gelling agents in food: a critical review. J Food Sci Technol 47(6):587–597. https://doi.org/10.1007/s13197-010-0162-6

Saldarriaga-Hernandez S, Hernandez-Vargas G, Iqbal HMN, Barceló D, Parra-Saldívar R (2020) Bioremediation potential of *Sargassum* sp. biomass to tackle pollution in coastal ecosystems: circular economy approach. Sci Total Environ 715:136978. https://doi.org/10.1016/j.scitotenv.2020.136978

Saleh RI, Kim M, Cha C (2021) Comprehensive enhancement of mechanical, water-repellent and antimicrobial properties of regenerated seaweed and plant-based paper with chitosan coating. Coatings 11(11):1384. https://doi.org/10.3390/coatings11111384

Salim BBM (2016) Influence of biochar and seaweed extract applications on growth, yield and mineral composition of wheat (*Triticum aestivum* L.) under sandy soil conditions. Ann Agric Sci 61(2):257–265. https://doi.org/10.1016/j.aoas.2016.06.001

Sani MNH, Yong JWH (2022) Harnessing synergistic biostimulatory processes: a plausible approach for enhanced crop growth and resilience in organic farming. Biology 11(2):41. https://doi.org/10.3390/biology11010041

Sanjivkumar M, Chandran MN, Suganya AM, Immanuel G (2020) Investigation on bio-properties and in-vivo antioxidant potential of carrageenans against alloxan induced oxidative stress in Wistar albino rats. Int J Biol Macromol 151:650–662. https://doi.org/10.1016/j.ijbiomac.2020.02.227

Santana I, Félix M, Guerrero A, Bengoechea C (2022) Processing and characterization of bioplastics from the invasive seaweed *Rugulopteryx okamurae*. Polymers 14(2):355. https://doi.org/10.3390/polym14020355

Sekkal M, Legrand P, Huvenne JP, Verdus MC (1993) The use of FTIR microspectrometry as a new tool for the identification in situ of polygalactanes in red seaweeds. J Mol Struct 294:227–230. https://doi.org/10.1016/0022-2860(93)80356-z

Şen M, Avcı EN (2005) Radiation synthesis of poly(N-vinyl-2-pyrrolidone)-κ-carrageenan hydrogels and their use in wound dressing applications. I. Preliminary laboratory tests. J Biomed Mater Res Part A 74A(2):187–196. https://doi.org/10.1002/jbm.a.30308

Senthilkumar R, Prasad DMR, Govindarajan L, Saravanakumar K, Prasad BSN (2019) Green alga-mediated treatment process for removal of zinc from synthetic solution and industrial effluent. Environ Technol 40(10):1262–1270. https://doi.org/10.1080/09593330.2017.1420696

Shafie MH, Kamal ML, Zulkiflee FF, Hasan S, Uyup NH, Abdullah S, Mohamed Hussin NA, Tan YC, Zafarina Z (2022) Application of carrageenan extract from red seaweed (Rhodophyta) in cosmetic products: a review. J Indian Chem Soc 99(9):100613. https://doi.org/10.1016/j.jics.2022.100613

Shah MT, Zodape ST, Chaudhary DR, Eswaran K, Chikara J (2013) Seaweed sap as an alternative liquid fertilizer for yield and quality improvement of wheat. J Plant Nutr 36(2):192–200. https://doi.org/10.1080/01904167.2012.737886

Shannon E, Abu-Ghannam N (2019) Seaweeds as nutraceuticals for health and nutrition. Phycologia 58(5):563–577. https://doi.org/10.1080/00318884.2019.1640533

Sharma S, Chen C, Khatri K, Rathore MS, Pandey SP (2019) *Gracilaria dura* extract confers drought tolerance in wheat by modulating abscisic acid homeostasis. Plant Physiol Biochem 136:143–154. https://doi.org/10.1016/j.plaphy.2019.01.015

Sharma S, Jaiswal AK, Duffy B, Jaiswal S (2020) Ferulic acid incorporated active films based on poly(lactide)/poly(butylene adipate-co-terephthalate) blend for food packaging. Food Packag Shelf Life 24:100491. https://doi.org/10.1016/j.fpsl.2020.100491

Shin D-H, Lee S-H, Huang Y, Lim HW, Lee Y, Jang K, Cho Y-W, Park SM, Kim D-D, Lim C-J (2018) Protective properties of geniposide against UV-B-induced photooxidative stress in human dermal fibroblasts. Pharm Biol 56(1):176–182. https://doi.org/10.1080/13880209.2018.1446029

Shukla PS, Shotton K, Norman E, Neily W, Critchley AT, Prithiviraj B (2018) Seaweed extract improve drought tolerance of soybean by regulating stress-response genes. AoB Plants 10(1):plx051. https://doi.org/10.1093/aobpla/plx051

Shukla PS, Mantin EG, Adil M, Bajpai S, Critchley AT, Prithiviraj B (2019) *Ascophyllum nodosum*-based biostimulants: sustainable applications in agriculture for the stimulation of plant growth, stress tolerance, and disease management. Front Plant Sci 10:655. https://doi.org/10.3389/fpls.2019.00655

Singh S, Singh MK, Pal SK, Perween S, Kumari J, Zodape ST, Ghosh A (2015) Seaweed sap as productivity booster of maize. Bioscan 10(3):1303–1305

Sivasankari S, Venkatesalu V, Anantharaj M, Chandrasekaran M (2006) Effect of seaweed extracts on the growth and biochemical constituents of *Vigna sinensis*. Bioresour Technol 97(14):1745–1751. https://doi.org/10.1016/j.biortech.2005.06.016

Song JH, Murphy RJ, Narayan R, Davies GBH (2009) Biodegradable and compostable alternatives to conventional plastics. Philos Trans R Soc B Biol Sci 364(1526):2127–2139. https://doi.org/10.1098/rstb.2008.0289

Sood A, Gupta A, Agrawal G (2021) Recent advances in polysaccharides based biomaterials for drug delivery and tissue engineering applications. Carbohydr Polym Technol Appl 2:100067. https://doi.org/10.1016/j.carpta.2021.100067

Soppelsa S, Kelderer M, Casera C, Bassi M, Robatscher P, Andreotti C (2018) Use of biostimulants for organic apple production: effects on tree growth, yield, and fruit quality at harvest and during storage. Front Plant Sci 9:1342. https://doi.org/10.3389/fpls.2018.01342

Soumya S, Sajesh KM, Jayakumar R, Nair SV, Chennazhi KP (2012) Development of a phytochemical scaffold for bone tissue engineering using *Cissus quadrangularis* extract. Carbohydr Polym 87(2):1787–1795. https://doi.org/10.1016/j.carbpol.2011.09.094

Souza RB, Frota AF, Silva J, Alves C, Neugebauer AZ, Pinteus S, Rodrigues JAG, Cordeiro EMS, de Almeida RR, Pedrosa R, Benevides NMB (2018) In vitro activities of kappa-carrageenan isolated from red marine alga *Hypnea musciformis*: antimicrobial, anticancer and neuroprotective potential. Int J Biol Macromol 112:1248–1256. https://doi.org/10.1016/j.ijbiomac.2018.02.029

Spinelli F, Fiori G, Noferini M, Sprocatti M, Costa G (2009) Perspectives on the use of a seaweed extract to moderate the negative effects of alternate bearing in apple trees. J Hortic Sci Biotechnol 84(6):131–137. https://doi.org/10.1080/14620316.2009.11512610

Stirk WA, Novák O, Strnad M, van Staden J (2003) Cytokinins in macroalgae. Plant Growth Regul 41(1):13–24. https://doi.org/10.1023/A:1027376507197

Sudhakar MP, Venkatnarayanan S, Dharani G (2022) Fabrication and characterization of bio-nanocomposite films using κ-carrageenan and *Kappaphycus alvarezii* seaweed for multiple industrial applications. Int J Biol Macromol 219:138–149. https://doi.org/10.1016/j.ijbiomac.2022.07.230

Sugiura Y, Matsuda K, Yamada Y, Nishikawa M, Shioya K, Katsuzaki H, Imai K, Amano H (2006a) Isolation of a new anti-allergic phlorotannin, phlorofucofuroeckol-B, from an edible brown alga, *Eisenia arborea*. Biosci Biotechnol Biochem 70(11):2807–2811. https://doi.org/10.1271/bbb.60417

Sugiura Y, Takeuchi Y, Kakinuma M, Amano H (2006b) Inhibitory effects of seaweeds on histamine release from rat basophile leukemia cells (RBL-2H3). Fish Sci 72(6):1286–1291. https://doi.org/10.1111/j.1444-2906.2006.01287.x

Sugiura Y, Matsuda K, Yamada Y, Nishikawa M, Shioya K, Katsuzaki H, Imai K, Amano H (2007) Anti-allergic phlorotannins from the edible brown alga, *Eisenia Arborea*. Food Sci Technol Res 13(1):54–60. https://doi.org/10.3136/fstr.13.54

Sugiura Y, Kinoshita Y, Misumi S, Yamatani H, Katsuzaki H, Hayashi Y, Murase N (2021) Correlation between the seasonal variations in phlorotannin content and the antiallergic effects of the brown alga *Ecklonia cava* subsp. *stolonifera*. Algal Res 58:102398. https://doi.org/10.1016/j.algal.2021.102398

Sultana F, Wahab MA, Nahiduzzaman M, Mohiuddin M, Iqbal MZ, Shakil A, Mamun A-A, Khan MSR, Wong L, Asaduzzaman M (2023) Seaweed farming for food and nutritional security, climate change mitigation and adaptation, and women empowerment: a review. Aquac Fish 8(5):463–480. https://doi.org/10.1016/j.aaf.2022.09.001

Sun J, Tan H (2013) Alginate-based biomaterials for regenerative medicine applications. Materials 6(4):1285–1309. https://doi.org/10.3390/ma6041285

Sun J, Norouzi O, Mašek O (2022) A state-of-the-art review on algae pyrolysis for bioenergy and biochar production. Bioresour Technol 346:126258. https://doi.org/10.1016/j.biortech.2021.126258

Sunarpi S, Jupri A, Kurnianingsih R, Julisaniah NI, Nikmatullah A (2010) Effect of seaweed extracts on growth and yield of rice plants. Nus Biosci 2(2):73–77. https://doi.org/10.13057/nusbiosci/n020204

Tabassum MR, Xia A, Murphy JD (2017) Potential of seaweed as a feedstock for renewable gaseous fuel production in Ireland. Renew Sustain Energ Rev 68:136–146. https://doi.org/10.1016/j.rser.2016.09.111

Takano R, Iwane-Sakata H, Hayashi K, Hara S, Hirase S (1998) Concurrence of agaroid and carrageenan chains in funoran from the red seaweed *Gloiopeltis furcata* post. et ruprecht (cryptonemiales, rhodophyta). Carbohydr Polym 35(1–2):81–87. https://doi.org/10.1016/s0144-8617(97)00230-0

Tanaka Y, Ashaari A, Mohamad FS, Lamit N (2020) Bioremediation potential of tropical seaweeds in aquaculture: low-salinity tolerance, phosphorus content, and production of UV-absorbing compounds. Aquaculture 518:734853. https://doi.org/10.1016/j.aquaculture.2019.734853

Tanna B, Choudhary B, Mishra A (2018) Metabolite profiling, antioxidant, scavenging and anti-proliferative activities of selected tropical green seaweeds reveal the nutraceutical potential of *Caulerpa* spp. Algal Res 36:96–105. https://doi.org/10.1016/j.algal.2018.10.019

Tannoury MY, Saab AM, Elia JM, Harb NN, Makhlouf HY, Diab-Assaf M (2017) In vitro cytotoxic activity of *Laurencia papillosa*, marine red algae from the Lebanese coast. J Appl Pharm Sci 7(3):175–179. https://doi.org/10.7324/JAPS.2017.70328

Teixeira-Guedes C, Gomes-Dias JS, Cunha SA, Pintado ME, Pereira RN, Teixeira JA, Rocha CMR (2023) Enzymatic approach for the extraction of bioactive fractions from red, green and brown seaweeds. Food Bioprod Process 138:25–39. https://doi.org/10.1016/j.fbp.2022.12.005

Tennakoon P, Chandika P, Yi M, Jung W-K (2023) Marine-derived biopolymers as potential bioplastics, an eco-friendly alternative. iScience 26(4):106404. https://doi.org/10.1016/j.isci.2023.106404

Thakur N, Salama ES, Sharma M, Sharma P, Sharma D, Li X (2022) Efficient utilization and management of seaweed biomass for biogas production. Mater Today Sustain 18:100120. https://doi.org/10.1016/j.mtsust.2022.100120

Thiruchelvi R, Das A, Sikdar E (2021) Bioplastics as better alternative to petro plastic. Mater Today Proc 37:1634–1639. https://doi.org/10.1016/j.matpr.2020.07.176

Thiviya P, Gamage A, Gama-Arachchige NS, Merah O, Madhujith T (2022) Seaweeds as a source of functional proteins. Phycology 2(2):216–243. https://doi.org/10.3390/phycology2020012

Toskas G, Heinemann S, Heinemann C, Cherif C, Hund R-D, Roussis V, Hanke T (2012) Ulvan and ulvan/chitosan polyelectrolyte nanofibrous membranes as a potential substrate material for the cultivation of osteoblasts. Carbohydr Polym 89(3):997–1002. https://doi.org/10.1016/j.carbpol.2012.04.045

Udayan A, Arumugam M, Pandey A (2017) Nutraceuticals from algae and cyanobacteria. In: Rastogi RP, Madamwar D, Pandey A (eds) Algal green chemistry. Elsevier, Amsterdam, pp 65–89. https://doi.org/10.1016/B978-0-444-63784-0.00004-7

Ullmann J, Grimm D (2021) Algae and their potential for a future bioeconomy, landless food production, and the socio-economic impact of an algae industry. Org Agric 11(2):261–267. https://doi.org/10.1007/s13165-020-00337-9

Vafajoo L, Cheraghi R, Dabbagh R, McKay G (2018) Removal of cobalt (II) ions from aqueous solutions utilizing the pre-treated 2-*Hypnea valentiae* algae: equilibrium, thermodynamic, and dynamic studies. Chem Eng J 331:39–47. https://doi.org/10.1016/j.cej.2017.08.019

Valencia RT, Acosta LS, Hernández MF, Rangel PP, Gallegos Robles MÁ, Antonio Cruz RC, Vázquez CV (2018) Effect of seaweed aqueous extracts and compost on vegetative growth, yield, and nutraceutical quality of cucumber (*Cucumis sativus* L.) fruit. Agronomy 8(11):264. https://doi.org/10.3390/agronomy8110264

Vasantharaja R, Abraham LS, Inbakandan D, Thirugnanasambandam R, Senthilvelan T, Jabeen SKA, Prakash P (2019) Influence of seaweed extracts on growth, phytochemical contents and antioxidant capacity of cowpea (*Vigna unguiculata* L. Walp). Biocatal Agric Biotechnol 17:589–594. https://doi.org/10.1016/j.bcab.2019.01.021

Venkatesan J, Bhatnagar I, Manivasagan P, Kang K-H, Kim S-K (2015) Alginate composites for bone tissue engineering: a review. Int J Biol Macromol 72:269–281. https://doi.org/10.1016/j.ijbiomac.2014.07.008

Venkatesan J, Anil S, Kim S-K, Shim M (2016) Seaweed polysaccharide-based nanoparticles: preparation and applications for drug delivery. Polymers 8(2):30. https://doi.org/10.3390/polym8020030

Verdy C, Branka J-E, Mekideche N (2011) Quantitative assessment of lactate and progerin production in normal human cutaneous cells during normal ageing: effect of an *Alaria esculenta* extract. Int J Cosmet Sci 33(5):462–466. https://doi.org/10.1111/j.1468-2494.2011.00656.x

Vieira BRC, Pintor AMA, Boaventura RAR, Botelho CMS, Santos SCR (2017) Arsenic removal from water using iron-coated seaweeds. J Environ Manag 192:224–233. https://doi.org/10.1016/j.jenvman.2017.01.054

Vinoth S, Sundari S, Gurusaravanan P, Sivakumar S, Siva G, Kumar GP, Manju M, Velmurugan K, Lakshminarayana V, Jayabalan N (2017) Evaluation of seagrass liquid extract on salt stress alleviation in tomato plants. Asian J Plant Sci 16(4):172–183. https://doi.org/10.3923/ajps.2017.172.183

Wadduwage J, Liu H, Egidi E, Singh BK, Macdonald CA (2023) Effects of biostimulant application on soil biological and physicochemical properties: a field study. J Sustain Agric Environ 2:285. https://doi.org/10.1002/sae2.12057

Wakhet S, Singh V, Sahoo S, Sagiri SS, Kulanthaivel S, Bhattacharya M, Kumar N, Banerjee I, Pal K (2015) Characterization of gelatin–agar based phase separated hydrogel, emulgel and bigel: a comparative study. J Mater Sci Mater Med 26(2):118. https://doi.org/10.1007/s10856-015-5434-2

Wan M-C, Qin W, Lei C, Li Q-H, Meng M, Fang M, Song W, Chen J-H, Tay F, Niu L-N (2021) Biomaterials from the sea: future building blocks for biomedical applications. Bioactive Mater 6(12):4255–4285. https://doi.org/10.1016/j.bioactmat.2021.04.028

Wang S, Li Y, White W, Lu J (2014) Extracts from New Zealand *Undaria pinnatifida* containing fucoxanthin as potential functional biomaterials against cancer in vitro. J Funct Biomater 5(2):29–42. https://doi.org/10.3390/jfb5020029

Wang Y, Fu F, Li J, Wang G, Wu M, Zhan J, Chen X, Mao Z (2016) Effects of seaweed fertilizer on the growth of *Malus hupehensis* Rehd. seedlings, soil enzyme activities and fungal communities under replant condition. Eur J Soil Biol 75:1–7. https://doi.org/10.1016/j.ejsobi.2016.04.003

Warkoyo W, Saati EA (2011) The solvent effectiveness on extraction process of seaweed pigment. Makara J Technol 15(1):5–8. https://doi.org/10.7454/mst.v15i1.850

Washington WS, Engleitner S, Boontjes G, Shanmuganathan N (1999) Effect of fungicides, seaweed extracts, tea tree oil, and fungal agents on fruit rot and yield in strawberry. Aust J Exp Agric 39(4):487–494. https://doi.org/10.1071/EA98164

Weill P, Plissonneau C, Legrand P, Rioux V, Thibault R (2020) May omega-3 fatty acid dietary supplementation help reduce severe complications in Covid-19 patients? Biochimie 179:275–280. https://doi.org/10.1016/j.biochi.2020.09.003

Wen Y, Wang S, Shi Z, Jin Y, Thomas J-B, Azzi ES, Franzén D, Gröndahl F, Martin A, Tang C, Mu W, Jönsson PG, Yang W (2022) Pyrolysis of engineered beach-cast seaweed: performances and life cycle assessment. Water Res 222:118875. https://doi.org/10.1016/j.watres.2022.118875

Wijesinghe WAJP, Jeon Y-J (2011) Biological activities and potential cosmeceutical applications of bioactive components from brown seaweeds: a review. Phytochem Rev 10(3):431–443. https://doi.org/10.1007/s11101-011-9214-4

Wu Z, Song Y, Shen H, Jiang X, Li B, Xiong Z (2019a) Biochar can mitigate methane emissions by improving methanotrophs for prolonged period in fertilized paddy soils. Environ Pollut 253:1038–1046. https://doi.org/10.1016/j.envpol.2019.07.073

Wu Z, Zhang X, Dong Y, Li B, Xiong Z (2019b) Biochar amendment reduced greenhouse gas intensities in the rice-wheat rotation system: six-year field observation and meta-analysis. Agric For Meteorol 278:107625. https://doi.org/10.1016/j.agrformet.2019.107625

Xu C, Leskovar DI (2015) Effects of *A. nodosum* seaweed extracts on spinach growth, physiology and nutrition value under drought stress. Sci Hortic 183:39–47. https://doi.org/10.1016/j.scienta.2014.12.004

Xu S-Y, Kan J, Hu Z, Liu Y, Du H, Pang G-C, Cheong K-L (2018) Quantification of neoagaro-oligosaccharide production through enzymatic hydrolysis and its anti-oxidant activities. Molecules 23(6):1354. https://doi.org/10.3390/molecules23061354

Yaich H, Amira AB, Abbes F, Bouaziz M, Besbes S, Richel A, Blecker C, Attia H, Garna H (2017) Effect of extraction procedures on structural, thermal and antioxidant properties of ulvan from *Ulva lactuca* collected in Monastir coast. Int J Biol Macromol 105:1430–1439. https://doi.org/10.1016/j.ijbiomac.2017.07.141

Yalçın S, Özyürek M (2018) Biosorption potential of two brown seaweeds in the removal of chromium. Water Sci Technol 78(12):2564–2576. https://doi.org/10.2166/wst.2019.007

Yang S, Wang H, Wang G, Wang J, Gu A, Xue X, Chen R (2023) Effects of seaweed-extract-based organic fertilizers on the levels of mineral elements, sugar-acid components and hormones in Fuji apples. Agronomy 13(4):969. https://doi.org/10.3390/agronomy13040969

Yantovski EI (2008) Solar energy conversion through seaweed photosynthesis and zero emissions power generation. Surf Eng Appl Electrochem 44(2):138–145. https://doi.org/10.3103/S1068375508020117

Yao Y, Wang X, Chen B, Zhang M, Ma J (2020) Seaweed extract improved yields, leaf photosynthesis, ripening time, and net returns of tomato (*Solanum lycopersicum* Mill.). ACS Omega 5(8):4242–4249. https://doi.org/10.1021/acsomega.9b04155

Yong WTL, Thien VY, Rupert R, Rodrigues KF (2022) Seaweed: a potential climate change solution. Renew Sustain Energ Rev 159:112222. https://doi.org/10.1016/j.rser.2022.112222

Yoon S-J, Pyun Y-R, Hwang J-K, Mourão PAS (2007) A sulfated fucan from the brown alga *Laminaria cichorioides* has mainly heparin cofactor II-dependent anticoagulant activity. Carbohydr Res 342(15):2326–2330. https://doi.org/10.1016/j.carres.2007.06.019

Yu X, Mou H, Zhang Y (2013) Research on alginate fiber and its application to textiles. Adv Mater Res 821–822:68–71. https://doi.org/10.4028/www.scientific.net/AMR.821-822.68

Yuan H, Song J (2005) Preparation, structural characterization and in vitro antitumor activity of kappa-carrageenan oligosaccharide fraction from *Kappaphycus striatum*. J Appl Phycol 17(1):7–13. https://doi.org/10.1007/s10811-005-5513-8

Yusuf R, Kristiansen P, Warwick N (2019) Effect of two seaweed products and equivalent mineral treatments on lettuce (*Lactuca sativa* L.) growth. J Agron 18(3):100–106. https://doi.org/10.3923/ja.2019.100.106

Yusuf R, Mahfudz, Muhardi, Syakur A, Mas'ud H, Latarang B, Kartika D, Kristiansen P (2021) Application of local seaweed extracts to increase the growth and yield eggplant (*Solanum melongena* L.). IOP Conf Ser Earth Environ Sci 681:012019. https://doi.org/10.1088/1755-1315/681/1/012019

Zahid PB (1999) Preparation of organic fertilizer from seaweed and its effect on the growth of some vegetable and ornamental plants. Pak J Biol Sci 2(4):1274–1277. https://doi.org/10.3923/pjbs.1999.1274.1277

Zhang H, Mao W, Fang F, Li H, Sun H, Chen Y, Qi X (2008) Chemical characteristics and anticoagulant activities of a sulfated polysaccharide and its fragments from *Monostroma latissimum*. Carbohydr Polym 71(3):428–434. https://doi.org/10.1016/j.carbpol.2007.06.012

Zhang C, Wang W, Wang C, Tian Q, Huang W, Yuan Z, Chen X (2009) Cytotoxicity of liver targeted drug-loaded alginate nanoparticles. Sci China Ser B Chem 52(9):1382–1387. https://doi.org/10.1007/s11426-009-0178-9

Zhang C, Wang W, Liu T, Wu Y, Guo H, Wang P, Tian Q, Wang Y, Yuan Z (2012) Doxorubicin-loaded glycyrrhetinic acid-modified alginate nanoparticles for liver tumor chemotherapy. Biomaterials 33(7):2187–2196. https://doi.org/10.1016/j.biomaterials.2011.11.045

Zhang K, Kim W-J, Park A-HA (2020) Alkaline thermal treatment of seaweed for high-purity hydrogen production with carbon capture and storage potential. Nat Commun 11:3783. https://doi.org/10.1038/s41467-020-17627-1

Zhao Y, Bourgougnon N, Lanoisellé J-L, Lendormi T (2022) Biofuel production from seaweeds: a comprehensive review. Energies 15(24):9395. https://doi.org/10.3390/en15249395

Zhao C, Liu G, Tan Q, Gao M, Chen G, Huang X, Xu X, Li L, Wang J, Zhang Y, Xu D (2023) Polysaccharide-based biopolymer hydrogels for heavy metal detection and adsorption. J Adv Res 44:53–70. https://doi.org/10.1016/j.jare.2022.04.005

Zhou G, Qiu X, Zhang J, Tao C (2019) Effects of seaweed fertilizer on enzyme activities, metabolic characteristics, and bacterial communities during maize straw composting. Bioresour Technol 286:121375. https://doi.org/10.1016/j.biortech.2019.121375

Zia KM, Tabasum S, Nasif M, Sultan N, Aslam N, Noreen A, Zuber M (2017) A review on synthesis, properties and applications of natural polymer based carrageenan blends and composites. Int J Biol Macromol 96:282–301. https://doi.org/10.1016/j.ijbiomac.2016.11.095

Znad H, Awual MR, Martini S (2022) The utilization of algae and seaweed biomass for bioremediation of heavy metal-contaminated wastewater. Molecules 27(4):1275. https://doi.org/10.3390/molecules27041275

Zumla A, Chan JFW, Azhar EI, Hui DSC, Yuen K-Y (2016) Coronaviruses — drug discovery and therapeutic options. Nat Rev Drug Discov 15(5):327–347. https://doi.org/10.1038/nrd.2015.37

Chapter 5
Harnessing the Potential of Seaweeds and Their Associated Bacteria for Polyhydroxyalkanoates Biosynthesis

Rima Gnaim, Razan Unis, Jallal Gnaim, and Alexander Golberg

Abstract Seaweed qualifies as a valuable and abundant resource for producing polyhydroxyalkanoates (PHAs), which are microbial polyesters that serve as renewable and biodegradable alternatives to conventional plastics derived from fossil fuels. PHAs are biosynthesized and accumulated intracellularly by various microorganisms, which act as carbon and energy storage compounds. However, the high production cost has hindered the widespread industrial production of PHAs. Extensive research has been dedicated to utilizing seaweeds and their by-products as affordable and sustainable substrates for producing PHAs in order to address this challenge. This chapter aims to provide a comprehensive overview from 2014 to 2023 of the recent advancements, challenges, and prospects associated with the microbial biosynthesis of PHAs using different seaweeds as carbon/nitrogen sources. Additionally, it examines in-depth seaweed-associated bacteria as novel bioreactors for PHA synthesis, shedding light on the symbiotic relationships between seaweed hosts and microorganisms that contribute to enhanced PHA yields with desirable properties.

Keywords Seaweed holobiont · Microbial fermentation · Polyhydroxyalkanoates · Bioplastics

---

R. Gnaim (✉) · R. Unis
Porter School of Environment and Earth Sciences, Tel Aviv University, Tel Aviv, Israel

The Triangle Regional R&D Center (TRDC), Kfar Qari, Israel
e-mail: rimagnaim@mail.tau.ac.il

J. Gnaim
The Triangle Regional R&D Center (TRDC), Kfar Qari, Israel

A. Golberg
Porter School of Environment and Earth Sciences, Tel Aviv University, Tel Aviv, Israel

N. Trivedi et al. (eds.), *Recent Advances in Seaweed Biotechnology*,
https://doi.org/10.1007/978-981-96-0519-4_5

5.1 Introduction

The increasing global concerns about the environmental impact of conventional plastics have sparked a quest for sustainable and environmentally friendly alternatives (Kumar et al. 2020). One promising alternative is polyhydroxyalkanoates (PHAs), a group of microbial polyesters offering renewable and biodegradable properties comparable to their petroleum-based counterparts (Dalton et al. 2022). Despite their immense potential, the high production cost of PHAs has hindered their widespread industrial adoption. However, researchers have increasingly turned their attention to seaweed, an abundant and valuable marine resource, as a promising substrate for PHA manufacturing (Dong et al. 2023). Seaweeds encompass diverse multicellular organisms that thrive in marine ecosystems worldwide. Their rapid growth rates, efficient nutrient uptake, and remarkable ability to absorb atmospheric carbon dioxide make them excellent candidates for PHA production (Han et al. 2014).

Additionally, seaweeds possess abundant polysaccharides like alginate, agar, and carrageenan, which can be hydrolyzed through chemical, hydrothermal, or enzymatic processes into fermentable sugars such as xylose, galactose, and glucose (Khambhaty and Reena 2023; Offei et al. 2018; Michalak and Chojnacka 2015; Lin et al. 2019). These sugars are one of the fundamental building blocks for PHA biosynthesis by numerous microorganisms, including bacteria, archaea, and yeasts (Koller et al. 2010). Seaweeds possess inherent versatility, allowing the production of PHA with a diverse array of chemical, mechanical, and physical properties. This expansive range of properties opens up a wide range of potential applications for PHA across various industries. Furthermore, seaweed harbors a wide range of bacteria on its surface, creating a diverse ecosystem. Among these bacteria, certain strains have the ability to synthesize PHAs (Rosenboom et al. 2022). Understanding the intricate interactions between seaweed hosts and their associated microbiota holds excellent promise for introducing novel bacterial-based bioreactors, improving PHA yields, and developing tailored bioprocessing strategies.

This chapter endeavors to present a thorough and encompassing analysis of the latest advancements, challenges, and prospects in the microbial biosynthesis of PHAs utilizing various seaweeds as carbon sources. Furthermore, it delves into a detailed examination of seaweed-associated bacteria as innovative bioreactors for PHA synthesis, illuminating the symbiotic associations between seaweed hosts and microorganisms that foster improved PHA production with desirable characteristics.

5.2 Seaweed Holobiont: The Interaction Between Seaweeds and Their Associated Bacteria

Seaweeds play a crucial and irreplaceable role within ecosystems by providing vital resources such as nourishment, protection, and suitable living environments for organisms at higher trophic levels. Moreover, seaweeds play a crucial role in the primary productivity of various ecosystems worldwide, including temperate, arctic, and tropical systems (Van der Loos et al. 2019). Apart from their ecological significance, seaweeds also hold considerable economic value due to their aquaculture and biotechnology applications (Leandro et al. 2019).

Seaweeds and their associated microbiota establish a cohesive and functional entity referred to as the "holobiont," characterized by intricate and symbiotic relationships (Wood et al. 2022). Contemporary understanding acknowledges the crucial involvement of microorganisms in the well-being, functionality, and development of seaweeds throughout various stages of their life cycle via mutualistic interactions. The dynamic interplay among seaweeds and their microbial counterparts encompasses multiple aspects, including settlement, reproduction, morphology, defense mechanisms, and nutrient exchange (Egan et al. 2013; Bengtsson et al. 2011). Evidently, the optimal functioning of seaweed heavily relies on its associated microbiota, emphasizing the influential role of this partnership in providing valuable ecosystem services. Notably, the seaweed surface is a highly conducive habitat for microbiotal proliferation, benefiting from abundant organic carbon, oxygen, and nutrients (Bengtsson et al. 2011; Lage et al. 2018).

Consequently, microbiota, such as nitrogen-fixing and heterotrophic bacteria, reciprocate by furnishing seaweed hosts with essential CO_2 and N_2 (Saha and Weinberger 2019). Moreover, they form a protective biofilm that shields seaweed from harmful factors, including UV radiation, heavy metals, and toxins. For example, seaweed can adsorb or chemically transform toxins into less harmful compounds (Wahl et al. 2020). Certain bacteria that inhabit seaweed play an active role in the host's defense mechanisms against unwanted microbes and biofouling. They outperform other bacteria by producing antifouling chemicals and antibiotics, thus safeguarding their host's health (Egan et al. 2013).

In addition, seaweed-associated bacteria have specific functions that involve the production of molecules (e.g., vitamins) that impact various aspects of seaweed, including growth, development, morphology, as well as zoospore release and settlement (Ihua et al. 2019).

This intricate relationship between seaweeds and their associated microbiota provides a fertile ground for exploring their potential to produce PHA polymers. These bacteria, residing on the seaweed surface, play a pivotal role in overall seaweed health and possess the enzymatic capabilities required for PHA synthesis (Han et al. 2014). By harnessing the metabolic potential of these seaweed-associated bacteria, researchers have begun to explore the potential utilization of seaweed as a viable substrate for PHA fabrication (Gnaim et al. 2021). By adopting this approach, we can tap into the abundant carbon content of seaweed and harness the mutually beneficial bond between seaweed and its associated bacteria. This approach has the potential to unlock a sustainable and renewable pathway for the synthesis of PHAs.

5.3 Polyhydroxyalkanoates

PHAs represent biodegradable biopolymers that microorganisms produce by incorporating lipid inclusions within their cellular structure. These lipid inclusions, formed in granular configurations, serve as a means of storing energy (Dalton et al. 2022). In 1926, the French microbiologist Maurice Lemoigne reported the discovery of PHAs in *Bacillus megaterium* (Keshavarz and Roy 2010). The specific class of compounds, referred to as poly(3-hydroxybutyrate) (PHB), was identified as part of this novel group of compounds (Kumar et al. 2020). PHAs are thermoplastic biodegradable polyesters derived from hydroxyalkanoic acids (Raza and Abid 2018). They are synthesized by a wide range of microorganisms, encompassing approximately 90 genera of bacteria (Aragosa et al. 2021). These bacteria can synthesize PHAs under varying environmental conditions, including oxygen-rich and oxygen-limited settings (Koller et al. 2022; Phanse et al. 2011). Bacteria store PHAs in the cytoplasm, forming granules between 0.2 and 0.5 μm in diameter (Fig. 5.1). Importantly, they synthesize PHAs when there is a limited supply of essential nutrients for bacterial growth in the cultivation medium (Chee et al. 2010). However, certain bacteria, such as *Escherichia coli*, can produce PHAs even under normal growth conditions without needing growth-limiting factors (Behera et al. 2023; Vicente et al. 2023).

5.3.1 *Biosynthesis of PHAs*

Four distinct categories of enzymes play a direct role in the pathways associated with the production and consumption of PHAs: PHA synthases, phasins (PhaPs), PHA depolymerases, and protein regulators such as PhaR (Mitra et al. 2022). Among these

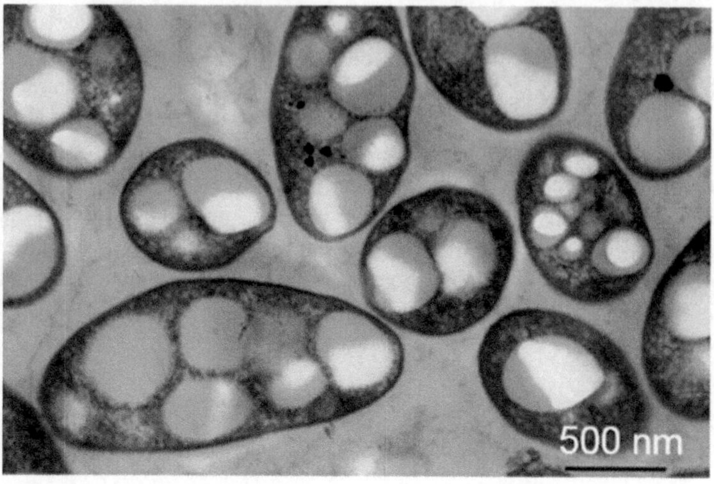

Fig. 5.1 Intracellular PHA granules. (Source: Giubilini et al. (2021))

enzymes, phasins are the most prevalent group and fulfill a fundamental function by facilitating PHA synthesis and regulating the exact localization of the granules to prevent their fusion. A study by York et al. (2001) revealed a clear linear correlation between the concentration of PhaPs and the production of PHAs. Researchers have identified three types of PHA synthases: type I, represented by *Ralstonia eutropha*, type II, represented by *Pseudomonas aeruginosa*, and type III, represented by *Allochromatium vinosum*. These enzymes' activity, manufacturing ability, and degradation rates display notable variations (Neoh et al. 2022). Figure 5.2 illustrates three distinct pathways for producing PHA (Meng and Chen 2018): Pathway I: The glycolysis pathway is utilized as the direct synthesis route, using acetyl-CoA, which is commonly employed for saccharides. Additionally, this pathway can integrate CO_2 and CH_4 via the serine pathway to produce acetyl-CoA. The resulting polymer predominantly consists of PHB. Pathway II: The predominant approach to generate medium-chain-length PHAs (*mcl*-PHAs) involves utilizing the acyl-CoA-dependent fatty acid (FA) β-oxidation pathway. Nonetheless, certain substrates like cyclohexanol and specific lactones can undergo conversion into *mcl*-PHAs. Another pathway, Pathway III, involves employing malonyl-CoA in the FA synthesis pathway for PHA production. All three pathways converge at the intermediate 3-hydroxyacyl-CoA, the PHA polymerase/synthase substrate. Pathway type I involves a series of three consecutive steps to produce *scl*-PHA: A pair of acetyl-CoA molecules undergo condensation into acetoacetyl-CoA catalyzed by a 3-ketothiolase enzyme, resulting in the release of CoA. The enzyme acetoacetyl-CoA reductase, which relies on NADPH, introduces a 3-hydroxy group to acetoacetyl-CoA. The PHA synthase then merges

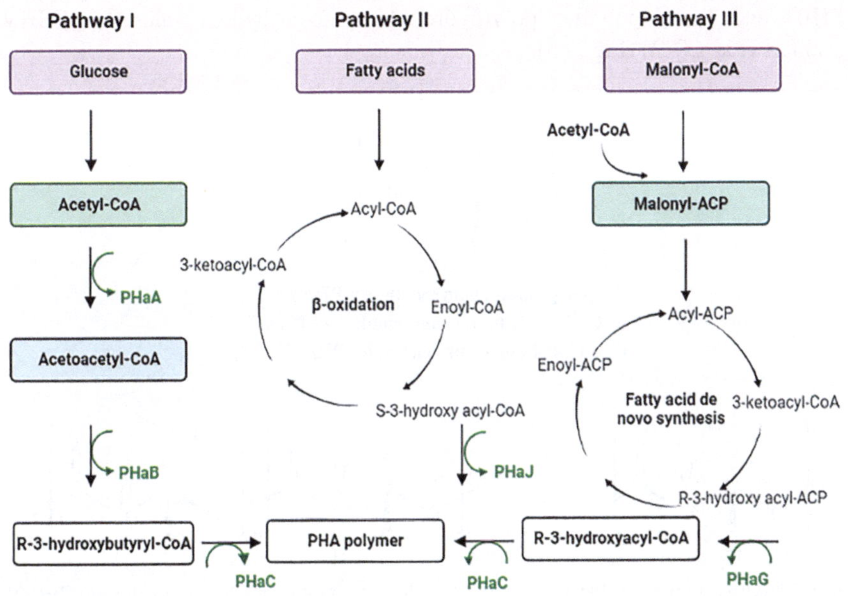

Fig. 5.2 Biosynthesis pathways of PHAs

the monomers acquired in the previous stage, elongating the chain by forming ester linkages. The genes phaA, phaB, and phaC encode these three enzymes, respectively (Meng and Chen 2018). A comprehensive exploration of the metabolic pathways for producing PHA shows the existence of numerous potential substrates, with varying convergences into these three PHA production pathways among different microorganisms. However, these possibilities are ultimately associated with the central metabolic processes, including β-oxidation, glycolysis, the Krebs cycle, the Calvin cycle, and catabolic pathways for amino acids (Tan et al. 2014; Bellini et al. 2022).

5.3.2 *Structure of PHAs*

PHAs exhibit three structural classes of short, medium, or long chain length (*scl*-, *mcl*-, and *lcl*-PHA, respectively) (Fig. 5.3). The side chains of these classes exhibit different numbers of carbon atoms (Wang et al. 2022). Different carbon chain lengths contribute to the variability in the mechanical attributes of PHAs. Their mechanical properties are influenced by the size of the carbon chain, leading to distinct characteristics. *scl*-PHA, composed of 3–5 carbon atoms, often exhibits rigidity and brittleness. In contrast, *mcl*-PHA (6–14 carbon chains) and *lcl*-PHA (more than 15 carbon in their chains) display enhanced flexibility and elastomeric properties (Dwivedi et al. 2020). This polymer's chemical structure divergence is directly associated with the microorganism species, the carbon source utilized, and the growth conditions employed (Alves et al. 2023). The two most commonly encountered variants within the PHA family are the homopolymer polyhydroxybutyrate (PHB) and the copolymer poly(hydroxybutyrate-co-hydroxyvalerate) (PHBV) (Ladhari et al. 2023) (Fig. 5.3).

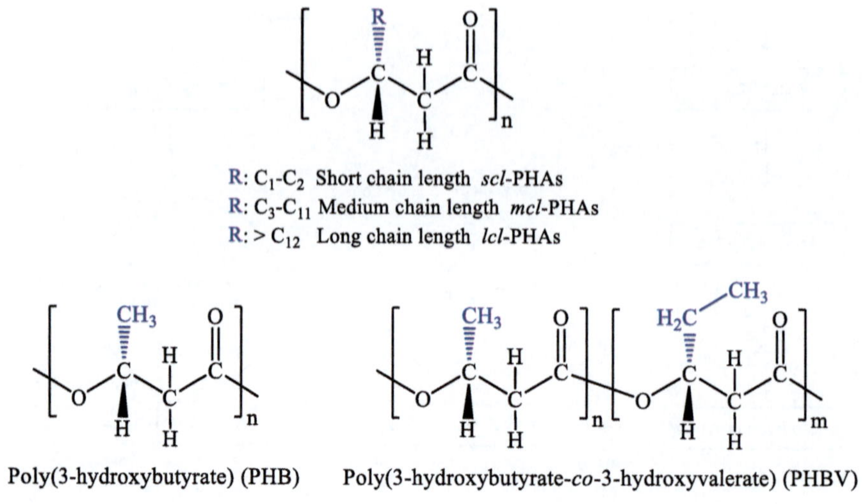

Fig. 5.3 Chemical structure of *scl*-PHAs, *mcl*-PHAs, *lcl*-PHAs, PHB, and PHBV

5.3.3 Properties of PHAs

Based on their structural composition, PHA can be divided into two main categories, namely homopolymers and copolymers. Homopolymers consist of a single type of monomer, while copolymers are composed of two or more different monomers.

The most common homopolymer is PHB, a highly crystalline polyester that shares similarities with conventional plastics like polypropylene (Turco et al. 2022; Altaee et al. 2016). However, compared to petroleum-based plastics, the homopolymer exhibits certain drawbacks, such as brittleness and a narrow processing window (Gigante et al. 2020). These limitations arise due to the high melting and low decomposition temperatures of PHB (Yeo et al. 2018). In addition, PHB exhibits characteristics comparable to polypropylene, including strong moisture resistance, and provides exceptional barrier properties to gases (Wu et al. 2021). In contrast, PHBV exhibits enhanced ductility compared to PHB, resulting in increased flexibility and reduced brittleness. In addition, it can be effectively processed at reduced temperatures (Eraslan et al. 2022). PHBV tends to have a slow crystallization rate and lower degrees of crystallinity (Muthuraj et al. 2021).

PHAs are only slightly soluble in water; therefore, they are suitable for applications where exposure to water or moisture is expected, such as in packaging materials. They are not easily degraded or deteriorated by exposure to sunlight, making them suitable for outdoor applications. Owing to their high density, PHAs readily settle and accumulate in aquatic sediments, enabling anaerobic biodegradation, where microorganisms in the sediment break down PHAs without oxygen (Raza and Abid 2018; Edaes and de Souza 2021).

5.3.4 Applications of PHAs

PHA polymers have attracted considerable interest due to their biocompatibility, biodegradability, and environmentally friendly properties (Gregory et al. 2022). These applications encompass a wide range of medical uses in soft tissue engineering, including cardiac patches for the repair of myocardial infarction (Dhania et al. 2022), stents and vessels (Jian and Zhu 2022), heart valves to create a scaffold or a framework that mimics the structure and function of natural heart valves (Guo et al. 2022), ear restoration, suture materials, and wound healing (Balcucho et al. 2023; Taylor et al. 2023). They also serve as excellent matrices for the in vitro proliferation of human cells and cartilage tissue (Kalia et al. 2023). The use of PHB composite materials for bone tissue regeneration has demonstrated remarkable outcomes in hard tissue engineering, with accelerated bone formation comparable to that of the implanted constructs. (Porter et al. 2013). PHAs can also be used as a matrix or coating material in drug delivery applications (Kumari and Singh 2023).

5.3.5 Costs Associated with PHA Production

Microbial plastics offer unique advantages over traditional petroleum-based plastics, positioning them as promising alternatives to conventional plastics derived from fossil fuels. In addition to their potential to mitigate environmental pollution, microbial plastics contribute to sustainable development and effective waste management practices, thereby offering long-term benefits for a more environmentally conscious future; nevertheless, the industrial scale of PHAs remains constrained by the significant expenses associated with the downstream processing, purification, and fermentation of these biopolymers (Dong et al. 2023; de Melo et al. 2022). Numerous approaches were applied to enhance the economic viability of PHA processes, including modifying carbon sources (Saratale et al. 2021) and bacterial strains (Wang et al. 2023). By manipulating production variables, it becomes possible to reduce production costs. Furthermore, incorporating enhanced properties into biopolymers, for instance, incorporating advancements in mechanical strength and environmental degradability (Kumar et al. 2023), can improve the performance and desirability of PHA, creating new possibilities for various industries and ecological initiatives (Gregory et al. 2022).

5.4 Seaweed-Associated Bacteria: A Potential Bioreactor for PHA Synthesis

Harnessing seaweed-associated bacteria for PHA production is a promising pathway toward sustainable bioplastics (Yamaguchi et al. 2019). Seaweed harbors a diverse microbial community that includes bacteria capable of synthesizing PHAs. These bacteria possess the necessary enzymes, such as PHA synthases, to convert carbon sources into PHAs and offer numerous advantages (Gnaim et al. 2023): (*i*) These bacteria are abundant and their diverse nature provides a wide range of potential microbial strains for screening and selection. This increases the likelihood of identifying strains that possess the desired traits and that are highly efficient in producing PHAs. This diversity also facilitates ongoing efforts to improve strains, enhancing overall efficiency and optimizing PHA productivity (Ramya and Manikandan 2017). (*ii*) Seaweed-associated bacteria are classified as marine halophilic bacteria, capable of surviving in challenging environments such as a high-salt alkaline medium with limited nutrients and low temperatures. This unique characteristic facilitates continuous open fermentation and minimizes the energy demand for equipment sterilization (Gunny et al. 2014; Ghosh et al. 2019). (*iii*) These bacteria align with sustainability goals because they are associated with seaweed, a rapidly renewable resource contributing to carbon sequestration. Consequently, the process becomes carbon-neutral, supporting environmentally friendly production practices. Therefore, these bacteria have garnered substantial interest as a promising and cost-effective approach for PHA production (Fig. 5.4) (Blandón et al. 2020).

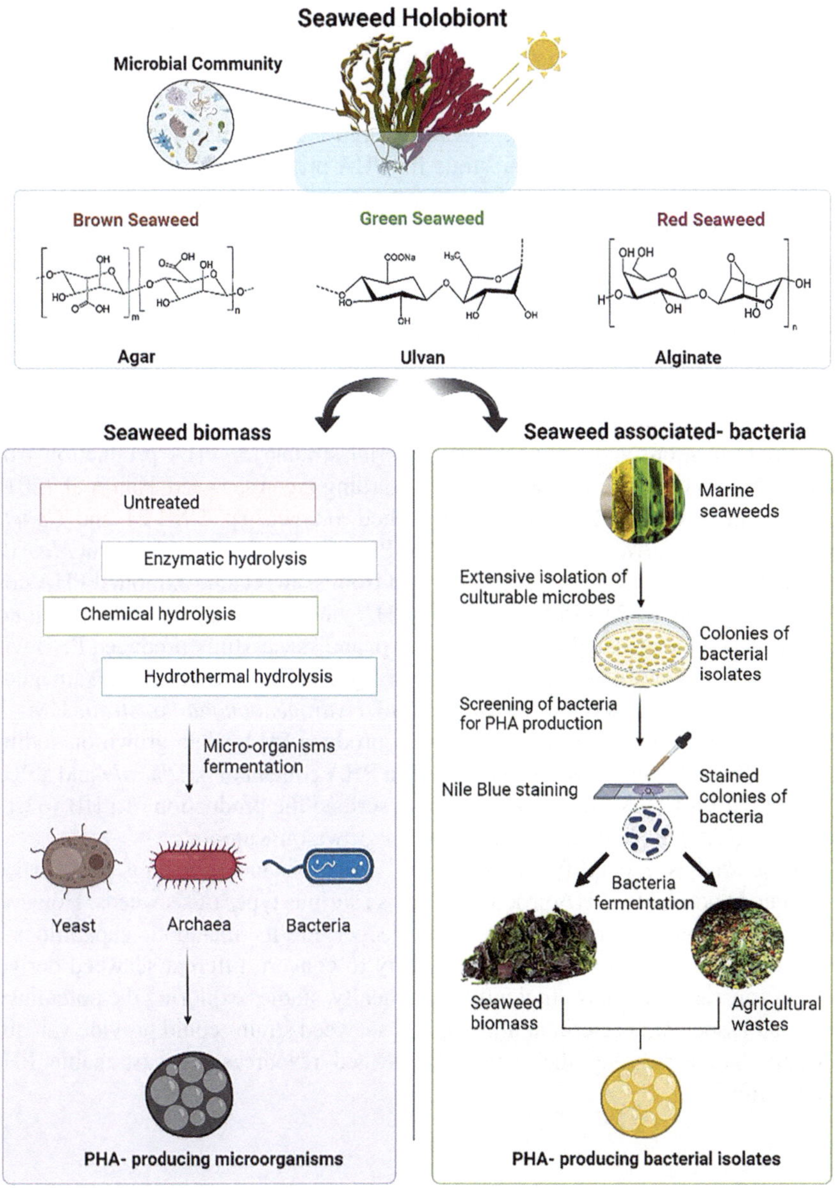

Fig. 5.4 Seaweeds and their associated bacteria for polyhydroxyalkanoate biosynthesis

5.4.1 Seaweed-Associated Bacteria Utilize Seaweed for PHA Production

Seaweeds, including red, green, and brown types, are a highly abundant marine resource and serve as an ideal substrate for PHA production due to their rich composition of polysaccharides and other complex carbon compounds (Table 5.1) (Sawant et al. 2016). These polysaccharides can be utilized by the seaweed bacterial isolates, which possess the enzymatic machinery to convert these carbon sources into PHAs efficiently (Han et al. 2014). The isolation and cultivation of seaweed bacterial isolates have shown great potential in unlocking the PHA-producing capabilities of these microorganisms. Researchers have identified bacterial strains with high PHA production yields and desirable properties through screening and selection processes (Moriya et al. 2020). However, few studies have reported the isolation, identification, and screening of bacterial strains for PHA production from different types of seaweed. For example, regarding green seaweed, Han et al. (2014) and Gnaim et al. (2021) isolated and studied *Massilia* sp. UMI-21 and *Cobetia amphilecti* from *Ulva* sp., respectively, for PHA production. The bacteria *Massilia* sp. UMI-21 was cultivated on starch derived from seaweed and exhibited PHA content ranging from 27.2 to 45.5% w/v and a PHA yield of 0.9–1.2 g L^{-1}. *C. amphilecti* was grown on the acid hydrolysate of *Ulva* sp. and successfully produced PHB with a content of 12% w/v and a PHB yield of 0.167 g L^{-1} (Gnaim et al. 2021). Yamaguchi et al. (2019) have described the isolation of *Hydrogenophaga* sp. *strain* UMI-18 strain from *Sargassum* sp.; it was found to produce PHA when grown on sodium alginate derived from brown seaweed, with a PHA content of 43.2% w/v and a PHA yield of 0.75 g L^{-1}. Moriya et al. (2020) described the production of PHB (62.1% w/v, 3.11 g L^{-1}) by *Cobetia* sp. strains when grown on Kumbo.

These studies exemplify the successful isolation and screening of seaweed-associated bacteria for PHA production across various types of seaweeds. However, further investigations are needed to delve deeper into the metabolic capabilities of seaweed-associated bacteria and their ability to convert different seaweed-derived carbon sources into PHAs efficiently. Specifically, studies exploring the potential of seaweed-associated bacteria in utilizing red seaweed strains could provide valuable insights into expanding the range of seaweed resources for sustainable PHA production.

5.4.2 Seaweed-Associated Bacteria Utilize Wastes for PHA Production

In addition to using seaweed as a substrate for PHA production by seaweed-associated bacteria, the isolated bacteria also have emerged as a promising means of producing PHA by utilizing agricultural waste (Table 5.2) (Ojha and Das 2020). Crop residues, food processing by-products, and animal manure are abundant and

Table 5.1 Seaweeds and their extracts as a substrate for PHA production

Seaweed type	Seaweed sp.	Bacterial strain	Carbon source	Fermentation conditions	Recovery/analysis method	PHA type	CDW (g/L)	PHA content (%)	Yield (g/L)	References
Green	*Ulva* sp.	*Haloferax mediterranei*	Subcritical hydrolysate of *Ulva* (8% load), at 180 °C, for 30 min	Hv-YPC, at 42 °C, for 72 h, 180 rpm	Solvent extraction, chloroform	PHBV	3.8	55	2.2	Ghosh et al. (2019)
	Ulva sp.	*Haloferax mediterranei*	Seaweed hydrolysis (5% load) at 170 °C, for 20 min, 38 g L^{-1} salinity	Hv-YPC, at 42 °C, for 72 h, 100 rpm	Chemical digestion, SDS	PHBV	3.9	53	2.1	Ghosh et al. (2021)
	Ulva sp.	*Cobetia amphilecti* (isolated)	Acid (2% v/v) hydrolysate of *Ulva* (9.5% load) autoclaved at 121 °C for 30 min	Marine Broth medium, at 32 °C, for 4 days, 90 rpm under aerobic conditions	Solvent extraction, chloroform	PHB	1.4	12	0.167	Gnaim et al. (2020)
	Ulva sp.	*Massilia* sp. UMI-21 (isolated)	Starch, 2% w/v	M.S. medium, at 30 °C, for 72 h, 150 rpm	Solvent extraction, chloroform	PHB	3.31–8.29	27.2–45.5	0.9–1.2	Han et al. (2014)

(continued)

Table 5.1 (continuued)

Seaweed type	Seaweed sp.	Bacterial strain	Carbon source	Fermentation conditions	Recovery/analysis method	PHA type	CDW (g/L)	PHA content (%)	Yield (g/L)	References
Red	*Kappaphycus alvarezii*	*Halomonas hydrothermalis*	Crude levulinic acid and biodiesel residues	Zobell marine broth, at 37 °C, for 96 h, 120 rpm	Solvent extraction, chloroform	PHBV	0.54–1.55	61.3–74.1	0.4–1.07	Bera et al. (2015)
	Gelidium amansii	*Bacillus megaterium* strains	*G. amansii* treated with H_2SO_4 at 121 °C for 60 min	LB medium with acid-treated *G. amansii,* at 30 °C, for 4 days, 200 rpm	Quantified by GCMS	PHB	3.1–4.1	30–55	0.93–2.56	Alkotaini et al. (2015)
	Gelidium amansii	*Bacillus megaterium* strains	Untreated *G. Amansii*	LB medium, Solidified seaweed G. amansii medium, at 30 °C, for 4 days, 200 rpm	Quantified by GCMS	PHB	1.0–2.0	19–28	0.19–0.56	Alkotaini et al. (2015)
	Gelidium amansii	*Bacillus megaterium* KCTC 2194	10% (w/v) acid-treated G. *Amansii*		Quantified using GCMS	PHB	3.3–10.1	51–54	1.68–4.04	Alkotaini et al. (2016)
	Gelidium sesquipedale	*Halomonas boliviensis*	Acid and enzymatic hydrolysis of *G. Amansii* (2%)	Salt-rich medium, at 30 °C, for 90 h, 170 rpm	Solvent extraction, chloroform	PHB	8.3	41	3.4	Tůma et al. (2020)
	Red algae (Rhodophyta) such as *Gelidium* and *Gracilaria*	*Bacillus cereus* & *Saccharophagus degradans*	Xylan or agarose	Minimal medium, at 30 °C, for 100 h, 200 rpm	Solvent extraction, chloroform	PHB	1.00–0.94	34.5	0.30–0.33	Sawant et al. (2017)
	Gelidium amansii	S. *degradans* or co-culture of *S. degradans* & *Bacillus cereus*	Untreated *Gelidium amansii*	Minimal medium supplemented with *G. amansii,* at 30 °C, pH 7.5, for 195 h	Analyzed by GCMS	PHB	0.74–1.47	17–27	0.25–0.42	Sawant et al. (2018)

	Corallina mediterranea	*Halomonas pacifica* ASL10	Algal acid hydrolysis 10% (w/v)	Mineral salt medium, at 37 °C, for 72 h, 160 rpm	Solvent extraction, cold acetone, then chloroform	PHBV	4.2	67	2.8	Abd El-Malek et al. (2021)
	Corallina mediterranea	*Halomonas salifodiane* ASL11	Algal acid hydrolysis 10% (w/v)	Mineral Salt Medium, at 37 °C, for 72 h, 160 rpm	Solvent extraction, cold acetone, then chloroform	PHBV	4.5	63	3.0	Abd El-Malek et al. (2021)
	Pterocladia capillacea	*Halomonas pacifica* ASL10	Algal acid hydrolysis 10% (w/v)	Mineral Salt Medium, at 37 °C, for 72 h, 160 rpm	Solvent extraction, cold acetone, then chloroform	PHBV	2.5	40	1.0	Abd El-Malek et al. (2021)
	Pterocladia capillacea	*Halomonas salifodiane* ASL11	Algal acid hydrolysis 10% (w/v)	Mineral Salt Medium, at 37 °C, for 72 h, 160 rpm	Solvent extraction, cold acetone, then chloroform	PHBV	4.9	41	1.2	Abd El-Malek et al. (2021)
	Eucheuma spinosum	*Halomonas* sp. YLGW01	*Eucheuma spinosum* Hydrolysate	Marine broth, and incubated at 30 °C	Analyzed by GCMS	PHB	6.32	62	3.88	Bhatia et al. (2023)

(continued)

Table 5.1 (continuued)

Seaweed type	Seaweed sp.	Bacterial strain	Carbon source	Fermentation conditions	Recovery/analysis method	PHA type	CDW (g/L)	PHA content (%)	Yield (g/L)	References
Brown	*Laminaria japonica*	*Cupriavidus necator NCIMB 11599*	Acid hydrolysis 10% (w/v)	At 30 °C, pH 7.5, for 96 h, 1300 rpm	–	PHBV	4.5–8.3	32.0–49.4	1.58–4.10	Muhammad et al. (2020)
	Laminaria japonica	*Paracoccus* sp. *LL1*	Acid hydrolysis 10% (w/v)	At 30 °C, pH 7.5, for 96 h, 1300 rpm	–	PHBV	3.6–13.4	29–37	1.05–4.98	Muhammad et al. (2020)
	Laminaria japonica	*megaterium ALA2*	Acid hydrolysis 10% (w/v)	At 30 °C, pH 7.5, for 96 h, 1300 rpm	–	PHBV	1.8	19.5–32.0	0.35–0.58	Muhammad et al. (2020)
	Wakame and kombu	*Burkholderia* sp.	10% Mannitol	MM medium with mannitol, at 30 °C, for 4 days, 120 rpm	Solvent extraction, chloroform	PHB	3.75	40	1.5	Yamada et al. (2018)
	Kombu (dried Laminaria spp.) & Wakame (Undaria pinnatifida)	*Cobetia* sp. *Strains* (isolated)	3% (w/v) Alginate	MM medium containing alginate, at 30 °C, for 2 days, 120 rpm	Solvent extraction, chloroform	PHB	5.01	62.1	3.11	Moriya et al. (2020)
	Kombu (dried Laminaria spp.) & Wakame (Undaria pinnatifida)	*Cobetia* sp. *Strains* (isolated)	5% (w/v) Kombu	MM medium, at 30 °C, for 4 days, 120 rpm	Solvent extraction, chloroform	PHB	0.7	13.5	0.09	Moriya et al. (2020)
	Sargassum sp.	*Cupriavidus necator PTCC 1615*	Acid hydrolysis 10% (w/v)	Mineral Medium, at 30 °C, for 60 h, 200 rpm	A solution of chloroform and 3% sulfuric acid solution of methanol, 100 °C, 3.5 h	PHB	5.36	74.4	3.93	Azizi et al. (2017)

	Sargassum sp.	*Hydrogenophaga* sp. strain UMI-18 (isolated)	1% (w/v) sodium alginate isolated from a decayed brown seaweed litter	1% (w/v) AMS medium, at 30 °C, for 72 h, 150 rpm	Sulfuric acid and chloroform	PHB	1.73	43.2	0.75	Yamaguchi et al. (2019)
	Algal wastes	*Halomonas ventosae* and *Halomonas daqingensis*	Algal biodiesel waste residue	Seed culture was grown on a Dry Sea Mix (DSM) medium. Halomonas strains were grown on 3% ABWR (algal biodiesel waste residues) supplemented with 5% NaCl, at 35 °C, for 48 h, 200 rpm	Sodium hypochlorite (4%)	PHB	0.362	65.2	0.236	Dubey and Mishra (2021)

Table 5.2 Seaweed-associated bacteria as a source of PHA production utilizing agro-industrial wastes

Seaweed type	Seaweed sp.	Bacterial strain	Carbon source	Fermentation conditions	Recovery/ analysis method	PHA type	CDW (g/L)	PHA content (%)	Yield (g/L)	References
Brown	*Kappaphycus alvarezii*	Seaweed isolates: *Bacillus cereus* and *Pseudomonas pseudoalcaligens*.	Palm jaggery and green gram flour	LB medium, at 35–37 °C for Bacillus and at 25–30 °C for Pseudomonas, for 48 h, 250 rpm	4% Sodium hypochlorite solution	PHB	–	66.45–75.68	–	Ramya and Manikandan (2017)
Green	*Ulva* sp.	*Cobetia amphilecti*	Olive leaves, olive mill waste and celery wastes	Marine Broth medium, 90 rpm at 32 °C for 4 days	Methyl-Levulinate	PHB	4.2–6	12–76.3	0.63–3.2	Gnaim et al. (2022)
			Avocado seed waste	Marine Broth medium, 90 rpm at 32 °C for 4 days	Ethyl-Levulinate	PHB	7.1	29.7	2.1	Gnaim et al. (2023)
	Seaweed associated with red mangroves	*Pichia kudriavzevii*	Hydrolysate of chicken feathers and banana peels	Mineral salts medium, 120 rpm, at 32 °C for 4 days, pH 10	Chloroform	PHBV	29.53	79.68	23.53	Ojha and Das (2020)

underutilized carbon sources that can be transformed into valuable substrates for PHA synthesis (Fernández et al. 2005). This strategy offers a dual benefit: effective waste management and bioplastic production. This approach contributes to waste valorization, reducing the environmental burden associated with waste disposal while simultaneously addressing the need for sustainable feedstock sources (Awasthi et al. 2022). Using waste as a substrate for PHA production has several advantages: (*i*) It provides a cost-effective solution because these waste materials are often readily available and may even have a low cost, since they may otherwise incur disposal charges. (*ii*) It eliminates the reliance on dedicated agricultural land for cultivating feedstocks specifically for PHA production, reducing land use and the associated environmental impacts (Razzaq et al. 2022). (*iii*) It helps mitigate the ecological footprint of PHA production by minimizing the reliance on virgin resources (Naitam et al. 2022).

Limited research has been conducted on utilizing different agro-industrial wastes for producing PHA from seaweed-associated bacteria. Gnaim et al. (2022, 2023) reported the cultivation of *Cobetia amphilecti*, an isolate from *Ulva* sp., on olive leaf waste, olive mill waste, celery waste, and avocado seed waste. *C. amphilecti* exhibited PHA content ranging from 12 to 76.3% w/v and a PHA yield of 0.63–3.2 g L^{-1}. In another study, Ramya and Manikandan (2017) documented the synthesis of PHA by *Bacillus cereus* and *Pseudomonas pseudoalcaligens,* two isolates from *Kappaphycus alvarezii* and *Ulva lactuca* when grown on palm jaggery and green gram flour. The PHA content was 66.45–75.68% w/v. Furthermore, Ojha and Das (2020) reported the production of PHBV from *Pichia kudriavzevii* VIT-NN02, a yeast isolated from seaweed associated with red mangroves. The PHA content was 79.68% w/v and the PHA yield was 23.53 g L^{-1}.

Utilizing waste materials as substrates enhances the sustainability aspects associated with PHA production. However, despite the potential of using seaweed-associated bacteria for PHA production, there are still many challenges to be addressed, which include optimizing fermentation conditions, selecting the best bacterial strains, developing efficient downstream processing methods, and scaling up the production process to meet industrial demands. By addressing these challenges and advancing the technology, this strategy holds significant potential to contribute to a more sustainable and environmentally friendly future for the plastics industry.

5.4.3 Seaweed as a Substrate for PHA Synthesis from Various Microorganisms

Seaweeds are utilized not only by seaweed-associated bacteria for PHA synthesis but also by other microorganisms (Dubey and Mishra 2021). Green seaweed, specifically *Ulva* sp. hydrolysate (Ghosh et al. 2019), has been employed as a substrate for PHBV production by *Haloferax mediterranei*, resulting in a yield of ~2 g L^{-1}.

Red seaweeds such as *Kappaphycus alvarezii*, *Gelidium amansii*, *Gelidium sesquipedale*, *Gracilaria*, *Corallina mediterranea*, and *Pterocladia capillacea* have been exploited as a substrate for *Halomonas hydrothermalis*, *Bacillus megaterium* strains, *Halomonas boliviensis*, *Saccharophagus degradans 2–40*, *Halomonas pacifica ASL10*, and *Halomonas salifodiane ASL11*, respectively. Bera et al. (2015) reported PHBV production (0.4–1.07 g L^{-1}) by *H. hydrothermalis* cultivated on *Kappaphycus alvarezii*-derived crude levulinic acid and biodiesel residues. Although the PHA yields and productivities were low, these findings offer promising opportunities for process optimization, since no pretreatment or enzymatic hydrolysis is required. *B. megaterium* strains were cultivated on both slurry hydrolyzed and untreated *G. amansii*, resulting in PHB yields of 0.93–2.56 g L^{-1} and 0.19–0.56 g L^{-1}, respectively (Alkotaini et al. 2015; Alkotaini et al. 2016). Tůma et al. (2020) achieved PHB production (3.4 g L^{-1}) from *H. boliviensis* when cultivated on *G. amansii*. Abd El-Malek et al. (2021) reported the growth of *H. pacifica ASL10* and *H. salifodiane ASL11*, with yields of 2.8 and 3 g L^{-1}, on the acid hydrolysate of *C. mediterranea*, and 1.0 and 1.2 g L^{-1}, and the acid hydrolysate of *P. capillacea*, respectively. A recent work (Bhatia et al. 2023) has demonstrated the accumulation of PHA (62% w/v, 3.88 g L^{-1}) by *Halomonas* sp. YLGW01 valorizing red seaweed *Eucheuma spinosum* under unsterilized conditions. Regarding brown seaweed, Azizi et al. (2017) focused on producing PHB from *Sargassum* sp. hydrolysates. This brown seaweed underwent pretreatment through acid hydrolysis, followed by an enzymatic treatment to produce monomeric sugars. The hydrolysate that ensued, containing ammonium sulfate as a nitrogen supply, underwent varying levels of NaCl as a stressor to enhance the synthesis of polymers by *Cupriavidus necator PTCC 1615*. The most substantial amount of PHA achieved was 3.9 g L^{-1} of PHB and an overall biomass of 5.4 g L^{-1} (Azizi et al. 2017). In another study, *Laminaria japonica* acid hydrolysate served as a substrate for *Paracoccus* sp. *LL1*, *Cupriavidus necator NCIMB 11599*, and *Bacillus megaterium ALA2*, resulting in PHBV yields ranging from 0.35 to 4.98 g L^{-1} (Muhammad et al. 2020). Furthermore, Yamaguchi et al. (2019) demonstrated the production of PHB from sodium alginate-based decayed brown seaweed litter, and Dubey and Mishra (2021) showcased the production of PHB from algal biodiesel waste residue (Yamaguchi et al. 2019; Dubey and Mishra 2021).

Various analytical techniques are employed to characterize the properties of PHA. One commonly used technique is Fourier-transform infrared spectroscopy (FTIR). FTIR allows for identifying functional groups in the polymer structure by measuring the absorption and transmission of infrared light. It provides valuable information regarding PHA's chemical composition and molecular structure, including the presence of ester bonds and other relevant functional groups (Godbole 2016). Gas chromatography (GC) coupled with mass spectrometry (MS) is widely used to analyze PHA monomers. GC-MS enables the separation and identification of individual monomers, providing insight into the monomeric composition (Gnaim et al. 2022). In addition to spectroscopic and chromatographic techniques, thermal analysis methods play a crucial role in characterizing PHA materials. Differential scanning calorimetry (DSC) is commonly employed to investigate the thermal properties of PHA, including glass transition temperature (Tg), melting temperature (Tm), and heat capacity. DSC measures the heat flow associated with phase transitions, providing valuable information about PHA's crystallinity, thermal stability, and processing behavior (Roja et al. 2019). Another widely used technique is thermogravimetric analysis (TGA), which assesses PHA's thermal stability and decomposition behavior. TGA measures the change in sample weight as a function of temperature, offering insights into degradation temperatures and decomposition rates (Ghosh et al. 2019). These thermal analysis techniques are instrumental in understanding the thermal behavior and stability of PHA, contributing to the design and application of these materials in various fields.

5.5 Conclusions and Future Perspectives

The growing demand for sustainable alternatives to conventional plastics has sparked significant interest in bioplastic production, particularly PHAs. Despite its potential, the scalability of PHA production faces a major challenge due to the prohibitively high expenses involved in using pure substrates. Recent research has focused on exploiting seaweed as an economically viable and promising alternative substrate, addressing the cost issue and creating new opportunities for sustainable bioplastic production. PHA production by utilizing seaweed offers several advantages, including the ability to produce PHA without consuming fresh water, aligning with the goal of sustainable production processes. Seaweeds have a wide range of polysaccharides, including alginate, agar, carrageenan, starch, and Ulvan, making them highly desirable substrates for bacteria involved in PHA biosynthesis. Studies have demonstrated the feasibility of using seaweed as a marine biomass source to produce monomeric sugars suitable for PHA production. However, further research is needed to optimize seaweed hydrolysis, increase sugar yields, and remove inhibitors from the cultivation medium to enhance PHA production efficiency. Tailoring the functional properties of PHA for specific applications is crucial and, once achieved, it enables a feasible scale-up of the production process. In the future, the proximity of marine microalgal cultivation for fuel additives could generate by-product streams that efficiently contribute to PHA manufacturing. This integration would enhance the overall sustainability and practicality of PHA production. Careful consideration should be given to the design of the cultivation infrastructure, aiming to minimize emissions while optimizing biomass yields. To fully assess the performance of the process and facilitate future enhancements, comprehensive techno-economic analysis and life cycle assessments are essential, as highlighted by greenhouse gas estimations.

Furthermore, exploring the co-production of PHA, carotenoids, and other valuable biomaterials from residual algal biomass can reduce the overall production cost of biopolymers while addressing challenges associated with petrochemical waste disposal. This approach aligns with circular economy principles by adding value to residual biomass. Finally, investigating the recyclability and scalability of this process using different microbial cultures can provide valuable insights. Overall, producing PHA from seaweed offers an exciting avenue for sustainable bioplastic production, potentially reducing production costs, conserving petroleum reserves, and contributing to a circular economy approach.

Acknowledgments A.G. and J.G. thank the Ministry of Science & Technology, the Ministry of Agriculture, and the Good Food Institute (GFI). R.G. thanks the Porter Excellence Scholarships, the Arianne de Rothchild Women's Doctoral Program Scholarship, the Lewis & Martin Whitman Scholarship, and the TRDC-TAU research grant. R.U. thanks the Eda and Shlomo Zalman Najman research fund. This work was supported by Aaron Frenkel's Air Pollution Initiative at Tel Aviv University.

References

Abd El-Malek F, Rofeal M, Farag A, Omar S, Khairy H (2021) Polyhydroxyalkanoate nanoparticles produced by marine bacteria cultivated on cost-effective Mediterranean algal hydrolysate media. J Biotechnol 328:95–105

Alkotaini B, Sathiyamoorthi E, Kim BS (2015) Potential of Bacillus megaterium for production of polyhydroxyalkanoates using the red algae *Gelidium amansii*. Biotechnol Bioprocess Eng 20:856–860

Alkotaini B, Koo H, Kim BS (2016) Production of polyhydroxyalkanoates by batch and fed-batch cultivations of *Bacillus megaterium* from acid-treated red algae. Korean J Chem Eng 33:1669–1673

Altaee N, Fahdil A, Yousif E, Sudesh K (2016) Recovery and subsequent characterization of polyhydroxybutyrate from *Rhodococcus equi* cells grown on crude palm kernel oil. J Taibah Univ Med Sci 10:543–550

Alves AA, Siqueira EC, Barros MP, Silva PE, Houllou LM (2023) Polyhydroxyalkanoates: a review of microbial production and technology application. Int J Environ Sci Technol 20:3409–3420

Aragosa A, Specchia V, Frigione M (2021) Isolation of two bacterial species from argan soil in Morocco associated with polyhydroxybutyrate (PHB) accumulation: current potential and future prospects for the bio-based polymer production. Polym J 13:1870

Awasthi MK, Sindhu R, Sirohi R, Kumar V, Ahluwalia V, Binod P, Juneja A, Kumar D, Yan B, Sarsaiya S, Zhang Z (2022) Agricultural waste biorefinery development towards circular bioeconomy. Renew Sust Energ Rev 158:112122

Azizi N, Najafpour G, Younesi H (2017) Acid pretreatment and enzymatic saccharification of brown seaweed for polyhydroxybutyrate (PHB) production using *Cupriavidus necator*. Int J Biol Macromol 101:1029–1040

Balcucho J, Narváez DM, Tarazona NA, Castro-Mayorga JL (2023) Microbially synthesized polymer-metal nanoparticles composites as promising wound dressings to overcome methicillin-resistance *Staphylococcus aureus* infections. Polymers 15:920

Behera S, Priyadarshanee M, Das S (2023) Properties, biochemical synthesis, and their applications. Chemosphere 24:133723

Bellini S, Tommasi T, Fino D (2022) Poly (3-hydroxybutyrate) biosynthesis by *Cupriavidus necator*: a review on waste substrates utilization for a circular economy approach. Bioresour Technol 17:100985

Bengtsson MM, Sjøtun K, Storesund JE, Øvreås L (2011) Utilization of kelp-derived carbon sources by kelp surface-associated bacteria. Aquat Microb Ecol 62:191–199

Bera A, Dubey S, Bhayani K, Mondal D, Mishra S, Ghosh PK (2015) Microbial synthesis of polyhydroxyalkanoate using seaweed-derived crude levulinic acid as co-nutrient. Int J Biol Macromol 72:487–494

Bhatia SK, Hwang JH, Oh SJ, Kim HJ, Shin N, Choi TR, Yang YH (2023) Macroalgae as a source of sugar and detoxifier biochar for polyhydroxyalkanoates production by *Halomonas sp.* YLGW01 under the unsterile condition. Bioresour Technol 384:129290

Blandón L, Alvarado-Campo KL, Patiño AD, Jiménez-Vergara E, Quintero M, Montoya-Giraldo M, Jutinico-Shubach LM, Santos-Acevedo M, Gómez-León J (2020) Polyhydroxyalkanoate production from two species of marine bacteria: a comparative study. J Polym Environ 28:2324–2334

Chee JY, Yoga SS, Lau NS, Ling SC, Abed RM, Sudesh K (2010) Bacterially produced polyhydroxyalkanoate (PHA): converting renewable resources into bioplastics. In: Current research, technology and education topics in applied microbiology and microbial biotechnology, vol 2. Formatex Research Center, Spain, pp 1395–1404

Dalton B, Bhagabati P, De Micco J, Padamati RB, O'Connor K (2022) A review on biological synthesis of the biodegradable polymers polyhydroxyalkanoates and the development of multiple applications. Catalysts 12:319

de Melo RN, de Souza HG, Steffens J, Junges A, Valduga E (2022) Recent updates to microbial production and recovery of polyhydroxyalkanoates. 3 Biotech 13:204

Dhania S, Bernela M, Rani R, Parsad M, Grewal S, Kumari S, Thakur R (2022) Scaffolds the backbone of tissue engineering: advancements in use of polyhydroxyalkanoates (PHA). Int J Biol Macromol 208:243–259

Dong H, Yang X, Shi J, Xiao C, Zhang Y (2023) Exploring the feasibility of cell-free synthesis as a platform for polyhydroxyalkanoate (PHA) production: opportunities and challenges. Polymers 15:2333

Dubey S, Mishra S (2021) Efficient production of polyhydroxyalkanoate through halophilic bacteria utilizing algal biodiesel waste residue. Front Bioeng Biotechnol 9:624859

Dwivedi R, Pandey R, Kumar S, Mehrotra D (2020) Polyhydroxyalkanoates (PHA): role in bone scaffolds. J Oral Biol Craniofac Res 10:389–392

Edaes FS, de Souza CB (2021) General aspects of biodegradable biopolymers and the polyhydroxyalkanoates' family. Curr Biotechnol 10:178–190

Egan S, Harder T, Burke C, Steinberg P, Kjelleberg S, Thomas T (2013) The seaweed holobiont: understanding seaweed–bacteria interactions. FEMS Microbiol Rev 37:462–476

Eraslan K, Aversa C, Nofar M, Barletta M, Gisario A, Salehiyan R, Goksu YA (2022) Poly (3-hydroxybutyrate-co-3-hydroxyhexanoate) (PHBH): synthesis, properties, and applications-a review. Eur Polym J 167:111044

Fernández D, Rodríguez E, Bassas M, Viñas M, Solanas AM, Llorens J, Manresa A (2005) Agro-industrial oily wastes as substrates for PHA production by the new strain *Pseudomonas aeruginosa* NCIB 40045: effect of culture conditions. Biochem Eng J 26:159–167

Ghosh S, Gnaim R, Greiserman S, Fadeev L, Gozin M, Golberg A (2019) Macroalgal biomass subcritical hydrolysates for the production of polyhydroxyalkanoate (PHA) by *Haloferax mediterranei*. Bioresour Technol 271:166–173

Ghosh S, Greiserman S, Chemodanov A, Slegers PM, Belgorodsky B, Epstein M, Golberg A (2021) Polyhydroxyalkanoates and biochar from green macroalgal *Ulva* sp. biomass subcritical hydrolysates: process optimization and a priori economic and greenhouse emissions break-even analysis. Sci Total Environ 770:145281

Gigante V, Cinelli P, Seggiani M, Alavarez VA, Lazzeri A (2020) Processing and thermomechanical properties of PHA. In: The handbook of polyhydroxyalkanoates. CRC Press, Boca Raton, FL, pp 91–118

Giubilini A, Bondioli F, Messori M, Nyström G, Siqueira G (2021) Advantages of additive manufacturing for biomedical applications of polyhydroxyalkanoates. Bioengineering 8:29

Gnaim R, Polikovsky M, Unis R, Sheviryov J, Gozin M, Golberg A (2021) Marine bacteria associated with the green seaweed *Ulva* sp. for the production of polyhydroxyalkanoates. Bioresour Technol 328:124815

Gnaim R, Unis R, Gnayem N, Das J, Gozin M, Gnaim J, Golberg A (2022) Turning mannitol-rich agricultural waste to poly(3-hydroxybutyrate) with *Cobetia amphilecti* fermentation and recovery with methyl levulinate as a green solvent. Bioresour Technol 352:127075

Gnaim R, Unis R, Gnayem N, Das J, Shamis O, Gozin M, Gnaim J, Golberg A (2023) Avocado seed waste bioconversion into poly (3-hydroxybutyrate) by using *Cobetia amphilecti* and ethyl levulinate as a green extractant. Int J Biol Macromol 239:124371

Godbole S (2016) Methods for identification, quantification and characterization of polyhydroxyalkanoates-a review. Int J Bioassays 5:2016

Gregory DA, Taylor CS, Fricker AT, Asare E, Tetali SS, Haycock JW, Roy I (2022) Polyhydroxyalkanoates and their advances for biomedical applications. Trends Mol Med 28:331–342

Gunny AA, Arbain D, Gumba RE, Jong BC, Jamal P (2014) Potential halophilic cellulases for in situ enzymatic saccharification of ionic liquids pretreated lignocelluloses. Bioresour Technol 155:177–181

Guo W, Yang K, Qin X, Luo R, Wang H, Huang R (2022) Polyhydroxyalkanoates in tissue repair and regeneration. Eng Regen 3:24–40

Han X, Satoh Y, Kuriki Y, Seino T, Fujita S, Suda T, Kobayashi T, Tajima K (2014) Polyhydroxyalkanoate production by a novel bacterium *Massilia* sp. UMI-21 isolated from seaweed, and molecular cloning of its polyhydroxyalkanoate synthase gene. J Biosci Bioeng 118:514–519

Ihua MW, Guihéneuf F, Mohammed H, Margassery LM, Jackson SA, Stengel DB, Dobson AD (2019) Microbial population changes in decaying *Ascophyllum nodosum* result in macroalgal-polysaccharide-degrading bacteria with potential applicability in enzyme-assisted extraction technologies. Mar Drugs 17:200

Jian Y, Zhu Y (2022) Poly 3-hydroxybutyrate 4-hydroxybutyrate (P34HB) as a potential polymer for drug-eluting coatings on metal coronary stents. Polymers 14:994

Kalia VC, Patel SK, Lee JK (2023) Exploiting polyhydroxyalkanoates for biomedical applications. Polymers 15:1937

Keshavarz T, Roy I (2010) Polyhydroxyalkanoates: bioplastics with a green agenda. Curr Opin Microbiol 13:321–326

Khambhaty Y, Reena R (2023) Macroalgal polysaccharides: biocatalysts in biofuel/bioenergy production. In: Polysaccharide degrading biocatalysts. Academic Press, Elsevier, Cambridge, MA, pp 227–273

Koller M, Salerno A, Dias M, Reiterer A, Braunegg G (2010) Modern biotechnological polymer synthesis: a review. Food Technol Biotechnol 48:255–269

Koller M, Mukherjee A, Obruca S, Zinn M (2022) Polyhydroxyalkanoates (PHA): microbial synthesis of natural polyesters. In: Microbial production of high-value products, vol 6. Springer, Cham, pp 185–236

Kumar M, Rathour R, Singh R, Sun Y, Pandey A, Gnansounou E, Lin KY, Tsang DC, Thakur IS (2020) Bacterial polyhydroxyalkanoates: opportunities, challenges, and prospects. J Clean Prod 263:121500

Kumar V, Lakkaboyana SK, Tsouko E, Maina S, Pandey M, Umesh M, Singhal B, Sharma N, Awasthi MK, Andler R, Jayaraj I (2023) Commercialization potential of agro-based polyhydroxyalkanoates biorefinery: a technical perspective on advances and critical barriers. Int J Biol Macromol 234:123733

Kumari P, Singh A (2023) Polyhydroxyalkanoates (PHA): production, properties, and packaging applications. In: Biopolymerbased films and coatings. Taylor & Francis, Abingdon, Oxfordshire, pp 241–264

Ladhari S, Vu NN, Boisvert C, Saidi A, Nguyen-Tri P (2023) Recent development of polyhydroxyalkanoates (PHA)-based materials for antibacterial applications: a review. ACS Appl Bio Mater 6:1398–1430

Lage S, Gojkovic Z, Funk C, Gentili FG (2018) Algal biomass from wastewater and flue gases as a source of bioenergy. Energies 11:664

Leandro A, Pereira L, Gonçalves AM (2019) Diverse applications of marine macroalgae. Mar Drugs 18:17

Lin R, Deng C, Ding L, Bose A, Murphy JD (2019) Improving gaseous biofuel production from seaweed *Saccharina latissima*: the effect of hydrothermal pretreatment on energy efficiency. Energy Convers Manag 196:1385–1394

Meng DC, Chen GQ (2018) Synthetic biology of polyhydroxyalkanoates (PHA). Adv Biochem Eng Biotechnol 162:147–174

Michalak I, Chojnacka K (2015) Production of seaweed extracts by biological and chemical methods. In: Marine algae extracts: processes, products, and applications. Wiley-VCH, Weinheim, pp 121–144

Mitra R, Xu T, Chen GQ, Xiang H, Han J (2022) An updated overview on the regulatory circuits of polyhydroxyalkanoates synthesis. Microb Biotechnol 15:1446–1470

Moriya H, Takita Y, Matsumoto A, Yamahata Y, Nishimukai M, Miyazaki M, Yamada M (2020) *Cobetia* sp. bacteria, which are capable of utilizing alginate or waste *Laminaria* sp. for poly (3-hydroxybutyrate) synthesis, isolated from a marine environment. Front Bioeng Biotechnol 8:974

Muhammad M, Aloui H, Khomlaem C, Hou CT, Kim BS (2020) Production of polyhydroxyalkanoates and carotenoids through cultivation of different bacterial strains using brown algae hydrolysate as a carbon source. Biocatal Agric Biotechnol. 30:101852

Muthuraj R, Valerio O, Mekonnen TH (2021) Recent developments in short-and medium-chain-length Polyhydroxyalkanoates: production, properties, and applications. Int J Biol Macromol 187:422–440

Naitam MG, Tomar GS, Kaushik R, Singh S, Nain L (2022) Agro-industrial waste as potential renewable feedstock for biopolymer polyhydroxyalkanoates (PHA) production. Enzym Eng 11:190–206

Neoh SZ, Chek MF, Tan HT, Linares-Pastén JA, Nandakumar A, Hakoshima T, Sudesh K (2022) Polyhydroxyalkanoate synthase (PhaC): the key enzyme for biopolyester synthesis. Curr Res Biotechnol 4:87–101

Offei F, Mensah M, Thygesen A, Kemausuor F (2018) Seaweed bioethanol production: a process selection review on hydrolysis and fermentation. Fermentation 4:99

Ojha N, Das N (2020) Process optimization and characterization of polyhydroxyalkanoate copolymers produced by marine *Pichia kudriavzevii* VIT-NN02 using banana peels and chicken feather hydrolysate. Biocatal Agric Biotechnol. 27:101616

Phanse N, Chincholikar A, Patel B, Rathore P, Vyas P, Patel M (2011) Screening of PHA (poly hydroxyalkanoate) producing bacteria from diverse sources. Int J Biosci 1:27–32

Porter MM, Lee S, Tanadchangsaeng N, Jaremko MJ, Yu J, Meyers M, McKittrick J (2013) Porous hydroxyapatitepolyhydroxybutyrate composites fabricated by a novel method via centrifugation. In: Mechanics of biological systems and materials, volume 5: proceedings of the 2012 annual conference on experimental and applied mechanics, pp 63–71. Springer, Cham, Switzerland

Ramya R, Manikandan A (2017) Standardization of biopolymer production from seaweed associative bacteria. Int J Biol Macromol 102:550–564

Raza ZA, Abid S, Banat IM (2018) Polyhydroxyalkanoates: characteristics, production, recent developments and applications. Int Biodeterior Biodegrad 126:45–56

Razzaq S, Shahid S, Farooq R, Noreen S, Perveen S, Bilal M (2022) Sustainable bioconversion of agricultural waste substrates into poly (3-hydroxyhexanoate) (mcl-PHA) by *Cupriavidus necator DSM 428*. Biomass Convers Biorefinery 14:1–11

Roja K, Sudhakar DR, Anto S, Mathimani T (2019) Extraction and characterization of polyhydroxyalkanoates from marine green alga and cyanobacteria. Biocatal Agric Biotechnol 22:101358

Rosenboom JG, Langer R, Traverso G (2022) Bioplastics for a circular economy. Nat Rev Mater 7:117–137

Saha M, Weinberger F (2019) Microbial "gardening" by a seaweed holobiont: surface metabolites attract protective and deter pathogenic epibacterial settlement. J Ecol 107:2255–2265

Saratale RG, Cho SK, Saratale GD, Kadam AA, Ghodake GS, Kumar M, Bharagava RN, Kumar G, Kim DS, Mulla SI, Shin HS (2021) A comprehensive overview and recent advances on polyhydroxyalkanoates (PHA) production using various organic waste streams. Bioresour Technol 325:124685

Sawant SS, Salunke BK, Tran TK, Kim BS (2016) Lignocellulosic and marine biomass as resource for production of polyhydroxyalkanoates. Korean J Chem Eng 33:1505–1513

Sawant SS, Salunke BK, Taylor LE, Kim BS (2017) Enhanced agarose and xylan degradation for production of polyhydroxyalkanoates by co-culture of marine bacterium, *Saccharophagus degradans* and its contaminant, Bacillus cereus. Appl Sci 7:225

Sawant SS, Salunke BK, Kim BS (2018) Consolidated bioprocessing for production of polyhydroxyalkanoates from red algae *Gelidium amansii*. Int J Biol Macromol 109:1012–1018

Tan GYA, Chen CL, Li L, Ge L, Wang L, Razaad IMN, Wang JY (2014) Start a research on biopolymer polyhydroxyalkanoate (PHA): a review. Polym J 6:706–754

Taylor CS, Behbehani M, Glen A, Basnett P, Gregory DA, Lukasiewicz BB, Nigmatullin R, Claeyssens F, Roy I, Haycock JW (2023) Aligned polyhydroxyalkanoate blend electrospun fibers as intraluminal guidance scaffolds for peripheral nerve repair. ACS Biomater Sci Eng 9:1472–1485

Tůma S, Izaguirre JK, Bondar M, Marques MM, Fernandes P, da Fonseca MM, Cesário MT (2020) Upgrading end-of-line residues of the red seaweed *Gelidium sesquipedale* to polyhydroxyalkanoates using *Halomonas boliviensis*. Biotechnol Rep 27:e00491

Turco R, Corrado I, Zannini D, Gargiulo L, Di Serio M, Pezzella C, Santagata G (2022) Upgrading cardoon biomass into Polyhydroxybutyrate based blends: a holistic approach for the synthesis of biopolymers and additives. Bioresour Technol 363:127954

van der Loos LM, Eriksson BK, Salles JF (2019) The macroalgal holobiont in a changing sea. Trends Microbiol 27:635–650

Vicente D, Proença DN, Morais PV (2023) The role of bacterial polyhydroxyalkanoate (PHA) in a sustainable future: a review on the biological diversity. Int J Environ Res Public Health 4:2959

Wahl M, Werner FJ, Buchholz B, Raddatz S, Graiff A, Matthiessen B, Karsten U, Hiebenthal C, Hamer J, Ito M, Gülzow E (2020) Season affects strength and direction of the interactive impacts of ocean warming and biotic stress in a coastal seaweed ecosystem. Limnol Oceanogr 65:807–827

Wang Q, Xu Y, Xu P, Yang W, Chen M, Dong W, Ma P (2022) Crystallization of microbial polyhydroxyalkanoates: a review. Int J Biol Macromol 209:330–343

Wang J, Liu S, Huang J, Cui R, Xu Y, Song Z (2023) Genetic engineering strategies for sustainable polyhydroxyalkanoate (PHA) production from carbon-rich wastes. Environ Technol Innov 30:103069

Wood G, Steinberg PD, Campbell AH, Vergés A, Coleman MA, Marzinelli EM (2022) Host genetics, phenotype and geography structure the microbiome of a foundational seaweed. Mol Ecol 31:2189–2206

Wu F, Misra M, Mohanty AK (2021) Challenges and new opportunities on barrier performance of biodegradable polymers for sustainable packaging. Prog Polym Sci 117:101395

Yamada M, Yukita A, Hanazumi Y, Yamahata Y, Moriya H, Miyazaki M, Yamashita T, Shimoi H (2018) Poly (3-hydroxybutyrate) production using mannitol as a sole carbon source by *Burkholderia* sp. AIU M5M02 isolated from a marine environment. Fish Sci 84:405–412

Yamaguchi T, Narsico J, Kobayashi T, Inoue A, Ojima T (2019) Production of poly (3-hydroxybutyrate) by a novel alginolytic bacterium *Hydrogenophaga* sp. strain UMI-18 using alginate as a sole carbon source. J Biosci Bioeng 128:203–208

Yeo JC, Muiruri JK, Thitsartarn W, Li Z, He C (2018) Recent advances in the development of biodegradable PHB-based toughening materials: approaches, advantages and applications. Mater Sci Eng C 92:1092–1116

York GM, Junker BH, Stubbe J, Sinskey AJ (2001) Accumulation of the PhaP phasin of *Ralstonia eutropha* is dependent on production of polyhydroxybutyrate in cells. J Bacteriol 183:4217–4226

Chapter 6
Bioprospecting of Seaweed Secondary Metabolites as Emerging Functional Ingredients

Dhara Dixit

Abstract Functional ingredients are described as standardized and characterized formulations, portions or extracts that consist of bioactive compounds with diverse levels of purity. These components serve as ingredients for manufacturers in the production of functional foods. The oceanic setting serves as an underexplored reservoir of functional components applicable to diverse facets of food processing, preservation and enhancement. Presently, the significance of biologically active compounds in the market is on the rise, and there is a noteworthy commercial interest in incorporating seaweeds into nutritious foods, cosmetics, agricultural products and pharmaceuticals. This is due to the outstanding potential of seaweeds as functional ingredients, possessing medicinal, antioxidant, antimicrobial properties and additional health advantages. Omics technologies (Genomics, Transcriptomics, Proteomics and Metabolomics) present a fundamental opportunity in the bioprospecting of seaweeds as functional ingredients by providing comprehensive insights into their genetic, metabolic and functional characteristics. Integrating data from genomics, transcriptomics, proteomics and metabolomics provides a holistic understanding of seaweed biology. Systems biology approaches further enable the modelling and prediction of the behaviour of biological systems, aiding in the optimization of seaweed cultivation and processing for functional ingredient production.

Keywords Bioprospecting · Omics · Genomics · Transcriptomics · Proteomics · Metabolomics

D. Dixit (✉)
Coastal and Marine Ecology Division, Gujarat Institute of Desert Ecology (GUIDE), Bhuj, Kachchh, Gujarat, India

N. Trivedi et al. (eds.), *Recent Advances in Seaweed Biotechnology*,
https://doi.org/10.1007/978-981-96-0519-4_6

6.1 Introduction

Bioprospecting/biodiversity prospecting or bioprospection, represents a scientific approach that systematically searches and explores biological resources (plants, microorganisms, animals and their genetic material) to uncover novel compounds, genes, or organisms in diverse ecosystems for various commercial applications (Dixit et al. 2021). The objective is to reveal new compounds, enzymes, genes and other biological entities with potential applications in various fields that can be harnessed for scientific, commercial and conservation purposes. Bioprospecting has its roots in the traditional knowledge systems of indigenous communities that have long recognized and utilized the benefits of nature's offerings (Efferth et al. 2016). Bioprospecting has emerged as a promising strategy for discovering valuable resources that can revolutionize several industries, such as pharmaceuticals, agriculture, biotechnology and environmental conservation. By tapping into the immense biological diversity in ecosystems worldwide, scientists and researchers can identify unique bioactive compounds, genetic sequences and even entire organisms with remarkable properties. These discoveries hold immense potential for developing new drugs, improving crop yields and advancing various technological applications.

Bioprospecting involves a multidisciplinary approach, integrating elements of biology, chemistry, ecology and genetics. Scientists employ various techniques to identify potential resources, including field surveys, botanical collections, DNA sequencing and biochemical screening. Researchers can unveil nature's hidden treasures by cataloguing and analysing the vast array of organisms and their genetic information. Several phases employed during the bioprospecting process include:

(a) Targeted Sampling: Scientists identify specific habitats, such as rainforests, marine environments or extreme ecosystems, likely to harbour novel organisms or compounds of interest. Samples, including plants, animals, microbes, or environmental materials, are collected, cleaned and stored for further analysis.
(b) Bioassays: Extracts derived from collected samples are subjected to various bioassays for characterization to determine their potential biological activity. To assess their efficacy, these assays involve testing the extracts against specific disease targets, pests or other relevant indicators.
(c) Chemical Analysis: Extracts exhibiting promising bioactivity are further analysed using chromatography, spectroscopy or mass spectrometry to separate and categorize the dynamic complexes responsible for the observed bioactive effects.
(d) Genetic Analysis: Modern bioprospecting methodologies also employ genetic analysis methods, like metagenomics, to study the genomic material present in samples. This allows scientists to determine novel genes and genetic pathways with potential applications for further product development for commercial market/s.

One significant area of focus in bioprospecting is the investigation of secondary metabolites (Pereira and Costa-Lotufo 2012) that represent a set of bioactive amalgams produced by living entities. These complex amalgams, however, do not straightaway hold a role in their growth or reproduction. Secondary metabolites often serve specialized functions, such as defence against predators, competition for resources, or communication within the organism's ecosystem. Secondary metabolites from terrestrial and marine sources reveal distinctive characteristics in terms of structural diversity, bioactivities and ecological roles. Examples of secondary metabolites include alkaloids, terpenes, phenols and flavonoids. Bioprospecting studies aim to identify, isolate, and characterize these secondary metabolites, unveiling their chemical assemblages and understanding their mode of action. Terrestrial organisms yield an extensive array of bioactives to defend against herbivores and pathogens, while marine organisms synthesize structurally diverse compounds to deter predation and fouling. In recent years, secondary metabolites have emerged as a primary target for bioprospecting owing to their diverse chemical makeup and potential biochemical activities (Marco and Abram 2019; Hemaiswarya et al. 2013). Aquatic organisms (bacteria, fish, algae, and invertebrates) have long captivated the scientific community due to their remarkable diversity and unique adaptations to the marine environment. Marine biodiversity is one of the most underexplored natural resources of bioactives that can potentially deliver novel compounds (Malve 2016). Therefore, there has been a rising emphasis on studying marine secondary metabolites as they exhibit an impressive array of biological activities, viz, antimicrobial, antifungal, anticancer, antiviral, anti-inflammatory and neuroprotective properties.

Seaweeds, or macroalgae, are an assorted assemblage of marine beings that portray a significant ecological role in marine ecosystems. Besides their ecological importance, seaweeds have gained attention recently due to their secondary metabolites. Marine macroalgae are segregated into three categories depending on their pigmentation: red (Rhodophyceae), brown (Phaeophyceae) and green (Chlorophyceae). These organisms contribute to the evolvement of a remarkable ability to produce extensive sorts of secondary metabolites, attracting significant attention from researchers and industries alike. The extraordinary diversity and unique chemical structures of seaweed secondary metabolites have led to their designation as functional ingredients.

6.2 Omics Approaches Underlying Bioprospecting

Omics approaches have revolutionized seaweed bioprospecting by providing comprehensive insights into seaweed species' genetic, molecular and metabolic aspects. These technologies have identified and characterized diverse bioactive molecules with significant potential for applications in various industries. Traditional bioprospecting methods involve the collection of samples from various ecosystems, then isolating and characterizing bioactive compounds through bioassays and chemical analysis. However, these conventional approaches are time-consuming, cumbersome, expensive and often limited by the availability of samples. Omics technologies (genomics, transcriptomics, proteomics, lipidomics, metabolomics and *in silico* bioprospecting) offer a paradigm shift by enabling the efficient and targeted exploration of natural resources.

6.2.1 Genomics

Genomics comprehensively explains an organism's genetic composition and potential functions (Liu et al. 2015). By sequencing and analysing the genomes of organisms, researchers can identify genes encoding enzymes, secondary metabolites or other bioactive compounds. Comparative genomics can also aid in identifying gene clusters associated with the biosynthesis of specific compounds, facilitating the prediction of novel pathways and the discovery of previously unknown molecules. Omics technologies offer powerful tools for deciphering seaweeds' complex biochemical pathways and genetic diversity. Genomic studies provide insights into the genetic makeup of seaweed species and offer valuable information on their evolutionary history, physiological adaptations and potential biosynthetic capabilities. Whole-genome sequencing of seaweed species has led to identifying genes involved in synthesizing bioactive compounds, such as agarose in red algae and fucoidan in brown algae. Comparative genomics enables researchers to uncover genomic variations and identify candidate genes responsible for producing specific bioactive molecules, aiding in targeted bioprospecting efforts. The dawn of the next-generation sequencing (NGS) tools, comprehensive sequencing of entire genomes through genomics, offers crucial understandings into the biology and evolutionary advancement of various organisms (Fort et al. 2018). Metagenomics, in particular, has proven to be a potent tool for investigating microbial communities linked to seaweeds, which play a significant role in aquatic biomes.

6.2.2 Transcriptomics

Transcriptomics encompasses studying a complete RNA molecule set within a cell/tissue at a particular interval. Researchers can identify and quantify gene expression levels through techniques such as RNA sequencing (RNA-seq). Transcriptomic analyses help identify genes and pathways involved in synthesizing bioactive compounds and provide insights into the regulatory mechanisms underlying their production. Transcriptomics involves studying an organism's complete RNA transcripts, providing information on gene expression patterns and regulation. Transcriptomic analyses of seaweeds under different environmental conditions or during specific developmental stages have shed light on the biosynthetic pathways involved in producing bioactive compounds. By identifying differentially expressed genes, transcriptomics facilitates the discovery of key regulatory factors and potential targets for enhancing the production of specific bioactive molecules. Transcriptomics involves the study of all RNA transcripts in a given organism, shedding light on gene expression patterns and regulatory mechanisms (Dong and Chen 2013). In seaweed bioprospecting, transcriptomics has been employed to detect genes playing a role in the biosynthesis of valuable compounds.

6.2.3 Proteomics

Proteomics focuses on studying proteins directly responsible for most biological functions (Pierce et al. 2007). By analysing the comprehensive set of proteins expressed in a living entity or a precise tissue, researchers can identify proteins involved in synthesizing, transporting and modifying bioactive compounds. Proteomics can also reveal post-translational modifications and protein-protein interactions that are crucial for the production of bioactive molecules. Proteomic approaches enable the identification and characterization of proteins expressed by seaweed species. By analysing the proteome, researchers can identify enzymes involved in synthesizing bioactive compounds and gain insights into post-translational modifications and protein-protein interactions. Proteomics also allows for comparing protein profiles among different seaweed species or under different environmental conditions, aiding in identifying unique and potentially valuable proteinaceous compounds.

6.2.4 Lipidomics

Lipids constitute a multifaceted class of biomolecules that exhibit diverse chemical structures insoluble in water. Lipids act as an edifice for cellular membranes, play a role as signalling molecules and are storage houses for energy (Zullig et al. 2020). A separate branch called ‘Lipidomics’ refers to undertaking the studies associated with the ‘lipidome’ (complete set of lipids) has emerged. Analytically, lipidomics is based on mass spectrometry that enables the qualitative and quantitative study of complex lipid systems by targeting the differences in the charge and mass of the moieties (Swinnen and Dehairs 2022). Recent developments in this direction have driven the advancement of spatial lipidomics and tracer lipidomics, which offer an unparalleled understanding of these biomolecules. As lipids are at the centre of many physiological and pathological roles, studying lipids and their involvement in metabolic pathways triggers immense possibilities for novel biomarker detections and therapeutic solutions.

6.2.5 Metabolomics

Metabolomics involves studying the broad set of minor moieties or metabolites within a biological system. Researchers can identify and quantify metabolites associated with bioactive compounds by employing mass spectrometry (MS) techniques and nuclear magnetic resonance (NMR). Metabolomics enables the characterization of metabolic pathways and the identification of novel compounds that may possess valuable biological activities (Dixit et al. 2020). Metabolomics involves the comprehensive analysis of small molecules or metabolites produced by seaweeds. This approach provides a snapshot of the metabolic profile of seaweeds under specific conditions and facilitates the discovery of bioactive compounds. Metabolomic studies have revealed the presence of diverse metabolites, including pigments, polyphenols and secondary metabolites, with potential pharmaceutical applications. Metabolomics can be coupled with other omics techniques to identify metabolic pathways associated with synthesizing specific bioactive molecules.

6.2.6 In Silico *Bioprospecting*

A computational method employed for the virtual screening of the bioactivity of compounds is defined as *in silico* bioprospecting. This computer-aided screening utilizes bioinformatics applications and tools followed by the database search to envisage the probable bioactivity of amalgams. *In silico* bioprospecting assists and accelerates the identification of potential candidates through screening the database/libraries against precise targets (Kamble and Singh 2022). The conventional methods of screening and identifying bioactive compounds are cumbersome, costly and time-consuming. *In Silico* bioprospecting allows for predicting the binding capacity and the bioactivity of compounds of interest via molecular docking, computational modelling and machine learning.

In order to overcome several analytical shortcomings, technologies like high-resolution mass spectroscopy (HR-MS) and NMR are coupled with omics techniques. A general overview of bioprospecting of seaweeds is shown in Fig. 6.1.

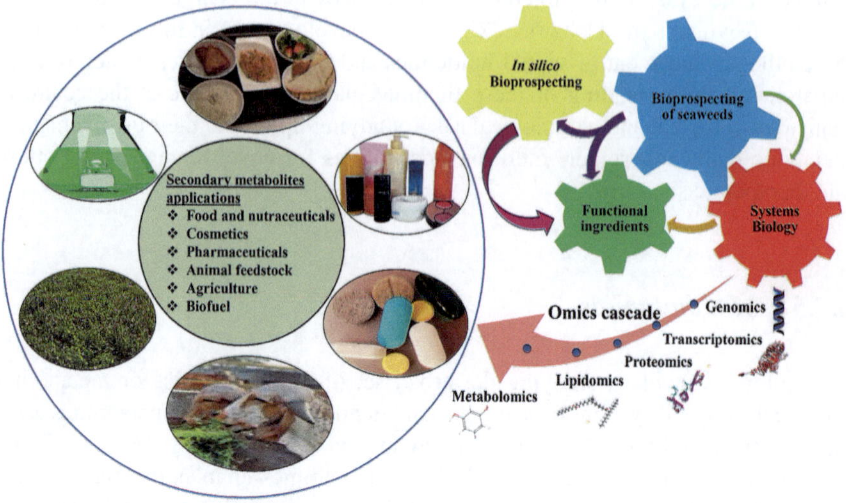

Fig. 6.1 A general overview of bioprospecting of seaweeds

6.3 Marine Biotechnology

The utilization of Marine Biotechnology or Blue Biotechnology, involves the usage of molecular biology procedures to explore and harness aquatic bioresources. This sector aims to significantly address critical societal challenges related to food and energy security, animal/human health drug development, sustainable practices associated with oceans and other waterbodies, and industrial materials and processes (Hurst et al. 2016). Within the realm of marine biotechnology, the focus is on studying the genes involved in synthesizing bioactive compounds. However, the genomic studies of marine microalgae and seaweeds, progresses slowly due to their complex and huge genomic size. Genomic sequencing and annotation contribute to a deeper understanding of non-explored strains, facilitating the identification of genes associated with the biosynthesis of specific compounds through exploring metabolic pathways.

Metabolic flux modelling, based on the stoichiometry of biochemical reactions and mass balances over intracellular metabolites, enables the connection of metabolic fluxes to substrate consumption, biomass production, and the generation of bioactive compounds (Ibanez and Cifuentes 2013). Through metabolic engineering approaches, strains can be genetically enhanced, and transgenic species developed to overproduce targeted compounds. Nevertheless, challenges arise when the overproduction of a specific metabolite is low or challenging to control in marine organisms, necessitating the introduction of responsible genes into a more easily cultivable host organism. This approach, although practical, is complex as it requires the identification and transfer of all genes involved in a particular metabolic pathway, thereby increasing the system's complexity.

6.4 Improvising the Synthesis of Secondary Metabolites

The secondary metabolite/bioactive production relies on the algal strain and its respective cultivation approaches. This offers the possibility to optimize the formation of different selective metabolites of choice using selected methodologies viz; metabolic engineering, genetic selection and metabolic flux modelling. The selection of suitable optimization processes, in turn, enhances the productivity/recovery of desired metabolites. Such an approach, when integrated with the biorefinery concept, can deliver sustainable and maximum utilization of the algal strain/s for producing food, feed, energy and high-value chemicals.

6.5 Manufacturing

Optimizing varied tactics for producing macroalgae-based bioactives of interest is imperative for commercial-scale production. Omics knowledge and tools combined with novel designs can be employed to optimize the culture conditions of seaweeds. Approaches encompassing the culturing of seaweeds using simulated enclosures (chambers, ponds, tanks, containers) may offer desired culture concentrations. Closed systems such as photobioreactors prove better compared to open ponds/tank cultures as any kind of exchange is restricted with the external surroundings, thereby resulting in quality biomass. Land-based seaweed cultivation serves few plusses when compared to the sea-based cultivation approaches. For example, the culturing conditions viz; light control, pH adjustment, temperature optimization, day length, oxygen content, CO_2 level, nutrient supplementation) can be conveniently manipulated based on the desired output and the productivity can be controlled throughout the year (Ibanez and Cifuentes 2013). However, the limitation underlying this system includes the possibility of infestation caused by predatory species thereby resulting in competition with the desired algal species impeding their flourishment. Additionally, the cost effectiveness of designing such photobioreactors is also a concern while scaling up at an industrial level. Different algal species require different optimal conditions of their growth and culturing, making the designing very specific and challenging. Targeting these practical issues and addressing them using scientific knowhow and technological tools thereby making the design and manufacturing process affordable with high productivity is indeed the need of the hour.

6.6 Collection, Rupturing and Extraction from Cells

Downstream processes like collection, cell rupturing and extraction may also be improved to augment the convalescence of priced chemicals. Innovative and species-specific harvesting methods, besides sedimentation, centrifugation and membrane filtration, need to be augmented for desired outputs (Barta et al. 2021). Novel strategies for post-harvest measures also should be focused on to make the final product feasible. Cell rupturing is undertaken using processes like microwave or ultrasound-assisted breakdown, enzymatic rupturing or high-pressure homogenization to facilitate the release of metabolites, as most of them are present within the cell. Concentration procedures like freeze-drying, fluidized bed-drying, spray-drying, etc., must also be enhanced. Post-rupturing and releasing the metabolites of interest, extracting them using economical, ecologically benign and effective protocols is another crucial facet of this chain. Traditional extraction protocols employ harmful organic solvents, while green extraction approaches involving microwave-assisted, ionic liquid, supercritical fluid, ultrasound-assisted, and pressurized liquid extractions depend upon minimal usage of solvents (Kadam et al. 2015). The solvents used in these procedures are usually metabolite-specific food-grade solvents.

Another method known as acoustic cavitation is also used for breaking down the cells, thereby reducing their size and increasing the contact surface area with the extraction solvents. These extraction procedures exhibit flexibility as they can be altered or curated by adjusting the viscosity, density, dielectric constant and diffusion rate that determine the solvating potential of the solvent used and its overall selectivity. These recent developments in secondary metabolite extraction from seaweeds have paved the way for furthering their exploration for sustainable applications.

6.7 Notion of Algal Biorefinery

The algal biorefinery notion promises to offer multiple assemblages of utility products ranging from food additives, natural food flavourings, bulk chemicals, nutraceuticals, and cosmeceuticals to biofuel (Griffiths et al. 2021). This concept depends upon the inventive utilization of algal biomass by targeting metabolites of interest using innovative and practical fractionation and isolation techniques for industrial-scale applications. Integrating the algal biorefinery concept with the existing industrial platforms and optimizing the processes while maintaining and preserving the product's bioactive properties and desired features is what needs to be addressed. Other potential metabolite applications include greenhouse gas emissions, fossil fuels, crop protection and resistance and bioplastic development. However, for all of the mentioned bioprospecting aspects of seaweed metabolites to work commercially, the analysis, optimization and practical implementation using viable technology shall have to work collectively.

6.8 Role of Bioprospecting Secondary Metabolites from Seaweeds as Functional Ingredients

In bioprospecting, there have been notable strides in technologies related to single-molecule and single-cell analysis. Current developments demonstrate a preference for computational approaches due to their ease, reliability and time efficiency. This shift has resulted in the transformation of the multidisciplinary approach, incorporating various scientific disciplines to unravel the mechanisms of individual biological entities. The abundance of extensive databases containing diverse metabolomic and proteomic information enhances the likelihood of discovering novel drugs or commercially valuable molecules through disseminating this information. Some of the top seaweed companies across the globe striding towards bioprospecting include Acadian Seaplants Limited (ASL): Nova Scotia, Canada; Cargill Incorporated: Minnesota, United States; DuPont de Nemours, Inc.: Delaware, United States; Irish Seaweeds: Northern Ireland, United Kingdom, Beijing Leili Marine Bioindustry

Inc. (Leili Group): Beijing, China, Mara Seaweed (Celtic Sea Spice Company): Edinburgh, Scotland, and Qingdao Gather Great Ocean Algae Industry Group (GGOG): Qingdao, China. A detailed list of seaweed companies at global scale is given in Table 6.1.

6.9 Cosmetics Industries

The shift from synthetic composites to organic goods has garnered attention from researchers due to seaweed's multifarious pharmacological actions with minimal cytotoxic implications on humans (Jayawardhana et al. 2023). Contemporarily, significant attention has grown for seaweeds as an ecological, safe and effective resource. It finds versatile applications in cosmetics, serving as an additive and as active ingredient (emulsifier, thickener and wetting agent) (Lopez-Hortas et al. 2021). Chlorophytes, in particular, contain active forms of vitamin B_3, such as niacinamide, nicotinate esters, and nicotinic acid which are incorporated into skincare products (El-Beltagi et al. 2022).

Seaweeds contain carotenoids and phycobilins, serving as natural colour enhancers known for their anti-allergic properties (Lopez-Hortas et al. 2021). One significant carotenoid, astaxanthin, displays a vital role as a potent antioxidant by scavenging free radicals. The presence of sulphuric compounds and terpenoids in seaweeds imparts unique odours to cosmetics, effectually enhancing the sensory characteristics of these compounds (El Hattab 2020). The mineral composition of seaweed closely resembles that of human skin, facilitating easy absorption. This quality makes skincare products derived from seaweed highly desirable for their numerous benefits, including moisturization, promotion of metabolism, reduction of acne and anti-ageing effects. Seaweed minerals and trace elements provide defence against ultraviolet rays. Phycocyanin, a pigment of the Rhodophytes, has received a nod from the Food and Drug Administration (FDA) as a natural colourant for foodstuffs and makeup products. Its acceptance is attributed to its safe, natural and decomposable properties, making it a focal point in the United States market.

The market has witnessed the introduction of diverse skincare products incorporating algae-based ingredients, highlighting the adaptability and effectiveness of such natural components. A prominent product is Helionori® by Gelyma and Helioguard365®, a sunscreen enriched with mycosporine-like amino acids (MAAs) sourced from *Porphyra umbilicalis*. The *Gracilaria* species has gained significant traction in the cosmetics industry owing to its stabilizing and gel formation properties. Commercial goods harnessing the benefits of *Gracilaria* species include a facial mask by Balinique®, a hydrogel soap by Sea Laria®, and a hydrating cream by Thalasso® (Torres et al. 2019). Consequently, obtaining bioactive combinations from algal origin emerges as a promising avenue for developing cosmeceutical agents in skincare, providing a diverse range of benefits across various cosmetic formulations. A list of the seaweed-based cosmetics companies and their products has been given in Table 6.2.

Table 6.1 Seaweed-based companies globally and their value-added products (Anilkumar 2022)

Name of the company/industry/ firm	Value-added product
Australia	
Alg Seaweed	Seaweed condiments and snacks
Australian Kelp Products Pty Ltd	Seaweed fertilizer, animal feed supplement (Sea Biscuits Stock Aid), carrageenan, alginate, agar
FutureFeed Pty Ltd	Animal feed supplement
NatraSol Sea Products	Plant biostimulant, animal feed supplement
Phyco Health	Food supplements (seaweed fibre and protein), cosmetics (skin care products and food (seaweed snacks, pasta))
Purify Co Pty Ltd	Cosmetics
Seasol International Pty Ltd	Liquid seaweed fertilizer
SeaStock Pty Ltd	Livestock feed supplement
Venus Shell Systems	Functional food ingredient, nutraceuticals, wound dressing
YelpKelp Pty Ltd	Dog feed supplement
China	
Andi-Johnson Group	Agar, carrageenan
Beijing Leili Marine Bioindustry Inc.	Seaweed fertilizer
Foodmate Co. Ltd.	Carrageenan
Greenfresh Food Co. Ltd.	Agar, carrageenan
High Hope Foods (Nanjing) Co. Ltd.	Carrageenan, sodium alginate, agar-agar
Hong Kong Sheli Ltd.	Carrageenan, agar, sodium alginate
Lauta Ltd.	Carrageenan
Leili Agrochemistry Co. Ltd.	Seaweed fertilizer
Mingfu Fujian Agar Co Ltd.	Agar
Nantong Oceans Flavor Food Co. Ltd.	Seaweed snacks
Qingdao Bright Moon Seaweed Group Co. Ltd.	Sodium alginate, calcium alginate, alginic acid, functional food ingredients
Qingdao Gather Great Ocean Algae Industry Group Co. Ltd.	Sodium alginate, carrageenan, agar, propylene glycol alginate, seaweed feeds, seaweed fertilizers, seaweed foods
Qingdao Ocean Algae Industry Group Co. Ltd.	Alginate, alginic acid, carrageenan, agar, seaweed fertilizer
Qingdao Seawin Biotech Group Co. Ltd.	Seaweed extract, seaweed-based water soluble NPK fertilizer, seaweed granular fertilizer
Shandong Topsea Seaweed Industrial Co., Ltd.	Alginic acid, sodium alginates
Yan Cheng Hairui Food Co. Ltd.	Roasted seaweed, seasoned seaweed, seaweed roll
Yantai Jiate Bio-Tech. Co., Ltd.	Seaweed foliar fertilizer, plant growth regulator
Yantai Sheli Hydrocolloids Co. Ltd.	Carrageenan, agar, sodium alginate
Fiji	

(continued)

Table 6.1 (continued)

Name of the company/industry/firm	Value-added product
Sea & Soil	Seaweed fertilizer, cosmetics
Israel	
Haifa Group	Plant biostimulant, soluble NPK fertilizers enhanced with seaweed extracts
Seakura	Seaweed cracker, nutritional supplements
Sealaria Ltd.	Cosmeceuticals (skin care products)
Japan	
Asahi & Co., Ltd.	Agar
Green Science Alliance Co. Ltd.	Bioplastics
KIMICA Corporation	Sodium alginate, ammonium alginate, potassium alginate, alginic acid
Maruesu Corporation	Seaweed snacks
Nisshinbo Holdings Inc.	Seaweed-based bioplastic
SWF Co. Ltd.	Cosmetics
Ventuno Co. Ltd.	Health food, cosmetics
New Zealand	
Abron Ltd.	Fertilizer
AgriSea New Zealand Ltd.	Animal feed supplement, plant biostimulant, seaweed salt
BioSea	Plant biostimulant
CH4 Global Inc.	Animal feed supplements
Eco Ora Ltd.	Dietary supplement
Enzalg	Dietary supplement
Farmlands Co-operative Society Ltd.	Plant biostimulant
Kereru Brewing Company	Seaweed beer (Kereru Karengose Salty)
KiwiWakame	Food condiments
New Zealand Southern Pacific Seaweed Company	Health supplement
Pacific Harvest Seaweeds	Seaweed flakes, seaweed powder, seaweed strips, agar
Syrene Skincare NZ Limited	Cosmetics
VOYA NZ	Cosmetics
Waikaitu Ltd.	Plant biostimulant
Republic of Korea	
Dysskorea Co., Ltd.	Seaweed snacks, roasted seaweed
Botamedi Inc.	Nutraceuticals, cosmetics, personal hygiene products etc. (prepared from polyphenol extract SEANOL®)
Manjun Seaweed	Seaweed snacks
The Garden of Naturalsolution	Cosmetic ingredients (Seactive™)
Marine Innovation Co., Ltd.	Seaweed-based plates, cups, egg cartons/ trays, fruit trays, coffee carriers
Haewon Biotech Inc.	Mannitol, alginic acid, fucoidan, fucoxanthin, nutraceutical
Singapore	

(continued)

Table 6.1 (continued)

Name of the company/industry/firm	Value-added product
Nutri-San Pte Ltd.	Animal feed supplement
Seychelles	
Seaweed Seychelles	Plant growth promoter, soil conditioner
South Africa	
Afrikelp	Plant biostimulant
Kelp Products International (Pty) Ltd	Liquid fertilizer, plant biostimulant (Kelpak®)
Malaysia	
Tacara Sdn Bhd	Kappa and iota carrageenan products for food & nonfood industries
Indonesia	
PT. Seaweedtama Biopac Indonesia	Bioplastics
Denmark	
FermBiotics	Pet food supplement (Petferm™)
Jens Møller Products ApS	Seaweed caviar (Cavi·Art® &Tosago®), seaweed pearls (Food art ®)
Melissa Organic Skincare	Cosmetics (skin care products)
Wavy Wonders	Seaweed snacks
Nordisk Tang	Seaweed beer, rum, snacks, truffle, seaweed salt, seaweed spice
Seaman Seaweed Chips	Seaweed chips
Finland	
Origin by Ocean	Sodium alginate (Caerulo), fucoxanthin (Axureo), fucoidan (Cumatilo), laminarin (Livido), sunscreen
Estonia	
Vetik OÜ	Phycoerythrin
Est-Agar AS	Furcellaran
France	
Bretagne Cosmétiques Marins	Cosmetics
Algotherm	Cosmetics
Gelyma	Cosmeceuticals (skin care products)
Nexira	Dietary supplements (ID-alGTM—weight loss solution)
Algopack	Bioplastics
Eranova	Bioplastics
Penn Ar Bed Company – Marine Biotechnologies	Seaweed biostimulant, seaweed fertilizer
Agrimer	Liquid fertilizer, animal feed supplement, cosmetics etc.
Thalion	Cosmetics
Phytomer	Cosmetics

(continued)

Table 6.1 (continued)

Name of the company/industry/firm	Value-added product
Seppic SA	Cosmetics (hair care—BIORESTORER™ PF, BIOENERGIZER™ P BG PF; skin care—AMBRE OCEANE™ SPE, ASPAR'AGE etc.)
The Plantin Company	Seaweed liquid fertilizer
Timac Agro International	Plant biostimulant
Germany	
Viva Maris GmbH	Seaweed sauces, pasta, bread spread, sausage alternatives
NOF Europe GmbH	Cosmetics ingredient (ECKLEXT® BG)
Smartfiber AG	Seaweed yarn (SeaCell™, Smartcel™)
Nordic Oceanfruit	Seaweed salads, toppings, burger patties
Neomed Pharma GmbH	Plant biostimulant, animal feed supplement, cosmetics etc.
Compo Expert GmbH	Plant biostimulant, liquid NK fertilizer, organic-mineral NK fertilizer
Ireland	
Donegal Seaweed	Seaweed fertilizer, plant biostimulant
Brandon Products Ltd.	Plant biostimulant
BioAtlantis Ltd.	Plant biostimulant
C-Bio Limited	Plant biostimulant
AlgAran Seaweed Products	Seaweed capsules, flakes, anti-ageing serum, shampoo, face cream, body lotion, cleansing lotion, hand cream, lip balm
Emerald Isle Seaweed	Seaweed flake capsules, seaweed granules, seaweed fertilizer, animal feed
Irish Seaweeds	Seaweed facial wrap, powder bath, smoked seaweed snacks, seaweed herbs, seaweed capsules
Wild Irish Seaweeds Ltd.	Seaweed sprinkles, smoothie, powder, cosmetics
Seavite Bodycare Ltd.	Cosmetics
Nutramara Ltd.	Cosmeceutical & nutraceutical formulations, fucoidan
Origin Spirits Ireland Ltd	Seaweed whisky
Ocean Harvest Technologies Limited	Animal nutritional supplement
Shamrock Farm Enterprises	Seaweed fertilizer, animal feed supplements
Sealac Ltd	Seaweed-based electrolyte for pigs & livestock, seaweed meal for equines, seaweed powder for calf
Fruit Hill Farm	Seaweed fertilizer, animal feed supplement
Oilean Glas Teoranta	Seaweed meal, plant biostimulant
Plantruption	Seaweed burger
RÍ NA MARA Irish Seaweed Cosmetics	Cosmetics
The Netherlands	
Koppert Biological Systems	Plant biostimulant
The Dutch Weed Burger BV	Seaweed burger
Olijck	Seaweed ravioli and tagliatelle, seaweed burgers, etc.

(continued)

Table 6.1 (continued)

Name of the company/industry/firm	Value-added product
UmaMeats	Seaweed burger, sausage
Italy	
Valagro SpA	Plant biostimulant
Biolchim SPA	Plant biostimulant
Norway	
Algea AS	Plant biostimulant, animal feed ingredient
AlgiPharma AS	Pharmaceuticals
SJY Seaweed	Seaweed snacks
B'ZEOS AS	Seaweed-based biolplastic
Portugal	
Bodyocean	Cosmetics (skin care)
Lusalgae	Cosmetics (Sealgae®)
Inclita Seaweed Solutions	Seaweed-based bioactive ingredients
Poland	
Biotrem	Seaweed-based tableware and packaging
Sweden	
Seatrients	Seaweed smoothies
Algine Products Sweden	Food supplements
Saint Lucia	
Algas Organics	Plant biostimulant
Mexico	
Palau BioQuim, S.A. de C.V.	Seaweed fertilizer, plant biostimulant
Tecniprocesos Biologicos, S.A. de C.V.	Seaweed fertilizer, plant biostimulant
Química Sagal, S.A. de C.V	Seaweed fertilizer, plant biostimulant
Salgax Biotecnologia Marina Aplicada	Seaweed fertilizer, cosmetics (skin care)
Dianco México, SAPI de CV	Biofertilizers, soil improver
Chile	
Amarea	Seaweed snacks
Quelp	Seaweed burger, nuggets and meatball
Nun Spa	Seaweed pasta
India	
Marine Hydrocolloids	Agar, agarose, carrageenan, sodium alginate, pulverized seaweed
Biostadt India Limited	Plant growth stimulant, fish growth stimulant
Acadian Seaplants India Private Limited	Plant growth stimulant
Dhanuka Agritech Limited	Plant growth regulator
PI Industries Ltd.	Plant growth stimulant, biofertilizer
AquaRev	Carrageenan, stabilizer for ice cream, khoya & paneer
Sathyam Kissan Care	Plant growth stimulant

(continued)

Table 6.1 (continued)

Name of the company/industry/ firm	Value-added product
Agri Life – SOM Phytopharma (India) Limited	Biopesticides, biofertilizers, biostimulants
Vikas Crop Care	Biofertilizers, biostimulants
Karma Chemicals	Biofertilizers, biostimulants
Wellspring Bio-Organics	Biofertilizers
Srinivasa Marine Chemicals	Agar-agar
Jumat Agar-Agar Industries	Agar-agar
East Coast Seaweed Inc.	Biofertilizers, biostimulants
Prathibha Biotech	Biofertilizers, biostimulants
Nitesh Agro Industries	Biostimulants
Saosis Biotech Private Limited	Biofertilizers
Humates and Seaweeds Pvt. Ltd.	Biofertilizers
Sujay Biotech Pvt. Ltd.	Biofertilizers
Indorama India Private Limited	Plant growth promoter
Dew Specialty Chemicals (P) Ltd.	Plant growth promoter
Sarda Biopolymers Pvt. Ltd.	Carrageenan
Hifield – Ag Chem. India Private Limited	Seaweed powder, flakes, plant growth promoter
AquaAgri Processing Private Ltd.	Carrageenan, plant growth stimulant, animal feed ingredients
Prasmo Agri	Biostimulants
Pioneer Pharmaceuticals	Seaweed-based nutraceuticals
Ideal Biosciences	Seaweed-based shrimp and fish feed supplements
Maziq Farm Technologies (P) Ltd.	Biostimulants
VLCC	Nutraceuticals
Tata Chemicals	Plant growth regulator
Lucid Colloids Ltd.	Carrageenan (Carraluc Ô)
Exotic Biosolutions Private Limited	Seaweed-based poultry and cattle feed supplements
SUBONEYO Chemicals Pharmaceuticals (P) Limited	Seaweed extract-based organic fertilizer
Pinnacle Biosciences	Seaweed-based fertilizer, animal feed supplements, nutraceuticals, cosmetics
Gayatri Herbals Pvt. Ltd.	Seaweed-based biostimulants
Parshv Chem Industries	Seaweed granules
HiMedia Laboratories	Agar, alginic acid
Redox Industries Limited	Seaweed extract flake
Aushadh Limited	Seaweed extract flake, liquid seaweed fertilizer
Sikko Industries Ltd.	Seaweed plant growth promoter

(continued)

Table 6.1 (continued)

Name of the company/industry/ firm	Value-added product
Manidharma Biotech Private Limited	Liquid fertilizer (biostimulant) (organic six)
Aquatic Chemicals	Biostimulant
VNET	Biostimulant
Agrocare India Private Limited	Seaweed-based enzymes, seaweed granules
Seaweed Fertilizer Co.	Semi-refined carrageenan, seaweed fertilizers
AK Seaweeds Company	Alginate, seaweed fertilizers
Purvraj Agro Industries Private Limited	Seaweed fertilizer, soil conditioner
Agrocart	Seaweed fertilizer
Laford Agrotech Limited	Seaweed extract flakes, granules, gel
AB Agritech Private Limited	Seaweed extract
Scarlet Marine Chemicals	Seaweed extract, fertilizer
Excel Crop Care Limited	Foliar plant nutritional supplement (excel AeROS)
AURA Biotechnologies Private Limited	Agar-agar, seaweed fertilizer
SNAP Natural and Alginate Products Pvt. Ltd.	Alginates, carrageenan, plant growth stimulant, aquaculture inputs, etc.

6.10 Agriculture Industries

Bioprospecting secondary metabolites from seaweeds for use as functional ingredients in agriculture industries involves exploring the prospective profits of these compounds in enhancing plant development, health and overall agricultural productivity. Extracts derived from various seaweed species have gained wide acceptance in the field of plant biostimulants. Marine macroalgae, in particular, are considered treasured sources for enhancing plant growth owing to high levels of amino acids, minerals, vitamins, and growth regulators (auxins, cytokinins, gibberellins) (Sharma et al. 2014). The use of seaweeds in plant biotechnology is associated with cultivating robust and healthy plants, resulting in a noteworthy increase in the amount and mass of fruits. Additionally, these seaweed-based solutions present a non-hazardous alternate for disease control (Mondal et al. 2014).

Seaweed-based fertilizers have proven effective in reducing seed germination time while simultaneously improving various aspects of plant development, including physical, genetic and biochemical properties (Shukla et al. 2019). The positive impact extends to enhanced asexual growth, encompassing leaf count, root and shoot development, leaf basal area, etc., with an improvement in the fresh weight (FW) as well as dry weight (DW). Furthermore, multiplicative features viz; floral set-up, fruit mass, and harvest are positively influenced using seaweed-based fertilizers (Mzibra et al. 2021). The incorporation of seaweed-based infusion into plants augments the resistance capability, thereby providing protection against plant pathogens due to the specific secondary metabolites (Mrid et al. 2021).

Table 6.2 Indian seaweed-based cosmetics companies and products

Name of company	Value-added products
Blossom Kochhar Beauty Products Pvt. Ltd., Roorkee, Uttarakhand	Hydrating face pack
Myra Veda Organics Pvt. Ltd., Mumbai	Seaweed scrubs
Bio Veda Action Research Company, Sirmaur, Himachal Pradesh	Bio Seaweed revitalizing anti-fatigue eye gel
Shahnaz Ayurvedics, Roorkee, Uttarakhand	Seaweed under-eye mask
Kaya Limited, Mumbai	Skin hydrating night mask face gel
Kulsum's Kaya Kalp Herbals Pvt Ltd, Jaipur	Under-eye dark circle seaweed gel
Good Vibes, Palghar, Maharashtra	Hydrating and age defying night gel
AM Enterprises, Solan, Himachal Pradesh	Anti-tan face mask
O3+ Skin Solution, Noida	Face emulsion, skin cleansing gel, face mask, face cream, acne control serum etc.
Cosmic Nutracos Solutions Private Limited, New Delhi	Insta glow face mask
Indus Cosmeceuticals, Shimla, Himachal Pradesh	Acne gel
Indrani Cosmetics, Pune, Maharashtra	Face/skin gel
VLCC, Gurgaon	Skin bleach

Seaweed-based fertilizers contribute to increased resilience to heat and salt stress, enabling plants to withstand adverse environmental conditions (Battacharyya et al. 2015). Additionally, the organic compounds within these fertilizers promote prolonged moisture retention and nutrient availability in the soil. Commercially available seaweed-based fertilizers are typically derived from various seaweed species viz; *Ulva* sp., *Kappaphycus alvarezii*, *Sargassum* sp., *Ascophyllum nodosum*, *Ecklonima maxima*, *Laminaria* sp., etc. The commercial cultivation of seaweeds for catering to industrial demands is interlinked. Some examples of the macroalgae-based products used commercially are as follows (Khan et al. 2009).

Product name	*Seaweed species*	*Use*
Fartum	*Eclonia nodosum*	Biofertilizer
Kelpak	*Durvillea antarctica*	Plant growth stimulant
Acadian	*Ascophllum nodosum*	Plant growth stimulant
Seasol	Unspecified	Plant growth stimulant
Profert	Unspecified	Plant biostimulant
Agri-Grow Ultra	*Nodosum macrocystis*	Plant growth stimulant
Sea winner	*Durvillea potatorum*	Plant growth stimulant
Seanure	Unspecified	Plant growth stimulant
AgroKelp	*Ascophllum nodosum*	Plant growth stimulant

Some examples of the commercial seaweed products of Indian origin used in agriculture comprise the following:

1. MicroLife Super Seaweed Fertilizer

2. OrganicDews Liquid Seaweed Concentrate
3. Garden Essentials Seaweed Fertilizer
4. Sea Secret—Seaweed Extract Liquid Concentrate
5. Casa De Amor Essential Seaweed Extract Liquid Fertilizer
6. PI Industries Organic BIOVITA Granules Seaweed Concentrate Fertilizer
7. Naturegoon Concentrate Liquid Seaweed Fertilizer
8. TrustBasket Seaweed Extract Liquid Concentrate
9. Shiviproducts Seaweed Flakes Powder
10. Holy Green—Nurture the Future Liquid Virgin Seaweed Fertilizer
11. Sea Ruby (liquid) green seaweed extract
12. Down to Earth organic Kelp meal
13. Simple grow solution seaweed fertilizer with humic acid
14. Sea magic dry soluble seaweed extract
15. Neptune's Harvest seaweed fertilizer
16. Feromones organic seaweed liquid fertilizer
17. Black amrut liquid organic fertilizer
18. Homecrop Biosmart
19. Gold Dust Seaweed liquid fertilizer
20. Aryabio Balada Seaweed Concentrate
21. Botanicare Seaplex
22. Medina organic liquid seaweed
23. Sagarika
24. Shelbysbe organic micronutrient seaweed fertilizer
25. Katyayani seaweed extract fertilizer
26. Gacil seaweed extract liquid fertilizer

6.11 Food/Health/Nutraceutical Industries

The edible seaweed food products industry is witnessing substantial growth projections, prompting a shift towards sustainable practices in seaweed cultivation to meet the rising demand (Rogel-Castillo et al. 2023). Seaweeds, abundant and economically viable, have emerged as a promising resource for use in food products, offering nutritional benefits and bioactive compounds with desirable technological properties. The chemical composition of seaweeds encompasses a diverse array of bioactives like amino acids, polyphenols, unsaturated fats, sulphated polysaccharides, etc., exhibiting antioxidant characteristics. Notably, phlorotannins, fucoidans and carotenoids stand out as significant components.

The precise control of the concentration of these compounds is crucial, as sensory quality may not always be favoured. Therefore, the incorporation of seaweed into food products presents an interesting challenge. In recent years, seaweeds have undergone re-evaluation and gradual integration into diets, especially among vegans and vegetarians. With their high polyphenolic secondary metabolite content, seaweeds offer phlorotannins as potential substitutes for synthetic antioxidants in the food industry due to their noteworthy antioxidant capacity linked to their structural characteristics (Fernando et al. 2022). Carotenoids, including xanthophylls and tocopherols, are also efficient antioxidants found in seaweeds, with fucoxanthin being a notable quencher of singlet oxygen. Tocopherols, widely utilized in the food industry, exhibit effective radical scavenging activity.

Seaweeds have found application in various meat products such as burgers, frankfurters, steaks, patties and sausages, serving a dual purpose as antioxidants to preserve meat quality and as fat replacers to create low-fat alternatives (Espinosa-Ramirez et al. 2023). Ingredients derived from seaweed extracts have emerged in the market as viable alternatives to traditional salt in various food formulations. A notable example of such a commercial ingredient is Algysalt®, a blend of seaweed components that includes *Ulva lactuca*, *Ascophyllum nodosum*, *Lithothamnium calcareum*, *Palmaria palmata*, *Himanthalia elongata*, *Porphyra umbilicalis*, *Fucus vesiculosus*, *Laminaria saccharina*, *Chondrus crispus* and *Undaria pinnatifida*. The antioxidation features of macroalgae are attributed to the presence of tocopherols, alkaloids, polyphenols, carotenoids and terpenes. Phaeophytes like *Fucus vesiculosus*, *Ascophyllum nodosum* and *Bifurcaria bifurcate* have demonstrated their ability to replace synthetic antioxidants and provide stability to lipids and proteins in the resultant product (Penalver et al. 2020). These positive effects were similarly observed in pork patties, explaining the favourable impact of phlorotannin content in *Fucus vesiculosus* on oxidation stability and sensory attributes.

6.12 Pharmaceutical Industries

Synthetic generation and/or alteration represent prevalent alternative methods for obtaining bioactive compounds, a category that encompasses a majority of the marine-originated medicines presently undergoing clinical trials and advancement. Certain composites can undergo total synthesis, while others may undergo structural modifications to enhance their effectiveness and/or minimize toxicity. Secondary metabolites display a wide array of functional characteristics, primarily revolving around pharmaceutical applications, including drug exploration and advancement (Kiruba et al. 2018), the creation of antiviral (Lomartire and Gonçalves 2022) and antibacterial agents (Fakee et al. 2023), as well as possessing anti-inflammatory (Vairappan et al. 2013), neuroprotective (Nho et al. 2020), antioxidant (Dixit 2023), cancer therapeutic (Lee et al. 2021), curative properties (Premarathna et al. 2023), anti-herpes virus activity (Ana et al. 2021), and anti-obesity and hypertension (Seca and Pinto 2018).

Pharmaceutical companies and investigators focusing on medicine are currently centring on various substances for drug development and medical research. These substances include antifouling compounds such as halogenated furanones derived from *Delisea pulchra*, and kahalalide F from a *Bryopsis* species, explored for their potential in treating lung cancer, tumours and AIDS (Qin 2020). Additionally, algal-derived components such as fucoidans, lectins and kainoids are regularly employed in biomedical studies. Numerous additional constituents with documented biological virtues are also under investigation, contributing to the diverse landscape of ongoing research in the pharmaceutical and medical fields.

6.13 Seaweed-Based Startups and Their Products in India

A few seaweed startups are operating in India, focussing on value addition to seaweed products. Here, we highlight some significant Indian startups in this sector and details about their respective products.

6.13.1 Zerocircle Alternatives Pvt. Ltd

Zerocircle Alternatives Private Limited is dedicated to developing eco-friendly alternatives to everyday products, utilizing renewable resources such as seaweeds to contribute to a circular economy. Zerocircle is venturing into various sectors, including food packaging (films, cups, bottles, and laminates), shoe soles (completely compostable soles for high-performance shoes), medical applications (materials to replace synthetic gloves, masks, and medical sutures), future food (vegan protein formulations incorporating plant-based B-12, omega-3 fatty acids, minerals, and vitamins), textiles (fabrics based on bioyarn), and courier packaging (durable materials for E-tailers to withstand long-distance transport). The biodegradable products are crafted from both edible and non-edible seaweeds.

6.13.2 Zaara Biotech

Zaara Biotech, located in Kerala, is a biotechnology startup with a mission to establish a brand that offers nutritionally superior, high-quality and affordable products to enhance the lifestyles of the general population. The company has introduced a seaweed-based cookie known as 'B-lite Cookies' as part of its product line. Distinguished by its rapid energy-boosting capabilities and elevated protein and dietary fibre levels, this cookie stands out among other biscuit products. Additionally, the cookie is enriched with essential minerals like iron and calcium and is free from preservatives, trans fat and cholesterol. The presence of superoxide dismutase in the cookie is anticipated to contribute to slowing down the ageing process in humans.

6.13.3 Agastya Marine Pvt. Ltd

Agastya Marine Pvt. Ltd. is a startup enterprise focused on cultivating seaweed and its subsequent processing. The startup's primary goal is to cultivate seaweed to produce various value-added products, including agar-agar, agarose, biostimulants and organic fertilizers.

6.13.4 Sea6 Energy Pvt. Ltd

Sea6 Energy Pvt. Ltd., headquartered in Bangalore, is driven by the vision of harnessing the oceans sustainably and ethically to establish a future for humanity that is not reliant on conventional energy sources. The company's manufacturing facilities are engineered to operate with minimal freshwater and chemical usage, and its production processes are certified organic according to standards equivalent to those in India and the EU. The diverse range of products offered by Sea6 Energy includes plant biostimulants and liquid fertilizers for agriculture, immunity boosters for aquatic animals and poultry in the animal feed sector, food gelling agents as a food ingredient, and biopolymers such as biofilms and bioplastics. Additionally, the company is involved in the production of biofuels.

6.13.5 Oceanic Organics

Oceanic Organics, a pioneering startup, aims to revolutionize and implement sustainable business models by integrating the blue and green economies. The startup is committed to developing non-chemically processed products and creating alternative sources of income for coastal and rural communities in India. Oceanic Organics produces a natural growth promoter derived from seaweeds, marketed under the brand name 'O-Plants'. This product is composed of plant growth hormones and various beneficial microbes, fostering the comprehensive and healthy growth of plants.

6.14 Challenges in the Trajectory

The prospective of algae as a basis for bioactive compounds for their usage as pharmaceutical products and/or functional food components is well-established. Algae's metabolic plasticity, allowing the tuning of metabolite production through cultivation conditions or metabolic engineering, positions them as natural reactors. They are also pivotal in a biorefinery platform as a sustainable foundation of biomass for biodiesel production, food and a rich source of secondary metabolites. Exploring marine secondary metabolites encounters challenges due to the vastness and complexity of marine ecosystems, necessitating extensive sampling efforts and efficient extraction techniques. Sustainable harvesting of marine organisms and conserving fragile ecosystems are pivotal considerations. However, recent technological strides, including omics and synthetic biology, offer novel opportunities to surmount these challenges and expedite the discovery and production of marine-derived compounds. Despite the tremendous progress, challenges remain in omics and *in silico* bioprospecting. Integration of multiple omics datasets, standardization of data analysis pipelines, and improvement in computational algorithms are ongoing efforts. Additionally, ethical considerations regarding admittance to molecular information, intellectual property rights and rational benefit allocation need to be addressed to ensure the sustainable utilization of biodiversity.

Numerous challenges must be tackled to attain economic viability and genuine sustainability within the industry. A primary concern is the industry's heavy reliance on biomass availability. National regulations governing coastal areas limit the industrial application of macroalgae exclusively to highly available or invasive species. This approach leaves significant biodiversity untapped and raises concerns about potential over-exploitation. The scientific methodology employed in macroalgae harvesting for sustainable energy manufacturing needs to be revised, as the current methods may result in product inconsistencies due to ecological disparities. Addressing the critical query of establishing a robust industry that fosters the progress of reliable, effective and frugally sustainable products becomes paramount. Additionally, the price range involving manufacturers and distributors poses affordability challenges for various categories of agriculturalists (El Boukhari et al. 2020). Thus, determining the optimal strategy for overcoming these hurdles and ensuring widespread accessibility to sustainable products remains crucial.

6.15 Future Prospective

Exploration into bioprospecting holds promise for fostering sustainable innovation. This practice diminishes dependence on finite resources by unveiling valuable compounds and offers alternative solutions to address prevalent societal issues. Furthermore, bioprospecting can create economic incentives for preserving natural habitats by showcasing their fundamental value for scientific and commercial purposes. Additionally, on the scientific front, comprehensive research is requisite for screening secondary metabolites produced by diverse marine entities, genetic understanding and sequencing, annotating, understanding metabolic synthesis pathways and strategically using this information for sustainable food, health and energy production. Optimization of various processes, including biomass production through advanced photobioreactor design, bioactive recovery via downstream operation optimization, and further processing of residual biomass for bulk chemicals and biofuels, is essential.

6.16 Conclusion

Contemporary bioprospecting has multiple goals, encompassing biodiversity protection, management of natural assets and sustainable economic expansion. Integrating omics approaches in seaweed bioprospecting allows for a systems-level understanding of seaweed biology and identifying critical molecular players in synthesizing bioactive compounds. Bioinformatics tools and computational approaches play a crucial role in analysing and integrating omics data, facilitating the discovery of novel candidate molecules and predicting their potential biological activities. It is crucial to conduct bioprospecting activities responsibly and ethically, ensuring fair benefit-sharing and respecting the rights of local communities. By embracing bioprospecting, we can unlock nature's secrets, developing groundbreaking solutions to global challenges while safeguarding the environment for future generations.

References

Ana P, Nathalie B, Gilles B, Daniel R, Tomas MS, Yolanda FP (2021) Anti-Herpes simplex virus (HSV-1) activity and antioxidant capacity of carrageenan-rich enzymatic extracts from *Solieria filiformis* (Gigartinales, Rhodophyta). Int J Biol Macromol 168:322–330

Anilkumar PK (2022) Mapping of technologies for value-addition of seaweeds in India. Technology Information, Forecasting and Assessment Council, New Delhi

Barta DG, Coman V, Vodnar DC (2021) Microalgae as sources of omega-3 polyunsaturated fatty acids: biotechnological aspects. Algal Res 58:102410

Battacharyya D, Babgohari MZ, Rathor P, Prithiviraj B (2015) Seaweed extracts as biostimulants in horticulture. Sci Hortic 196(30):39–48

Dixit D (2023) Untargeted secondary metabolite profiling and putative pathway deduction of a marine Ochrophyte – *Iyengaria stellata* (Børgesen) using liquid chromatography-mass spectrometry. Food Biosci 51:102211

Dixit D, Reddy CRK, Trivedi MH, Gadhavi DK (2020) Non-targeted metabolomics approach to assess the brown marine macroalga *Dictyota dichotoma* as a functional food using liquid chromatography with mass spectrometry. Sep Sci Plus 3(5):140–149

Dixit S, Shukla A, Singh V, Upadhyay SK (2021) Bioprospecting of natural compounds for industrial and medical applications current scenario and bottleneck. In: Upadhyay SK, Singh SP (eds) Bio prospecting of plant biodiversity for industrial molecules, pp 53–71

Dong Z, Chen Y (2013) Transcriptomics: advances and approaches. Sci China Life Sci 56:960–967

Efferth T, Banerjee M, Paul NW, Abdelfatah S, Arend J, Elhassan G, Hamdoun S, Hamm R, Hong C, Kadioglu O, Naß J, Ochwangi D, Ooko E, Ozenver N, Saeed ME, Schneider M, Seo EJ, Wu CF, Yan G, Zeino M, Zhao Q, Abu-Darwish MS, Andersch K, Alexie G, Bessarab D, Bhakta-Guha D, Bolzani V, Dapat E, Donenko FV, Efferth M, Greten HJ, Gunatilaka L, Hussein AA, Karadeniz A, Khalid HE, Kuete V, Lee IS, Liu L, Midiwo J, Mora R, Nakagawa H, Ngassapa O, Noysang C, Omosa LK, Roland FH, Shahat AA, Saab A, Saeed EM, Shan L, Titinchi SJ (2016) Biopiracy of natural products and good bioprospecting practice. Phytomedicine 23(2):166–173

El Boukhari MEM, Barakate M, Bouhia Y, Lyamlouli K (2020) Trends in seaweed extract based biostimulants: manufacturing process and beneficial effect on soil-plant systems. Plan Theory 9(3):359

El Hattab M (2020) Algae essential oils: chemistry, ecology, and biological activities. In: de Oliveira MS, Silva S, Da Costa WA (eds) Essential oils: bioactive compounds, new perspectives and applications. IntechOpen, London, pp 1–27

El-Beltagi HS, Mohamed AA, Mohamed HI, Ramadan KMA, Barqawi AA, Mansour AT (2022) Phytochemical and potential properties of seaweeds and their recent applications: a review. Mar Drugs 20(6):342

Espinosa-Ramirez J, Mondragon-Portocarrero AC, Rodríguez JA, Lorenzo JM, Santos EM (2023) Algae as a potential source of protein meat alternatives. Front Nutr 10:1254300

Fakee J, Bolton JJ, Le Roes-Hill M, Durrell KA, Antunes E, Beukes DR (2023) Antimicrobial activity of the secondary metabolites isolated from a South African red seaweed, *Laurencia corymbosa*. Molecules 28(5):2063

Fernando IPS, Lee WW, Ahn G (2022) Marine algal flavonoids and phlorotannins; an intriguing frontier of biofunctional secondary metabolites. Crit Rev Biotechnol 42(1):23–45

Fort A, Guiry MD, Sulpice R (2018) Magnetic beads, a particularly effective novel method for extraction of NGS-ready DNA from macroalgae. Algal Res 32:308–313

Griffiths G, Hossain AK, Sharma V, Duraisamy G (2021) Key targets for improving algal biofuel production. Clean Technol 3(4):711–742

Hemaiswarya S, Raja R, Ravikumar R, Kumar AY, Carvalho IS (2013) Microalgal omics and their applications. In: Barh D, Zambare V, Azevedo VO (eds) Applications in biomedical, agricultural, and environmental sciences. CRC Press, Taylor and Francis, Boca Raton, London, p 439

Hurst D, Børresen T, Almesjö L, Raedemaecker F, Bergseth S (2016) Marine biotechnology strategic research and innovation roadmap: insights to the future direction of European marine biotechnology. ERA-NET Marine Biotechnology

Ibanez E, Cifuentes A (2013) Benefits of using algae as natural sources of functional ingredients. J Sci Food Agric 93(4):703–709

Jayawardhana HHACK, Jayawardena TU, Sanjeewa KKA, Liyanage NM, Nagahawatta DP, Lee HG, Kim JI, Jeon YJ (2023) Marine algal polyphenols as skin protective agents: current status and future prospectives. Mar Drugs 21(5):285

Kadam SU, Alvarez C, Tiwari BK, O'Donnell CP (2015) Extraction of biomolecules from seaweeds. In: Tiwari BK, Troy DJ (eds) Seaweed sustainability food and non-food applications. Academic Press, pp 243–269

Kamble AD, Singh H (2022) Finding novel enzymes by *in silico* bioprospecting approach. In: Kuddus M, Aguilar CN (eds) Value-addition in food products and processing through enzyme technology. Academic Press, pp 347–364

Khan W, Rayirath UP, Subramanian S, Jithesh MN, Rayorath P, Hodges DM, Critchley AT, Craigie JS, Norrie J, Prithiviraj B (2009) Seaweed extracts as biostimulants of plant growth and development. J Plant Growth Regul 28:386–399

Kiruba NJM, Pradeep MA, Thatheyus AJ (2018) Discovering promising anti-cancer drug candidates from marine algae. Sci Int 6:44–50

Lee H, Selvaraj B, Lee JW (2021) Anticancer effects of seaweed-derived bioactive compounds. Appl Sci 11(23):11261

Liu J, Li J, Qu J, Yan S (2015) Development of genome-wide insertion and deletion polymorphism markers from next-generation sequencing data in Rice. Rice 8(1):63

Lomartire S, Gonçalves AMM (2022) Antiviral activity and mechanisms of seaweeds bioactive compounds on enveloped viruses—a review. Mar Drugs 20(6):385

Lopez-Hortas L, Florez-Fernandez N, Torres MD, Ferreira-Anta T, Casas MP, Balboa EM, Falque E, Dominguez H (2021) Applying seaweed compounds in cosmetics, cosmeceuticals and nutricosmetics. Mar Drugs 19(10):552

Malve H (2016) Exploring the ocean for new drug developments: marine pharmacology. J Pharm Bioallied Sci 8:83–91

Marco DE, Abram F (2019) Using genomics, metagenomics and other "omics" to assess valuable microbial ecosystem services and novel biotechnological applications. Front Microbiol 10:151

Mondal D, Ghosh A, Prasad K, Singh S, Bhatt N, Zodape ST, Chaudhary JP, Chaudhari J, Chatterjee PB, Seth A, Ghosh PK (2014) Elimination of gibberellin from *Kappaphycus alvarezii* seaweed sap foliar spray enhances corn stover production without compromising the grain yield advantage. Plant Growth Regul 75:657–666

Mrid RB, Benmrid B, Hafsa J, Boukcim H, Sobeh M, Yasri A (2021) Secondary metabolites as biostimulant and bioprotectant agents: a review. Sci Total Environ 777:146204

Mzibra A, Aasfar A, Khouloud M, Farrie Y, Boulif R, Kadmiri IM, Bamouh A, Douira A (2021) Improving growth, yield, and quality of tomato plants (*Solanum lycopersicum* L) by the application of Moroccan seaweed-based biostimulants under greenhouse conditions. Agronomy 11(7):1373

Nho JA, Shin YS, Jeong HR, Cho S, Heo HJ, Kim GH, Kim DO (2020) Neuroprotective effects of phlorotannin-rich extract from brown seaweed *Ecklonia cava* on neuronal PC-12 and SH-SY5Y cells with oxidative stress. J Microbiol Biotechnol 30(3):359–367

Penalver R, Lorenzo JM, Ros G, Amarowicz R, Pateiro M, Nieto G (2020) Seaweeds as a functional ingredient for a healthy diet. Mar Drugs 18(6):301

Pereira RC, Costa-Lotufo LV (2012) Bioprospecting for bioactives from seaweeds: potential, obstacles and alternatives. Rev Bras Farmacogn 22(4):894–905

Pierce JD, Fakhari M, Works KV, Pierce JT, Clancy RL (2007) Understanding proteomics. Nurs Health Sci 9(1):54–60

Premarathna AD, Tuvikene R, Somasiri M, De Silva M, Adhikari R, Ranahewa TH, Wijesundara R, Wijesekera SK, Dissanayake I, Wangchuk P, Rjabovs V, Jayasooriya AP, Rajapakse R (2023) A novel therapeutic effect of mannitol-rich extract from the brown seaweed *Sargassum ilicifolium* using in vitro and in vivo models. BMC Complement Med Ther 23(1):26

Qin Y (2020) Research, development and commercial applications of bioactive seaweed substances in health products. Ann Pub Health Epidemiol 1(4). https://doi.org/10.33552/APHE.2021.01.000516

Rogel-Castillo C, Latorre-Castaneda M, Munoz-Munoz C, Agurto-Munoz C (2023) Seaweeds in food: current trends. Plants 12(12):2287

Seca AML, Pinto DCGA (2018) Overview on the antihypertensive and anti-obesity effects of secondary metabolites from seaweeds. Mar Drugs 16(7):237

Sharma HSS, Fleming C, Selby C, Rao JR, Martin T (2014) Plant biostimulants: a review on the processing of macroalgae and use of extracts for crop management to reduce abiotic and biotic stresses. J Appl Phycol 26(1):465–490

Shukla PS, Mantin EG, Adil M, Bajpai S, Critchley AT, Prithiviraj B (2019) *Ascophyllum nodosum*-based biostimulants: sustainable applications in agriculture for the stimulation of plant growth, stress tolerance, and disease management. Front Plant Sci 10:655

Swinnen JV, Dehairs J (2022) A beginner's guide to lipidomics. Biochem (Lond) 44(1):20–24

Torres P, Santos JP, Chow F, dos Santos DYAC (2019) A comprehensive review of traditional uses, bioactivity potential, and chemical diversity of the genus *Gracilaria* (Gracilariales, Rhodophyta). Algal Res 37:288–306

Vairappan CS, Kamada T, Lee WW, Jeon YJ (2013) Anti-inflammatory activity of halogenated secondary metabolites of *Laurencia snackeyi* (Weber-van Bosse) Masuda in LPS-stimulated RAW 264.7 macrophages. J Appl Phycol 25:1805–1813

Zullig T, Trotzmuller M, Kofeler HC (2020) Lipidomics from sample preparation to data analysis: a primer. Anal Bioanal Chem 412:2191–2209

Chapter 7
A Blueprint of Global Seaweed Industry: Challenges and Future Perspectives

Cheng-gang Ren, Zheng-yi Liu, Zhi-hai Zhong, Chun-lin Chen, Ya-jun Shi, and Song Qin

Abstract The seaweed industry is a vital component of the marine economy in East and Southeast Asian coastal nations and is witnessing remarkable growth over the years. Global seaweed production surged from 10.6 million tons in 2000 to 35.82 million tons (fresh weight) in 2019, with a value exceeding $11 billion. Traditionally, seaweed finds applications in food (45–75%), hydrophilic colloids, and food additives. However, emerging markets in molecular gastronomy, dietary fiber, and prebiotics are paving the way for seaweed active substances like polysaccharides and oligosaccharides to become a prominent player in the future functional seaweed industry. Recent scientific research highlights the exceptional potential of algal polysaccharides as dietary fiber supplements due to their structural characteristics and indigestibility. Additionally, algal oligosaccharides possess unique antioxidant and probiotic effects, making them ideal for the growing health market. Technological advancements in polysaccharide and oligosaccharide production can lead to the production of diverse products catering to the emerging market needs. Guided by scientific support, the sustainable development of the seaweed industry thrives through functional advancements, guaranteeing the responsible utilization of seaweed resources.

Keywords Seaweed industry · Innovative products · Seaweed polysaccharides and oligosaccharides · Structure-activity relationship · Dietary fiber · Prebiotics

---

C.-g. Ren (✉) · Z.-y. Liu · Z.-h. Zhong · C.-l. Chen · Y.-j. Shi · S. Qin (✉)
Yantai Institute of Coastal Zone Research, Chinese Academy of Sciences, Laishan, Yantai, P. R. China
e-mail: cgren@yic.ac.cn; sqin@yic.ac.cn

N. Trivedi et al. (eds.), *Recent Advances in Seaweed Biotechnology*,
https://doi.org/10.1007/978-981-96-0519-4_7

7.1 Introduction

The seaweed industry encompasses various types of seaweeds, including farmed, wild harvested, and salvaged varieties of brown, red, and green ones. Over the past two decades (2000–2018), seaweed production has tripled, with significant consumption and production areas concentrated in Asian countries like China, Indonesia, the Philippines, Japan, and South Korea. These countries, with well-established seaweed industries, predominantly process seaweed into various food forms such as salads, sushi, nori rice, condiments, and miso soup (dashi). Additionally, seaweed is used as a food additive, including alginate, agar, and carrageenan. However, the seaweed industry and the range of seaweed-derived products have undergone significant diversification. On one hand, seaweed processing has entered the realm of Molecular Gastronomy, a modernist cuisine. On the other hand, seaweed-based medicinal foods have gained increasing importance in promoting human health.

The global population suffering from metabolic diseases, such as hyperglycemia, cardiovascular problems, overweight, and obesity, has witnessed a dramatic increase. Seaweed polysaccharides, with their satiating, indigestible, low-calorie, and intestinal fermentation properties, along with the antioxidant and probiotic properties of seaweed oligosaccharides, have demonstrated positive impacts in reducing the risk of lifestyle-related diseases and promoting intestinal health (Jiménez-Escrig and Sánchez-Muniz 2000; Shannon and Abu-Ghannam 2019). Concurrently, economic growth, technological innovation, market dynamics, and shifts in demand have brought profound changes to the global and regional seaweed industry leading to the development of new supply chains disrupting traceability.

In summary, the seaweed industry has transcended its traditional role as a primary food source. It has reshaped different industries, including natural additives, dietary fiber, and prebiotics. A new seaweed value chain, encompassing the comprehensive development of marine resources, supporting the human health industry, and targeting high-value-added markets, is gradually taking shape.

7.2 Global Seaweed Industry Pattern and Traditional Value Chain

7.2.1 Production of Raw Materials

Natural seaweed boasts a rich historical culinary tradition, with well-documented evidence of its consumption dating back over 2000 years in East Asian countries such as China and Japan. According to FAO statistics, recent years (2010–2017) have witnessed natural seaweed collection accounting for a mere 2–3% of the total global seaweed production, peaking at 1.1 million tons. The majority of these natural algae are utilized directly as feed, fertilizers, and chemical raw materials.

Examining the global distribution pattern of natural seaweed, countries like Norway, Ireland, and Canada along the Atlantic coast, the United States, Chile, and Peru along the Pacific coast, and Australia and New Zealand along the Southern Ocean coast possess abundant natural seaweed biomass. These regions contribute to 30–50% of the global annual seaweed production and encompass a wide variety of edible species, such as *Laminaria* and *Fucus*. However, due to factors such as food culture restrictions, ecological protection, labor costs, and transportation challenges, these countries have yet to establish mature seaweed food industry chains. Considering the concept of protective utilization of seaweed resources, developing a seaweed food industry represents a crucial expansion area for the formation of a new value chain.

In 1971, seaweed became one of the first aquatic species to achieve the milestone of being "more cultivated than salvaged" (Chopin 2018). Today, seaweed cultivation accounts for 51.3% of the total mariculture production, with eight genera dominating the world's seaweed aquaculture production (increasing from 72.3% in 2000 to 96% in 2011 and 96.8% in 2018). These genera include red algae species like *Eucheuma* spp. (29.0%), *Kappaphycus* spp. (4.9%), *Gracilaria* spp. (10.7%), *Porphyra/Pyropia* spp. (nori) (8.9%), as well as *Laminaria japonica* (35.3%), *Undaria pinnatifida* (wakame) (7.2%), *Sargassum* spp. (0.8%), and other *Phaeophyceae*. In 2010, the estimated annual global seaweed production reached 19 million tonnes of fresh weight, valued at approximately US $5.7 billion (FAO 2012; Vairappan et al. 2014).

By 2017, the world production of aquatic plants, primarily macroalgae, had reached 32.9 million tons, with 96.6% (31.8 million tons) obtained through cultivation (FAO 2017). In 2018, the global production of cultivated aquatic algae reached 32.4 million tons (97.1% of seaweed production) with a market value of $13.3 billion (FAO 2018). Seaweed cultivation, also known as mariculture, is primarily concentrated in Asia. In 2011, Asian countries accounted for 99.05% of the total seaweed production and 99.36% of the total output value (Chopin and Tacon 2021). Notably, China accounted for 63% of global production, while Indonesia and the Philippines contributed 14% and 11% respectively (Hardouin et al. 2014). The remaining regions of the world combined produced approximately 0.5% of seaweed.

In addition, seaweed cultivation witnessed a threefold increase between 2000 and 2018, with the development of the seaweed industry varying across countries. For instance, Indonesia's production of carannas (*Eucheuma* and *Kappaphycus* species) contributed to 28.8% of the world's seaweed aquaculture production, which also resulted in a doubling of seaweed production in China. However, China's share declined from 77.7% in 2000 to 57.1% in 2018, primarily due to various factors (Chopin and Tacon 2021). Furthermore, the significant growth in genera like *Eucheuma* and *Gracilaria*, known for their hydrocolloids production, led to red algae production (17.8 million tonnes) surpassing brown algae production (1320 tonnes) for the first time in 2015 (Chopin 2018). Nevertheless, in the past 3 years, seaweed production has stabilized due to a slowdown in Southeast Asia, while the cultivation of temperate and cold water species continues to rise. Looking ahead, North America, Europe, and South America can facilitate the multidimensional development of the seaweed industry through the implementation of integrated multi-trophic aquaculture (IMTA).

7.2.2 *Major Producing Country*

China stands as one of the leading producers of seaweed. According to the statistical year book of China (2015–2019) (source: https://www.cafs.ac.cn/info/1397/25066.htm), the cultivation of seaweed and the trade of algae and its products are depicted in Fig. 7.1. Figure 7.1a illustrates that between 2014 and 2018, seaweed cultivation in China primarily encompassed kelp, wakame, laver, and *Gracilaria*, among others. The cultivated area of seaweed exhibited a gradual expansion at a rate of 3.15–7.86% annually, after which the growth rate slowed down. Concurrently, the annual yield displayed a steady and modest increase. Additionally, China engages in significant seaweed trade with other regions. The country predominantly imports hundreds of thousands of tons of edible seaweed raw materials from other Asian countries (97.71–99.76%) each year, while its exports are primarily directed toward Asia (65.03–81.04%) and Europe (15.17–29.78%), as demonstrated in Fig. 7.1b. Notably, China's algin exports are predominantly to Asia (64.08–70.79%) and Europe (17.89–23.11%). Over the past decade, the export volume of algin has increased by 1.43 times, from 20,010,104 tons in 2010 to 28,626,533 tons in 2019. The production value has also witnessed a growth of 1.58 times, amounting to $167 million in 2019.

Indonesia plays a significant role in the production of ten red algae species, including *Kappaphycus alvarezii*, *Eucheuma denticulatum*, and various *Gracilaria* spp. Around 70% of the seaweed is exported in the form of dried algae to mainland China, Hong Kong, Denmark, and the Philippines (Kambey et al. 2020). The remaining 30% undergoes processing into seaweed gum, with an annual output of 1000–2000 tons (source: https://www.fao.org/fishery/en/collection/countrysector). Notably, *Kappaphycus alvarezii* and Eucheuma species account for approximately 88% of the carrageenan processing raw materials, resulting in an annual production of approximately 120,000 tons (dry weight) (FAO 2020). The Philippines also serves as a major source of seaweed raw materials, with dried seaweed, semi-refined carrageenan (SRC), and refined carrageenan being the primary seaweed products. In 2017, the Philippines produced 1.4 million tons (wet weight) of seaweed, valued at $174 million (Mateo et al. 2020). The main export markets for these products include Europe, North America, Asia, among others. In Japan, the main cultivated algae species are *Porphyra pseudolinearis* and Porphyra spp., which account for approximately 27% of the volume of production and 15% of the value of the Japanese aquaculture industry. *Undaria pinnatifida*, commonly known as wakame, contributes to 5% of the total aquaculture production in Japan and 1% of the total output value (source: http://www.fao.org/fishery/countrysector/naso_japan/en). Lastly, the seaweed production in South American countries like Peru and Chile should not be overlooked due to its significance in the industry.

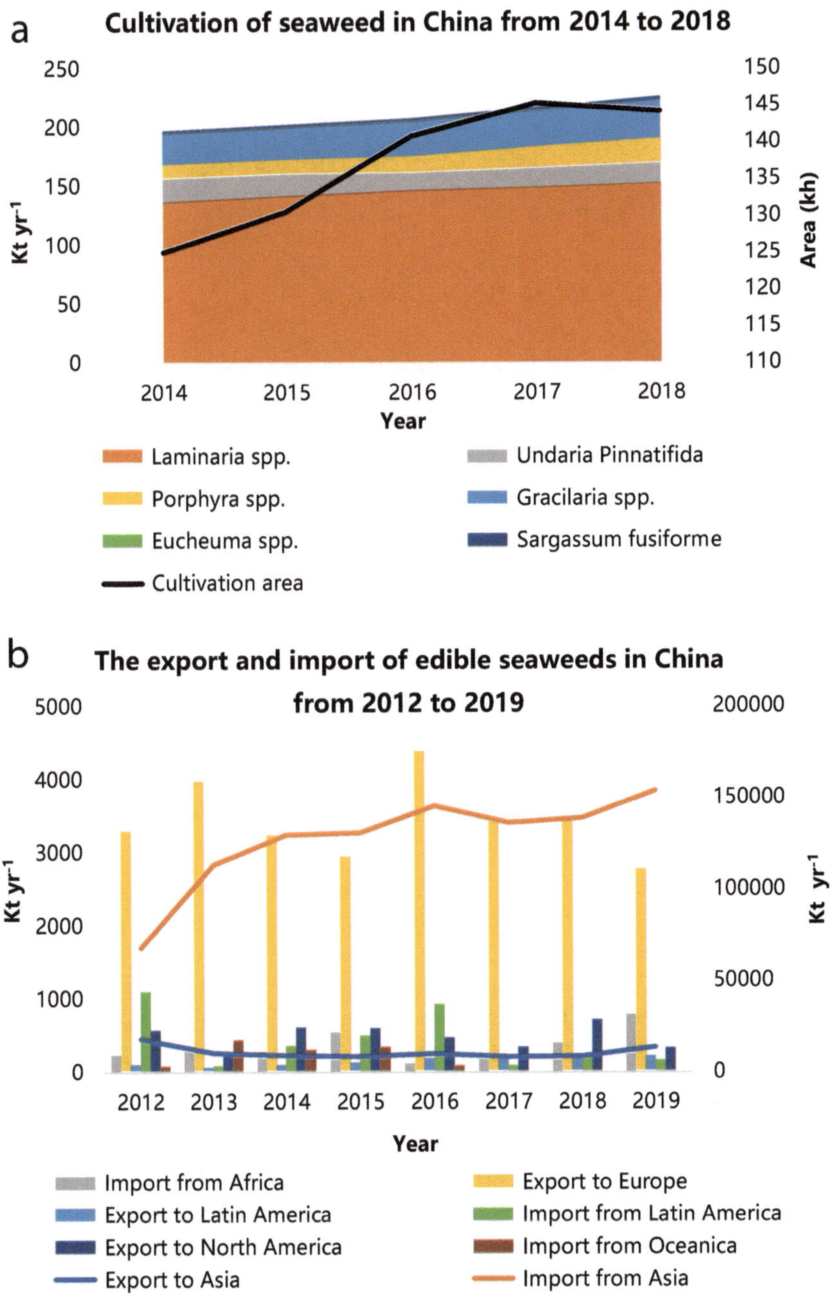

Fig. 7.1 Seaweed cultivation, production, import, and export in China. (**a**) Cultivation of seaweed in China from 2014 to 2018. (**b**) The export and import of edible seaweeds in China from 2012 to 2019. (**c**) Export of Alginate from China to Asia from 2010 to 2019

C

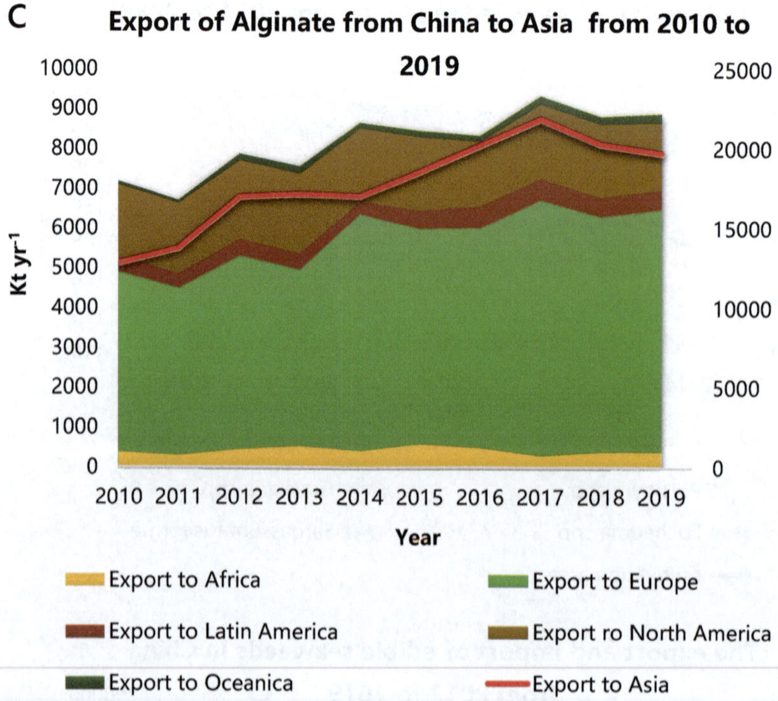

Fig. 7.1 (continued)

7.2.3 *Traditional Seaweed Products and Derivatives*

Approximately 75% of the global seaweed production is directly consumed in food products, while 13% is utilized in the production of hydrophilic colloids (polysaccharides) such as agar, alginate, and carrageenan. These substances find extensive application in the food industry as clarifying agents, gelling agents, emulsifiers, stabilizers, thickeners, and flocculants for various food items like ice cream, yogurt, candy, meat products, and beverages (refer to Fig. 7.2). The remaining 12% of seaweed is processed for agricultural products (Hardouin et al. 2014; Mouritsen et al. 2019).

Among the algae directly consumed by humans, five dominant genera include *Saccharina* and *Laminaria* (kombu), *Undaria* (wakame), and *Porphyra* and *Pyropia* (nori). The iron content in red algae *Palmaria palmata* averages 0.8 mg/g, while lean meat only reaches an average of 1.2–3.1 mg/100 g. Seaweed protein content ranges from 5% to 47% of the dry weight, with red algae exhibiting the highest protein content. Approximately half of the free amino acids in seaweed are essential amino acids. Furthermore, seaweed polysaccharides play a vital role in the development of plant-based foods, artificial meat, and other globally popular food items. Most seaweeds are rich in polysaccharides, with alginate content by dry weight in

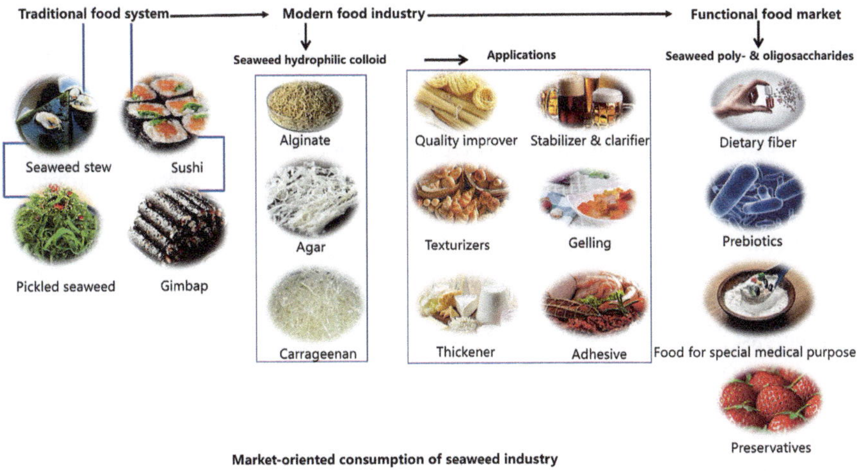

Fig. 7.2 Market-oriented consumption of seaweed industry

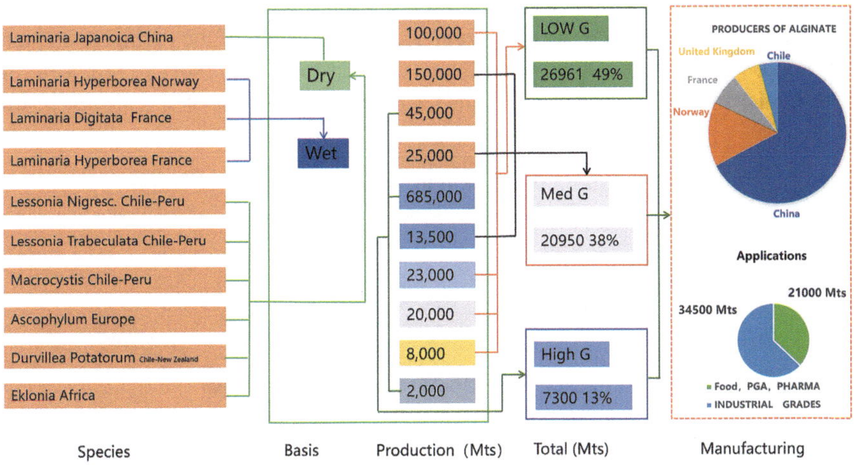

Fig. 7.3 World production of Alginate in 2014. Abbreviations in the figure: G guluronate, Mts Megaton

Ascophylum nodosum is 22–30%, *Laminaria digitata* fronds 25–44%, *Laminaria digitata* stipes 35–47%, *Laminaria hyperborea* fronds 17–33%, *Laminaria hyperborea* stipes 25–38%, and *Sargassum* spp. 17–45% (O'Sullivan et al. 2010).

Alginate, agar, carrageenan, and other seaweed polysaccharides have long been recognized as essential natural food additives, playing crucial roles in gelling, emulsification, water retention, and thickening properties. The global production and utilization of alginate in 2014 are depicted in Fig. 7.3. According to the Danish Seaweed Congress report (2014), the worldwide alginate production estimated at 55,500 tons, with China having the production share of 74% followed by Norway

(14%), France (6%), and Chile (3%). Among them, 21,000 tons (37%) are used for food purposes, while the remaining 34,500 tons (63%) utilized for industrial applications, as shown in Fig. 7.3.

Agar is primarily derived from *Gracilaria* (53%) and *Gelidium* (44%), with a small portion representing the genus *Gelidiella* and *Pterocladia* (Marinho-Soriano and Bourret 2003). Gracilaria-derived agar gel exhibits lower strength, but alkali treatment can enhance its properties, making it an important commercial source. The International Seaweed Congress in Denmark reports that *Gracilaria* production increased to 88,000 tons in the Asia-Pacific region in 2015, while *Gelidium* decreased by almost half during the same period. Nevertheless, overall agar production increased from 7500 to 14,500 tons. The Asia-Pacific region demonstrates substantial growth, as depicted in Fig. 7.4.

Statistical analysis indicates that 70–80% of carrageenan production is utilized in the food industry, with an estimated annual consumption of around 45,000 metric tons. Approximately 45% of carrageenan is used in dairy products, and 30% in meat and meat products, contributing to a total estimated market value of $300 million per year (FAO 2013). Commercial Carrageenan is primarily sourced from *Eucheuma cottonii* and *Eucheuma spinosum*, mainly produced in the Philippines and Indonesia. κ-Carrageenan extracted from *E. cottonii* accounts for 68% of the carrageenan market, while ι-carrageenan is the main type extracted from *E. spinosum*, representing 19% of the market. Between 1999 and 2015, the production of *E. cottonii* in the

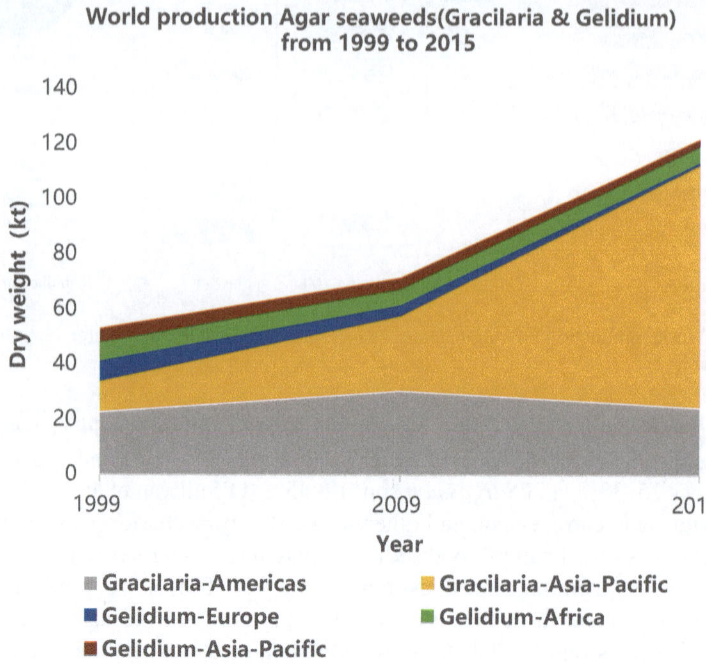

Fig. 7.4 World production of *Gracilaria*, *Gelidium*, and Agar production from 1999 to 2015

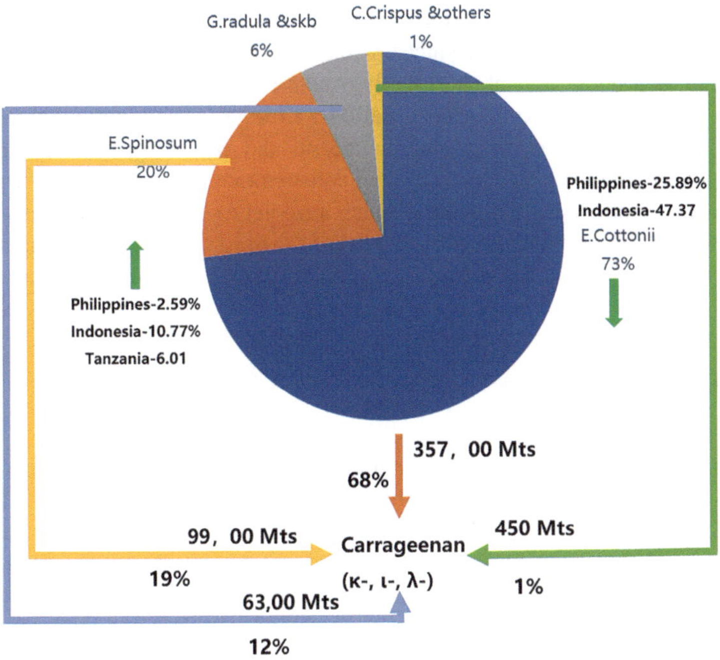

Fig. 7.5 World production of carrageenan in 2015

Philippines nearly halved. By 2015, the Philippines accounted for only 25.89% (60,000 tons, dry weight) of the *E. cottonii* market, while Indonesia experienced a significant increase, capturing 47.37% of the market share (Fig. 7.5).

7.2.4 *The Growth Potential of the Seaweed Industry*

The seaweed food value chain consists of four key links: seaweed cultivation, primary processing, food production, and consumption. These links span the agriculture, industry, and service sectors, encompassing labor-intensive industries with high energy and water consumption, primarily concentrated in developing Asian countries. Although these regions possess abundant labor resources, their environmental protection systems often lag behind economic demands, reflecting the characteristics of traditional coastal biological industries. Seaweed food products usually have a simple form, posing challenges for preserving long-term quality and efficiency of active ingredients like polysaccharides and oligosaccharides. The global seaweed food market is concentrated in East Asia, Southeast Asia, and South Asia, where traditional consumption of fresh, seaweed and primary products limits the demand for research and development of new offerings. Consumer demand for seaweed food primarily stems from taste preferences rather than the precise health benefits it offers. Developed countries like Japan, South Korea, and Taiwan tend to consume high-end seaweed products, while in other regions, seaweed is mainly consumed as a daily vegetable. Notably, Asian countries generally exhibit lower obesity rates compared to developed nations. However, developing countries face challenges in health food development concepts and market awareness, complying with regulatory regimes leading to limited consumption levels. Consequently, high-end seaweed products with precise efficacy have yet to become the primary choice for consumers. The potential health benefits of seaweed consumption, demonstrated by lower obesity rates in Asian countries, highlight its significance for human health.

7.3 Functional Food

Metabolic syndromes closely linked to lifestyle habits include obesity, uncontrolled blood sugar, dyslipidemia, and high blood pressure. However, the precise mechanisms underlying each symptom are not fully understood. Dietary fibers have garnered significant interest due to their beneficial effects on body weight, food intake, glucose homeostasis, and insulin sensitivity. This is likely because dietary fiber intake profoundly influences nutrient bioavailability, microbial composition, and gastrointestinal function, thus regulating nutrient absorption mechanisms in human and animal diets (Fig. 7.6).

Seaweed polysaccharides, which constitute 25–75% of the dry weight of seaweed, play a crucial role as dietary fiber, with water-soluble fiber accounting for approximately 50–85% (Jiménez-Escrig and Sánchez-Muniz 2000). In 2018, the

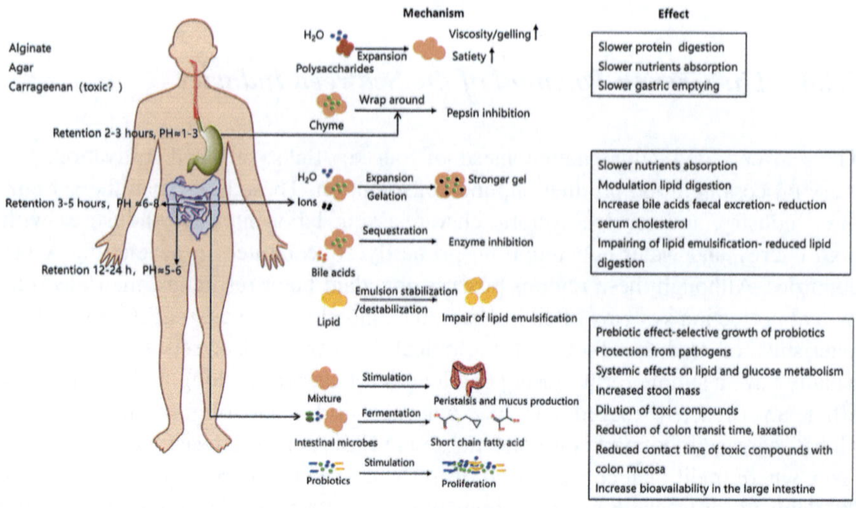

Fig. 7.6 Intervention and improvement of dietary fiber from seaweed on human metabolic syndrome

U.S. Food and Drug Administration (FDA) officially recognized alginate as a dietary fiber (den Besten et al. 2013). The American Dietetic Association recommends adults consume 25–30 g of dietary fiber per day (Das et al. 2020). Alginate forms a gel by chelating with ions like Ca^{2+} and Mg^{2+}. This gel exhibits water absorption and swelling properties, leading to a slower gastric emptying rate and enhanced feelings of fullness and satisfaction. Seaweed dietary fiber has also demonstrated the ability to reduce serum cholesterol levels in rats, bind cholesterol and related active compounds, and to some extent inhibit lipid absorption by the gastrointestinal tract (Brownlee et al. 2005; Jiménez-Escrig and Sánchez-Muniz 2000).

Agar gel, a food additive with a high soluble crude fiber content (94%), has long been consumed as a traditional ingredient (80%) in various foods such as frozen noodles (tokoroten), jelly (mitsumame jelly), and others (Armisén and Gaiatas 2009; Nishinari and Fang 2017). Agar can form a gel without the need for sugar, and its digestibility in the human body is low, contributing only 10% of intake (Armisén and Gaiatas 2009), making it a negligible source of calories. As a soluble dietary fiber (hydrophilic colloid), agar exhibits high water retention, forms low viscosity mixtures in the intestines, and remains undegraded by coliform bacteria in the colon. Consequently, it finds application in fat-reducing foods, hypoglycemic foods, and other specialized food products. Moreover, agar has a dilatant effect and is utilized as a laxative ingredient in the United States. In fact, it is listed on the national prescription to enhance the slow-release efficacy of drugs (Armisén and Gaiatas 2009).

However, it is important to note that not all seaweed dietary fiber is equally safe. Animal studies have demonstrated that low molecular weight carrageenan is associated with the development and promotion of intestinal tumors and ulcers (Bhattacharyya et al. 2008; Tobacman 2001). Following the recommendation of the Joint Food and Agriculture Organization/World Health Organization Expert Committee on Food Additives (JECFA) at its 57th meeting, which stated that only degraded carrageenan exhibits adverse effects and should not be used as a food additive, JECFA announced in 2007 that the use of carrageenan in infant formula is not recommended (JECFA 2007). This highlights the urgent need for the development of safer and healthier processing technologies for algal dietary fiber.

7.4 The New Demand Under the Market Orientation

Function and safety are crucial focus in the future of the food industry. Concerning seaweed food, there is a growing demand for its accessibility among diverse consumer groups, coupled with more precise requirements for product efficacy. With the rapid development of new seaweed food products such as additives, dietary fiber, and prebiotics, the global landscape of the seaweed food industry is poised for profound transformations. The consumption market for seaweed food will expand beyond Asia, where traditional eating habits prevail, and reach coastal and even inland countries worldwide.

In 2014, the number of people living with type 2 diabetes worldwide reached 422 million, and approximately 40% of adults over the age of 25 were affected by high blood pressure (WHO). Furthermore, in 2015, 38.2% of adults in the United States were obese, and in 2016, nearly 40% of adults and 18% of children (aged 5–19) in the UK were overweight. Cardiovascular disease costs an estimated $200 billion annually in terms of healthcare and productivity losses, according to the Centers for Disease Control and Prevention (source: https://www.nutritioninsight.com/news/obesity-related-cancer-cases-on-the-rise-study-finds.html). Hypercholesterolemia is responsible for 45% of heart attacks in Western Europe and 35% in Central and Eastern Europe, while South Asians experience the highest incidence of coronary heart disease (CAD) worldwide. Consequently, a substantial market for functional and specialized medical foods has emerged. Nutritional intervention studies focusing on seaweed polysaccharides and oligosaccharides have demonstrated their potential to prevent and alleviate such metabolic disorders, thereby highlighting the foreseeable high-value utilization of seaweed in the future.

On the other hand, the preservation and storage of food are crucial aspects of modern food processing. Algal polysaccharides, such as Enteromorpha polysaccharides (ulvan?), exhibit significant antibacterial properties against both gram-positive bacteria (such as *Bacillus subtilis* and *Staphylococcus aureus*) and gram-negative bacteria (including *Salmonella*, *Pseudomonas aeruginosa*, and *Escherichia coli*). This bacteriostatic effect is likely attributed to the iron-binding ability and electron-donating properties of isohydroxamic acid. *Enteromorpha* polysaccharide demonstrates strong iron (III) chelating ability, which inhibits the absorption of Fe(III) (Zhong et al. 2020), a vital element for *Enteromorpha* growth. Brown algal oligosaccharides (2–8 DP) can effectively inhibit gray mold, blue mold, and black rot in kiwifruit (Liu et al. 2020). They also suppress pectin dissolution, pectin methylase, polygalacturonase gene expression, and the corresponding enzyme activities encoded in kiwifruit. In terms of antioxidant capacity, particularly for lipid oxidation, algal polysaccharides exhibit promising performance. For instance, when green algal polysaccharide extracted from *Chaetomorpha linum* is added to beef sausage stored at 4 °C, samples treated with 0.25% of the green algal polysaccharide during a 12-day storage period show notable improvements in lipid antioxidant capacity, color characteristics, and lipid oxidation compared to the standard formula (with added vitamin C) (Das et al. 2020; Hamzaoui et al. 2020). Moreover, seaweed polysaccharides and oligosaccharides are naturally sourced and have fewer side effects compared to common antioxidants used in the modern food industry, such as BHA (butylated hydroxyanisole), BHT (butylated hydroxytoluene), PG (propyl gallate), and TBHQ (tert-butylhydroquinone).

7.5 Concluding Remarks

With the world population expected to exceed 9 billion by the middle of the twenty-first century, providing adequate food in the face of climate change and environmental degradation has become a major challenge for human society. Seaweed is a sustainable, low-cost crop, does not occupy arable land, does not increase the burden of global warming, and the annual biomass per unit area of seaweed is higher than land crops, is an excellent alternative food source. Seaweed and its processed products are also more in line with modern society's requirements for healthy food. From fat substitutes to antioxidants, dietary fiber to low-calorie foods to antimicrobial agents, the physical and chemical properties of seaweed and its sugars allow it to meet the needs of healthy foods. China, Japan, South Korea, Southeast Asia, and other Asian countries have a high degree of acceptance of seaweed products, while Western countries have a low degree of acceptance of seaweed and trehalose due to different dietary habits, which can be solved through reasonable formulation, in-depth processing or fermentation. In addition, further development of seaweed polysaccharide and seaweed oligosaccharide resources and establishment of sound standards and management systems. To promote the wide application of seaweed polysaccharides and oligosaccharides.

The health awareness of consumers, especially post COVID-19 times, has increased, and they are looking forward to the market of high-quality functional foods with good flavor and immune-boosting properties. Seaweed polysaccharides and oligosaccharides have nutritional and functional properties, as well as good processing characteristics, which can meet the nutritional and sensory needs of food, and meet the needs of high-value-added seaweed products now and in the future. According to GI Insight's report, the global dietary fiber market was worth more than $5 billion in 2018 and is expected to grow at a CAGR >10% through 2025. Among these dietary fiber ingredients, inulin has the largest market share and may account for more than 30% of the global dietary fiber market share by 2024 (source: https://www.gminsights.com/industry-analysis/dietary-fibers-market). It can be seen that seaweed food is still an under-developed natural food resource. Changing people's dietary preferences is not easy, so it is necessary to develop diverse and acceptable seaweed products to stimulate the development of a thriving seaweed industry.

The seaweed market from low value to high value requires the joint efforts of research institutions, government forces and financial capital. The booming demand of the global health food market, the development of seaweed seedling technology, cultivation technology and harvesting technology will drive the vigorous development of the global seaweed industry. In the future, the more production areas or processing areas that can catch up with the wave of science and technology will gradually occupy a larger market and lead to the renewal of the seaweed industry chain. Basic research and scientific validation will lay a good foundation for the development of more new products, promote innovation in the downstream seaweed value chain, and react to the upstream seaweed value chain. Ultimately, the collective effort of technology, science and marketing will lead to the reshaping of the global seaweed food value chain.

References

Armisén R, Gaiatas F (2009) Agar. In: Phillips GO, Williams PA (eds) Handbook of hydrocolloids, 2nd edn. Woodhead Publishing, Cambridge, pp 82–107

Bhattacharyya S, Borthakur A, Dudeja PK, Tobacman JK (2008) Carrageenan induces cell cycle arrest in human intestinal epithelial cells in vitro. J Nutr 138(3):469–475

Brownlee IA, Allen A, Pearson JP, Dettmar PW, Havler ME, Atherton MR, Onsøyen E (2005) Alginate as a source of dietary fiber. Crit Rev Food Sci Nutr 45(6):497–510

Chopin T (2018) Seaweeds: the world's largest mariculture crop. Int Aquafeed 21:14–15

Chopin T, Tacon AGJ (2021) Importance of seaweeds and extractive species in global aquaculture production. Rev Fish Sci Aquac 29(2):139–148

Das AK, Nanda PK, Madane P, Biswas S, Das A, Zhang W, Lorenzo JM (2020) A comprehensive review on antioxidant dietary fibre enriched meat-based functional foods. Trends Food Sci Technol 99:323–336

den Besten G, van Eunen K, Groen AK, Venema K, Reijngoud DJ, Bakker BM (2013) The role of short-chain fatty acids in the interplay between diet, gut microbiota, and host energy metabolism. J Lipid Res 54(9):2325–2340

FAO (2012) The state of the world fisheries & aquaculture. FAO Fisheries Department, Rome, Italy

FAO (2013) FAO fisheries technical paper 105

FAO (2017) Fishery and aquaculture statistics 2017/FAO annuaire

FAO (2018) Meeting the sustainable development goals. FAO

FAO (2020) Sustainability in action

Hamzaoui A, Ghariani M, Sellem I, Hamdi M, Feki A, Jaballi I, Nasri M, Amara IB (2020) Extraction, characterization and biological properties of polysaccharide derived from green seaweed "*Chaetomorpha linum*" and its potential application in Tunisian beef sausages. Int J Biol Macromol 148:1156–1168

Hardouin K, Bedoux G, Burlot A-S, Nyvall-Collen P, Bourgougnon N (2014) Enzymatic recovery of metabolites from seaweeds: potential applications. Adv Bot Res 71:279–320

JECFA (2007) Joint FAO/WHO Expert Committee on Food Additives

Jiménez-Escrig A, Sánchez-Muniz FJ (2000) Dietary fibre from edible seaweeds: chemical structure, physicochemical properties and effects on cholesterol metabolism. Nutr Res 20(4):585–598

Kambey CSB, Campbell I, Sondak CFA, Nor ARM, Lim PE, Cottier-Cook EJ (2020) An analysis of the current status and future of biosecurity frameworks for the Indonesian seaweed industry. J Appl Phycol 32(4):2147–2160

Liu J, Kennedy JF, Zhang X, Heng Y, Chen W, Chen Z, Wu X, Wu X (2020) Preparation of alginate oligosaccharide and its effects on decay control and quality maintenance of harvested kiwifruit. Carbohydr Polym 242:116462

Marinho-Soriano E, Bourret E (2003) Effects of season on the yield and quality of agar from *Gracilaria* species (Gracilariaceae, Rhodophyta). Bioresour Technol 90(3):329–333

Mateo JP, Campbell I, Cottier-Cook EJ, Luhan MRJ, Ferriols VMEN, Hurtado AQ (2020) Analysis of biosecurity-related policies governing the seaweed industry of The Philippines. J Appl Phycol 32(3):2009–2022

Mouritsen OG, Rhatigan P, Pérez-Lloréns JL (2019) The rise of seaweed gastronomy: phycogastronomy. Bot Mar 62(3):195–209

Nishinari K, Fang Y (2017) Relation between structure and rheological/thermal properties of agar. A mini-review on the effect of alkali treatment and the role of agaropectin. Food Struct 13:24–34

O'Sullivan L, Murphy B, McLoughlin P, Duggan P, Lawlor PG, Hughes H, Gardiner GE (2010) Prebiotics from marine macroalgae for human and animal health applications. Mar Drugs 8(7):2038–2064

Shannon E, Abu-Ghannam N (2019) Seaweeds as nutraceuticals for health and nutrition. Phycologia 58(5):563–577

Tobacman JK (2001) Review of harmful gastrointestinal effects of carrageenan in animal experiments. Environ Health Perspect 109(10):983–994

Vairappan CS, Razalie R, Elias UM, Ramachandram T (2014) Effects of improved post-harvest handling on the chemical constituents and quality of carrageenan in red alga, *Kappaphycus alvarezii* Doty. J Appl Phycol 26(2):909–916

Zhong R, Wan X, Wang D, Zhao C, Liu D, Gao L, Wang M, Wu C, Nabavid SM, Daglia M, Capanoglu E, Xiao J, Cao H (2020) Polysaccharides from marine *Enteromorpha*: structure and function. Trends Food Sci Technol 99:11–20

Chapter 8
An Eco-sustainable and Circular Economy—Benefits and Constraints

João Cotas, Silvia Lomartire, Madalena Mendes, Leonel Pereira, and Ana Marta M. Gonçalves

Abstract Although there are various valorisation routes for seaweeds, the preferential use is for food, cosmetic, and pharmaceutical areas, considering the amounts (e.g. scalable by cultivation), nutritional profile, and presence of bioactive compounds. Given the characteristics of seaweeds, the primary recovery methods to be developed should be combined with the obtainment of value-added goods (e.g. pigments for developing bioplastics). Seaweeds can thus be used to directly acquire such molecules of interest, and the biomass can then be transformed into energy or bioproducts. The biorefinery idea takes a multi-solution strategy, obtaining low-volume, high-value-added goods and vice versa, completing the cycle. The cascade valorisation method is used in the biorefinery idea. In this chapter, we focus on the seaweed biorefinery concept from an eco-sustainable and circular economy point of view for a Scale-up to the industrial level, focusing on the benefits and constraints.

Keywords Seaweeds · Bioactive compounds · Biorefinery concept · Circular economy · Scale-up

J. Cotas · S. Lomartire · M. Mendes · L. Pereira
Marine Resources, Conservation and Technology, Marine Algae Lab, CFE—Centre for Functional Ecology—Science for People & Planet, Department of Life Sciences, University of Coimbra, Coimbra, Portugal
e-mail: jcotas@uc.pt; silvia.lomartire@uc.pt; leonel.pereira@uc.pt

A. M. M. Gonçalves (✉)
Marine Resources, Conservation and Technology, Marine Algae Lab, CFE—Centre for Functional Ecology—Science for People & Planet, Department of Life Sciences, University of Coimbra, Coimbra, Portugal

Department of Biology and CESAM, University of Aveiro, Aveiro, Portugal
e-mail: amgoncalves@uc.pt

N. Trivedi et al. (eds.), *Recent Advances in Seaweed Biotechnology*,
https://doi.org/10.1007/978-981-96-0519-4_8

8.1 Introduction

Seaweeds are multicellular, photoautotrophic organisms (aquatic primary producers) that exist by using oxygen and water dissolved carbon dioxide (CO_2), light, and nutrients. They constitute the foundation of various food chains in aquatic ecosystems. The primary groupings (phyla) of seaweeds are members of the Chlorophyta (green), Rhodophyta (red), and Ochrophyta-Phaeophyceae (brown) (Rodrigues et al. 2015; Barbot et al. 2016).

Despite having varied valorisation paths, harvested, wild, and cultivated seaweeds have been involved in food, cosmetic, and pharmaceutical industries (Rodrigues et al. 2015). However, seaweeds metabolise many structural molecules, such as proteins, lipids, and carbohydrates (primary metabolites), which are necessary required as a food source, as well as other interesting bioactive compounds (secondary metabolites), which can be used in a variety of industries (food, feed, agriculture, cosmetics, pharmaceutical, and biotechnological) (Leandro, et al. 2020). As a result, seaweeds own a lot of revolutionary uses, and seaweeds have been shown to be a renewable resource for the future, resulting in increasing demand for the processing of these species and, consequently, for their production (García-Poza et al. 2020). Furthermore, the biological productivity of seaweed contributes to photosynthetic carbon sequestration, mostly the seaweeds that have high quantities of monosaccharides and polysaccharides (carbohydrates form). Harvesting and using macroalgae for biofuels and other sectors (food, feed, medicines, and fertilisers) can also help to reduce CO_2 emissions due to reducing fossil fuel usage (Duarte et al. 2017).

Seaweeds constitute the foundation of the coastal ecosystem, sustaining a diverse range of aquatic animals both directly and indirectly (Reisewitz et al. 2006). Because seaweeds are a key player in the foundation of several aquatic food chains that provide a variety of ecological resources, such as habitats, food, and shelter to various trophic levels (herbivores, omnivores, and carnivores, from invertebrates to vertebrate species), seaweeds are essential to the support of aquatic ecosystems (Almanza and Buschmann 2013; Vásquez et al. 2014).

Due to the limitations of wild harvest, the global demand for seaweeds and their by-products has spurred interest in their production and cultivation. This is quite problematic from an ecological standpoint, as unchecked wild gathering can have a harmful influence on ecosystems (Jung et al. 2013) because seaweed is the basis of the food chain; however, too much biomass can create blooms of pathogenic bacteria and microalgae. As a result, stakeholders (seaweed industry and associations) are more likely to invest in the development of various algae species that may meet various economic sector demands (Ashkenazi et al. 2019). The use of seaweeds might be critical to reduce the need to feed a growing human population on a world with insufficient area for agricultural crops, because seaweed production does not compete for inland arable land, freshwater, or agricultural fertilisers (Charrier et al. 2017; Hermans 2022; Hermans 2023).

The polysaccharides portion, which percentage per dry weight may vary from species to species (Filote et al. 2021), is the principal economic utilisation of seaweed compounds. In a study of Kraan (2012), the investigation of total polysaccharide concentrations in different seaweed species showed a total polysaccharide concentration in a range between 4–76% of the dry weight (DW). The species analysed were *Laminaria* sp., *Saccharina* sp., *Fucus* sp., *Ascophyllum* sp., *Undaria* sp., *Sargassum* sp., for brown seaweeds; *Chondrus* sp., *Porphyra* sp., *Gracilaria* sp., *Palmaria* sp., for red seaweeds, and *Ulva* sp., for green seaweeds.

The highest contents are found in species such as *Ascophyllum*, *Porphyra*, and *Palmaria*; however, green seaweed species such as *Ulva* also have a high content, up to 65% of DW. The polysaccharides of the brown group, with alginate being the predominant constituent, following by laminarin and fucoidan. Despite this, these polymers are widely utilised in a variety of sectors, and their use is increasing as synthetic polymer manufacturing declines. The primary polysaccharide in red seaweed is either agar (agarophyte seaweed) or carrageenan (carragenophyte seaweed) (Imeson 2009; Filote et al. 2021). The extraction of pigments is another major business, where seaweed biomass must be properly gathered and treated before extraction at an industrial level. Based on their pigment contents, seaweeds can be categorised into brown, red, and green algae, referred to as Phaeophyceae, Rhodophyta, and Chlorophyta. These algae contain three main classes of pigments: chlorophylls, carotenoids, and phycobiliproteins. Chlorophylls are essential for photosynthesis and are greenish lipid-soluble pigments featuring a porphyrin ring. They are present in all types of algae, higher plants, and cyanobacteria. Among the four types of chlorophyll found in marine algae, chlorophyll a is the most significant, absorbing energy from violet-blue and orange-red light. Chlorophylls b and c are the second and third types, and chlorophyll d, discovered in red algae, is the fourth. They can be extracted with ethanol and isolated using column chromatography (Freitas et al. 2022).

Carotenoids, considered accessory pigments, enhance light absorption by transferring light excitation to chlorophyll. Carotenoids can be classified into two categories: carotenes, which are unsaturated hydrocarbons, and xanthophylls, containing oxygen-containing functional groups. One prominent carotenoid in algae is fucoxanthin, which is responsible for the brown colour of kelps, brown algae, and diatoms. Fucoxanthin is a major contributor, accounting for approximately 10% of the total carotenoid production in nature. They can be extracted with ethanol:acetone and isolated using column chromatography (Freitas et al. 2022).

Phycobiliproteins are water-soluble fluorescent proteins that serve as accessory pigments for capturing light during photosynthesis. They absorb energy in specific portions of the visible spectrum (450–650 nm) and are the primary photoreceptor for photosynthesis in Cyanobacteria, red algae, and cryptomonads. Three primary categories of phycobiliproteins exist: phycocyanins, allophycocyanins, and phycoerythrins, with phycoerythrins being the most abundant among many red algae species (Pangestuti and Kim 2011). The phycobiliproteins can be extracted with aqueous solutions with a phosphate buffer (Freitas et al. 2022).

The research of Saati (2011) demonstrated the contents of pigments obtained for green, brown, and red seaweeds respectively. The results showed that green, brown, and red seaweed had pigment content of 0.98 mg/100 g (DW), 1.28 mg/100 g (DW), 1.35 mg/100 g (DW), and yield of pigment for each group was 4.85%, 6.24% and 6.65% respectively.

Some of the pigments described are considered high-value pigments that are utilised in the biomedical and food industries after passing safety requirements (Freitas et al. 2022; Gomes et al. 2022). Although certain pigment extraction technologies, such as fucoxanthin and phycoerythrin, have been established for commercial exploitation, industrial extraction methods are currently being developed (Leandro et al. 2020; Freitas et al. 2022).

Pigments are linked to proteins found within algae in various forms and locations, including components of the cell wall, enzymes, and associations with pigments and carbohydrates. The extraction of these compounds can be impeded by their specific locations and the composition of the cell wall. Traditionally, the extraction of proteins and pigments from fresh, dried, or freeze-dried algae involved maceration in a buffer or aqueous solvent, sometimes followed by a secondary alkaline maceration if necessary. In the present day, multiple innovative extraction techniques have been developed to improve the extraction of bioactive substances from seaweeds, such as the utilisation of enzymes, ultrasound, microwaves, or supercritical fluids. Most contemporary methods focus on breaking down the cell wall. To achieve this, seaweeds are ground in liquid nitrogen using a mortar to facilitate cell wall disruption, which is the primary barrier to access and extract algal pigments.

However, this approach is not entirely effective for cell wall degradation and is also expensive when applied on an industrial scale. Most chlorophylls and carotenoids are hydrophobic pigments, making them soluble in non-polar solvents. Nevertheless, extracting these pigments from biological materials necessitates the use of a solvent that can mix with water. The most employed solvents, either pure or with added water, include acetone, dimethylformamide, dimethyl sulfoxide (DMSO), diethylether, methanol, or various mixtures. For the extraction of phycobiliproteins, water can be utilised for one or several days, extracting proteins through osmotic shock. However, this method is time-consuming, and its primary drawback is the partial degradation of phycobiliproteins by proteases. Various buffers are employed, such as sodium phosphate (ranging from 5 to 100 mM, with a pH between 6.5 and 7.4) and 0.1 M acetate buffer with a pH of 5.5.

Alternative methods for extracting pigments and seaweed proteins have been explored, with enzymatic hydrolysis of the cell wall suggesting as another means to access algal protein and pigments (Dumay and Morançais 2016).

However, these principal exploitation routes for seaweed are being commercially explored with industrial well-known seaweed (such as kelps and *Ascophyllum nodosum*). Although, there are seaweeds that do not attract industry. One of the major problems, ecological and economic, are the seaweed blooms and seaweed biomass beach drifted (who can damage the ecosystem and anthropogenic activities, such as tourism and fish farming). Therefore, this chapter will focus on the biorefinery concept from the biochemical preliminary profile up to the applications of the biomass.

8.2 Valorisation Chains for Seaweed Pre-step

Seaweeds are a source of raw material for sectors such as food, feed, energy, biomolecules, and human livelihood (Charrier et al. 2017). The utilisation of seaweed biomasses provides several uses of alternative raw material sources, which have some advantages, including decreased production costs and increased product value. Seaweeds can also be used as a standalone finished product (Pardilhó et al. 2022a). There two types of biorefineries that work based in seaweeds, the polysaccharide (carrageenan and alginate) extraction in Oceania, Canada, and South Africa, where the industry try a full usage of the biomass, mostly for polysaccharide obtention, and after the waste is decomposed to soil fertiliser. Also, this industry uses seaweed drifted-ashore using to alleviate a negative environmental impact (Pardilhó et al. 2022a, b). This example is followed by other extraction industries nowadays, where the polymer product is already on the market. Although, there are two value products on a simplistic biorefinery approach, where the decomposition of the waste can be useful for production of bioethanol or biogas (Pardilhó et al. 2023a, b).

It is critical to determine whether a given molecule is present in economically exploitable proportions to make it interesting from the costs-efficacy point of view. Seaweeds have a high carbohydrate content, a variable ash and protein content, and a low lipid content that varies depending on the species and seasonal abiotic and biotic variations (Rodrigues et al. 2015; Barbot et al. 2016; Pardilhó et al. 2022a). For example, in the summer, tropical seaweeds produce more sugars and photoprotective pigments, whereas in the spring, they accumulate more minerals and produce more proteins due to the more mineral-enriched seawater; as a result, natural seaweed growth increases in the spring and early summer (Jard et al. 2013; Barbot et al. 2016).

For example, the study of Zubia et al. (2014) showed that *Gracilariopsis tenuifrons*, a tropical red alga living in the upper intertidal zone in Mexico, has an increase in chlorophyll a, carotenoids content and antioxidant activity in response to stress-induced by UV-B radiation coupled with high photosynthetically active radiation (PAR). Another study showed that the light regime conditions typical of tropical environment greatly impacted photosynthetic pigment production, as it demonstrated the investigation on the tropical marine algae *Chlorella* sp. (Nurachman et al. 2015). Tropical seaweeds such as *Gracilaria arcuata* and *Ulva lactuca* have been investigated for their pigment contents and results showed that pigment content of *Gracilaria arcuata* extracted with ethyl-acetate represent for carotenoid 0.528 ± 0.009 μmol/g, while chlorophyll a 2.845 ± 0.069 mg/g. *U. lactuca* carotenoid content was 0.442 ± 0.046 μmol/g and chlorophyll a 9.216 ± 0.103 mg/g, confirming the high content of pigments in tropical seaweeds (Hidayati et al. 2020).

Therefore, seaweed quality will vary greatly between seasons, being a problem for the industry. Thus, the seaweed needs to have preliminary characterisation step, which can be bottleneck, as it might be hard to find rapid and good techniques (Pardilhó et al. 2022a, 2023b). Before the biorefinery approach, there is a need to develop rapid seaweed characterisation assessment assays to evaluate the best technique to exploit the biomass in a more efficient way and with a higher economic output.

8.2.1 Seaweed Rapid Assessment Assay

This theory or concept is a preliminary step before the initiation of industrial process. Thin layer chromatography (TLC) is the most basic approach for determining the presence of a specific type/class of substance, such as detecting, identifying, and isolating seaweed pigments/phenolic compounds from various types of seaweed samples (Cotas et al. 2019) (Fig. 8.1). This approach has the benefit of using just little amounts of sample, being simple to apply, and requiring no costly equipment. It enables qualitative (using proper reagents and visualisation methods) and quantitative (using devices that detect stain density) examination.

UV/V spectrophotometry is used to ensure seaweed quality (aqueous fractions), primarily by evaluating the oxidation of seaweed pigments, phenolics, and proteins, and by using spectrum analysis with seaweed extracts to reduce complexity and noise in the analysis (Cotas et al. 2019). This approach is already used to test phosphorus and can be used to examine the oxidation of biomass and pigment concentration with or without TLC prior to pigment separation analysis (with pigment standards analysis). Colorimetric techniques (CT) employing colorimetric reagents, such as the Bradford or Kjeldahl procedures, can be used to determine protein concentration (Mæhre et al. 2018). The phenolic content may be determined using the

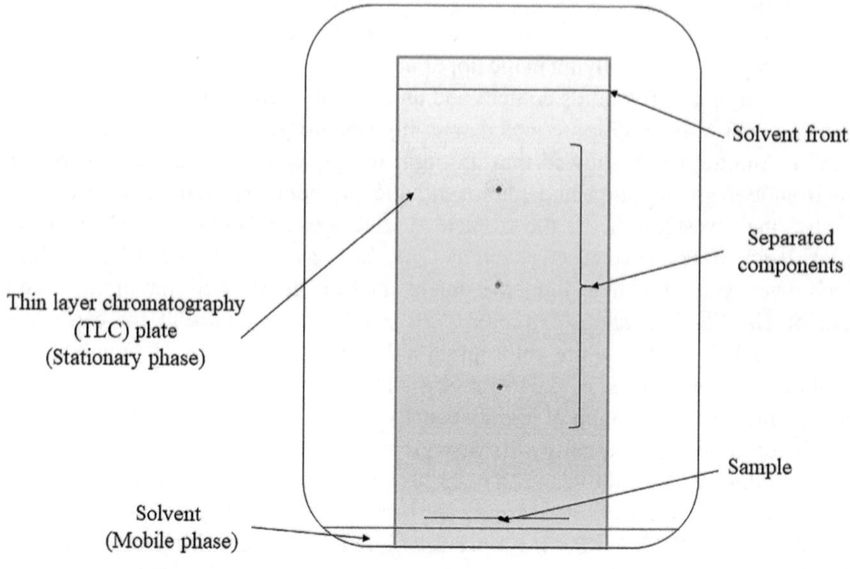

Fig. 8.1 Schematic representation of TLC technique

CT method and the Folin-Ciocalteau test (Ford et al. 2019); however, due to the involvement of other seaweed components, this approach might produce erratic results (Sumampouw et al. 2021).

Even dangerous minerals (lead, mercury, arsenic, cobalt, chromium, lithium, bromine, aluminium, iron, cobalt, nickel, copper, zinc, rubidium, strontium, cadmium, barium) may be detected using atomic absorption spectrophotometry using dry, digested, or seaweed ashes, which is the standard method for analysing and quantifying total and particular mineral content and heavy metals (Assubaie 2015).

IR spectroscopy employs infrared light frequencies to analyse light absorption by the sample, as well as vibration by chemical bonds, which is the main target (Unknown). The Fourier-transform infrared spectroscopy (FTIR) technique is a low-cost approach for analysing the chemical bonding of dried materials in a liquid extract solution (Pereira et al. 2003). Infrared light is used in this approach to cause chemical bonds to vibrate (Coates 2004). Before commercialisation, it may be used to examine polysaccharides, pigments, phenolic fractions, compound oxidation, and microplastics in seaweed (Cotas et al. 2019; Li et al. 2020; Bogolitsyn et al. 2020). When compared to chromatography, the FTIR approach is less expensive and easier to use, although it is less sensitive for biochemical quantification and quality analysis. Nonetheless, when combined with chromatographic analysis, this approach can improve seaweed quality characterisation (Sim et al. 2014).

However, due to the complexity of seaweed composition, these analytical approaches for certifying seaweed quality are continually advancing. Characterisation of seaweeds is critical to ensure quality control and customer safety. Safety frameworks and regulations are still being developed in order to govern seaweed food quality control, employing particular analyses for seaweeds, taking into account their chemical diversity and compound complexity (Leandro et al. 2020; Monteiro et al. 2022).

The evaluation of integrated solutions based on a seaweed characterisation, biorefinery-based concept to analyse the potential seaweed biomass-derived products/compounds is of high interest (allowing scale-up process, and reducing the error and miscalculations). Integrated approaches that involve the use of seaweeds, will incentive the use and exploitation of the seaweed biomass using an economically feasible cascade process to obtain valuable and less valuable (biofuels and soil conditioners) products, contributing to a circular economy, and adopting a zero-waste approach (Cotas et al. 2019; Pardilhó et al. 2023b).

In their study, Muscat et al. (2021) formulated five ecological principles aimed at steering biomass utilisation towards a circular economy. However, it was not until the work of Chary et al. (2023) that these principles were specifically investigated in the context of aquaculture. This research offers a narrative review designed to identify the key strategies for enhancing the circularity of aquaculture, along with their potential impacts on primary species and production systems. One illustrative approach involves prioritising the cultivation of species that enhance food and nutrition security, while also minimises their environmental footprint and considering the broader food system perspective to prevent competition between feed and food resources in aquaculture. As a result, integrating seaweed cultivation into aquaculture systems can contribute to a more comprehensive circular economy. This integration can lead to increased production and demand for crucial species, reduced losses and waste during farm and post-harvest phases, and the promotion of nutrient recycling practices across multiple scales.

8.3 Biorefinery Approach for Seaweed

The utilisation of a seaweed-based biorefinery idea is gaining popularity due to the need to address the economic and environmental limitations associated with traditional terrestrial biomass-based biorefineries, such as the high energy requirement, high quantity of biomass required for a low final product, lack of proper techno-economic evaluations, and limited number of biofuel products (Banu et al. 2021; Kokossis and Yang 2010). Recent research shown that cascade extraction might be a viable procedure for obtaining high-value goods without generating waste, hence contributing to the beneficial economic effect of putting into service and implementing the utilisation of such underused biomass (Pardilhó et al. 2022a, 2023b). In the same small scale, using seaweed and other examples from European companies, it is possible biorefinery facility be sustainable ecologically and economically, although there is a keystone to do a step-by-step biorefinery growth module from a small scale to a large scale.

Currently, raw seaweed biomass employed in current industrial processes results in the production of a low-value by-product, such as polymers and biofertilisers. This underutilised residue (mainly for the polysaccharide extraction industry) by-product can be used to produce biofuels or other economically valuable goods, bridging the gap towards the creation of an economically viable biorefinery with zero-waste output (circular economy), which is being developed by the extraction industry to reduce the waste decomposition into a new add-value product, biogas, or bioethanol. In agreement, the cascade strategy should be explored to enable full utilisation of biomass in a circular economy approach, thus, reducing the use of toxic cleaning agents to clean polysaccharides, normally. Thus, doing a downstream extraction sequence diminishes the cleaning process and adds a subproduct with economic value.

8.3.1 Non-polar Compounds

The principal non-polar chemicals in seaweed are lipids, which can be an essential basis for animal feed, particularly for aquatic species; however, the seaweed must be shown to be safe before being used in animal feed. Despite the fact that seaweed has a low lipid content (usually less than 1% of lipid content), another commercial use might be biodiesel synthesis, which will be further discussed in this section (Morais et al. 2020; Pardilhó et al. 2022a).

8.3.1.1 Lipidic Compounds

The non-polar fraction is removed first in the sequential extraction starting from washed, dried, and milled seaweed biomass, yielding mostly lipids (oils/FA). Solvents are the most often used method for extracting oil from algae (or other products). Normally, there are main industrial ways, with acetone (which is approved by EFSA) or using the traditional Soxhlet extraction using water vapour to extract the oils, which is already applied to the essential oils extraction in plants (Maceiras et al. 2011).

Even though seaweeds are not a traditional source of energy (based on lipid content), their polyunsaturated fatty acid (PUFA) content (such as omega 3 and 6) might be higher than that of terrestrial crops (Leandro et al. 2020). As a result, algal oils are regarded as a source of omega-3 fatty acid (FA). Two of the most essential omega-3 fatty acids present in algae oil are docosahexaenoic acid (DHA) and eicosapentaenoic acid (EPA). Oil derived from seaweed is now utilised in foods and medication (Leandro et al. 2020). There are few investigations on the lipid fraction of seaweed; however, some inferences concerning seaweed lipid fraction may be drawn from previous studies on non-residual biomass (Pardilhó et al. 2022a). The scarce studies with seaweeds are due to seaweed having normally low quantity of fatty acids and, thus, not being attractive to exploit them, even for studies. This can change if the seaweed fatty acids are one of the sub-products. Although, there are companies specialised in kelps oils which can be a key element for the brown seaweed extraction industry, because alginate and fucoidan gained the major commercial interests (Repêchage 1980; Cardamo 1936; Connemara Organic Seaweed Company 2013).

Variations in lipid content and FA profiles can be linked to environmental factors as well as species differences. Lipid levels are always less than 5% (FA per seaweed dry weight), with the majority of species having lipid values of less than 1% (FA per seaweed dry weight). Individually, saturated fatty acids (SFA) are found in larger concentrations, but when total unsaturated FA are examined, they comprise the majority (MUFA: monounsaturated fatty acids plus PUFA), except for some red seaweeds which have abundance of SFA. Total lipid content is generally assessed in the context of the species' nutritional profile, and it is an indicator of the possible applications of seaweed. However, the FA concentration and quality (usually up to 24-carbon (C24)) are regarded as essential parameters of yield and the acceptability of using these oils as feedstock for biofuel or bioproduct manufacture (Rocha et al. 2021; Pardilhó et al. 2022a).

Biodiesel, derived from vegetable oils and fats, is the most often used biofuel to replace fossil diesel. Compared to fossil fuels, biofuel is a renewable and green energy that is biodegradable and helps to the reduction of greenhouse gases, mainly CO_2, as well as other pollutants such as particles (Costa et al. 2012).

Transesterification is a chemical process in which a vegetable oil or animal fat (triglycerides) interacts with an alcohol (e.g. methanol, ethanol) in the presence of a catalyst to create biodiesel and a by-product (glycerol). Purified biodiesel can be utilised in diesel engines alone or in combination with regular diesel, at the end, biodiesel can be renewable, non-toxic, and eco-friendly solution for combustion engines. Biodiesel is typically made from edible oils, which creates several issues linked to food versus fuel rivalry. It is critical to find alternative raw materials different from oils to develop biodiesel to overcome these issues (Maceiras et al. 2011; Pardilhó et al. 2022a).

Non-polar extraction may be required in the remaining cascade process to eliminate pre-treatment (PT) or associated impurities (mostly co-extraction of lipids with other seaweed components of interest, such as pigments and phenolic compounds). In other words, there are expenses connected with the seaweed simple biorefinery sector, and non-polar extractions may be employed as PT to reduce the likelihood of multi-compound extraction, hence lowering residues (impurities) creation and increasing profitability (Pardilhó et al. 2022a, 2023b).

8.3.2 Polar Compounds

Seaweed polar compounds are used as a substitute for synthetic compounds in a variety of sectors; however, the high extraction costs for high-quality compounds have an impact on companies that may employ low-cost natural resources, such as the paper and textile industries. Thus, the biorefinery technique may be useful in exploring seaweed molecules in a variety of applications at a lower cost. There is a scarcity of knowledge on the extraction of polar compounds from seaweed; yet a robust business exists that uses fresh seaweed as raw material (Cotas et al. 2020; Freitas et al. 2022; Pardilhó et al. 2022a, 2023b).

8.3.2.1 Pigments

Brown, red, and green seaweeds can provide several pigments with different colours. Nonetheless, several chromatographic methods exist to extract and quantify the most important pigments, as well as to separate all the seaweed pigments (chromatographic separation methods) to obtain various commercial pigments from the same species used for the extraction (Leandro et al. 2020; Freitas et al. 2022; Gomes et al. 2022).

The fundamental difference between fresh seaweed pigment extraction and drifted seaweed pigment extraction is the heterogeneity of the raw dried drifted seaweed biomass employed, which is more heterogeneous and is a blend of seaweed species, and the projected lower pigment quality output (when compared to fresh seaweed pigment). However, given their low cost and abundance, as well as the industries' desire for natural pigments, it is recommended to support pigment extraction and seek new extraction methods to increase the pigment yield, hence the profit (Leandro et al. 2020; Freitas et al. 2022; Gomes et al. 2022).

There is a need to analyse the quality of the pigments in order to target the optimal use, which may be animal feed or the textile sector. Since seaweed can be a low-cost material to explore (normally, 1 kilo is below 10€), after the pigment extraction, the residual biomass can be re-utilised for polysaccharides extraction; another possibility is that the residual biomass with no pigments might be used for fermentation process (e.g. to obtain energy products) or agricultural application, promoting full exploitation of seaweed (Pardilhó et al. 2022a, 2023b).

8.3.2.2 Phenolic Compounds

Seaweed phenolic compounds have essential roles in seaweed protection against predators and have a wide spectrum of actions even when oxidised. These chemicals can be used as a supplement in agricultural crops and animal feed, benefiting both crops and animal health (Cotas et al. 2020).

The phenolic components of seaweed may be of interest for future investigation in order to get phenolic-enriched extracts; also, their separation aids in lowering the adverse impacts of their presence while aiming for further pigment extraction. This technique may be accomplished using ethanol at an industrial level due to its low cost and ability to be easily recovered (as an example of technology already applied in the carrageenan industry and also, in the pigment extraction using bacteria). The primary expenditures of this technique are related to the need to characterise the phenolic extracts further in relation to the chemical safety sheet, which is an expensive procedure. However, with recent technological advances, phenolic extraction can be a new complementary valorisation solution integrated into the seaweed cascade biorefinery approach. Phenolic extraction must be performed prior to obtaining the final compounds useful for biofuel production; in this way, phenolic compounds will not interfere with the extraction of other useful compounds. Even though information regarding the exploitation of seaweed phenolic compounds is rare in the literature, and even in fresh seaweed, the extraction of these compounds is being scaled up from R&D to small-scale manufacturing (Michalak and Chojnacka 2015; Lomartire et al. 2021; Pardilhó et al. 2022a; Cotas et al. 2023a, b; Cotas 2023). For example, SeapolynolTM (Botamedi Inc., Seoul, Korea) is a European Food Safety Authority-approved food supplement. This supplement is based on dieckol and other polyphenols derived from *E. cava* (also used in the extraction of alginate and fucoidan). The cosmetic industry is a major driver of the incorporation of seaweed phenolic compounds into commercial applications such as natural UV screening (Helioguard® 365, produced by Mibelle Biochemistry in Buchs, Switzerland; Aethic Sôvée® produced with Photamin, phenolic extracted from seaweeds, produced by AETHIC® in London, UK from *Porphyra umbilicalis*), or anti-ageing (ECKLEXT® BG, produced by NOF An extract of *Asparagopsis armata* containing MAAs is used in anti-ageing creams) (Cotas et al. 2020). Companies have been researching phlorotannins in order to develop new goods throughout the previous decade. The primary goals of phlorotannins' supplements in cardiovascular illness are to avoid arteriosclerosis and enhance protective high-density lipoprotein cholesterol (HDL-C). HealSea™ (made by Diana Naturals in Rennes, France), IdAlg™ (manufactured by Bio Serae in Bram, France), and Seanol™ (produced by

LiveChem in Jeju-do, South Korea and sold by Simple Health in Maitland, USA) are phlorotannin-containing products (Cotas et al. 2020). Although, the companies do not reveal what they do after the phenolic extraction, which can be used for polymer extraction, and the waste used for energy production.

This problematic exploitation is connected to the expected expenses as well as the difficulty in isolating phenolic compounds and ensuring industrial safe uses owing to the quick oxidation rate in the absence of the appropriate buffer (non-oxidative buffer, normally a semi-ethanolic solution with a polymeric ingredient, such as alginate or carboxymethylcellulose (Cotas et al. 2020; Cotas 2023; Cotas et al. 2023a, b). Ongoing research indicates that the content of phenolic compounds in seaweed (from northern Portugal) is low owing to deterioration under natural environmental circumstances (Pardilhó et al. 2023b).

8.3.3 Polysaccharides

Polysaccharides are the most stable sulphated compounds involved in seaweed-based industry; however, using selected species (rather than a mixture) to extract the most stable compounds in the seaweed, which normally are not affected in chemical purity (polysaccharides are stored within the cell wall and membrane; these compounds are really stable due to their position and chemical structures). Typically, seaweed must be picked at the collecting site or at the industrial location before being sent to the proper valorisation pathway, which is to get alginate, carrageenan, and agar. If used from the start, the cascade can be fully explored with higher value and lower-cost production; this is the case when it is considered a selection of seaweeds, thus it is possible to extract refined compounds based on the needs with a higher yield, while seaweed mixture reduces extraction yield and promotes polysaccharide mixture, as is typical in the traditional seaweed-based industry. Although the results of seaweed mixture polysaccharides extraction cannot be as pure or as high in quality as fresh seaweed polysaccharides, there is a market for these polysaccharides, which may be utilised in textiles, agriculture, paper manufacture, or animal feed (Leandro et al. 2020; Pardilhó et al. 2021, 2022a).

8.3.4 Biofermentation

Fermentation is the primary method for obtaining energy products from seaweed, either by anaerobic digestion (AD) to create biogas (rich in methane—CH_4) or ethanol fermentation. Other energy conversion methods include direct combustion, gasification, pyrolysis, and hydrothermal liquefaction; however, due to the decreased heating value associated with seaweed biomass's high moisture content (Murphy et al. 2013).

Due to seaweeds photosynthetic activity, carbon resources are stored in seaweed biomass and available to be extracted and used for biorefinery. Biofermentation is a popular method for exploring seaweed biomass since it involves little selection of the seaweed in industrial facilities. However, unlike other technologies in seaweed refineries, seaweed biofermentation is still in its early stages. This part includes revised research on the issue in order to comprehend the benefits and drawbacks from R&D studies to industrial scale-up (Monteiro et al. 2021; Pardilhó et al. 2022a).

The enhancement of sulphated polysaccharides productivity usually occurs through "algal-bacterial symbiosis". From another perspective, some Gram-negative bacterial strains such as *Nonlabens ulvanivorans* and *Alteromonas* sp. have shown a profound capability to use green algae in batch and fed-batch fermentation approaches, producing valuable microbial enzymes such as ulvan lyase (Qiao et al. 2020). Also, numerous microbial strains could ferment during algal-bacterial symbiosis, producing butyric acid and kainic acid, such as *Clostridium tyrobutyricum* for butyric acid fermentation and *Escherichia coli* to obtain kainic acid (Moore et al. 2020; Song et al. 2011). *Gracilaria* sp., *Sargassum siliquosum*, and *Ulva lactuca* hydrolysates have been suitable substrates for lactic acid bacteria such as *Lactobacillus bulgaricus*, *Lactobacillus delbreckii*, *Lactobacillus lactis*, and *Lactobacillus brevis* yielding lactic acid to develop biodegradable polylactic acid materials (Lin et al. 2020).

Several additional chemicals (acetic acid, lactic acid, methane, ethanol, methanol), however, may be generated by fermentative processes employing a diverse range of substrates and microbes. Due to a lack of quantitative data because the literature uses a mixture of seaweed with other residues (manure, plant residues), estimating the potential of seaweed as a fermentation feedstock for biogas and bioethanol is challenging. Ethanol fermentation procedures are commonly connected with the generation of ethanol from glucose by a single yeast species: *Saccharomyces cerevisiae*. AD is also a fermentation process in which a consortium of microorganisms begins by hydrolysing and then degrading organic substrates, yielding, in the end, CH_4 and CO_2 (biogas) (Barbot et al. 2016; Monteiro et al. 2021; Pardilhó et al. 2022a).

The high moisture content makes seaweed more suitable for microbial conversion than thermochemical processes. However, the high moisture content of seaweed could be an inconvenient in several processes, and for such reason biomass was generally dried prior to application of other PT methods, which is already used and applied by seaweed polysaccharide extraction industry. Sun-drying is the most used method to dry seaweed since it is not only effective but also the most economic (do not require fossil fuel energy); nevertheless, it depends on the weather and could vary from 2–3 days, on sunny days, to weeks, in rainy seasons, and it requires a huge surface to place the seaweed under the sun. A PT treatment can enhance the fermentation efficiency and efficacy, moreover, producing other high-value products reducing the biomass processing costs (Barbot et al. 2016; Monteiro et al. 2021; Pardilhó et al. 2022a, 2023b).

8.4 Advantages in Seaweed Biorefinery: A Next Step for Ecofriendly Solution

Seaweeds are flexible in a variety of ways, including their natural occurrence, composition, harvesting and growing methods, and uses. Important developments include the development of mechanised agriculture in temperate areas, as well as a wide range of valorisations via biorefinery for feed, food, chemicals, and fuels. Seaweed fast growth rates and composition make it ideal for hastening the transition to a bio-based, circular, sustainable, and equitable society. Seaweeds perform an ecological significance in aquatic habitats by generating oxygen, serving as a nursery habitat for a number of marine animals, and providing food for a variety of herbivores. They also support herbivorous animal groups (invertebrates like sea urchins and/or gastropods, as well as vertebrates like herbivorous fish) and give protection from carnivorous predators. They play an important role in the aquatic complex trophic web. Seaweed is important for coastal protection, fertiliser absorption, and carbon storage. Furthermore, seaweeds are significant for a variety of human activities since they are a food supply for many human populations, animal feed, and fertiliser in agriculture. Additionally, numerous seaweed chemicals are already being used commercially in human goods such as pharmaceuticals, orthopaedics, processed foods, and dentistry. Seaweed farming can be a low-cost way to reduce ecological burden. Seaweed may be used as an alternate source of subsistence by coastal people communities in coastal nations because to the low capital investment, short crop/harvest cycle, and simplicity of growing, resulting in a multifunctional primary product (Cotas et al. 2023a, b). Furthermore, they can be sustainable method to mitigate anthropogenic impacts and to be used to produce compounds needed by humans in an economic feasible pathway.

Many various biorefinery and valorisation schemes are being developed (mostly, as demonstrated above focussing in the polysaccharide extraction efficiency and reducing the cleaning costs), and depending on the local circumstances, some *of them can be feasible*. This is most difficult, step from small scale because the complexity of the process engineering. Clearly, many seaweed compounds must be valorised to ensure economic viability. Additional value from ecosystem services and maybe carbon sequestration could help to establish feasible processes employing seaweed as feedstock (Dussan et al. 2023). Due to seaweed growth is on aquatic system, they can provide coastal protection, and be a feeder for the aquatic species and also a key element to fight the acidity of the ocean. On other point, they can absorb the CO_2 into sugars and carbohydrates (polysaccharides), which, can substitute non-organic carbon (fuels and crude products), reducing CO_2 liberated to atmosphere using a circular economy approach (Cotas et al. 2023a, b).

The realisation of a global bio-economy is of great importance, and biomass utilisation plays an important role, where innovative solutions, such as biorefinery technologies, should be considered (Balina et al. 2017; Pardilhó et al. 2022a Cascade valorisation procedures are often used to monetise biomass by exposing it to several extraction operations to selectively acquire different components, allowing for a zero-waste strategy (Balina et al. 2017; Pardilhó et al. 2022a). The assessment of integrated solutions based on a biorefinery concept to get different products/compounds utilising the same initial biomass is of high interest (reducing the waste), allowing full utilisation of the seaweed biomass potential (more by-products from the same biomass). This should be accomplished using an economically possible procedure to extract desirable add-values components in low quantities (e.g. fatty acids (FA), proteins, phenolic compounds, and pigments) alongside lower valuable products in greater quantities (e.g. polymers, sugars for biofuel and residues to soil conditioners) (Pardilhó et al. 2022a, 2023b).

In the context of a sustainable and circular economy, the biorefinery final waste can be used as fertiliser or soil amendment for agriculture industry, taking into consideration the phosphorus and other nutrient levels, and heavy metal concentration (more important, lead and mercury). Thus, promoting a circular economy, where all the seaweed biomass is exploited (Fig. 8.2). This type of concept is being developed in small scale (Pardilhó et al. 2022b, 2023a, b), with good results; however, it is done by traditional techniques (using solvents methodology (hexane and methanol/ethanol), due to the usage of reagent that needs to be disposed or recovered with high percentage (the recovery yield is near 95% using the recovery method at laboratory scale, which is used for alginate and carrageenan extractions).

However, the great advantage of the biorefinery concept, even using traditional methods, it can replace the traditional industrial pre-treatment to extract extra-compounds sequentially, without the need of additional procedures that result in non-recovered residues. Indeed, it is an economic input for the seaweed-based industry, and reduces the waste produced by them.

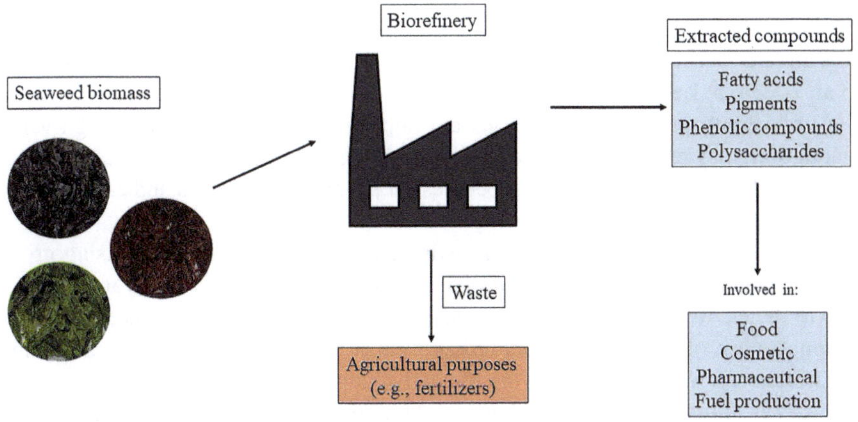

Fig. 8.2 Schematic representation of the production obtained from biorefinery

8.5 Bottlenecks in Seaweed Biorefinery Processes

Most of the research has focused on the effects of thermal processing on the concentration of bioactive components, with little attention paid to the bioavailability of these compounds following thermal processing. It is vital to understand how these nutrients, phytochemicals, and pollutants are affected by post-harvest processing, which includes storage, thermal processing, and cooking (Ho and Redan 2020).

The actual and official characterisation procedures in seaweeds are based on plant methodologies but have higher variability and mistakes since the same kind of components in plants, such as carbohydrates/dietary fibres, do not react the same way as similar compounds in seaweeds (Leandro et al. 2020). Although, in seaweeds' characterisation there are some techniques that are being developed for high economic and high efficiency to understand the best way to exploit the seaweed biomass (Cotas et al. 2019; Pardilhó et al. 2023b).

These biorefinery processes' restrictions can be addressed in the future by improving and standardising them, allowing for an accurate evaluation and control of seaweed advantages and risks (Pardilhó et al. 2022a); thus, seaweed producers and industry do the required analysis, which can give deviated results due to the methodology applied sensitivity, also because, in general until now, laboratory tests are standard for plants and not for seaweed (mostly in agriculture and biofermentation). Although, more recently, there is new supporting seaweed measurement standardisation by the National Institute of Standards and Technology IST Dietary Supplement Laboratory Quality Assurance Program (United States of America) (Nist 2023), however, European Quality Agency of Agriculture, food/feed, is still underdeveloped where some countries have more restrictive legislation than others; for example, Spain and France have a clear quality assay that is needed when, for example, east countries and Portugal do not have clear indications. Even the European Union is not assertive in the methodology to use in the certification of seaweed biostimulants (Pardilhó et al. 2021) (even certification laboratory differ in the method to be used) (New Food Magazine, 2021). One of the more evident cases of standard measures is in the agriculture and food/feed quality checkers (Pardilhó et al. 2022a, b; Leandro et al. 2020).

While multi-product biorefineries are appealing from a zero-waste/high-value standpoint, they rely on the very complicated integration of various extraction methods and a battery of downstream processes for product purification and conditioning (Dussan et al. 2023).

This cascade notion should be investigated further in the future utilising green solvents and/or procedures, as traditional solvents are mostly poisonous and destructive to the environment. On an industrial scale, it is important to emphasise the traditional reuse of organic solvents (ethanol, acetone, hexane, methanol) following their recovery, which can reduce their negative impact and process costs because the industry is almost based on non-green techniques when processing biomass which is a bottleneck when RD units are pulling greener techniques without support from industry.

In this biorefinery scenario, adequate laws and certifications about seaweeds safety and security application are required in order to certify the seaweeds using the best approach in a cost-effective manner (Salehi et al. 2019). During this time, it is critical that institutional frameworks encourage growth and commercialisation, therefore propelling the algal biomass sector to prominence as a source of high-value goods (Trentacoste et al. 2015). Furthermore, the seaweed industry is reliant on regulatory factors such as aquaculture/agriculture subsidies, marine spatial planning or land use spatial planning (for offshore or inshore growing, respectively), licensing, and other legislative agreements. As the biorefinery concept is only used in a small scale, their impact to change political sets is very hard (Lähteenmäki-Uutela et al. 2021), due to various forms of biorefinery concept (green or solvent downstream, or a mix between green and solvent) and biomass utilisation. Also, the technology is only now to be developed at industrial scale with a coupled pre-step to analyse the seaweed biomass and their economic rentability. The Umimare green biorefinery (Umimare 2022) concept, which is being developed to scale up to use cultivated low-quality *Gracilaria* sp. (not approved for food or feed), at this moment, passes by a biomass rapid characterisation (to determine the ratios and downstream) before being used for agar extraction or biofertiliser (a commercial product already) production, ethanolic fermentation (bioethanol production, using a new green technique with high efficiency), and residue as a soil conditioner. This concept is based on the biorefinery by Hemporama, which is at medium scale for Hemp biorefinery, to obtain fibres, Hurd and fermented liquid (Hemporama 2023). Also, TNO, a Norwegian-based company, is developing at a pilot scale new biorefineries technologies to process the seaweed biomass up to downstream add-value products (TNO Innovation for Life 2018).

The results of the seaweed cascade extraction processes demonstrate a viable hypothesis to be exploited in the future to use a raw material which can be applied in several industries; however, this seaweed can have seasonal and geographical variations, and thus, further studies should more deeply analyse such parameters to ensure feasibility of industrial-scale applications. Such studies should be specific for the targeted compounds, especially for high-value-added compounds (pigment and fatty acids), which might depend upon the species. The use of dedicated seaweed cultivation systems as a complement/support to guarantee constant availability of the seaweed biomass to obtain target compounds should also be carefully analysed. This, since generally, seaweed dedicated cultivation systems have different applications, usually incompatible with seaweed drifted-ashore (e.g. food and cosmetics). Thus, the minimum quantity of available seaweed drifted-ashore (e.g. *Rugulopterix okumorae* in the Mediterranean Sea) biomass should be ensured due to economic constraints. In the future, the traditional cascade concept should also be explored using green solvents and/or techniques, since the conventional solvents are mainly toxic and harmful to the environment. At an industrial level, it is relevant to mention the conventional reuse of the organic solvents (mostly alcohols and acetone) after their recovery, minimising their negative impact and process costs.

8.6 Conclusions

From seaweeds research conducted during the last two decades, it has been confirmed that the economic potential of seaweeds is immense, contributing to *revolutionary uses* in various industries and showing promise as a renewable resource for the future. By harvesting and utilising all the macroalgae biomass, they not only provide valuable bioactive compounds but also contribute to reduce carbon dioxide emissions through photosynthetic carbon buildup. Seaweeds significantly support coastal ecosystems, providing habitats, food, and shelter to a wide array of aquatic organisms, thus playing a pivotal role in sustaining aquatic biodiversity. They can be cultivated to protect the coastal area with advantage to reduce mechanical aggression from the wave but also reducing the carbon footprint. Seaweeds can be a key for new natural solutions for agriculture, polymers (for bioplastic) and new fuels (bioethanol), which can be obtained using the same raw biomass.

With growing demand and concerns over ecological impact such as overuse of terrestrial sources or increase of pollution due to biorefinery processes, there is a shift towards cultivated seaweed production. The economic utility of seaweeds largely stems from their polysaccharide content, particularly alginate, laminarin, and fucoidan in brown seaweeds, and agar or carrageenan in red seaweeds. Pigment extraction, including valuable compounds like chlorophylls, carotenes, and phycobiliproteins, further adds to their commercial potential. Despite these advancements, challenges remain, particularly in addressing negative environmental impacts like seaweed blooms and beach-drifted biomass.

The concept of biorefinery offers a comprehensive approach, encompassing biochemical profiles and applications of seaweed biomass.

The results of the seaweed cascade extraction processes show a viable hypothesis that can be exploited in the future to use seaweed raw material, which can be applied in several industries (agriculture, energy, feed and textile, as major examples). Furthermore, the development of new biorefinery approaches (not the traditional and more common) can be important to obtain interesting seaweed compounds (e.g. lipids, pigments, neutral sugars, polysaccharides, and minerals), contributing to technology development. These compounds can be valuable for several industries, such as animal feed, agriculture, textile industry or bioenergy production. Therefore, biorefinery concept could potentially unlock even more sustainable and innovative uses of seaweeds, contributing to both economic growth and ecological preservation.

Consequently, RD investigations should be tailored to the targeted molecules, particularly high-value-added chemicals that may vary depending on the species. The usage of specialised growing methods as a supplement/support to ensure consistent availability of biomass to extract target chemicals should also be further investigated.

Because traditional solvents, such as methanol, ethanol, hexane, and acetone, are mostly hazardous and destructive to the environment, this cascade idea can be approached with green solvents (ionic liquids) or new mechanical approaches (ultrasound or electric pulses) in the future. However, this technique can include solvent recovery, which the business can use to reduce the negative impact of organic solvents and process costs.

Acknowledgements This work was supported by FCT-Fundação para a Ciência e Tecnologia, I.P., in the framework of the Projects UIDB/04004/2020 and DOI identifier 10.54499/UIDB/04004/2020 (doi:10.54499/UIDB/04004/2020 accessed on 30 July 2024) granted to the Centre for Functional Ecology—Science for People and the Planet, (LA/P/0092/2020; https:/doi.org/10.54499/LA/P/0092/2020) granted to the Associate Laboratory TERRA and UIDP/50017/2020 + UIDB/50017/2020 + LA/P/0094/2020 granted to CESAM. Silvia Lomartire thanks the Foundation for Science and Technology (FCT) for the financial support granted through the doctoral grant 2021.05005.BD. Ana M. M. Gonçalves acknowledges the University of Coimbra for the contract IT057-18-7253.

References

Almanza V, Buschmann AH (2013) The ecological importance of *Macrocystis pyrifera* (Phaeophyta) forests towards a sustainable management and exploitation of Chilean coastal benthic co-management areas. Int J Environ Sustain Dev 12(4):341. https://doi.org/10.1504/IJESD.2013.056331

Ashkenazi DY, Israel A, Abelson A (2019) A novel two-stage seaweed integrated multi-trophic aquaculture. Rev Aquac 11(1):246–262. https://doi.org/10.1111/raq.12238

Assubaie FN (2015) Assessment of the levels of some heavy metals in water in Alahsa Oasis farms, Saudi Arabia, with analysis by atomic absorption spectrophotometry. Arab J Chem 8(2):240–245. https://doi.org/10.1016/j.arabjc.2011.08.018

Balina K, Romagnoli F, Blumberga D (2017) Seaweed biorefinery concept for sustainable use of marine resources. Energy Procedia 128:504–511. https://doi.org/10.1016/j.egypro.2017.09.067

Banu JR, Kavitha S, Tyagi VK, Gunasekaran M, Karthikeyan OP, Kumar G (2021) Lignocellulosic biomass based biorefinery: a successful platform towards circular bioeconomy. Fuel 302:121086

Barbot Y, Al-Ghaili H, Benz R (2016) A review on the valorization of macroalgal wastes for biomethane production. Mar Drugs 14(6):120. https://doi.org/10.3390/md14060120

Bogolitsyn K, Parshina A, Druzhinina A, Ovchinnikov D, Krasikov V, Khviyuzov S (2020) Physicochemical characteristics of the active fractions of polyphenols from Arctic macrophytes. J Appl Phycol 32(6):4277–4287. https://doi.org/10.1007/s10811-020-02226-w

Cardamo (1936) Seaweed oil. https://www.cardamo.gr/en/products/seaweed-oil_fykia_elaio. Access in date 19 Oct 2023

Charrier B, Abreu MH, Araujo R, Bruhn A, Coates JC, De Clerck O, Katsaros C, Robaina RR, Wichard T (2017) Furthering knowledge of seaweed growth and development to facilitate sustainable aquaculture. New Phytol 216(4):967–975. https://doi.org/10.1111/nph.14728

Chary K, van Riel AJ, Muscat A, Wilfart A, Harchaoui S, Verdegem M et al (2023) Transforming sustainable aquaculture by applying circularity principles. Rev Aquac 16:656

Coates J (2004) Encyclopedia of analytical chemistry -interpretation of infrared spectra, a practical approach. Encycl. Anal Chem:1–23

Connemara Organic Seaweed Company (2013) Organic seaweed body oil. https://www.connemaraseaweedcompany.ie/products/organic-seaweed-body-oil. Access in date 19 Oct 2023

Costa JC, Gonçalves PR, Nobre A, Alves MM (2012) Biomethanation potential of macroalgae *Ulva* spp. and *Gracilaria* spp. and in co-digestion with waste activated sludge. Bioresour Technol 114:320–326. https://doi.org/10.1016/j.biortech.2012.03.011

Cotas J (2023) Marine phenolics. In: Functional ingredients from algae for foods and nutraceuticals. Elsevier, pp 213–245

Cotas J, Figueirinha A, Pereira L, Batista T (2019) The effect of salinity on *Fucus ceranoides* (Ochrophyta, Phaeophyceae) in the Mondego River (Portugal). J Oceanol Limnol 37(3):881–891. https://doi.org/10.1007/s00343-019-8111-3

Cotas J, Leandro A, Monteiro P, Pacheco D, Figueirinha A, Gonçalves AMM, da Silva GJ, Pereira L (2020) Seaweed phenolics: from extraction to applications. Mar Drugs 18(8):384. https://doi.org/10.3390/md18080384

Cotas J, Gomes L, Pacheco D, Pereira L (2023a) Ecosystem services provided by seaweeds. Hydrobiology 2(1):75–96

Cotas J, Pacheco D, Monteiro P, Gonçalves AMM, Pereira L (2023b) Marine phenolics: extractions at low pressure. In: Marine phenolic compounds. Elsevier, pp 115–146

Duarte CM, Wu J, Xiao X, Bruhn A, Krause-Jensen D (2017) Can seaweed farming play a role in climate change mitigation and adaptation? Front Mar Sci 4(100):1–8. https://doi.org/10.3389/fmars.2017.00100

Dumay J, Morançais M (2016) Proteins and pigments. In: Seaweed in health and disease prevention. Academic, pp 275–318

Dussan K, Dijkstra JW, Luzzi S, van Zandvoort I, van Hal JW (2023) Seaweed versatility for biorefinery: blessing or burden? Curr Opin Green Sustain Chem 39:100728. https://doi.org/10.1016/j.cogsc.2022.100728

Filote C, Santos SCR, Popa VI, Botelho CMS, Volf I (2021) Biorefinery of marine macroalgae into high-tech bioproducts: a review. Environ Chem Lett 19(2):969–1000. https://doi.org/10.1007/s10311-020-01124-4

Ford L, Theodoridou K, Sheldrake GN, Walsh PJ (2019) A critical review of analytical methods used for the chemical characterisation and quantification of phlorotannin compounds in brown seaweeds. Phytochem Anal 30(6):587–599. https://doi.org/10.1002/pca.2851

Freitas MV, Pacheco D, Cotas J, Mouga T, Afonso C, Pereira L (2022) Red seaweed pigments from a biotechnological perspective. Phycology 2(1):1–29. https://doi.org/10.3390/phycology2010001

García-Poza S, Leandro A, Cotas C, Cotas J, Marques JC, Pereira L, Gonçalves AMM (2020) The evolution road of seaweed aquaculture: cultivation technologies and the industry 4.0. Int J Environ Res Public Health 17(18):6528. https://doi.org/10.3390/ijerph17186528

Gomes L, Monteiro P, Cotas J, Gonçalves AMM, Fernandes C, Gonçalves T, Pereira L (2022) Seaweeds' pigments and phenolic compounds with antimicrobial potential. Biomol Concepts 13(1):89–102. https://doi.org/10.1515/bmc-2022-0003

Hemporama (2023). https://www.hemporama.pt/. Access in date 17 Oct 2023

Hermans S (2022) Oceans 2050 results further seaweed carbon science. Phyconomy – tracking the seaweed economy. https://phyconomy.net/articles/oceans-2050-seaweed-carbon/. Access in date 20 Oct 2023

Hermans S (2023) 2023 seaweed state of the industry. Phyconomy – tracking the seaweed economy. https://phyconomy.net/articles/2022-seaweed-review/. Access in date 20 Oct 2023

Hidayati JR, Yudiati E, Pringgenies D, Oktaviyanti DT, Kusuma AP (2020) Comparative study on antioxidant activities, total phenolic compound and pigment contents of tropical *Spirulina platensis, Gracilaria arcuata* and *Ulva lactuca* extracted in different solvents polarity. In: E3S web of conferences, vol 147. EDP Sciences, p 03012

Ho KKHY, Redan BW (2020) Impact of thermal processing on the nutrients, phytochemicals, and metal contaminants in edible algae. Crit Rev Food Sci Nutr 0(0):1–19. https://doi.org/10.1080/10408398.2020.1821598

Imeson A (ed) (2009) Food stabilisers, thickeners and gelling agents. Wiley

Jard G, Marfaing H, Carrère H, Delgenes JP, Steyer JP, Dumas C (2013) French Brittany macroalgae screening: composition and methane potential for potential alternative sources of energy and products. Bioresour Technol 144:492–498. https://doi.org/10.1016/j.biortech.2013.06.114

Jung KA, Lim S-R, Kim Y, Park JM (2013) Potentials of macroalgae as feedstocks for biorefinery. Bioresour Technol 135:182–190. https://doi.org/10.1016/j.biortech.2012.10.025

Kokossis AC, Yang A (2010) On the use of systems technologies and a systematic approach for the synthesis and the design of future biorefineries. Comput Chem Eng 34(9):1397–1405

Kraan S (2012) Algal polysaccharides, novel applications and outlook. In: Carbohydrates-comprehensive studies on glycobiology and glycotechnology. IntechOpen

Lähteenmäki-Uutela A, Rahikainen M, Camarena-Gómez MT, Piiparinen J, Spilling K, Yang B (2021) European Union legislation on macroalgae products. Aquac Int 29:487. https://doi.org/10.1007/s10499-020-00633-x

Leandro A, Pacheco D, Cotas J, Marques JC, Pereira L, Gonçalves AMM (2020) Seaweed's bioactive candidate compounds to food industry and global food security. Life 10(8):140. https://doi.org/10.3390/life10080140

Li Q, Feng Z, Zhang T, Ma C, Shi H (2020) Microplastics in the commercial seaweed nori. J Hazard Mater 388:122060. https://doi.org/10.1016/j.jhazmat.2020.122060

Lin HTV, Huang MY, Kao TY, Lu WJ, Lin HJ, Pan CL (2020) Production of lactic acid from seaweed hydrolysates via lactic acid bacteria fermentation. Fermentation 6(1):37

Lomartire S, Cotas J, Pacheco D, Marques JC, Pereira L, Gonçalves AMM (2021) Environmental impact on seaweed phenolic production and activity: an important step for compound exploitation. Mar Drugs 19(5):245. https://doi.org/10.3390/md19050245

Maceiras R, Rodrı'guez M, Cancela A, Urréjola S, Sánchez A (2011) Macroalgae: raw material for biodiesel production. Appl Energy 88(10):3318–3323. https://doi.org/10.1016/j.apenergy.2010.11.027

Mæhre HK, Dalheim L, Edvinsen GK, Elvevoll EO, Jensen I-J (2018) Protein determination – method matters. Food Secur 7(1):5. https://doi.org/10.3390/foods7010005

Michalak I, Chojnacka K (2015) Algae as production systems of bioactive compounds. Eng Life Sci 15:160

Monteiro P, Lomartire S, Cotas J, Pacheco D, Marques JC, Pereira L, Gonçalves AMM (2021) Seaweeds as a fermentation substrate: a challenge for the food processing industry. PRO 9(11):1953. https://doi.org/10.3390/pr9111953

Monteiro P, Cotas J, Pacheco D, Figueirinha A, da Silva GJ, Pereira L, Gonçalves AMM (2022) Seaweed as food: how to guarantee their quality? In: Sustainable global resources of seaweeds, vol 2. Springer, Cham, pp 309–321

Moore BS, Chekan JR, McKinnie SM (2020) The regents of the University of California. In: Scalable biosynthesis of the seaweed neurochemical kainic acid

Morais T, Inácio A, Coutinho T, Ministro M, Cotas J, Pereira L, Bahcevandziev K (2020) Seaweed potential in the animal feed: a review. J Mar Sci Eng 8(8):559. https://doi.org/10.3390/jmse8080559

Murphy F, Devlin G, Deverell R, McDonnell K (2013) Biofuel production in Ireland – an approach to 2020 targets with a focus on algal biomass. Energies 6(12):6391–6412. https://doi.org/10.3390/en6126391

Muscat A, de Olde EM, Ripoll-Bosch R, Van Zanten HH, Metze TA, Termeer CJ et al (2021) Principles, drivers and opportunities of a circular bioeconomy. Nat Food 2(8):561–566

New Food Magazine (2021) Setting the standards for seaweed analysis. https://www.newfood-magazine.com/article/138149/setting-the-standards-for-seaweed-analysis/. Access in date 17 Oct 2023

Nist (2023) Supporting the standardization of seaweed measurements. https://www.nist.gov/news-events/events/2023/10/supporting-standardization-seaweed-measurements. Access in date 17 Oct 2023

Nurachman Z, Hartini H, Rahmaniyah WR, Kurnia D, Hidayat R, Prijamboedi B et al (2015) Tropical marine *Chlorella* sp. PP1 as a source of photosynthetic pigments for dye-sensitized solar cells. Algal Res 10:25–32

Pangestuti R, Kim S-K (2011) Biological activities and health benefit effects of natural pigments derived from marine algae. J Funct Foods 3(4):255–266. https://doi.org/10.1016/j.jff.2011.07.001

Pardilhó S, Cotas J, Gonçalves AMM, Dias JM, Pereira L (2021) Seaweeds used in wastewater treatment: steps to industrial commercialization. In: Phycology-based approaches for wastewater treatment and resource recovery. CRC Press, Boca Raton, pp 247–262

Pardilhó S, Cotas J, Pereira L, Oliveira MB, Dias JM (2022a) Marine macroalgae in a circular economy context: a comprehensive analysis focused on residual biomass. Biotechnol Adv 60:107987. https://doi.org/10.1016/j.biotechadv.2022.107987

Pardilhó S, Pires JC, Boaventura R, Almeida M, Maia Dias J (2022b) Biogas production from residual marine macroalgae biomass: kinetic modelling approach. Bioresour Technol 359:127473. https://doi.org/10.1016/j.biortech.2022.127473

Pardilhó S, Cotas J, Pacheco D, Bahcevandziev K, Pereira L, Oliveira MB, Maia Dias J (2023a) Back to the origins: potential of beach-cast macroalgae as biofertilizer. Waste Biomass Valoriz 14(4):1097–1111. https://doi.org/10.1007/s12649-022-01922-7

Pardilhó S, Cotas J, Pacheco D, Gonçalves AMM, Bahcevandziev K, Pereira L, Figueirinha A, Dias JM (2023b) Valorisation of marine macroalgae waste using a cascade biorefinery approach: exploratory study. J Clean Prod 385:135672. https://doi.org/10.1016/j.jclepro.2022.135672

Pereira L, Sousa A, Coelho H, Amado AM, Ribeiro-Claro PJA (2003) Use of FTIR, FT-Raman and 13C-NMR spectroscopy for identification of some seaweed phycocolloids. Biomol Eng 20(4–6):223–228. https://doi.org/10.1016/S1389-0344(03)00058-3

Qiao L, Yang X, Xie R, Du C, Chi Y, Zhang J, Wang P (2020) Efficient production of ulvan lyase from *Ulva prolifera* by *Catenovulum* sp. LP based on stage-controlled fermentation strategy. Algal Res 46:101812

Reisewitz SE, Estes JA, Simenstad CA (2006) Indirect food web interactions: sea otters and kelp forest fishes in the Aleutian archipelago. Oecologia 146(4):623–631. https://doi.org/10.1007/s00442-005-0230-1

Repêchage (1980) From the line "Beauty From The Sea": essential oil of seaweed. https://www.beautyfromthesea.co.uk/products/essential-oil-of-seaweed. Access in date 19 Oct 2023

Rocha CP, Pacheco D, Cotas J, Marques JC, Pereira L, Gonçalves AMM (2021) Seaweeds as valuable sources of essential fatty acids for human nutrition. Int J Environ Res Public Health 18(9):4968. https://doi.org/10.3390/ijerph18094968

Rodrigues D, Freitas AC, Pereira L, Rocha-Santos TAP, Vasconcelos MW, Roriz M, Rodríguez-Alcalá LM, Gomes AMP, Duarte AC (2015) Chemical composition of red, brown and green macroalgae from Buarcos bay in Central West Coast of Portugal. Food Chem 183:197–207. https://doi.org/10.1016/j.foodchem.2015.03.057

Saati EA (2011) The solvent effectiveness on extraction process of seaweed pigment. MAKARA of Technology Series. https://doi.org/10.7454/mst.v15i1.850

Salehi S-R, Seca P, Michalak T, Mishra N, Zam M (2019) Current trends on seaweeds: looking at chemical composition, phytopharmacology, and cosmetic applications. Molecules 24(22):4182. https://doi.org/10.3390/molecules24224182

Sim SF, Lee TZE, Mohd Irwan LNAL, Samling B (2014) Synchronized analysis of FTIR spectra and GCMS chromatograms for evaluation of the thermally degraded vegetable oils. J Anal Methods Chem 2014:1–9. https://doi.org/10.1155/2014/271970

Song JH, Ventura JRS, Lee CH, Jahng D (2011) Butyric acid production from brown algae using *Clostridium tyrobutyricum* ATCC 25755. Biotechnol Bioprocess Eng 16:42–49

Sumampouw GA, Jacobsen C, Getachew AT (2021) Optimization of phenolic antioxidants extraction from Fucus vesiculosus by pressurized liquid extraction. J Appl Phycol 33:1195. https://doi.org/10.1007/s10811-020-02362-3

TNO Innovation for life (2018). https://www.tno.nl/en/technology-science/labs/seaweed-lab/. Access in date 17 Oct 2023.

Trentacoste EM, Martinez AM, Zenk T (2015) The place of algae in agriculture: policies for algal biomass production. Photosynth Res 123(3):305–315. https://doi.org/10.1007/s11120-014-9985-8

Umimare (2022). https://umimare.com/. Access in date 17 Oct 2023

Vásquez JA, Zuñiga S, Tala F, Piaget N, Rodríguez DC, Vega JMA (2014) Economic valuation of kelp forests in northern Chile: values of goods and services of the ecosystem. J Appl Phycol 26(2):1081–1088. https://doi.org/10.1007/s10811-013-0173-6

Zubia M, Freile-Pelegrín Y, Robledo D (2014) Photosynthesis, pigment composition and antioxidant defences in the red alga *Gracilariopsis tenuifrons* (Gracilariales, Rhodophyta) under environmental stress. J Appl Phycol 26:2001–2010

Chapter 9
Seaweed-Based Proteins and Peptides: Processing and Development of Innovative Products

Armin Mirzapour-Kouhdasht, Samaneh Shaghaghian, and Marco Garcia-Vaquero

Abstract Seaweeds are currently an under-utilized source of proteins and peptides with great potential for the nutraceutical and functional food sectors. Due to the high cost of animal-sourced proteins, extensive research into alternative protein sources has been recently conducted. On the other hand, environmental and health issues caused by animal-sourced proteins, as well as animal welfare concerns, have drawn attention to alternative and sustainable sources of dietary protein. In terms of nutrition, seaweeds contain not only a significant amount of protein but also all of the essential amino acids (EAAs), which are vital for preserving good health. In addition, the processing of these proteins can result in the generation of bioactive peptides with a wide variety of health-promoting potentials, such as cardioprotective, antioxidative, antidiabetic, and immunomodulatory activities. This chapter analyses the literature available on the use of seaweeds as sources of proteins and EAAs, briefly summarizing current research available on the generation of bioactive peptides and their health-promoting activities. Key aspects of the use of seaweeds or seaweed protein ingredients, such as their techno-functional properties for the incorporation of seaweed proteins into foods, will also be covered in detail, together with the current challenges and bottlenecks hindering the full utilization of this biomass and future research needed to overcome them, helping to turn seaweed into a viable protein source for the processing and development of novel functional foods.

A. Mirzapour-Kouhdasht
School Agriculture and Food Science, University College Dublin, Dublin 4, Ireland

Department of Food Science, Purdue University, West Lafayette, United States

S. Shaghaghian
Department of Food Science and Technology, Shiraz University, Shiraz, Iran

Department of Food Science, University of Laval, Quebec, Canada

M. Garcia-Vaquero (✉)
School Agriculture and Food Science, University College Dublin, Dublin 4, Ireland
e-mail: marco.garciavaquero@ucd.ie

N. Trivedi et al. (eds.), *Recent Advances in Seaweed Biotechnology*,
https://doi.org/10.1007/978-981-96-0519-4_9

Keywords Seaweed · Proteins · Peptides · Bioactive peptides · Functional food · Techno-functional properties · Food formulation

9.1 Introduction

The demand for sustainable and eco-friendly protein sources is increasing, given the environmental and health concerns associated with animal-sourced proteins (Shaghaghian et al. 2022). Seaweeds have recently emerged as a promising source of proteins and peptides with great potential for the development of functional foods and nutraceuticals (Biparva et al. 2023). The incorporation of seaweed proteins in plant-based meat and fish analogs was shown to enhance the nutritional quality of the final product, while reducing the environmental impact of meat production (Nowacka et al. 2023).

Seaweeds have long been recognized as a valuable source of biologically active compounds, such as polysaccharides, minerals, vitamins, and antioxidants. However, seaweed proteins and peptides have received less attention than other compounds. Recent research has focused on seaweed proteins as a potential alternative to animal-sourced proteins. For example, several studies showed that seaweed proteins can be a sustainable and eco-friendly source of high-quality inputs with low environmental impact (Mouritsen 2013; Wijesekara et al. 2011).

One of the most significant advantages of seaweed proteins is their amino acid composition, which contains all essential amino acids (EAAs) in sufficient amounts to meet human dietary needs (Echave et al. 2022; Wijesekara et al. 2011). In addition, the processing of seaweed proteins can result in the generation of bioactive peptides with diverse health-promoting properties, such as antioxidative, antihypertensive, and anti-inflammatory activities (Cermeño et al. 2020; Echave et al. 2022). This makes seaweed proteins a valuable source of high-quality protein for the development of functional foods, which has been studied mainly in plant-based meat analogs, where the use of this protein source results in high-quality meat analogs with all EAAs and fewer environmental problems and health risks (Nadeeshani et al. 2022). For example, several studies have reported that seaweed proteins can be incorporated into plant-based meat analogs, such as burgers, sausages, and nuggets (Joshi and Kumar 2015; Kurek et al. 2022; Kyriakopoulou et al. 2021; Trottet et al. 2018).

Despite the potential of seaweed as a sustainable and eco-friendly protein source, there are still challenges and bottlenecks hindering the full utilization of this biomass. As an example, there is a lack of standardized methods for seaweed protein extraction, and the variability in their content and composition among different species can affect the quality and consistency of the final product (Pliego-Cortés et al. 2020). Therefore, future research is needed to overcome these challenges and to fully utilize the potential of seaweed biomass as a protein source for the processing and development of novel functional foods.

This chapter provides a comprehensive overview of the current state of the art of knowledge on seaweed proteins, including the generation of bioactive peptides and their potential for the development of functional foods. The techno-functional properties of seaweed proteins, current research on seaweed protein derivatives, including bioactive peptides, lectins, phycobiliproteins and free amino acids, and their health-promoting activities will also be covered together with the challenges and obstacles still impede the complete exploitation of seaweed biomass as a protein source. This chapter also discusses the benefits and limitations of seaweed proteins as an ingredient for incorporation in food products, as well as highlighting that future research is needed to realize the full potential of this valuable natural resource.

9.2 Use of Seaweeds as a Source of Protein

Population growth, climate change, and depletion of arable land and freshwater resources, in addition to the growing concerns about zoonotic diseases and antibiotic resistance, are all key drivers toward the movement and exploration of novel, and ideally sustainable, safe and nutritious foods. Algae may meet most of these criteria to be considered an ideal food for the future. Within the algae, marine macroalgae or seaweeds are versatile products that have been considered a source of minerals and for other food applications since ancient times (Kumar et al. 2023). Seaweeds are consumed fresh, dried, added as ingredients in foods, or for the generation of food supplements. Moreover, edible seaweeds can be found in supermarkets in multiple formats, such as canned (*Saccorhiza polyschides* and *Himanthalia elongata*) or dried (*H. elongata*, *Laminaria ochroleuca*, *Undaria pinnatifida*, *Palmaria* spp., and *Porphyra* spp.) (Kılınç et al. 2013).

Some seaweeds have been described as sources of protein, with contents of up to 45% dry weight (DW), this being comparable to eggs (53% DW), and other terrestrial plant sources such as soybean and legumes (30–40% DW). The inclusion of seaweeds with high levels of protein in the diet can contribute toward achieving the recommended dietary intake (RDI) of dietary protein for adults (i.e., 0.8 g of protein per kg of body weight per day) (Delimaris 2013; Puertas and Vázquez 2021). However, the protein content of seaweed is variable, and it has been described as ranging between 5 and 47% DW depending on the species, maturity, season, and geographical location of the biomass. Red seaweeds with 12.5–35.2% DW protein, followed by green seaweeds (9.6–23.3% DW), have been described as the macroalgal species with the highest protein contents, as compared to those of brown seaweeds, which ranged between 4.5 and 16.8% DW (Echave et al. 2022; Hentati et al. 2020; Ms et al. 2022). Table 9.1 shows the protein content of selected seaweed species currently consumed as human food and as described by (Fleurence et al. 2018).

Amino acids, as the building blocks of proteins, are precursors of low molecular-weight compounds (i.e., creatine, carnitine, carnosine, thyroid hormones, serotonin, melanin, melatonin, and heme groups), thus playing an essential role in regulating

Table 9.1 Protein content of some seaweed species consumed as food

	Phaeophyta (brown seaweed)			Chlorophyta (green seaweed)		Rhodophyta (red seaweed)		
Seaweed species	*Laminaria digitata*	*Ascophyllum nodosum*	*Undaria pinnatifida*	*Ulva lactuca*	*Ulva pertusa*	*Porphyra tenera*	*Chondrus crispus*	*Palmaria palmata*
Protein content (% DW)	8.0–15.0	3.0–15.0	11.0–24.0	8.7–32.7	17.5–26.0	33.0–47.0	21.4	8.0–35.0

The content of the table was originally published by (Fleurence et al. 2018) and reproduced with permission from Elsevier metabolism, gene expression, immunological activity, reproduction, tissue healing, neuro-transmission and other crucial physiological activities (Wu 2013). Seaweeds are an excellent source of amino acids that, in a similar way to the protein contents, can be variable depending on multiple factors affecting the biomass. The EAA concentration of seaweed reaches up to 24% DW in some species, including *U. pinnatifida* (48.4 g/100 g protein), *Ulva australis* (42 g/100 g protein), *Ascophyllum nodosum* (34.4 g/100 g protein), *Fucus vesiculosus* (25.1 g/100 g protein), and *Ulva* spp. (27.0 g/100 g protein). The high contents of these seaweeds in EAAs suggest that they may be considered good sources, as they can help to fulfill the nutritional requirements indicated by the FAO/WHO (Echave et al. 2022; World Health Organization 1991; Peñalver et al. 2020). The EAA contents of these selected seaweed are comparable to those of other traditional sources of these amino acids, such as casein (43.6% DW?), ovalbumin (52.4%), and legumes (45%) (Echave et al. 2022). Among the EAAs, leucine is the most predominant in *Alaria esculenta* and *Fucus serratus* (brown seaweeds), followed by valine in *Ulva lactuca* (green seaweed) and *Palmaria palmata* (red seaweed) (Thiviya et al. 2022). The content of arginine in *Chondrus crispus* was reported to be approximately 34% of the total amino acid fraction, which is much higher than its concentration in leguminous plants (i.e., 13–14% DW?) and ovalbumin (11.7% DW?) (Fleurence et al. 2018). The levels of isoleucine and threonine in *P. palmata* are similar to those in legumes, and the level of histidine in *U. pertusa* is close to that of egg proteins (Pangestuti and Kim 2015). Despite the abundance of leucine, valine, glutamic acid, taurine, threonine, arginine, alanine, glycine, proline, and aspartic acid, some seaweeds, such as *C. crispus*, *Gracilaria* spp., *Porphyra* spp. have a deficiency in phenylalanine and methionine. Most limiting amino acids in seaweed proteins consist of sulfur amino acids (methionine and cysteine), lysine, and tryptophan, whereas the mean values for each amino acid are reported to be relatively similar in brown, green, and red algae (Holdt and Kraan 2011; Lafarga et al. 2020). The amino acid score (AA score) and the EAA index are higher in red seaweeds than in brown and green seaweeds. The AA score of *Porphyra* spp. (91) and *Undaria* spp. (100) is comparable to those of animal-derived foods (Pangestuti and Kim 2015). Extracted amino acids have medical and nutraceutical benefits. For instance, glycine supplementation is prescribed for liver and lung injuries, arginine is prescribed to patients with brain ischemia and infertility, histidine is prescribed for hyper-insulinemia, and glutamic acid is prescribed for maintaining mental activities (Wu 2013).

Seaweeds have a high protein quality due to their high ratio of EAAs to total amino acids (approximately 50%) in comparison with rich protein sources such as soybean (39%) and egg protein (47%) (Thiviya et al. 2022). Furthermore, high values of EAA to non-essential amino acid (NEAA) recorded for red seaweeds (*Gracilaria* spp., i.e., 1.47–1.74) and brown seaweeds (*A. nodosum* 1.00–1.06) make them suitable candidates for animal-based protein replacements, such as egg substitute products, including mayonnaise, and alternatives to meat patties and burgers (Brien et al. 2022; Ganesan et al. 2019). Many seaweeds (*Caulerpa racemosa*, *U. intestinalis*, *U. lactuca*, and *Ulva rigida*) are now utilized as supplementary food ingredients to improve antioxidant properties in biscuits, bread, pork patties, and fish surimi (Brien et al. 2022; Kumar et al. 2023).

9.3 Bioactive Protein Derivatives of Seaweeds with Bio-functional Value

Seaweeds contain bioactive protein derivatives, such as free amino acids, mycosporine-like amino acids, peptides, glycoproteins, lectins, and phycobiliproteins (e.g., phycoerythrin, phycocyanin, allophycocyanin, and phycoerythrocyanin) with antihypertensive, antidiabetic, antioxidant, anti-inflammatory, antitumoral, antiviral, and antimicrobial activities (Ganesan et al. 2019). Figure 9.1 indicates the nutritional and techno-functional properties of selected seaweed proteins that can be useful in food and pharmaceutical applications.

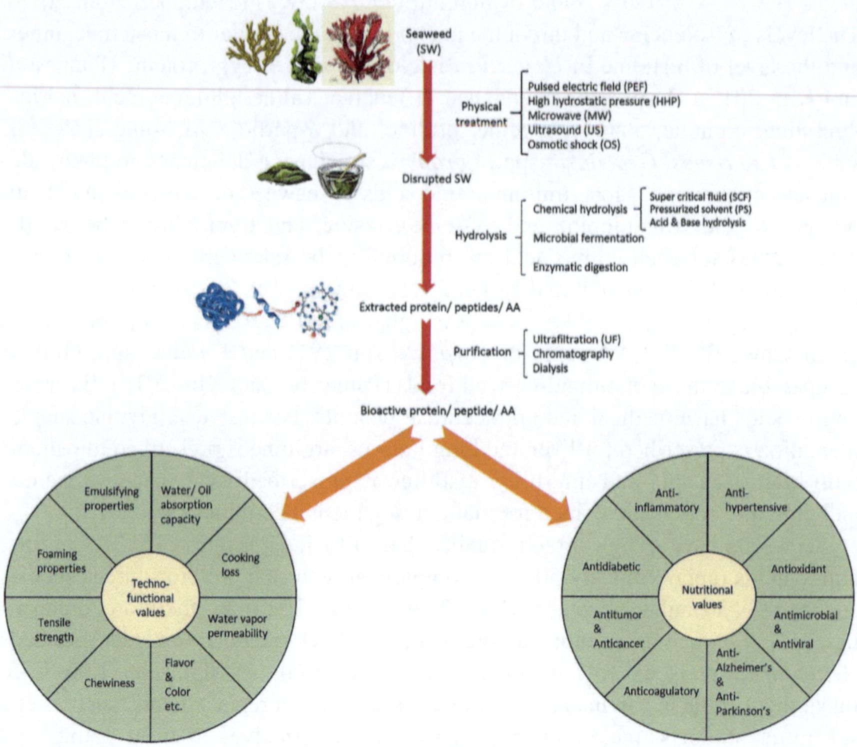

Fig. 9.1 Process of seaweed protein production and its nutritional and techno-functional properties valuable in food and pharmaceutical applications

9.3.1 Bioactive Peptides

These peptides usually consist of 2–30 amino acids. These short chains of amino acids are formed from protein degradation during food processing, fermentation, or hydrolysis. Once released, they may display hormone-like or drug-like biological activities, depending on their amino acid sequences (Lafarga et al. 2020; Pangestuti and Kim 2015). Low molecular weight (MW) peptides exhibit significant bioactivity due to their ability to pass through the gastrointestinal tract (GIT) wall, while long-chain peptides are more susceptible to being degraded by enzymes or microflora of the gastrointestinal tract (GIT) (Hosseini et al. 2022). Seaweeds are a natural source of bioactive peptides. Peptides isolated from various seaweeds show inhibitory effects on enzymes such as renin, angiotensin-I-converting enzyme (ACE-1), cyclooxygenases (COX), tyrosinase, α-amylase and α-glucosidase inhibitors associated with hormone-like properties, including antioxidant, antitumoral, antimicrobial, antiviral, opioid agonistic, prebiotic, anticoagulatory, and other inhibitory activities (Brien et al. 2022; Thiviya et al. 2022). Some of these seaweed-derived peptides showed thermally stable antioxidant potential (Lafarga et al. 2020). Carnosine and glutathione are two antioxidant peptides found in some seaweeds that exhibit the ability to combat tumors and cancer, diabetes, neurodegenerative disorders, and cardiovascular disease by binding reactive oxygen species (Peñalver et al.). This characteristic is attributed to the presence of hydrophobic amino acids (e.g., valine, leucine, glycine, methionine, alanine, tyrosine, histidine) in the structure of peptides (Harnedy et al. 2017). Carnosine (β-alanyl-L-histidine), generally found in animal muscles, can be extracted from *Acanthophora delilei* (red seaweed) and is involved in muscle activity; thus, it is suitable for athletes' diets (Fleurence et al. 2018; Pangestuti and Kim 2015). Zhang et al. (2019) purified an antioxidant peptide from *Gracilariopsis lemaneiformis* protein hydrolysates (i.e., Glu-Leu-Trp-Lys-Thr-Phe). The study concluded that smaller hydrophobic peptides (MW < 3 kDa) consisting of aromatic amino acids which showed higher antioxidant activities (Zhang et al. 2019). The presence of Glu and Asp in an extracted peptide resulted in high antioxidant properties, acting as strong electron donors, while Cys scavenged free radicals via sulfhydryl (-SH) groups (Mousaie et al. 2022). An extracted protein from *Sargassum ilicifolium* (brown seaweed) with low MW (≈4.1 kDa), mainly consisting of Asp-*Glu-Lys-Phe-Leu*, showed relatively good antioxidant, antibacterial (against *S. Typhimurium and Bacillus stearothermophilus*), and angiotensin-I-converting enzyme (ACE) inhibitory (IC_{50} of 1.28 mg/ml) activities. The hydrolysate protein from *S. ilicifolium* also had significant inhibitory activity (IC_{50} of 300.70 mg/ml) against the growth of carcinoma A549 cells in lung cancer (Mousaie et al. 2022). Moreover, a sixteen amino acid long peptide (MW ~1.77 kDa), composed of NH2-Asn-Met-Glu-Lys-Gly-Ser-Ser-Val-Val-Ser-Ser-Arg-Met-Lys-Gln-COOH, obtained from *Porphyra yezoensis* showed in vitro antithrombotic properties as the peptide interacted with clotting factors (Indumathi and Mehta 2016).

P. yezoensis (red seaweed) protein hydrolysates have been reported to have anti-pain properties due to their inhibitory activity against the COX enzyme (Lee et al.

2015). Thus, suppressing COX enables seaweed-derived peptides to treat pathogenic and inflammatory diseases (Brien et al. 2022). Admassu et al. (2018) reported the α-amylase inhibitory potential of *Porphyra* spp. (red seaweed) protein hydrolysates, with potential to decrease glucose absorption in the bloodstream and antidiabetic effects. The authors isolated and characterized the peptides Gly-Gly-Ser-Lys and Glu-Leu-Ser, with α-amylase inhibitory activities and IC_{50} values of 2.58 mM and 2.62 mM, respectively (Admassu et al. 2018).

9.3.2 Lectins and Phycobiliproteins

Recently, red and green seaweed-derived lectins and phycobiliproteins have gained the attention of researchers for their host–pathogen interactions, anti-inflammatory, and immune-modulation properties and used to develop functional food (Holdt and Kraan 2011; Mendis and Kim 2011). Lectins are glycoproteins or hemagglutinin proteins that exert antibacterial, anti-human immune-deficiency (anti-HIV), anti-hepatitis C, and anti-SARS-CoV effects by binding to the spike glycoprotein of these viruses (Lafarga et al. 2020; Nascimento da Silva et al. 2021). The antibacterial activity of lectins has been reported against pathogenic gram-negative bacteria, including *Serratia marcescens*, *Salmonella typhi, Klebsiella pneumoniae*, *Enterobacter aerogenes*, *Proteus* spp., and *Pseudomonas aeruginosa* (Pangestuti and Kim 2015). These antibacterial properties have been related to the ability of lectins to immobilize and deactivate bacteria by binding carbohydrates or glycoproteins on either the cell walls or plasma membranes of bacteria (Holdt and Kraan 2011). Lectins can recognize gametes and reproductive cell fusion to defend against pathogens (Echave et al. 2022). Grifithsin (GRFT) is a lectin protein isolated from *Griffithsia* spp. (red seaweed), which can decrease the mortality rate in lethal pulmonary SAR-CoV-infected patients by reducing pro-inflammatory cytokines in infected lung tissue (Nascimento da Silva et al. 2021). Seaweed-derived lectins also demonstrated red blood cell agglutination abilities in clinical blood typing recognition. These compounds can be isolated from *Eucheuma* spp., *Sagittaria filiformis*, *Enantiocladia duperreyi*, *Pterocladiella capillacea*, *Valerianella obtusiloba*, *Gracilaria cornea*, *Gracilaria ornate*, and *Codium fragile*. Lectins extracted from *C. fragile* (green seaweed) can detect specific antigens for blood group A (Pangestuti and Kim 2015).

Phycobiliproteins are water-soluble, fluorescent proteins of various colors, generally recognized as safe (GRAS). They are used in several food ingredients as colorants in chewing gums, popsicles, confectionery, dairy products, soft drinks, ice sherbets, jellies, and cosmetics (lipstick and eyeliners). Phycobiliproteins also have therapeutic value, such as anticancer, antiatherosclerosis, antihyperlipidemic, and anti-irradiative effects, in addition to Alzheimer's and Parkinson's prevention by reducing oxidative stress. These capabilities of phycobiliproteins are attributed to the potential of the peptide in chelating metals and scavenging free radicals (Peñalver et al. 2020; Román et al. 2002). Phycoerythrin, a red protein pigment complex produced by the light-harvesting phycobiliprotein family in *Gracilaria turuturu*, is a natural colorant and used as a fluorescent probe in diagnostics, as well as antioxidant, antitumoral, antidiabetic, and antiangiogenic compounds (Hentati et al. 2020).

9.3.3 Free Amino Acids

Selected seaweeds are rich sources of free amino acids and amino acid-like compounds, especially taurine and mycosporine-like amino acids (MAAs) (Pangestuti and Kim 2015; Thiviya et al. 2022). Taurine can lower cholesterol and fat digestibility in young, overweight adults by stimulating bile excretion and limiting serum lipids, thereby reducing the risk of atherosclerosis and coronary heart diseases. Taurine can also enhance heart-muscle contraction efficiency in congestive heart failure patients. Taurine has shown antihypertensive and antioxidant activities by chelating heavy metals, mainly lead and cadmium (Brien et al. 2022; Ganesan et al. 2019; Mendis and Kim 2011; Pangestuti and Kim 2015). Kanoids are unusual amino acids commercially isolated from *Digenea simplex* (red seaweed) with neuro-excitatory and antihelmintic activity to treat ascariasis. Kanoid amino acids also show desirable effects against neurophysiological diseases such as Alzheimer's, Parkinson's, and epilepsy (Holdt and Kraan 2011). Laminine (trimethyl (5-amino-5-carboxypentyl) ammonium oxalate) is another free amino acid extracted from *Saccharina angustata* (brown seaweed) and *Chondria armata* (red seaweed) with a choline-like base in its structure. Laminin monocitrate can have a transitory hypotensive impact by reducing smooth muscle contractions (Holdt and Kraan 2011).

Mycosporine-like amino acids (MAAs) are small secondary metabolites that absorb ultraviolet radiation. Macroalgae, especially red seaweeds, are a rich source of MAAs that can protect cells from oxidation and UV damage. Various MAAs, including palaythine, asterina, shinorine, porphyra-344, mycosporine-glycine, usujirene, and palythene, have been found in several seaweeds, including *U. intestinalis*, *A. esculenta*, *Pyropia columbina*, and *C. crispus* (Echave et al. 2022; Sun et al. 2020).

Some thermogenic proteins, such as uncoupling protein (UCP) families in the brown seaweed (*U. pinnatifida*), showed anti-obesity activity, which can reduce the risk of diabetes, hypercholesterolemia, and stroke with no variation in food intake of rats (Ganesan et al. 2019). Table 9.2 lists bioactive compounds in seaweeds with different biological activities.

Inhibiting enzymes involved in the renin-angiotensin system (Kumar et al. 2018), including angiotensin-converting enzyme-I (ACE-I), is a pharmacologically proven strategy to prevent cardiovascular diseases. The presence of hydrophobic amino acids at the C-terminus of the tripeptide sequence of the substrate has the potential to inhibit ACE (Pangestuti and Kim 2015). Cian et al. (2014) produced an extruded maize product fortified with *P. columbina* (red alga), showing 41% higher dialysability of ACE inhibitor compounds. In another study, Fitzgerald et al. reported

Table 9.2 Bioactive protein compounds in seaweeds

Seaweed	Bioactive compounds	Properties	References
Bryopsis spp. (green)	Cyclic depsipeptide	Antimicrobial activity against *Mycobacterium tuberculosis*	Dmitrenok et al. (2006)
Gracilariopsis lemaneiformis (red)	TGAPCR, FQIN [M(O)] CILR	Angiotensin-I-converting enzyme (ACE) inhibitory activity	Deng et al. (2018)
Mazzella japonica (red)	YRD, VSEGLD, TIMPHPR, GGPAT, SSNDYPI, SRIYNVKSNG, VDAHY, CPYDWV, YGDPDHY, NLGN, DFGVPGHEP, YRD, LDY, LRY, VY, LF, FY	ACE inhibitory activity	Kitade et al. (2018) and Kumagai et al. (2020)
Neopyropia yezoensis (red)	Di- and tripeptides	ACE inhibition/ antihypertensive activity	Jiao et al. (2019) and Qu et al. (2010)
	TPDSEAL	Antimicrobial activity against *Staphylococcus aureus*	
P. palmata (red)	Peptides derived from phycobiliproteins: YRD, AGGEY, VYRT, VDHY, IKGHY, LKNPG, LDY, LRY, FEQDWAS	ACE inhibition, antioxidant	Furuta et al. (2016)
	Alcalase, bromelain, and Promod-derived hydrolysates	Dipeptidyl peptidase IV (DPP-IV) inhibitory activities	McLaughlin et al. (2021)
	Alcalase/Flavourzyme hydrolysates	Antihyperglycemic/ antidiabetic potential	McLaughlin et al. (2020)
	Peptides: ILAP, LLAP, MAGVDHI	DPP-IV inhibitory activities	Harnedy et al. (2015)
	Papain hydrolysates: NIGK	Platelet-activating factor acetyl-hydrolase (PAF-AH) inhibitory peptides	Fitzgerald et al. (2013)
	IRLIIVLMPILMA	Renin-inhibitory activity	Fitzgerald et al. (2012)
	SDITRPGGQM	Antioxidant	Harnedy et al. (2017)
Porphyra dioica (red)	Peptides: DYYKR, YLVA	Antioxidant, ACE inhibition, DPP-IV inhibitory activities	Cermeño et al. (2019)

(continued)

Table 9.2 (continued)

Seaweed	Bioactive compounds	Properties	References
P. columbina (red)	Peptides	ACE inhibitory, immunosuppressive, antioxidant properties	Cian et al. (2012)
Saccharina longicruris (brown)	TITLDVEPSDTIDGVK, ISGLIYEETR, MALSSLPR, ILVLQSNQIR, ISAILPSR, IGNGGELPR, LPDAALNR, EAESSLTGGNGCAK, QVHPDTGISK	Antibacterial activities	Beaulieu et al. (2015)
Sargassum pallidum (brown)	Dipeptides (aurantiamide, aurantiamide acetate, dia-aurantiamide)	Antibiotic activity in vitro against *S. aureus*, *Staphylococcus epidermidis*, and *Pseudomonas aeruginosa*	Liu et al. (2009)
Sargassum thunbergii (brown)	Iodo-amino acids	Possibly helps in human thyroid metabolism	Dominguez (2013)
Ulva rigida (green)	Peptides	ACE inhibition	Paiva et al. (2016)
U. pinnatifida (brown)	Di-, tri-, and tetrapeptides: VYIY, AWFY, VW, IW, LW, YNKLKFYG, YKYY	ACE inhibition/ antihypertensive activity, antioxidant	Sato et al. (2002) and Suetsuna and Nakano (2000)
Boodlea coacta (green), *Griffithsia* spp. (red)	Lectins and griffithsin	Antiviral effects against human immunodeficiency virus (HIV), Hepatitis C virus and SARS-CoV/ eSARS-CoV-2 by preventing the entry into the host cells	Barre et al. (2021) and Cheung et al. (2015)
Caulerpa cupressoides (green)	Lectins	Antinociceptive and anti-inflammatory activities	Vanderlei et al. (2010)
C. fragile (green), *Eucheuma serra* (red)	Lectins	Mitogenic activities, lipogenic activity	Smit (2004)
Mimica amakusaensis (red), *Ulva* spp. (green)	Lectins	Induce apoptosis, metastasis, and cell differentiation in cancer cells, antibiotic, anti-inflammatory, anti-HIV activity, and human platelet aggregation inhibition	Mendis and Kim (2011)

(continued)

Table 9.2 (continued)

Seaweed	Bioactive compounds	Properties	References
N. yezoensis (red)	Taurine	Antioxidant	Torres et al. (2019) and Wang et al. (2015)
Saccharina angustata (brown), *Chondria armata* (red)	Laminin	Hypertensive effect, depress contraction of smooth muscles	Holdt and Kraan (2011)
C. crispus, Gelidium pusillum, Dasysiphonia japonica, P. palmata (red)	Phycobiliproteins	Antioxidant, antidiabetic, antitumour, anti-inflammatory, neuroprotective, and hepatoprotective properties	Mittal et al. (2017)
Gracilaria tikvahiae, P. palmata (red)	Phycobiliproteins (phycocyanins and allophycocyanin)	Anti-inflammatory, liver-protecting, antiviral, antitumour, antiatherosclerosis, lipase activity inhibitor, serum lipid reducing agent, and antioxidant	Mendis and Kim (2011)

Table was originally published by Thiviya et al. (2022)

renin-inhibitory peptide activity in *P. palmata* protein hydrolysate incorporated into bread (Fitzgerald et al. 2014).

9.4 Techno-functional Properties of Seaweed Proteins

Numerous studies have investigated incorporating seaweed into different food products and reported positive results in the protein content of products and their nutritional benefits. Adding seaweed to food product formulations can also enhance desirable techno-functional properties (Arufe et al. 2019; Cian et al. 2014; Kumar et al. 2018; O'sullivan et al. 2016; Rattaya et al. 2009). Different techno-functional qualities include water- and oil-binding, gelation, foaming, swelling, emulsifying, elasticity, cooking loss, firmness, and color, among others. Changes in the technological quality of products depend on the amount of a particular seaweed which is incorporated, e.g., low, supplemental amounts (i.e., 3%) of *U. lactuca, Porphyra tenera*, and *U. pinnatifida* in fortified pasta did not have noticeable effects on its technological properties, such as weight gain, swelling index, cooking loss, and moisture. Regarding textural characteristics, enriched pasta showed lower hardness, springiness, gumminess, and chewiness, whereas the adhesiveness of pasta incorporated with seaweed increased (Ainsa et al. 2022). Fitzgerald et al. (2014) reported that adding 4% *P. palmata* (red seaweed) protein hydrolysate to bread dough resulted in a specific volume reduction of the loaf. This may be attributed to the competitive water absorption of proteins and swelling restriction of wheat starch interfering with the leavening stage. The authors concluded that incorporating protein hydrolysate caused a harder crumb, less cohesive, and elastic bread texture, while seaweed proteins led to a higher degree of non-enzymatic Maillard reactions, resulting in a darker crumb appearance (Fitzgerald et al. 2014).

9.4.1 Water/Oil Absorption Capacity

One of the major techno-functional properties is the water/oil absorption capacity (WAC/OAC). These absorption capacities relate to the hydrophobicity and hydrophilicity of the food components, as they refer to the binding ability of the proteins to water/oil via various interactions between water/oil and constituents such as proteins, polysaccharides, or fibers. The water absorption capacity is a significant characteristic in viscous foods, specifying the machinability of the product (e.g., dough, sausage batter). High WAC provides greater viscosity, thickening, and mouthfeel in soup, gravies, dough, and bakery products. Various seaweeds show different WACs, including *Enteromorpha compressa*, *Kappaphycus alvarezii*, *Gracilaria fisheri*, and *Gracilaria tenuistipitata,* with respective values of 153, 222, 553, and 897% (Benjama and Masniyom 2012; Kandasamy et al. 2012; Kumar et al. 2018; Kumar et al. 2014).

Mohammed et al. (2021) compared two brown (i.e., *H. elongata* and *A. esculenta*) and two red (i.e., *Palmaria palmata* and *Porphyra umbilicalis*) seaweeds and concluded that *H. elongata* and *P. palmata* had the highest and lowest WAC (691 and 376%), respectively (Mohammed et al. 2021). Semi-sweet biscuits were utilized to carry the health benefits of *C. racemosa* (a green alga) (Kumar et al. 2018). The inclusion of *C. racemosa* incorporation in biscuits increased the protein and fiber content, improved the antioxidant activity, increased the water and oil absorption capacity of dough, and enhanced the taste-panel acceptability of the final product (Kumar et al. 2018). Bozdemir et al. (2022) estimated the WAC of protein extracted from powdered *Gracilaria dura* to be approximately 195%, which was significantly lower than that of soy protein isolate (447%) but higher than that of chickpea and lentil (119 and 133%) (Bozdemir et al. 2022). Moreover, the use of seaweed powder with high water affinity led to an increase in the WAC of seaweed-enriched chestnut flour doughs to produce modified gluten-free cookies (Arufe et al. 2019).

Oil absorption capacity plays a crucial role in several food formulations, such as bakeries, sausages, salad dressings, and mayonnaise, as these properties govern the flavor retention ability and palatability of the product. While the OACs of egg protein isolates, whey protein isolate, wheat protein isolate, and soy protein isolate are 210, 190, 260, and 154, respectively, the OAC of *G. dura* was reported to be approximately 568% (Bozdemir et al. 2022; Kumar et al. 2018; Lam et al. 2017). The high

OAC of seaweeds is due to the physical entrapment of oil via capillary attraction and the hydrophobicity of proteins. Therefore, different proportions of polar groups of the amino acids in seaweeds result in variation in the OAC. Accordingly, the high OHC of *P. umbilicalis* (359%) is probably related to its high protein content (i.e., 32.03%) (Mohammed et al. 2021).

The high WAC and OAC of various seaweeds make them potential candidates to enhance the texture and hold the moisture of food products. These characteristics are attributed to the protein conformation and amino acid composition of seaweeds (Bozdemir et al. 2022).

9.4.2 Foaming Properties

The high foaming capacity (FC) of proteins is used to improve smoothness and lightness through the uniform distribution of air bubbles in food formulations. Foaming capacity and stability (FS) determine the volatilization of flavors and develop food palatability (García-Vaquero et al. 2017). The FC of proteins depends on the net charges of proteins, amino acid compositions, ionic strength, and pH of the solution, resulting in rapid protein adsorption to the air–water interface and encapsulation of air bubbles. The FS of proteins greatly relies on time and the cohesiveness of a viscoelastic multilayer protein film around air cells formed via protein inter-molecular interactions (García-Vaquero et al. 2017; Kandasamy et al. 2012; Kumar et al. 2014; Lam et al. 2017). The green edible seaweed, *E. compressa*, showed a higher FC (55%) of proteins in acidic conditions (pH ~ 2), at zero ionic strength. Low salt concentrations, at pH ~ 6, improved the protein FC of *Enteromorpha* spp. due to an increase in protein solubility. The FS of *Enteromorpha* spp. protein concentrates, in acidic pH, increased because of a rise in the protein charge density (Kandasamy et al. 2012). Different foaming properties of proteins are due to protein concentration, pH, temperature, and methods for foam production (Yucetepe et al. 2022).

9.4.3 Emulsifying Properties

The emulsifying activity (EA) of a protein refers to its ability to form and stabilize an emulsion, which is a mixture of two immiscible liquids (i.e., oil and water) that are held together by the protein. This involves the protein adsorbing at the interface between the two liquids and reducing the interfacial tension, allowing the liquids to mix more easily (Bozdemir et al. 2022; Joshi et al. 2012). The emulsifying stability (ES) of a protein refers to its ability to maintain this emulsion over time without separating or breaking down. This involves factors such as the strength of the protein–ligand interactions at the interface, as well as any changes in pH or temperature that may affect these interactions (García-Vaquero et al. 2017; Joshi et al. 2012).

Overall, the hydrophobicity, net surface charge, and solubility define the EA and ES of the protein (Yucetepe et al. 2022).

The emulsifying properties of the protein extracts of *G. dura* were examined to investigate the techno-functional features that make it a low-cost and innovative source, suitable for being employed as functional food ingredients. The EA and ES of the protein extract from *G. dura* were 44 and 75%, respectively, whereas the hydrophobic proteins in an emulsion from *K. alvarezii* protein concentrate were found to be responsible for the formation of stable emulsions, with an ES of 62% (Bozdemir et al. 2022; Kumar et al. 2014). In a recent study investigating the techno-functional properties of *Caulerpa prolifera* protein hydrolysates, these hydrolysates had lower EA and ES compared to commercial emulsifiers, whereas the emulsion capacity of soybeans was approximately 70% (Yucetepe et al. 2022). Despite the lower protein content of seaweeds, the emulsifying characteristics of seaweeds are considerable in comparison with animal and plant sources, making them suitable functional ingredients. Table 9.3 summarizes the protein content as well as the main techno-functional properties of proteins from animal, plant, and seaweed origins for the purposes of comparison.

9.4.4 Other Properties and Uses of Seaweed and Seaweed-Derived Proteins

Seaweeds can also be used for the generation of edible films. Synthesizing fish skin gelatin films incorporated with protein extract of *Turbinaria ornata* exhibited higher elongation at break, while water vapor permeability and film solubility decreased. These changes in the film characteristics were attributed to non-disulfide interactions between gelatin molecules and oxidized phenols in seaweed isolates. Moreover, the protein cross-links formed in the film incorporated with seaweed extract resulted in a rough appearance of the film (Rattaya et al. 2009). In another study, the inclusion of *F. vesiculosus* extract into a whey protein concentrate active film not only inhibited lipid oxidation but also improved the thickness, strength, tensile strength, and elastic modulus of the produced edible film (Andrade et al. 2021). Bozdemir et al. (2022) extracted proteins of *G. dura* and examined their techno-functional properties, making them suitable for various food products. The authors determined the minimum (18.9%) and maximum (58.5%) solubility of *G. dura* proteins at pH 4.0 and pH 13.0, respectively (Bozdemir et al. 2022).

In another study conducted by Prabhasankar et al. (2009), the incorporation of 2.5% *Sargassum marginatum* with 6% DW protein into pasta was found to be useful to decrease cooking loss and increase the cooking weight of pasta. This fact was attributed to improvements to the gluten network by the addition of *Sargassum*, while incorporation levels beyond this amount resulted in weakness of the network and leaching of components. In fact, a thin film of the coating was formed on the

Table 9.3 Protein content and techno-functional properties of selected animal, plant, and seaweed protein sources

Protein source	Protein (% DW)	Techno-functional properties: Protein Solubility (%)	WAC (%)	OAC (%)	FC (%)	FS (%)[a]	EA (%)	ES (%)	References
Egg protein isolate	88.8	88.5	–	210	234.4	79.8	–	91.3	Lam et al. (2017)
Whey protein isolate	83.5	94.2	–	190	302	78.8	–	89.3	Lam et al. (2017)
Soy protein isolate	94.9	45	–	230	198.9	62.7	–	94.7	Lam et al. (2017)
Wheat protein isolate	83.4	8.4	–	260	192.2	9.8	–	63.3	Lam et al. (2017)
Protein extract from *Gracilaria dura*	21.2	29*	195	568	12.5	0.0	44	75	Bozdemir et al. (2022)
Gracilaria fisheri	11.6	–	553	179	–	–	–	–	Benjama and Masniyom (2012)
Gracilaria tenuistipitata	21.6	–	897	223	–	–	–	–	Benjama and Masniyom (2012)
Protein extract from *Enteromorpha compressa*	60.35	–	153	134	55**	37.5**	–	–	Kandasamy et al. (2012)
Protein extract from *Kappaphycus alvarezii*	62.3	–	222	129	53.33	45	–	62	Kumar et al. (2014)
Protein extract from *Caulerpa prolifera*	51.57	2*	54	21	26.5	26.7	41.4	41.7	Yucetepe et al. (2022)

Abbreviations in the table are as follows: *WAC* water absorption capacity, *OAC* oil absorption capacity, *FC* foaming capacity, *FS* foaming stability, *EA* emulsifying activity, *ES* emulsifying stability

[a]FS is measured as % at room temperature for 30 min

* at pH 7; ** at pH 2 and zero ionic strength

protein matrix of seaweed-incorporated pasta enveloping gelatinized starch granules (Prabhasankar et al. 2009).

Some seaweed amino acids, especially glutamate (glutamic acid) and aspartic acid (≈30% of the total amino acids), are responsible for an umami taste, while glycine and alanine induce a sweet taste associated with the particular taste of seaweeds (Echave et al. 2021; Echave et al. 2022; Thiviya et al. 2022). Brown seaweeds (e.g., cited as *Laminaria japonica*) have high amounts of glutamic acid, contributing to the formation of a flavor enhancer, monosodium glutamate (Pangestuti and Kim 2015).

9.5 Challenges and Bottlenecks of Seaweeds as a Source of Protein

The broad variability in algal proteins is a serious challenge to producing specific bioactive peptides commercially. The protein content of various seaweeds differs by location, temperature, light intensity, and nutrient concentration in water; e.g., it has been proven that the protein concentrations of *P. palmata*, *Gracilaria cervicornis*, and *Sargassum vulgare* were negatively correlated with seawater temperature and salinity. The protein content of seaweeds harvested between May and June in Scotland was lower than that of seaweeds collected in Brazil in autumn. Usually, during winter to early spring, seaweeds contain their highest content of protein. Similarly, the amino acid composition of proteins changes by season and consequently alters the yield of functional ingredients (Echave et al. 2022; Lafarga et al. 2020). Moreover, there are also analytical difficulties concerning the determination or estimation of proteins from a variety of seaweed biomass, with several reports considering modifications of the nitrogen-to-protein conversion factors used when determining the concentration of nitrogen to estimate protein accumulation. Several reports studied this analytical complication in multiple seaweed species and, in so doing, established multiple recommendations of conversion factors ranging from 3 to 5, which are lower than the conventional 6.25 used when estimating protein in previous studies (Ganesan et al. 2019; Thiviya et al. 2022).

The nutritional value of protein sources significantly depends on their digestibility and bioavailability. To digest proteins, various enzymes and hormones are utilized to degrade complex structures into simpler peptide fragments, which can then be reassembled. In general, algal proteins have high digestibility rates (>70%), comparable to terrestrial protein sources such as grains (69–84%), legumes (72–92%), fruits (72–92%), and vegetables (68–80%) (Echave et al. 2021). However, a major challenge in the application of seaweed proteins is their bioaccessibility. High amounts of phenolic compounds and polysaccharide fibers are present in seaweed cell walls, thereby limiting protein digestibility and absorption, ranging from 2 to 24% (Brien et al. 2022; Ganesan et al. 2019; Ms et al. 2022). Oxidized phenolic compounds in seaweeds, including catechins, flavanols, and phlorotannins, inactivate proteolytic enzymes and have a negative correlation with protein digestibility. The strong intercellular ionic bonds of proteins to seaweed cell walls have an adverse effect on protein digestibility (Brien et al. 2022). The complex cell wall of raw seaweed makes proteins unavailable and necessitates their extraction through different physical, chemical, enzymatic, and other novel methods (Ganesan et al. 2019; Thiviya et al. 2022). For instance, washing can be useful to remove soluble fibers, such as xylan, and minerals that contribute to improving the digestibility of protein from *P. palmata* from 1.5 to 58%, achieving similar digestibility levels to those of casein. Fermentation can also breakdown insoluble fibers and make the encapsulated cytoplasmic protein of seaweeds accessible to the hydrolytic enzymes responsible for gastrointestinal digestion, increasing the digestibility of samples to 45–65% (Marrion et al. 2003). After protein extraction, over 67% of protein from *U. pertusa* was digestible in the presence of pancreatic enzymes, increasing even further to 90% by adding pronase to the in vitro digestion enzymes. Moreover, the digestibility of seaweed proteins depends on their relative solubility, i.e., water-soluble proteins show a higher degree of degradation in their digestion by the enzymatic complexes of human intestinal juice, rather than trypsin or chymotrypsin separately (Fleurence et al. 2018).

Seaweed proteins have a suitable composition in terms of their essential amino acid composition to be considered a source of these compounds for human nutrition, and the generation of bioactive peptides offers excellent opportunities for the use of these proteins in the field of functional foods. However, further research is needed to evaluate these novel sources of protein, including studies on toxicity, allergenicity, microbial safety, and acceptability of the new(novel) ingredients. As an example, aside from their beneficial bioactivities, lectins can have toxic effects due to hemagglutinating properties and interference with the absorption of nutrients. It has also been shown that amphipods inhabiting *Porphyra* spp. may still be present at trace levels in the processed product (nori), resulting in food allergies in sensitive people consuming nori sheets. In addition, kainic acid (kainate) in red seaweeds (i.e., *P. palmata*) with glutamate-like structures has neurotoxic effects when consumed at excessive levels (Mendis and Kim 2011; Thiviya et al. 2022). Consumers should have a greater awareness of these issues.

9.6 Conclusions

In conclusion, seaweeds represent a promising alternative source of proteins and essential amino acids for the food industry. With increasing demands for sustainable and environmentally friendly protein sources, seaweed-derived proteins have emerged as a valuable resource that can help to address several health and environmental concerns of traditional protein sources. Moreover, seaweeds and their derived proteins can also be used for the development of functional foods. The generation of bioactive peptides from seaweed proteins adds to its appeal as a functional food ingredient with potential additional health benefits. Despite the many advantages of seaweed proteins, there are still some challenges and bottlenecks hindering their full utilization, including limited knowledge of variability in quality and quantity, technological treatments for standardized extraction, or their functionality and improved digestibility. Future research will be necessary to overcome these obstacles in order to fully realize the potential of selected seaweeds as a viable source of proteins for the food industry.

Acknowledgments The authors acknowledge the funding received from the project AMBROSIA, funded by the Department of Agriculture Food and the Marine (DAFM) under the umbrella of the European Joint Programming Initiative "A Healthy Diet for a Healthy Life" (JPI-HDHL) and of the ERA-NET Cofund ERA HDHL (GA No 696295 of the EU Horizon 2020 Research and Innovation Programme).

References

Admassu H, Gasmalla MA, Yang R, Zhao W (2018) Identification of bioactive peptides with α-amylase inhibitory potential from enzymatic protein hydrolysates of red seaweed (*Porphyra* spp). J Agric Food Chem 66(19):4872–4882. https://doi.org/10.1021/acs.jafc.8b00960

Ainsa A, Honrado A, Marquina P, Beltrán JA, Calanche J (2022) Influence of seaweeds on the quality of pasta as a plant-based innovative food. Food Secur 11(16):2525. https://doi.org/10.3390/foods11162525

Andrade MA, Barbosa CH, Souza VG, Coelhoso IM, Reboleira J, Bernardino S et al (2021) Novel active food packaging films based on whey protein incorporated with seaweed extract: development, characterization, and application in fresh poultry meat. Coatings 11(2):229. https://doi.org/10.3390/coatings11020229

Arufe S, Sineiro J, Moreira R (2019) Determination of thermal transitions of gluten-free chestnut flour doughs enriched with brown seaweed powders and antioxidant properties of baked cookies. Heliyon 5(6):e01805. https://doi.org/10.1016/j.heliyon.2019.e01805

Barre A, Van Damme EJ, Simplicien M, Le Poder S, Klonjkowski B, Benoist H et al (2021) Man-specific lectins from plants, fungi, algae and cyanobacteria, as potential blockers for SARS-CoV, MERS-CoV and SARS-CoV-2 (COVID-19) coronaviruses: biomedical perspectives. Cells 10(7):1619

Beaulieu L, Bondu S, Doiron K, Rioux L-E, Turgeon SL (2015) Characterization of antibacterial activity from protein hydrolysates of the macroalga *Saccharina longicruris* and identification of peptides implied in bioactivity. J Funct Foods 17:685–697

Benjama O, Masniyom P (2012) Biochemical composition and physicochemical properties of two red seaweeds (*Gracilaria fisheri* and *G. tenuistipitata*) from the Pattani Bay in Southern Thailand. *Sonklanakarin*. J Sci Technol 34(2):223

Biparva P, Mirzapour-Kouhdasht A, Valizadeh S, Garcia-Vaquero M (2023) Advanced processing of giant kelp (*Macrocystis pyrifera*) for protein extraction and generation of hydrolysates with anti-hypertensive and antioxidant activities in vitro and the thermal/ionic stability of these compounds. Antioxidants 12(3):775

Bozdemir A, Şensu E, Okudan EŞ, Özçelik B, Yücetepe A (2022) Ultrasound-assisted enzymatic extraction of proteins from *Gracilaria dura*: investigation of antioxidant activity and techno-functional properties. J Food Process Preserv 46(8):e16803. https://doi.org/10.1111/jfpp.16803

Brien RO, Hayes M, Sheldrake G, Tiwari B, Walsh P (2022) Macroalgal proteins: a review. Foods 11(4):571. https://doi.org/10.3390/foods11040571

Cermeño M, Stack J, Tobin PR, O'Keeffe MB, Harnedy PA, Stengel DB, FitzGerald RJ (2019) Peptide identification from a *Porphyra dioica* protein hydrolysate with antioxidant, angiotensin converting enzyme and dipeptidyl peptidase IV inhibitory activities. Food Funct 10(6):3421–3429

Cermeño M, Kleekayai T, Amigo-Benavent M, Harnedy-Rothwell P, FitzGerald RJ (2020) Current knowledge on the extraction, purification, identification, and validation of bioactive peptides from seaweed. Electrophoresis 41(20):1694–1717

Cheung RCF, Wong JH, Pan W, Chan YS, Yin C, Dan X, Ng TB (2015) Marine lectins and their medicinal applications. Appl Microbiol Biotechnol 99:3755–3773

Cian RE, Martínez-Augustin O, Drago SR (2012) Bioactive properties of peptides obtained by enzymatic hydrolysis from protein byproducts of *Porphyra columbina*. Food Res Int 49(1):364–372

Cian RE, Caballero MS, Sabbag N, González RJ, Drago SR (2014) Bio-accessibility of bioactive compounds (ACE inhibitors and antioxidants) from extruded maize products added with a red seaweed *Porphyra columbina*. LWT-Food Sci Technol 55(1):51–58. https://doi.org/10.1016/j.lwt.2013.08.011

Delimaris I (2013) Adverse effects associated with protein intake above the recommended dietary allowance for adults. Int Scholar Res Notice 2013. https://doi.org/10.5402/2013/126929

Deng Z, Liu Y, Wang J, Wu S, Geng L, Sui Z, Zhang Q (2018) Antihypertensive effects of two novel angiotensin I-converting enzyme (ACE) inhibitory peptides from *Gracilariopsis lemaneiformis* (Rhodophyta) in spontaneously hypertensive rats (SHRs). Mar Drugs 16(9):299

Dmitrenok A, Iwashita T, Nakajima T, Sakamoto B, Namikoshi M, Nagai H (2006) New cyclic depsipeptides from the green alga *Bryopsis* species; application of a carboxypeptidase hydrolysis reaction to the structure determination. Tetrahedron 62(6):1301–1308

Dominguez H (2013) Functional ingredients from algae for foods and nutraceuticals. Elsevier

Echave J, Fraga-Corral M, Garcia-Perez P, Popović-Djordjević J, Avdović HE, Radulović M et al (2021) Seaweed protein hydrolysates and bioactive peptides: extraction, purification, and applications. Mar Drugs 19(9):500. https://doi.org/10.3390/md19090500

Echave J, Otero P, Garcia-Oliveira P, Munekata PE, Pateiro M, Lorenzo JM et al (2022) Seaweed-derived proteins and peptides: promising marine bioactives. Antioxidants 11(1):176

Fitzgerald CN, Mora-Soler L, Gallagher E, O'Connor P, Prieto J, Soler-Vila A, Hayes M (2012) Isolation and characterization of bioactive pro-peptides with in vitro renin inhibitory activities from the macroalga Palmaria palmata. J Agric Food Chem 60(30):7421–7427

Fitzgerald C, Gallagher E, O'Connor P, Prieto J, Mora-Soler L, Grealy M, Hayes M (2013) Development of a seaweed derived platelet activating factor acetylhydrolase (PAF-AH) inhibitory hydrolysate, synthesis of inhibitory peptides and assessment of their toxicity using the Zebrafish larvae assay. Peptides 50:119–124

Fitzgerald C, Gallagher E, Doran L, Auty M, Prieto J, Hayes M (2014) Increasing the health benefits of bread: assessment of the physical and sensory qualities of bread formulated using a renin inhibitory *Palmaria palmata* protein hydrolysate. LWT-Food Sci Technol 56(2):398–405. https://doi.org/10.1016/j.lwt.2013.11.031

Fleurence J, Morançais M, Dumay J (2018) Seaweed proteins. In: Proteins in food processing. Elsevier, pp 245–262

Furuta T, Miyabe Y, Yasui H, Kinoshita Y, Kishimura H (2016) Angiotensin I converting enzyme inhibitory peptides derived from phycobiliproteins of dulse *Palmaria palmata*. Mar Drugs 14(2):32

Ganesan AR, Tiwari U, Rajauria G (2019) Seaweed nutraceuticals and their therapeutic role in disease prevention. Food Sci Human Wellness 8(3):252–263. https://doi.org/10.1016/j.fshw.2019.08.001

García-Vaquero M, López-Alonso M, Hayes M (2017) Assessment of the functional properties of protein extracted from the brown seaweed *Himanthalia elongata* (Linnaeus) SF Gray. Food Res Int 99:971–978. https://doi.org/10.1016/j.foodres.2016.06.023

Harnedy PA, O'Keeffe MB, FitzGerald RJ (2015) Purification and identification of dipeptidyl peptidase (DPP) IV inhibitory peptides from the macroalga *Palmaria palmata*. Food Chem 172:400–406

Harnedy PA, O'Keeffe MB, FitzGerald RJ (2017) Fractionation and identification of antioxidant peptides from an enzymatically hydrolysed *Palmaria palmata* protein isolate. Food Res Int 100:416–422

Hentati F, Tounsi L, Djomdi D, Pierre G, Delattre C, Ursu AV et al (2020) Bioactive polysaccharides from seaweeds. Molecules 25(14):3152. https://doi.org/10.3390/molecules25143152

Holdt SL, Kraan S (2011) Bioactive compounds in seaweed: functional food applications and legislation. J Appl Phycol 23:543–597

Hosseini SF, Rezaei M, McClements DJ (2022) Bioactive functional ingredients from aquatic origin: a review of recent progress in marine-derived nutraceuticals. Crit Rev Food Sci Nutr 62(5):1242–1269. https://doi.org/10.1080/10408398.2020.1839855

Indumathi P, Mehta A (2016) A novel anticoagulant peptide from the Nori hydrolysate. J Funct Foods 20:606–617. https://doi.org/10.1016/j.jff.2015.11.016

Jiao K, Gao J, Zhou T, Yu J, Song H, Wei Y, Gao X (2019) Isolation and purification of a novel antimicrobial peptide from *Porphyra yezoensis*. J Food Biochem 43(7):e12864

Joshi V, Kumar S (2015) Meat analogues: plant based alternatives to meat products – a review. Int J Food Ferment Technol 5(2):107–119

Joshi M, Adhikari B, Aldred P, Panozzo J, Kasapis S, Barrow C (2012) Interfacial and emulsifying properties of lentil protein isolate. Food Chem 134(3):1343–1353. https://doi.org/10.1016/j.foodchem.2012.03.029

Kandasamy G, Karuppiah SK, Subba Rao PV (2012) Salt-and pH-induced functional changes in protein concentrate of edible green seaweed *Enteromorpha* species. Fish Sci 78(1):169–176. https://doi.org/10.1007/s12562-011-0423-y

Kılınç B, Cirik S, Turan G, Tekogul H, Koru E (2013) Seaweeds for food and industrial applications. In: Food industry. IntechOpen

Kitade Y, Miyabe Y, Yamamoto Y, Takeda H, Shimizu T, Yasui H, Kishimura H (2018) Structural characteristics of phycobiliproteins from red alga *Mazzaella japonica*. J Food Biochem 42(1):e12436

Kumagai Y, Kitade Y, Kobayashi M, Watanabe K, Kurita H, Takeda H et al (2020) Identification of ACE inhibitory peptides from red alga *Mazzaella japonica*. Eur Food Res Technol 246:2225–2231

Kumar KS, Ganesan K, Selvaraj K, Rao PS (2014) Studies on the functional properties of protein concentrate of *Kappaphycus alvarezii* (Doty) Doty – an edible seaweed. Food Chem 153:353–360. https://doi.org/10.1016/j.foodchem.2013.12.058

Kumar A, Krishnamoorthy E, Devi HM, Uchoi D, Tejpal C, Ninan G, Zynudheen A (2018) Influence of sea grapes (*Caulerpa racemosa*) supplementation on physical, functional, and anti-oxidant properties of semi-sweet biscuits. J Appl Phycol 30:1393–1403. https://doi.org/10.1007/s10811-017-1310-4

Kumar A, Hanjabam MD, Kishore P, Uchoi D, Panda SK, Mohan CO et al (2023) Exploitation of seaweed functionality for the development of food products. Food Bioprocess Technol 1–31:1873. https://doi.org/10.1007/s11947-023-03023-2

Kurek MA, Onopiuk A, Pogorzelska-Nowicka E, Szpicer A, Zalewska M, Półtorak A (2022) Novel protein sources for applications in meat-alternative products – insight and challenges. Food Secur 11(7):957

Kyriakopoulou K, Keppler JK, van der Goot AJ (2021) Functionality of ingredients and additives in plant-based meat analogues. Food Secur 10(3):600

Lafarga T, Acién-Fernández FG, Garcia-Vaquero M (2020) Bioactive peptides and carbohydrates from seaweed for food applications: natural occurrence, isolation, purification, and identification. Algal Res 48:101909. https://doi.org/10.1016/j.algal.2020.101909

Lam ACY, Warkentin TD, Tyler RT, Nickerson MT (2017) Physicochemical and functional properties of protein isolates obtained from several pea cultivars. Cereal Chem 94(1):89–97. https://doi.org/10.1094/CCHEM-04-16-0097-FI

Lee HA, Kim IH, Nam TJ (2015) Bioactive peptide from *Pyropia yezoensis* and its anti-inflammatory activities. Int J Mol Med 36(6):1701–1706. https://doi.org/10.3892/ijmm.2015.2386

Liu X, Wang C-Y, Shao C-L, Wei Y-X, Wang B-G, Sun L-L et al (2009) Chemical constituents from *Sargassum pallidum* (Turn.) C. Agardh. Biochem Systemat Ecol 2(37):127–129

Marrion O, Schwertz A, Fleurence J, Guéant JL, Villaume C (2003) Improvement of the digestibility of the proteins of the red alga *Palmaria palmata* by physical processes and fermentation. Food Nahrung 47(5):339–344

McLaughlin CM, Sharkey SJ, Harnedy-Rothwell P, Parthsarathy V, Allsopp PJ, McSorley EM et al (2020) Twice daily oral administration of *Palmaria palmata* protein hydrolysate reduces food intake in streptozotocin induced diabetic mice, improving glycaemic control and lipid profiles. J Funct Foods 73:104101

McLaughlin C, Harnedy-Rothwell PA, Lafferty R, Sharkey S, Parthsarathy V, Allsopp PJ et al (2021) Macroalgal protein hydrolysates from *Palmaria palmata* influence the 'incretin effect' in vitro via DPP-4 inhibition and upregulation of insulin, GLP-1 and GIP secretion. Eur J Nutr 60(8):4439–4452

Mendis E, Kim S-K (2011) Present and future prospects of seaweeds in developing functional foods. Adv Food Nutr Res 64:1–15

Mittal R, Tavanandi HA, Mantri VA, Raghavarao K (2017) Ultrasound assisted methods for enhanced extraction of phycobiliproteins from marine macro-algae, *Gelidium pusillum* (Rhodophyta). Ultrason Sonochem 38:92–103

Mohammed HO, O'Grady MN, O'Sullivan MG, Hamill RM, Kilcawley KN, Kerry JP (2021) An assessment of selected nutritional, bioactive, thermal and technological properties of Brown and red Irish seaweed species. Food Secur 10(11):2784. https://doi.org/10.3390/foods10112784

Mouritsen OG (2013) Seaweeds: edible, available, and sustainable. University of Chicago Press

Mousaie M, Khodadadi M, Tadayoni M (2022) Hydrolysate protein from brown macroalgae (*Sargassum ilicifolium*): antioxidant, antitumor, antibacterial, and ACE inhibitory activities. J Food Process Preserv 46(11):e17020. https://doi.org/10.1111/jfpp.17020

Ms VQ, Miss MS, Miss GG-V, Ms PJ (2022) Seaweeds in bakery and farinaceous foods: a mini-review. Int J Gastron Food Sci 28:100403. https://doi.org/10.1016/j.ijgfs.2021.100403

Nadeeshani H, Hassouna A, Lu J (2022) Proteins extracted from seaweed Undaria pinnatifida and their potential uses as foods and nutraceuticals. Crit Rev Food Sci Nutr 62(22):6187–6203

Nascimento da Silva LC, Mendonça JSP, de Oliveira WF, Batista KLR, Zagmignan A, Viana IFT, dos Santos Correia MT (2021) Exploring lectin–glycan interactions to combat COVID-19: lessons acquired from other enveloped viruses. Glycobiology 31(4):358–371. https://doi.org/10.1093/glycob/cwaa099

Nowacka M, Trusinska M, Chraniuk P, Drudi F, Lukasiewicz J, Nguyen NP et al (2023) Developments in plant proteins production for meat and fish analogues. Molecules 28(7):2966

O'sullivan A, O'grady M, O'callaghan Y, Smyth T, O'brien N, Kerry J (2016) Seaweed extracts as potential functional ingredients in yogurt. Innovative Food Sci Emerg Technol 37:293–299. https://doi.org/10.1016/j.ifset.2016.07.031

Paiva L, Lima E, Neto AI, Baptista J (2016) Isolation and characterization of angiotensin I-converting enzyme (ACE) inhibitory peptides from *Ulva rigida* C. Agardh protein hydrolysate. J Funct Foods 26:65–76

Pangestuti R, Kim S-K (2015) Seaweed proteins, peptides, and amino acids. In: Seaweed sustainability. Elsevier, pp 125–140

Peñalver R, Lorenzo JM, Ros G, Amarowicz R, Pateiro M, Nieto G (2020) Seaweeds as a functional ingredient for a healthy diet. Mar Drugs 18(6):301. https://doi.org/10.3390/md18060301

Pliego-Cortés H, Wijesekara I, Lang M, Bourgougnon N, Bedoux G (2020) Current knowledge and challenges in extraction, characterization and bioactivity of seaweed protein and seaweed-derived proteins. Adv Bot Res 95:289–326

Prabhasankar P, Ganesan P, Bhaskar N (2009) Influence of Indian brown seaweed (*Sargassum marginatum*) as an ingredient on quality, biofunctional, and microstructure characteristics of pasta. Food Sci Technol Int 15(5):471–479. https://doi.org/10.1177/1082013209350267

Puertas G, Vázquez M (2021) Evaluation of the composition and functional properties of whole egg plasma obtained by centrifugation. Int J Food Sci Technol 56(10):5268–5276. https://doi.org/10.1111/ijfs.15124

Qu W, Ma H, Pan Z, Luo L, Wang Z, He R (2010) Preparation and antihypertensive activity of peptides from *Porphyra yezoensis*. Food Chem 123(1):14–20

Rattaya S, Benjakul S, Prodpran T (2009) Properties of fish skin gelatin film incorporated with seaweed extract. J Food Eng 95(1):151–157. https://doi.org/10.1016/j.jfoodeng.2009.04.022

Román RB, Alvarez-Pez J, Fernández FA, Grima EM (2002) Recovery of pure B-phycoerythrin from the microalga *Porphyridium cruentum*. J Biotechnol 93(1):73–85

Sato M, Hosokawa T, Yamaguchi T, Nakano T, Muramoto K, Kahara T et al (2002) Angiotensin I-converting enzyme inhibitory peptides derived from wakame (*Undaria pinnatifida*) and their antihypertensive effect in spontaneously hypertensive rats. J Agric Food Chem 50(21):6245–6252

Shaghaghian S, McClements DJ, Khalesi M, Garcia-Vaquero M, Mirzapour-Kouhdasht A (2022) Digestibility and bioavailability of plant-based proteins intended for use in meat analogues: a review, vol 129. Trends Food Sci Technol, p 646

Smit AJ (2004) Medicinal and pharmaceutical uses of seaweed natural products: a review. J Appl Phycol 16(4):245–262

Suetsuna K, Nakano T (2000) Identification of an antihypertensive peptide from peptic digest of wakame (*Undaria pinnatifida*). J Nutr Biochem 11(9):450–454

Sun Y, Zhang N, Zhou J, Dong S, Zhang X, Guo L, Guo G (2020) Distribution, contents, and types of mycosporine-like amino acids (MAAs) in marine macroalgae and a database for MAAs based on these characteristics. Mar Drugs 18(1):43. https://doi.org/10.3390/md18010043

Thiviya P, Gamage A, Gama-Arachchige NS, Merah O, Madhujith T (2022) Seaweeds as a source of functional proteins. Phycology 2(2):216–243. https://doi.org/10.3390/phycology2020012

Torres MD, Flórez-Fernández N, Domínguez H (2019) Integral utilization of red seaweed for bioactive production. Mar Drugs 17(6):314

Trottet G, Fernandes S, Grunz G, Thoma JP, Brunner-Komorek K (2018) A process for preparing a meat-analogue food product. Google Patents

Vanderlei ESO, Patoilo KKNR, Lima NA, Lima APS, Rodrigues JAG, Silva LMCM et al (2010) Antinociceptive and anti-inflammatory activities of lectin from the marine green alga *Caulerpa cupressoides*. Int Immunopharmacol 10(9):1113–1118

Wang F, Guo X-Y, Zhang D-N, Wu Y, Wu T, Chen Z-G (2015) Ultrasound-assisted extraction and purification of taurine from the red algae *Porphyra yezoensis*. Ultrason Sonochem 24:36–42

Wijesekara I, Pangestuti R, Kim S-K (2011) Biological activities and potential health benefits of sulfated polysaccharides derived from marine algae. Carbohydr Polym 84(1):14–21

World Health Organization (1991) Protein quality evaluation: report of The Joint FAO/WHO Expert Consultation, Bethesda, Md, USA, 4–8 December 1989, vol 51. Food & Agriculture Org

Wu G (2013) Functional amino acids in nutrition and health, vol 45. Springer, pp 407–411

Yucetepe A, Okudan EŞ, Özçelik B (2022) Antioxidant activity and techno-functional properties of protein extracts from *Caulerpa prolifera*: optimization of enzyme-assisted extraction by response surface methodology. J Food Nutr Res 61(3)

Zhang X, Cao D, Sun X, Sun S, Xu N (2019) Preparation and identification of antioxidant peptides from protein hydrolysate of marine alga *Gracilariopsis lemaneiformis*. J Appl Phycol 31:2585–2596. https://doi.org/10.1007/s10811-019-1746-9

Chapter 10
Molecular Mechanism for Utilizing Macroalgal Polysaccharides by the Human Gut Commensal Bacteria

Jayashree Niharika and Ravindra Pal Singh

Abstract Macroalgae are profoundly present in the intertidal regions of oceans and have been consumed as human diet for several decades. Dietary glycans derived from macroalgae have incredible powers to keep the equilibrium of the human gut microbiome. So far, it has been confirmed that the human gut bacteria possess a unique cluster of genes that are dedicated for utilizing a specific type of glycan. Some genes of those clusters consist of endo-acting enzymes that yield a diversity of oligosaccharides in vivo. In vivo-produced oligosaccharides are often reckoned by immune cell receptors and play a crucial role in gut homeostasis. Some molecular mechanisms by which human gut bacteria utilize macroalgal glycan have been discovered in the past couple of decades. Notwithstanding, studies based on human gut bacteria and nutraceuticals suggest that we should unravel glycan-utilizing mechanism of every microbial strain present in gut microbiota. This will devise innovative nutraceutical strategies to decisively manipulate the gut microbiota for improving human health. Thus, this chapter critically summarizes known molecular mechanisms of human gut bacteria by which they are utilizing macroalgal-derived glycans.

Keywords Macroalgal-derived glycans · Polysaccharide-utilizing locus · *Bacteroides uniformis* · Agarose · Porphyrin · Laminarin

10.1 Introduction

Macroalgae contribute greatly to the total carbon generated in the marine environment and the carbon chain for marine animals worldwide (Becker et al. 2020). Macroalgae fall into three distinct phyla: rhodophyta (red), phaeophyta (brown), and chlorophyte (green) where dietary fibers are found as food storage elements

J. Niharika · R. P. Singh (✉)
Department of Industrial Biotechnology, Gujarat Biotechnology University, Gandhinagar, Gujarat, India
e-mail: ravindra.singh@gbu.edu.in

N. Trivedi et al. (eds.), *Recent Advances in Seaweed Biotechnology*,
https://doi.org/10.1007/978-981-96-0519-4_10

(such as laminarin) and cell wall components, such as alginate and carrageenan (Xu et al. 2017). Edible macroalgae such as wakame (Undaria), nori (Porphyra), and kombu (Laminaria) are consumed as dietary fibers globally (Urbano and Goni, 2002). Macroalgal resources will contribute greatly to food security when the world population will exceed 9.1 billion by 2050. Given the significant role of macroalgae in a wide variety of industries, including food and fodder, the global market worth of consumption of macroalgae is projected to reach US $ 2492 million by 2028 from US $ 1501 million in 2021, with the compound annual growth rate of 7.51%, according to Fortune Business (https://www.fortunebusinessinsights.com/industry-reports/commercial-seaweed-market-100077). These dietary fibers have complicated structures and comprise of the variability of linkages with a heterogeneous mixture of monosaccharides and the bulk of them are adorned with sulfate groups (Yuan et al. 2006; Singh and Reddy 2014; Hwang et al. 2016; Yang et al. 2019). These polysaccharides have been associated with a variety of molecular and physiological functions, including anticancer, antioxidant, anti-diabetic, anti-viral, drug delivery vehicles, immunomodulatory, and prebiotic properties. As a result, they have a wide range of uses, including agricultural, food, and pharmaceutical fields.

A great deal of progress has been made to comprehend the importance of macroalgal oligo- and polysaccharides (collectively known as glycans) in enhancing the health of the gut by altering gut microbiota in recent years. For instance, the molecular mechanism of utilizing agarose by the human gut bacterium, *Bacteroides uniformis* NP1, has been demonstrated by Pluvinage et al. (2018). Glycan-utilizing genes have been discovered in *Bacteroides* in a cluster known as the polysaccharide-utilization locus (PUL). The PULs of *Bacteroides* express all genes that are required for glycan depolymerization (Reeves et al. 1997; Shipman et al. 2000). Similarly, laminarin utilization pathways (composed of β-1,3 linked linear main chain with 1,6- β linked single D glucan) have been demonstrated in *B. uniformis* ATCC 8492 and *B. uniformis* JCM 13288^T (Dejean et al. 2020; Singh et al. 2020).

Because of their better solubility and bioavailability, oligosaccharides derived from macroalgae have shown stronger prebiotic potential than their respective polysaccharides (Vera et al. 2011; Saigusa et al. 2015; Zhou et al. 2015; Park et al. 2016; Fang et al. 2017). For instance, alginate lyases of *Pseudomonas* sp. HZJ 216 (Li et al. 2011) and *Cellulophaga* sp. NJ-1 (Zhu et al. 2016) were used for yielding alginate-oligosaccharides (AOs) with degree of polymerization (DP) of 2–5 or 2–7. Those produced oligosaccharides were then utilized to encourage the growth of useful bacteria such as *Bifidobacterium* and *Lactobacillus.* AOs were observed to be more potent than plant-originated fructo-oligosaccharides in aiding growth of beneficial microflora (Wang et al. 2006).

Even though oligosaccharides can be produced using enzymatic, chemical, and physical methods, determining their monosaccharide composition and linkage heterogeneity remains difficult due to their complex nature (Lang et al. 2014). As a result, we expect that future research will first report the purification of specified DP of a used glycan with a thorough information on structural features, followed by an evaluation of biological activities to optimally utilize them.

Despite the numerous benefits of macroalgal glycans, they are not reaching to the majority of the microbial population as simple oligosaccharides are primarily processed in the proximal intestine and rarely reach the distal region (Duncan et al. 2003; Zoetendal et al. 2012). There is substantial evidence pointing out that microbial communities steadily expand throughout the intestine, with a significant population of microbes being present in the distal gut (Hillman et al. 2017; Kastl Jr. et al. 2020). This suggested that it is important for oligosaccharides to reach the distal section of the gut for encouraging the growth of beneficial microbial communities. Given the complexity of macroalgal glycans, they would be an excellent candidate for achieving this aim. This should be nailed down by further studies.

In this chapter, we summarize the discovered molecular mechanisms of human gut bacteria for utilizing macroalgal glycans.

10.2 Molecular Mechanism for Utilizing Macroalgal Glycans by Human Gut Bacteria

Prebiotics are one example of the nutraceutical strategy that uses the host-diet-microbe idea to change the gut microbiota and enhance the makeup of the gut's microbial population. According to epidemiological data gathered from Western and Japanese diets, macroalgal consumption in Japan may have reduced the prevalence of chronic non-communicable diseases like inflammatory bowel disease, obesity, metabolic diabetes, and cardiovascular disease (Chirumbolo and Bjorklund 2016; Kang et al. 2016). According to McDermid et al. (2005), the total dietary fiber content of macroalgae ranges from 20 to 60%. This may be a great source for fostering the growth of probiotics and other beneficial microbial communities.

Many in vivo and in vitro studies have been conducted on macroalgal glycans as carbon sources. However, most of those studies have not conferred the precise molecular mechanisms. It has been conclusively demonstrated that dietary fiber utilization is fully dependent on carbohydrate-active enzymes (CAZymes) present in gut microbial populations. Most glycan-utilizing enzymes used by human gut bacteria belong to polysaccharide lyase (PL) and glycoside hydrolase (GH) families (El Kaoutari et al. 2013). Indeed, it is believed that nutraceutical approaches to promoting gut health would be greatly aided if we knew which bacterium to target and what kind of arsenal they have for exploiting glycans.

Gram-negative bacterial PUL (gnPUL) was discovered in the gut commensal *Bacteroides thetaiotaomicron* (Anderson and Salyers 1989). An archetypal PUL architecture includes genes homologous to susC (a TonB-dependent receptor, which transports generated oligosaccharides into the periplasmic space), susD (an oligosaccharide binding transmembrane protein), genes encoding assorted enzymes (which hydrolyze glycans), and genes responsible for expressing sensory proteins (Larsbrink et al. 2014; Glenwright et al. 2017). It has been reported that *Bacteroides* have over 13,500 PULs (Lapebie et al. 2019). In *Bacteroides*, sensory proteins known as hybrid-two component systems (HTCSs) govern each PUL (Sonnenburg et al. 2006). HTCS can recognize oligosaccharides in periplasmic space and promotes the expression of PUL-encoded genes until the bacterial outer surface is not adequately saturated by the required protein that break down specific glycans (Larsbrink et al. 2014). While Gram-positive bacterial PUL (gpPUL) is less extensive and comprehensive than the gnPUL, we are aware that gpPUL is composed of an extracellular endolytic enzyme, sensory transcriptional regulator, exo-acting enzyme, and ATP-binding cassette (ABC) transporters (La Rosa et al. 2019; Singh et al. 2023).

Deville et al. (2004) proposed laminarin as a dietary fiber concept. Their following study has discovered that laminarin significantly promotes the growth of *Lactobacillus, Bifidobacteria*, and *Bacteroides*. After the growth experiment, the quantities of acetate, propionate, and butyrate were measured and were found to be higher than the control condition (Deville et al. 2007; Lam et al. 2020). It has been observed that *B. uniformis* ATCC 8492 (Salyers et al. 1977; Desai et al. 2016) and *Bacteroides cellulosilyticus* WH2 (McNulty et al. 2013) showed a significant growth on laminarin. Meanwhile, it was noted that *Bacteroides* dedicate about 20% of genes of their whole genome to carbohydrate utilization repertoire (Martens et al. 2008; Terrapon et al. 2015). Dejean et al. (2020) unveiled a molecular mechanism for laminarin utilization by *B. uniformis* ATCC 8492. Laminarin PUL of *B. uniformis* ATCC 8492 encodes surface glycan-binding proteins (SGBPs) and endo-acting glycoside hydrolases (GH158 and 16) at the outer membrane surface for efficiently utilizing laminarin. The GH158 and GH16 cleave laminarin and then generate oligosaccharides, which are then internalized by the SusCD complex in the periplasmic space. Imported β-1,3- linked oligosaccharides are then processed by exo-acting *Bu*GH3, which transforms the oligosaccharides to glucose before they enter the cytoplasm. Newly identified curdlan and laminarin-utilizing GH family belongs to 158 (Helbert et al. 2019). Further exploration of this enzyme was performed and observed that GH158 can cleave the soluble form of curdlan (Singh et al. 2020). The study demonstrated that GH158 generates oligosaccharides with β-1,3-linkage that support the growth of *Bifidobacteria* and *Lactobacillus*. The GH158 and *Bp*GH16$_{MLG}$ produce β-1,3 linked oligosaccharides, which can also promote the growth of *Blautia producta* JCM 1471^{T} (Singh et al. 2020, 2023). It has recently been observed that *B. producta* JCM 1471^{T} processes a gpPUL, consisting of cell wall membrane-bound extracellular (*Bp*GH16$_{MLG}$) and cytoplasmic enzymes (*Bp*GH94$_{MLG}$ and *Bp*GH3), a transcriptional regulator and ABC transporter. Although it has *Bp*GH16$_{MLG}$ and is highly active on laminarin, gpPUL does not support the growth on polysaccharides form of laminarin. It was postulated that the chemical structure of laminarin might not be an active transcriptional regulator in this bacterium. Furthermore, a solute binding protein (*Bp*GHSBP$_{MLG}$) is involved in utilizing β-1,3-linked oligosaccharides in *B. producta* JCM 1471^{T} (Singh et al. 2023).

Porphyran (galactans), agar, and agarose-cleaving enzymes were identified in Japanese gut bacteria- *B. plebeius* DSM 17135 (Hehemann et al. 2010) and *B. uniformis* JCM 13288^{T} (Singh et al. 2020) and marine bacterium—*Zobellia galactanivorans* (Hehemann et al. 2010). These red macroalgal-utilizing genes are often orchestrated in a PUL manner with outer surface tethered endo-acting enzymes, belonging to GH16 family, SusCD complex, SGBP, and endo-acting enzymes (GH86A and B) and exo-acting enzymes (GH2 and a sulphatase) in periplasmic space (Pluvinage et al. 2018). Further investigation of two, GH86A and B leads to the revelation that GH86A can cleave porphyran but not agarose and yield larger oligosaccharides than GH16-β-porphyranase as shown in Fig. 10.1. Given that GH86A generated larger oligosaccharides, it is unknown whether those oligosaccharides can be involved in glycan cross-feeding or just be utilized by *B. plebeius* DSM 17135. For example, a cross-feeding system was explored using *B. plebeius* and *Bifidobacterium longum* ssp. *infantis* in which GH16 of *B. plebeius* produces agarotriose. Produced agarotriose can be further cleaved by agarose-utilizing β-galactosidases (i.e., Bga2A and Bga42A) expressed by the *B. longum* ssp. *infantis* producing neoagarobiose. This neoagarobiose is not consumed by *B. longum* ssp. *Infantis*; therefore, *B. plebeius* can again use neoagarobiose to generate 3,6-anhydro-L-galactose (L-AHG) by the action of *Bp*GH117. L-AHG possesses anti-inflammatory activity (Yun et al. 2021). Only, produced galactose from agarotriose by the action of Bga2A and Bga42A is used by *B. longum* ssp. *infantis*.

Hehemann et al. (2012) postulated that *B. thetaiotaomicron* VP1–3731 and *B. uniformis* NP1 may use κ- carrageenan and agarose, respectively. Subsequently, identification of carrageenan and agarose-utilizing PUL was discovered in the marine bacterium *Z. galactanivorans* (Ficko-Blean et al. 2017) and biochemical characterization of PUL-associated protein was spotted in the human gut bacterium *B. uniformis* NP1 (Pluvinage et al. 2018). Concurrently, κ-carrageenan and agarose utilization capacity were also described in a human gut bacterium, *B. xylanisolvens* 38F6 and *B. uniformis* L8 (Li et al. 2017), respectively. The breakdown of agarose and porphyran initiated via GH16B and GH16C activities in *B. uniformis* NP1 (Pluvinage et al. 2018). Generated neoagaro-oligosaccharides (NAOs) are imported by orthologous SusCD complexes. NAOs are further cleaved by two different pathways in periplasmic space; (1) can be depolymerized cyclically by exo-acting enzymes complex, GH117B–GH2C and (2) periplasmic endo-acting enzyme, GH86, converts NAOs into neoagrobiose [L-AHG-β-1,4-galactose]. Generated monosaccharide (GAL) or (L-AHG) enters the cytoplasm by multifacilitator superfamily (MFS) proteins and the RhaT, respectively (Pluvinage et al. 2018). Similar to *B. uniformis* NP1, an exo-acting 3,6-anhydro-α-(1,3)-L-galactosidase (GH117) is recently characterized from *B. plebeius* and it eliminates AHG residue from the non-reducing end of NAO (Jin et al. 2021).

B. uniformis NP1 expresses 5 dedicated proteins, namely AHG isomerase, keto-deoxygluconate kinase, AHG dehydrogenase, galactose mutarotase, and a RhaT-like sugar importer for utilizing L-AHG. Particularly, RhaT-like sugar importer is suggested to import L-AHG into the cytoplasm. Rest four proteins are involved in metabolizing L-AHG through the standard metabolic pathway (Pluvinage et al. 2018). These proteins, particularly AHG isomerase and AHG dehydrogenase, share significant sequence similarity (>64%) to those proteins reported in the sea bacterium, *Vibrio* sp. EJY3 (Yun et al. 2015). Furthermore, it suggests that the en bloc horizontal gene transfer (HGT) has been involved in acquiring agarose PUL in

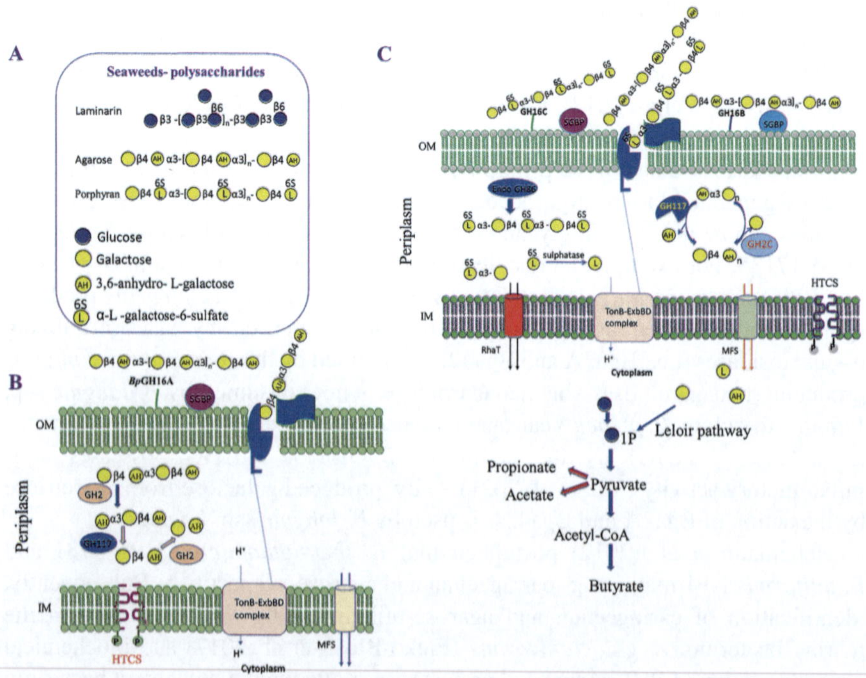

Fig. 10.1 Utilizing macroalgal glycans by human gut bacteria. (**a**) Representative structures of laminarin, agarose, and porphyran according to symbolic nomenclature for glycans (https://www.ncbi.nlm.nih.gov/glycans/snfg.html). (**b**) Utilization of agarose by *Bacteroides plebeius*. Agarose present in the human gut can first be recognized and trapped by surface glycan-binding protein (SGBP). SGBP allows trapped glycan to reach *Bp*GH16A, which cleaves polysaccharides into neoagarooligosaccharides, primarily neoagarotetraose (Park et al. 2021). The SusCD complex imports neoagarotetraose into the periplasmic region, where it is further transformed into monosaccharides by the GH117 and GH2 activities in a cyclic manner. (**c**) Utilization of agarose and porphyran by *Bacteroides uniformis* NP1. Polysaccharide forms of agarose and porphyrin are initially depolymerized by the activity of GH16B and GH16C, respectively. Alike *B. plebeius,* polysaccharide forms of agarose and porphyran are first trapped by SGBP and are then cleaved by GH16B and GH16C for the generation of oligosaccharides. Produced neoagarooligosaccharides imported into periplasmic space by the SusCD complex. Porphyran-derived neoagarooligosaccharides are further cleaved by the endolytic enzyme GH86, yielding disaccharides. GH117/GH2 cleaves this disaccharide to monosaccharides while agarose produced neoagarooligosaccharides that are cyclically converted into monosaccharides with the help of GH117 and GH2 (Pluvinage et al. 2018). These monosaccharides enter into glycolysis by the Leloir pathway (Holden et al. 2003) and participate in the formation of SCFAs such as acetate, propionate, and butyrate (Singh et al. 2022)

B. uniformis NP1 from marine bacteria. It has been further observed that diverse proceedings have taken place in the transfer of genes from marine bacteria to human gut bacteria for edible macroalgal digestion (Pudlo et al. 2022).

Alginate is a class of linear copolymers, which is composed of (1,4)-linked β-D-mannuronate (M) and α-L-guluronate (G) residues. These residues can be arranged in a form of successive M series (MMMMMM), successive G series (GGGGGG), and alternating G and M series (GMGMGM) (Urtuvia et al. 2017). Polysaccharide lyase (PL) is bacterial enzymes that cleave uronic acid-containing glycan such as alginate. PL families, belonging to 17, 7, and 6, cleave uronic acid-containing substrate through a β- elimination mechanism and release products with 4, 5-unsaturated residue from non-reducing end of a substrate (Lapebie et al. 2019). As compared to another glycan PUL, two distinct types of alginate PUL were observed (1) a PUL containing either PL6_1 or PL17_2 and (2) the one with PL17_2 and PL6_1 (Mathieu et al. 2018). This suggests that it could have occurred as a result of distinct HGT events in which PL6 had a shared ancestor. Stender et al. (2019) identified a polyM-specific PL6 from *B. cellulosilyticus*, which can generate unsaturated oligosaccharides of DPs 2 to 7. Unsaturated oligosaccharides produced by PL activity appear to be beneficial in boosting the growth of probiotic strains of *Lactobacillus* and *Bifidobacteria* species (Akiyama et al. 1992; Li et al. 2016). However, additional investigation needs to be done to comprehend their utilization mechanism. Those investigations should clearly show in terms of that distinct *Bacteroides* harbor different forms of PUL, which are more likely to maintain gut microbial equilibrium under diverse nutritional environmental situations.

It has been postulated that contaminated raw macroalgal material could have been a source of HGT event for transferring unique gene clusters from marine bacteria to human gut bacteria. It may be also possible that consistent consumption of a diet comprising macroalgae forces a selective pressure on certain gut bacteria to adapt new genes in their genome for efficient utilization of macroalgal dietary glycan. Subsequently, those PULs become evolutionary preserved when passing from one generation to the next. A better example was provided by a study of Hehemann et al. (2010). The study observed that a porphyran-utilizing locus was widely spread among Japanese gut microbiota and even was vertical transferred between mother and child. Similarly, bioinformatic analysis has observed that agarose-utilizing locus of *B. uniformis* NP1, laminarin-utilizing locus, and porphyran-utilizing locus of *B. plebeius* DSM 17135 are highly abundant in the Japanese population (Pluvinage et al. 2018; Dejean et al. 2020). Interestingly, agarose-utilizing locus of *B. uniformis* NP1 was observed in substantial amounts in North American metagenomic samples, implying that mass population and cultural movement can influence HGT incidences. A true example of HGT event was observed in our previous study in which agarose and porphyrin-utilizing locus of *B. uniformis* JCM 13288^{T} showed 100% sequence homology with a locus of another Japanese gut bacteria *B. plebeius* DSM 17135 (Singh et al. 2020). These bacteria are likely to take up this locus from *Zobellia galactanivorans* (Hehemann et al. 2010). Indeed, a dietary habit greatly influences the persistence of a locus. For example, β-1-3 linked glucan-utilizing locus identified in *B. uniformis, B. thetaiotaomicron*, and *B. fluxus* can be split into three distinct groups based on the architecture of the locus (Dejean et al. 2020). Among these, a *B. uniformis* type β-1-3 linked glucan using locus is widely distributed in Asian, European, and North American populations. While these three loci were not found in the Yanomami or Hadza tribes, this suggests that *Bacteroides* is not as abundant as *Prevotella* as their food habits are different (De Filippo et al.

2010; Smits et al. 2017). Although we gained some adaptive information, we still need to compare genomes of different bacterial taxa from different environmental situations to understand the adaptive genomic evolution of human gut bacteria.

Fucoidan and ulvan are complex glycans that require several enzymes, belonging to GHs and PL families, for their degradation in a marine environment (Foran et al. 2017; Ulaganathan et al. 2017; Sichert et al. 2020). To the best of our knowledge, a locus utilizing these glycans in human gut bacteria has not been identified yet. It would be interesting to dig out through further studies whether and how different genes that degrade fucoidan and ulvan have been transported from marine environments to human gut bacteria.

10.3 Roles of Bacterial-Produced Metabolites in Human Health

Overall, bacteria utilize different macroalgal glycans for their carbon source and survival. When glycans are used, they produce a variety of other compounds, including small-chain fatty acids (SCFAs, acetate, butyrate, propionate, fumarate, and lactate) (Macfarlane and Macfarlane, 2003), tryptophan compounds (tryptamine and indolic compounds) (Gao et al. 2020), and gamma-aminobutyric acid (Evrensel and Ceylan 2015). These SCFAs are produced by different pathways and some of the genes involved in these formations have already been previously summarized (Singh 2019; Zhao et al. 2019). Butyrate predominantly supplies energy to colon epithelial cells and exhibits a protective effect in colonic illnesses such as colon cancer and inflammatory bowel disease (Beauvieux et al. 2001; Zhang et al. 2021). Formate and propionate lower the pH of the gut lumen, inhibiting the growth of *Salmonella*—a pH-sensitive pathogenic bacteria (Jacobson et al. 2018; Gomez-Garcia et al. 2019). In a similar investigation, we discovered that butyrate can suppress the quorum-sensing mechanism in human pathogenic *Clostridium perfringens* (Adachi et al. 2018) and modulates mucin secretion along with bacterial adherence (Jung et al. 2022). Acetate is produced in high amounts and that can again transform into other forms of SCFAs (Reichardt et al. 2018). In addition, colonic acetate enters the brain after crossing the blood-brain barrier and intraperitoneal acetate suppresses appetite and patterns of hypothalamic neuronal activity (Frost et al. 2014). Produced lactic acid can control bacterial vaginosis (Tachedjian et al. 2017). The role of tryptophan metabolism has now been recognized in the gut-brain axis in which bacterial-produced tryptophan modulates brain functions. In particular, tryptophan metabolizes into tryptamine, kynurenine, serotonin, and indole, and these are known to modulate intestinal and neuroendocrine-immune responses (Gao et al. 2020). Tryptophan has centrally produced serotonin and is predominantly involved in connecting gut-brain signaling system (Layunta et al. 2021, 2022). Therefore, extensive research aimed at broadening the portfolio of glycan utilization by beneficial and commensal gut flora will reveal the secrets to modulate eubiosis and dysbiosis gut conditions.

10.4 Conclusion

Until now, we have exploited the molecular mechanism of utilizing macroalgal glycans by some human gut model bacteria. The bacterial strains not only rely on glycan utilization for their survival but also share generated oligosaccharides with other gut communities. Such trophic level glycan sharing system is essential for maintaining microbial homeostasis in the gut. Nevertheless, we need much more evidence about molecular mechanisms of macroalgal glycans utilization by other human gut bacteria. Such understanding will woo the beneficial roles of macroalgal glycans in the human diet and will promote their use in food industries.

Acknowledgments The authors would like to thank the Department of Biotechnology (DBT), India, for providing the Grant in-aid-number: BT/PR32876/PFN/20/1471/2020 and BT/PR38298/PFN/20/1558/2020. R. P. Singh would like to thank DBT, India, for providing the Ramalingaswami Re-entry Fellowship.

References

Adachi K, Ohtani K, Kawano M, Singh RP, Yousuf B, Sonomoto K, Shimizu T, Nakayama J (2018) Metabolic dependent and independent pH-drop shuts down VirSR quorum sensing in *Clostridium perfringens*. J Biosci Bioeng 125:525–531

Akiyama H, Endo T, Nakakita R, Murata K, Yonemoto Y, Okayama K (1992) Effect of depolymerized alginates on the growth of *Bifidobacteria*. Biosci Biotechnol Biochem 56:355–356

Anderson KL, Salyers AA (1989) Genetic evidence that outer membrane binding of starch is required for starch utilization by *Bacteroides thetaiotaomicron*. J Bacteriol 171:3199–3204

Beauvieux MC, Tissier P, Gin H, Canioni P, Gallis JL (2001) Butyrate impairs energy metabolism in isolated perfused liver of fed rats. J Nutr 131:1986–1992

Becker S, Tebben J, Coffinet S, Wiltshire K, Iversen MH, Harder T, Hinrichs KU, Hehemann JH (2020) Laminarin is a major molecule in the marine carbon cycle. Proc Natl Acad Sci USA 117:6599–6607

Chirumbolo S, Bjorklund G (2016) Red seaweeds for obesity prevention? Food Chem Toxicol 94:268–269

De Filippo C, Cavalieri D, Di Paola M, Ramazzotti M, Poullet JB, Massart S, Collini S, Pieraccini G, Lionetti P (2010) Impact of diet in shaping gut microbiota revealed by a comparative study in children from Europe and rural Africa. Proc Natl Acad Sci USA 107:14691–14696

Dejean G, Tamura K, Cabrera A, Jain N, Pudlo NA, Pereira G, Viborg AH, Van Petegem F, Martens EC, Brumer H (2020) Synergy between cell surface glycosidases and glycan-binding proteins dictates the utilization of specific beta(1,3)-glucans by human gut *Bacteroides*. MBio 11

Desai MS, Seekatz AM, Koropatkin NM, Kamada N, Hickey CA, Wolter M, Pudlo NA, Kitamoto S, Terrapon N, Muller A, Young VB, Henrissat B, Wilmes P, Stappenbeck TS, Nunez G, Martens EC (2016) A dietary fiber-deprived gut microbiota degrades the colonic mucus barrier and enhances pathogen. Cell 167:1339–1353 e1321

Deville C, Damas J, Forget P, Dandrifosse G, Peulen O (2004) Laminarin in the dietary fibre concept. J Sci Food Agric 84:1030–1038

Deville C, Gharbi M, Dandrifosse G, Peulen O (2007) Study on the effects of laminarin, a polysaccharide from seaweed, on gut characteristics. J Sci Food Agric 87:1717–1725

Duncan SH, Scott KP, Ramsay AG, Harmsen HJ, Welling GW, Stewart CS, Flint HJ (2003) Effects of alternative dietary substrates on competition between human colonic bacteria in an anaerobic fermentor system. Appl Environ Microbiol 69:1136–1142

El Kaoutari A, Armougom F, Gordon JI, Raoult D, Henrissat B (2013) The abundance and variety of carbohydrate-active enzymes in the human gut microbiota. Nat Rev Microbiol 11:497–504

Evrensel A, Ceylan ME (2015) The gut-brain axis: the missing link in depression. Clin Psychopharmacol Neurosci 13:239–244

Fang W, Bi D, Zheng R, Cai N, Xu H, Zhou R, Lu J, Wan M, Xu X (2017) Identification and activation of TLR4-mediated signalling pathways by alginate-derived guluronate oligosaccharide in RAW264.7 macrophages. Sci Rep 7:1663

Ficko-Blean E, Prechoux A, Thomas F, Rochat T, Larocque R, Zhu Y, Stam M, Genicot S, Jam M, Calteau A, Viart B, Ropartz D, Perez-Pascual D, Correc G, Matard-Mann M, Stubbs KA, Rogniaux H, Jeudy A, Barbeyron T, Medigue C, Czjzek M, Vallenet D, Mcbride MJ, Duchaud E, Michel G (2017) Carrageenan catabolism is encoded by a complex regulon in marine heterotrophic bacteria. Nat Commun 8:1685

Foran E, Buravenkov V, Kopel M, Mizrahi N, Shoshani S, Helbert W, Banin E (2017) Functional characterization of a novel "ulvan utilization loci" found in *Alteromonas* sp. LOR genome. Algal Res 25:39–46

Frost G, Sleeth ML, Sahuri-Arisoylu M, Lizarbe B, Cerdan S, Brody L, Anastasovska J, Ghourab S, Hankir M, Zhang S, Carling D, Swann JR, Gibson G, Viardot A, Morrison D, Louise Thomas E, Bell JD (2014) The short-chain fatty acid acetate reduces appetite via a central homeostatic mechanism. Nat Commun 5:3611

Gao K, Mu CL, Farzi A, Zhu WY (2020) Tryptophan metabolism: a link between the gut microbiota and brain. Adv Nutr 11:709–723

Glenwright AJ, Pothula KR, Bhamidimarri SP, Chorev DS, Basle A, Firbank SJ, Zheng H, Robinson CV, Winterhalter M, Kleinekathofer U, Bolam DN, Van Den Berg B (2017) Structural basis for nutrient acquisition by dominant members of the human gut microbiota. Nature 541:407–411

Gomez-Garcia M, Sol C, De Nova PJG, Puyalto M, Mesas L, Puente H, Mencia-Ares O, Miranda R, Arguello H, Rubio P, Carvajal A (2019) Antimicrobial activity of a selection of organic acids, their salts and essential oils against swine enteropathogenic bacteria. Porcine Health Manag 5:32

Hehemann JH, Correc G, Barbeyron T, Helbert W, Czjzek M, Michel G (2010) Transfer of carbohydrate-active enzymes from marine bacteria to Japanese gut microbiota. Nature 464:908–912

Hehemann JH, Kelly AG, Pudlo NA, Martens EC, Boraston AB (2012) Bacteria of the human gut microbiome catabolize red seaweed glycans with carbohydrate-active enzyme updates from extrinsic microbes. Proc Natl Acad Sci USA 109:19786–19791

Helbert W, Poulet L, Drouillard S, Mathieu S, Loiodice M, Couturier M, Lombard V, Terrapon N, Turchetto J, Vincentelli R, Henrissat B (2019) Discovery of novel carbohydrate-active enzymes through the rational exploration of the protein sequences space. Proc Natl Acad Sci USA 116:6063–6068

Hillman ET, Lu H, Yao T, Nakatsu CH (2017) Microbial ecology along the gastrointestinal tract. Microbes Environ 32:300–313

Holden HM, Rayment I, Thoden JB (2003) Structure and function of enzymes of the Leloir pathway for galactose metabolism. J Biol Chem 278:43885–43888

Hwang PA, Phan NN, Lu WJ, Ngoc Hieu BT, Lin YC (2016) Low-molecular-weight fucoidan and high-stability fucoxanthin from brown seaweed exert prebiotics and anti-inflammatory activities in Caco-2 cells. Food Nutr Res 60:32033

Jacobson A, Lam L, Rajendram M, Tamburini F, Honeycutt J, Pham T, Van Treuren W, Pruss K, Stabler SR, Lugo K, Bouley DM, Vilches-Moure JG, Smith M, Sonnenburg JL, Bhatt AS, Huang KC, Monack D (2018) A gut commensal-produced metabolite mediates colonization resistance to *Salmonella* infection. Cell Host Microbe 24:296–307 e297

Jin Y, Yu S, Kim DH, Yun EJ, Kim KH (2021) Characterization of neoagarooligosaccharide hydrolase *Bp*GH117 from a human gut bacterium *Bacteroides plebeius*. Mar Drugs 19

Jung TH, Han KS, Park JH, Hwang HJ (2022) Butyrate modulates mucin secretion and bacterial adherence in LoVo cells via MAPK signaling. PLoS One 17:e0269872

Kang MC, Kang N, Ko SC, Kim YB, Jeon YJ (2016) Anti-obesity effects of seaweeds of Jeju Island on the differentiation of 3T3-L1 preadipocytes and obese mice fed a high-fat diet. Food Chem Toxicol 90:36–44

Kastl AJ Jr, Terry NA, Wu GD, Albenberg LG (2020) The structure and function of the human small intestinal microbiota: current understanding and future directions. Cell Mol Gastroenterol Hepatol 9:33–45

La Rosa SL, Leth ML, Michalak L, Hansen ME, Pudlo NA, Glowacki R, Pereira G, Workman CT, Arntzen MO, Pope PB, Martens EC, Hachem MA, Westereng B (2019) The human gut Firmicute *Roseburia intestinalis* is a primary degrader of dietary beta-mannans. Nat Commun 10:905

Lam KL, Cheng WY, Su Y, Li X, Wu X, Wong KH, Kwan HS, Cheung PCK (2020) Use of random forest analysis to quantify the importance of the structural characteristics of beta-glucans for prebiotic development. Food Hydrocoll 108:106001

Lang Y, Zhao X, Liu L, Yu G (2014) Applications of mass spectrometry to structural analysis of marine oligosaccharides. Mar Drugs 12:4005–4030

Lapebie P, Lombard V, Drula E, Terrapon N, Henrissat B (2019) *Bacteroidetes* use thousands of enzyme combinations to break down glycans. Nat Commun 10:2043

Larsbrink J, Rogers TE, Hemsworth GR, Mckee LS, Tauzin AS, Spadiut O, Klinter S, Pudlo NA, Urs K, Koropatkin NM, Creagh AL, Haynes CA, Kelly AG, Cederholm SN, Davies GJ, Martens EC, Brumer H (2014) A discrete genetic locus confers xyloglucan metabolism in select human gut *Bacteroidetes*. Nature 506:498–502

Layunta E, Buey B, Mesonero JE, Latorre E (2021) Crosstalk between intestinal serotonergic system and pattern recognition receptors on the microbiota-gut-brain axis. Front Endocrinol (Lausanne) 12:748254

Layunta E, Latorre E, Grasa L, Arruebo MP, Buey B, Alcalde AI, Mesonero JE (2022) Intestinal serotonergic system is modulated by Toll-like receptor 9. J Physiol Biochem 78:689–701

Li L, Jiang X, Guan H, Wang P (2011) Preparation, purification and characterization of alginate oligosaccharides degraded by alginate lyase from *Pseudomonas* sp. HZJ 216. Carbohydr Res 346:794–800

Li M, Li G, Shang Q, Chen X, Liu W, Pi X, Zhu L, Yin Y, Yu G, Wang X (2016) In vitro fermentation of alginate and its derivatives by human gut microbiota. Anaerobe 39:19–25

Li M, Shang Q, Li G, Wang X, Yu G (2017) Degradation of marine algae-derived carbohydrates by *Bacteroidetes* isolated from human gut microbiota. Mar Drugs 15

Macfarlane S, Macfarlane GT (2003) Regulation of short-chain fatty acid production. Proc Nutr Soc 62:67–72

Martens EC, Chiang HC, Gordon JI (2008) Mucosal glycan foraging enhances fitness and transmission of a saccharolytic human gut bacterial symbiont. Cell Host Microbe 4:447–457

Mathieu S, Touvrey-Loiodice M, Poulet L, Drouillard S, Vincentelli R, Henrissat B, Skjak-Braek G, Helbert W (2018) Ancient acquisition of "alginate utilization loci" by human gut microbiota. Sci Rep 8:8075

Mcdermid KJ, Stuerckea B, Haleakala OJ (2005) Total dietary fiber content in Hawaiian marine algae. Bot Mar 48

Mcnulty NP, Wu M, Erickson AR, Pan C, Erickson BK, Martens EC, Pudlo NA, Muegge BD, Henrissat B, Hettich RL, Gordon JI (2013) Effects of diet on resource utilization by a model human gut microbiota containing *Bacteroides cellulosilyticus* WH2, a symbiont with an extensive glycobiome. PLoS Biol 11:e1001637

Park HJ, Ahn JM, Park RM, Lee SH, Sekhon SS, Kim SY, Wee JH, Kim YH, Min J (2016) Effects of alginate oligosaccharide mixture on the bioavailability of lysozyme as an antimicrobial agent. J Nanosci Nanotechnol 16:1445–1449

Park NJ, Yu S, Kim DH, Yun EJ, Kim KH (2021) Characterization of BpGH16A of *Bacteroides plebeius*, a key enzyme initiating the depolymerization of agarose in the human gut. Appl Microbiol Biotechnol 105:617–625

Pluvinage B, Grondin JM, Amundsen C, Klassen L, Moote PE, Xiao Y, Thomas D, Pudlo NA, Anele A, Martens EC, Inglis GD, Uwiera RER, Boraston AB, Abbott DW (2018) Molecular basis of an agarose metabolic pathway acquired by a human intestinal symbiont. Nat Commun 9:1043

Pudlo NA, Pereira GV, Parnami J, Cid M, Markert S, Tingley JP, Unfried F, Ali A, Varghese NJ, Kim KS, Campbell A, Urs K, Xiao Y, Adams R, Martin D, Bolam DN, Becher D, Eloe-Fadrosh EA, Schmidt TM, Abbott DW, Schweder T, Hehemann JH, Martens EC (2022) Diverse events have transferred genes for edible seaweed digestion from marine to human gut bacteria. Cell Host Microbe 30:314–328 e311

Reeves AR, Wang GR, Salyers AA (1997) Characterization of four outer membrane proteins that play a role in utilization of starch by *Bacteroides thetaiotaomicron*. J Bacteriol 179:643–649

Reichardt N, Vollmer M, Holtrop G, Farquharson FM, Wefers D, Bunzel M, Duncan SH, Drew JE, Williams LM, Milligan G, Preston T, Morrison D, Flint HJ, Louis P (2018) Specific substrate-driven changes in human faecal microbiota composition contrast with functional redundancy in short-chain fatty acid production. ISME J 12:610–622

Saigusa M, Nishizawa M, Shimizu Y, Saeki H (2015) In vitro and in vivo anti-inflammatory activity of digested peptides derived from salmon myofibrillar protein conjugated with a small quantity of alginate oligosaccharide. Biosci Biotechnol Biochem 79:1518–1527

Salyers AA, Palmer JK, Wilkins TD (1977) Laminarinase (beta-glucanase) activity in *Bacteroides* from the human colon. Appl Environ Microbiol 33:1118–1124

Shipman JA, Berleman JE, Salyers AA (2000) Characterization of four outer membrane proteins involved in binding starch to the cell surface of *Bacteroides thetaiotaomicron*. J Bacteriol 182:5365–5372

Sichert A, Corzett CH, Schechter MS, Unfried F, Markert S, Becher D, Fernandez-Guerra A, Liebeke M, Schweder T, Polz MF, Hehemann JH (2020) *Verrucomicrobia* use hundreds of enzymes to digest the algal polysaccharide fucoidan. Nat Microbiol 5:1026–1039

Singh RP (2019) Glycan utilisation system in *Bacteroides* and *Bifidobacteria* and their roles in gut stability and health. Appl Microbiol Biotechnol 103:7287–7315

Singh RP, Reddy CR (2014) Seaweed-microbial interactions: key functions of seaweed-associated bacteria. FEMS Microbiol Ecol 88:213–230

Singh RP, Rajarammohan S, Thakur R, Hassan M (2020) Linear and branched b-glucans degrading enzymes from versatile *Bacteroides uniformis* JCM 13288^{T} and their roles in cooperation with gut bacteria. Gut Microbes 12:1–18

Singh RP, Bhaiyya R, Khandare K, Tingirikari JMR (2022) Macroalgal dietary glycans: potential source for human gut bacteria and enhancing immune system for better health. Crit Rev Food Sci Nutr 62:1674–1695

Singh RP, Niharika J, Thakur R, Wagstaff BA, Kumar G, Kurata R, Patel D, Levy CW, Miyazaki T, Field RA (2023) Utilization of dietary mixed-linkage b-glucans by the Firmicute *Blautia producta*. J Biol Chem. Accepted on 04/05/2023

Smits SA, Leach J, Sonnenburg ED, Gonzalez CG, Lichtman JS, Reid G, Knight R, Manjurano A, Changalucha J, Elias JE, Dominguez-Bello MG, Sonnenburg JL (2017) Seasonal cycling in the gut microbiome of the Hadza hunter-gatherers of Tanzania. Science 357:802–806

Sonnenburg ED, Sonnenburg JL, Manchester JK, Hansen EE, Chiang HC, Gordon JI (2006) A hybrid two-component system protein of a prominent human gut symbiont couples glycan sensing *in vivo* to carbohydrate metabolism. Proc Natl Acad Sci USA 103:8834–8839

Stender EGP, Dybdahl Andersen C, Fredslund F, Holck J, Solberg A, Teze D, Peters GHJ, Christensen BE, Aachmann FL, Welner DH, Svensson B (2019) Structural and functional aspects of mannuronic acid-specific PL6 alginate lyase from the human gut microbe *Bacteroides cellulosilyticus*. J Biol Chem 294:17915–17930
Tachedjian G, Aldunate M, Bradshaw CS, Cone RA (2017) The role of lactic acid production by probiotic *Lactobacillus* species in vaginal health. Res Microbiol 168:782–792
Terrapon N, Lombard V, Gilbert HJ, Henrissat B (2015) Automatic prediction of polysaccharide utilization loci in *Bacteroidetes* species. Bioinformatics 31:647–655
Ulaganathan T, Boniecki MT, Foran E, Buravenkov V, Mizrachi N, Banin E, Helbert W, Cygler M (2017) New ulvan-degrading polysaccharide lyase family: structure and catalytic mechanism suggests convergent evolution of active site architecture. ACS Chem Biol 12:1269–1280
Urbano MG, Goni I (2002) Bioavailability of nutrients in rats fed on edible seaweeds, Nori (*Porphyra tenera*) and Wakame (*Undaria pinnatifida*), as a source of dietary fibre. Food Chem 76:281–286
Urtuvia V, Maturana N, Acevedo F, Pena C, Diaz-Barrera A (2017) Bacterial alginate production: an overview of its biosynthesis and potential industrial production. World J Microbiol Biotechnol 33:198
Vera J, Castro J, Gonzalez A, Moenne A (2011) Seaweed polysaccharides and derived oligosaccharides stimulate defense responses and protection against pathogens in plants. Mar Drugs 9:2514–2525
Wang Y, Han F, Hu B, Li J, Yu W (2006) *In vivo* prebiotic properties of alginate oligosaccharides prepared through enzymatic hydrolysis of alginate. Nutr Res 26:597–603
Xu SY, Huang X, Cheong KL (2017) Recent advances in marine algae polysaccharides: isolation, structure, and activities. Mar Drugs 15
Yang CF, Lai SS, Chen YH, Liu D, Liu B, Ai C, Wan XZ, Gao LY, Chen XH, Zhao C (2019) Anti-diabetic effect of oligosaccharides from seaweed *Sargassum confusum* via JNK-IRS1/PI3K signalling pathways and regulation of gut microbiota. Food Chem Toxicol 131:110562
Yuan H, Song J, Li X, Li N, Dai J (2006) Immunomodulation and antitumor activity of kappa-carrageenan oligosaccharides. Cancer Lett 243:228–234
Yun EJ, Lee S, Kim HT, Pelton JG, Kim S, Ko HJ, Choi IG, Kim KH (2015) The novel catabolic pathway of 3,6-anhydro-L-galactose, the main component of red macroalgae, in a marine bacterium. Environ Microbiol 17:1677–1688
Yun EJ, Yu S, Park NJ, Cho Y, Han NR, Jin YS, Kim KH (2021) Metabolic and enzymatic elucidation of cooperative degradation of red seaweed agarose by two human gut bacteria. Sci Rep 11:13955
Zhang L, Liu C, Jiang Q, Yin Y (2021) Butyrate in energy metabolism: there is still more to learn. Trends Endocrinol Metab 32:159–169
Zhao C, Dong H, Zhang Y, Li Y (2019) Discovery of potential genes contributing to the biosynthesis of short-chain fatty acids and lactate in gut microbiota from systematic investigation in E. Coli. NPJ Biofilms Microbiomes 5:19
Zhou R, Shi XY, Bi DC, Fang WS, Wei GB, Xu X (2015) Alginate-derived oligosaccharide inhibits neuroinflammation and promotes microglial phagocytosis of beta-amyloid. Mar Drugs 13:5828–5846
Zhu B, Chen M, Yin H, Du Y, Ning L (2016) Enzymatic hydrolysis of alginate to produce oligosaccharides by a new purified endo-type alginate lyase. Mar Drugs 14
Zoetendal EG, Raes J, Van Den Bogert B, Arumugam M, Booijink CC, Troost FJ, Bork P, Wels M, De Vos WM, Kleerebezem M (2012) The human small intestinal microbiota is driven by rapid uptake and conversion of simple carbohydrates. ISME J 6:1415–1426

Chapter 11
Phytoremediation Potential of Seaweeds: Exploring Their Environmental Cleanup Capabilities

Sarah H. Rashedy and Amany G. Madkour

Abstract The phytoremediation potential of seaweeds has gained significant attention in recent years due to their remarkable ability to clean up and restore polluted environments. They are useful tools in phytoremediation because of their distinct characteristics and diverse mechanisms. This chapter aims to explore the environmental cleanup abilities of seaweeds and shed light on their potential applications in phytoremediation. Seaweeds have the ability to remove a variety of contaminants, including nutrients, heavy metals, organic pollutants, and radioactive materials. Furthermore, they have demonstrated efficacy in the removal and detoxification of contaminants from aquatic ecosystems, coastal areas, and industrial sites. There are many benefits to using seaweeds in phytoremediation, including their vast availability, low cultivation costs, and capacity to restore ecosystem balance. However, challenges such as proper species selection, site-specific considerations, and long-term monitoring need to be addressed for successful implementation.

Keywords Phytoremediation · Seaweeds · Nutrients · Heavy metals · Environmental contamination

Abbreviations

CFs	Concentration factors
IMTA	Integrated multitrophic aquaculture
PAHs	Polycyclic aromatic hydrocarbons
REEs	Rare earth elements

S. H. Rashedy (✉) · A. G. Madkour
National Institute of Oceanography and Fisheries (NIOF), Cairo, Egypt

N. Trivedi et al. (eds.), *Recent Advances in Seaweed Biotechnology*,
https://doi.org/10.1007/978-981-96-0519-4_11

11.1 Introduction

Bioremediation, as defined by Zhou et al. (2008), is a scientific technique that involves the assessment of the environment, including human exposure to both natural and synthetic chemicals. This approach relies on the sampling and analysis of tissues and fluids from individual organisms. By utilizing this technique, researchers can leverage the understanding that chemicals present in organisms can leave markers that indicate the extent of exposure. Phytoremediation is a natural, cost-effective, and environmentally friendly approach used to clean up contaminated soils and water (Kafle et al. 2022). It involves the use of plants and their associated microorganisms to remove, degrade, or immobilize pollutants from the environment. While phytoremediation has been widely studied for terrestrial plants, there is growing interest in exploring the potential of seaweeds for this purpose. Furthermore, algae are a prominent part of biological monitoring programs and play a crucial ecological role in many aquatic habitats (Omar 2010).

Seaweeds are prominent for their global presence and adaptability to various environments and hold immense potential as a natural resource for mitigating water pollution. They can serve as valuable source material for the removal of harmful contaminants from polluted water, offering notable advantages. Dried or fresh seaweeds are widely used for their capability to remediate different types of contamination, either in the field or in the laboratory (Mansour et al. 2022). The use of live organisms instead of dry biomass is advantageous because seaweeds exhibit rapid growth rates, ensuring a continuous and renewable supply of material for efficient pollutant removal (Baghour 2017). Seaweeds in their live form are recognized as highly effective biosorbents for various pollutants, they have confirmed proficiency in removing a range of contaminants, including nutrients (Madkour et al. 2020), polycyclic aromatic hydrocarbons (PAHs) (Nisha et al. 2019; Zhang et al. 2019), pesticides (Neveux et al. 2018), antimicrobials (Rosa et al. 2020), and heavy metals (Rahhou et al. 2023), among others (Arumugam et al. 2018). In this chapter, we give an overview of seaweed phytoremediation as an environmentally favorable method of environmental cleanup, advance knowledge of seaweed phytoremediation potential, and promote the use of seaweeds in sustainable cleanup efforts.

11.2 Mechanism of Phytoremediation by Seaweeds

Phytoremediation facilitated by seaweeds is a two-stage process (Fig. 11.1) that enables them to efficiently remove and mitigate pollutants from the environment. The process involves the following steps:

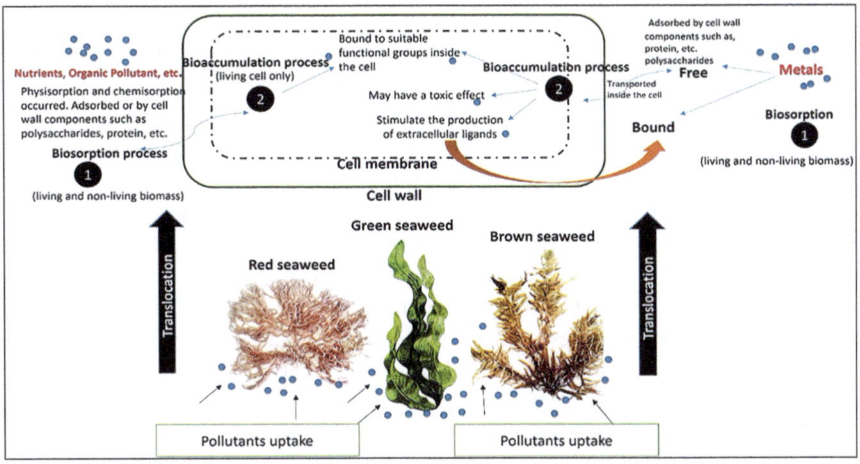

Fig. 11.1 Mechanism of phytoremediation by seaweed cell through biosorption and bioaccumulation stages

11.2.1 Biosorption (The First Stage)

Biosorption is a process by which seaweeds (living and non-living biomass) can remove and bind pollutants from their surrounding through passive concentration and bind contaminants onto their surfaces, primarily through physical and chemical interactions. (Neveux et al. 2018). According to Velásquez and Dussan (2009), biosorption is a passive process that does not require energy expenditure by the organism. The efficiency of biosorption depends on achieving kinetic equilibrium and the composition of the sorbent's cellular surface and the amount of contaminants that can be removed by a sorbent is determined by these factors.

Seaweeds possess various mechanisms that contribute to biosorption. The complex cell wall structure of seaweeds, composed of polysaccharides, proteins, and other biomolecules, provides numerous binding sites for pollutants. These binding sites can interact with pollutants through mechanisms such as adsorption, electrostatic forces, ion exchange, complexation, surface precipitation, hydrogen bonding, chelation, and van der Waals attraction (Arumugam et al. 2018). Among these, ion exchange is the most significant mechanism of biosorption (Sadhukhan et al. 2019).

Studies have demonstrated an adequate fit of experimental kinetic data to the pseudo-first-order model for the removal of pollutants (nitrogen, phosphorus, or metal) concluding that the biosorption process is made up of extracellular and intracellular transfer. The attraction of contaminants by active sites on the surface of biosorbents (physisorption) and chemical bonding (chemisorption) are categorically included in the extracellular transfer, whereas biotransformation and intracellular accumulation are included in intracellular transfer (Rathod et al. 2014) The data that the pseudo-second-order model fits the best also provide supporting evidence. The chemisorption including valence forces through the sharing or exchange of electrons between the biosorbent and adsorbate is described by a pseudo-second-order equation. Thus, data demonstrate that the process of pollutants biosorption onto seaweed may entail multiple steps, including chemisorptions (Ugurlu and Karaoglu 2011).

The biosorption of pollutants by seaweed biomass is influenced by various factors. These factors include the size and morphology of the cells, pH of the solution, and duration of contact between the biomass and pollutants, concentrations of cations and anions, temperature, metal speciation, and the physiological characteristics of the biomass. Additionally, the quantity of functional groups present in the algae matrix plays a significant role in determining the biosorption capacity (Bulgariu et al. 2015). Moreover, different macroalgal strains exhibit varying metal uptake capacities due to the differences in cell wall constituents among various species of macroalgae (Arumugam et al. 2018).

11.2.2 Bioaccumulation (The Second Stage)

Bioaccumulation is a cellular process (living biomass only) that involves the binding of toxic metals or organic substances within the internal structure of cells (Zabochnicka-Świątek and Krzywonos 2014). Seaweeds can bioaccumulate pollutants by actively concentrating and storing toxins inside their tissues. Pollutants are taken up and stored during this process in different cellular compartments such as cell walls and vacuoles. Velásquez and Dussan 2009 confirmed that bioaccumulation is an active metabolic process driven by energy from a living organism and requires respiration. Leitenmaier and Küpper (2013) findings indicate that in hyperaccumulators, stronger ligands are utilized for transient binding during the transport of metals to storage sites, whereas in non-hyperaccumulators, these ligands are known to complex metals. This confirms that hyperaccumulation is primarily driven by enhanced active metal transport rather than metal complexation. Hyperaccumulating plants store metals in vacuoles, which are devoid of enzymes such as phosphatases, lipases, and proteinases that are susceptible to heavy metal toxicity (Baghour 2017). Seaweeds exhibit the ability to accumulate trace metals, achieving concentrations thousands of times higher than those found in seawater. However, according to Henriques et al. (2015), bioaccumulation serves as an effective remediation process only if the contaminated water provides a suitable growth medium and does not exert critical toxic effects on cells. Following bioaccumulation, these pollutants undergo detoxification processes, either through their removal, transformation into non-toxic forms, or their complexation with other compounds. Detoxification can occur through various mechanisms such as compartmentalization, complexation, or other phenomena that facilitate the conversion of pollutants into less harmful substances.

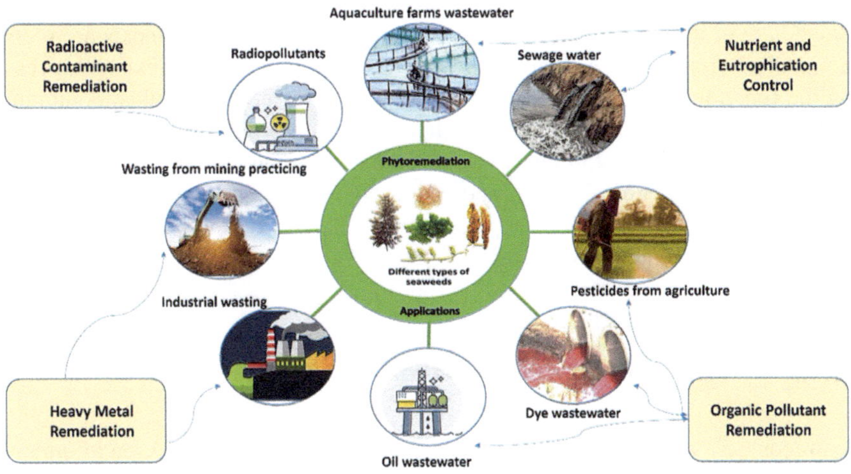

Fig. 11.2 Phytoremediation applications by seaweeds

11.3 Phytoremediation Applications of Seaweeds

Seaweeds possess remarkable abilities to absorb and assimilate a wide range of pollutants from water bodies, including nutrients, heavy metals, organic pollutants, and even radioactive contaminants (Fig. 11.2). Through processes like bioaccumulation, biosorption, and enzymatic degradation, they can effectively sequester and immobilize these harmful substances, reducing their impact on aquatic ecosystems and human health. Additionally, seaweeds play a crucial role in mitigating eutrophication, a phenomenon caused by excessive nutrient levels in water bodies, by efficiently absorbing nitrogen and phosphorus.

Phytoremediation applications can be summarized as follows:

11.3.1 Heavy Metal Remediation

Algae play a crucial ecological role in various aquatic ecosystems and are essential components of biological monitoring programs (Wan Maznah 2010). According to Baghour (2017), macroalgal species are effective biomonitors for contaminants in the dissolved phases, particularly heavy metals, while lipophilic contaminants are less readily absorbed due to their low lipid content. Seaweeds are utilized as bioindicators because of their widespread distribution, size, longevity, presence in pollution-affected areas, and their ability to accumulate metals efficiently, making them easily identifiable and valuable for water quality assessment (Stengel et al. 2004). In addition, Wan Maznah (2010) pointed out that seaweeds are especially well suited for water quality assessment as they exhibit rapid reproduction rates and short life cycles, allowing them to indicate short-term impacts. Furthermore, macroalgae are easily cultured in laboratories, and their sampling is inexpensive and minimally disruptive to resident biota. Standard evaluation methods exist for assessing the functional and non-taxonomic structural characteristics of algal communities (Baghour 2017). As shown above, the mechanism of removal of heavy metals by seaweeds can occur through two processes: bioaccumulation and biosorption (Fig. 11.1). Many brown, green, and red seaweeds (as listed in Table 11.1) are known as efficient heavy metal hyperaccumulators, making them valuable for bioremediation of polluted ecosystems.

11.3.2 Organic Pollutant Remediation

Seaweeds have demonstrated remarkable capabilities in the remediation of organic pollutants, showcasing their potential as efficient biofilters in aquatic environments. Organic pollutants, such as polycyclic aromatic hydrocarbons (PAHs), pesticides, dyes, phenolic compound, and various industrial contaminants, pose significant threats to marine ecosystems and human health. In recent decades, there has been a growing interest in studying the bioaccumulation of pesticide residues in seaweed. Researchers have found compelling evidence supporting the effectiveness of various seaweed species as biomonitors for this purpose (Sundhar et al. 2019; Banach et al. 2020). Nisha et al. (2019) evaluated the ability of certain seaweed species (*Sargassum wightii* (Greville) and *Gracilara corticata* (J. Agardh)) to sequester PAHs from contaminated waters, offering an eco-friendly approach to remediation. Similarly, Zokm et al. (2022) found that seaweeds can act as bioaccumulators of pesticides, reducing their concentrations and mitigating their impact on aquatic life. Besides, the active metabolic processes within seaweeds enable the breakdown and detoxification of organic pollutants, making them essential components in the restoration of polluted marine habitats. These findings underscore the importance of exploring seaweeds as natural tools in the ongoing efforts to combat organic pollutant contamination in our oceans.

11.3.3 Nutrient and Eutrophication Control

Seaweeds play a crucial role in nutrient and eutrophication control in aquatic ecosystems, offering valuable ecological services. As primary producers, they have the ability to absorb and utilize nutrients, including nitrogen and phosphorus, from the surrounding water. This process, known as nutrient uptake, helps to reduce nutrient levels in the water column, which is vital in mitigating eutrophication. Studies have demonstrated that seaweeds can effectively assimilate and store excess nutrients, preventing their release into the water and thus mitigating the negative impacts of eutrophication (Eriksson et al. 2018; Uku and Mohamed 2019). By removing

Table 11.1 Different seaweeds species for heavy metal removal

Seaweed	Type of algae	Metal ions	References
Sargassum sp.	Brown algae	Cd^{2+}	Sheng et al. (2004) Mahmood et al. (2017)
		Cu^{2+}	Karthikeyan et al. (2007) Barquilha et al. (2017)
		Ni^{2+}	Sheng et al. (2004) Barquilha et al. (2017)
		Pb^{2+}	Sheng et al. (2004)
		Zn^{2+}	Sheng et al. (2004) Mahmood et al. (2017)
		Cr	Yang and Chen (2008)
Sargassum filipendula	Brown algae	Ag^{+} and Cr^{3+}	Cardoso et al. (2017)
		Pb^{2+}	Verma et al. (2016) Cardoso et al. (2017)
		Cu^{2+}	Davis et al. (2000) Kleinübing et al. (2011) Cardoso et al. (2017)
		Cd^{2+}	Luna et al. (2010) Cardoso et al. (2017)
		Zn^{2+}	Luna et al. (2010) Cardoso et al. (2017)
		Ni^{2+}	Kleinübing et al. (2011) Cardoso et al. (2017)
Sargassum hystrix	Brown algae	Mn^{2+}	Ghasemi et al. (2016)
		Pb^{2+}	Jalali et al. (2002)
Sargassum muticum	Brown algae	Sb^{3+}	Ungureanu et al. (2015)
		Cd^{2+}	Davis et al. (2000)
		Cr	Bermúdez et al. (2012)
Sargassum ilicifolium	Brown algae	Pb^{2+}	Tabaraki et al. (2014)
Sargassum baccularia	Brown algae	Cd^{2+}	Hashim (2004)
Sargassum fluitans	Brown algae	Cu^{2+}	Davis et al. (2000)
		Cd^{2+}	Davis et al. (2000)
		Ni^{2+}	Holan and Volesky (1994)
Sargassum glaucescens	Brown algae	Ni^{2+}	Pahlavanzadeh et al. (2010)
Sargassum natans	Brown algae	Pb^{2+}	Jalali et al. (2002)
		Ni^{2+}	Holan and Volesky (1994)
Sargassum siliquosum	Brown algae	Cd^{2+}	Hashim (2004)

(continued)

Table 11.1 (continued)

Seaweed	Type of algae	Metal ions	References
Sargassum vulgare	Brown algae	Pb^{2+} and Ni^{2+}	Holan and Volesky (1994)
		Cu^{2+} and Cd^{2+}	Davis et al. (2000)
Aschophyllum nodosum	Brown algae	Sb^{3+}	Ungureanu et al. (2015)
		Pb^{2+}	Holan and Volesky (1994) Romera et al. (2007)
		Cu^{2+} and Zn^{2+}	Romera et al. (2007)
		Cd^{2+}	Romera et al. (2007) Lodeiro et al. (2005)
		Ni^{2+}	Romera et al. (2007) Holan and Volesky (1994)
Fucus vesiculosus	Brown algae	Pb^{2+}	Holan and Volesky (1994) Mata et al. (2008)
		Cu^{2+} and Cd^{2+}	Mata et al. (2008)
		Ni^{2+}	Holan and Volesky (1994)
		Cr	Murphy et al. (2008)
Fucus spiralis	Brown algae	Pb^{2+}, Cu^{2+}, Cd^{2+}, Ni^{2+}, and Zn^{2+}	Romera et al. (2007)
		Cr	Murphy et al. (2008)
Padina pavonia	Brown algae	Pb^{2+}	Jalali et al. (2002)
Padina sp.	Brown algae	Pb^{2+}, Cu^{2+}, Cd^{2+}, Zn^{2+}, and Ni^{2+}	Sheng et al. (2004)
Fucus serratus	Brown algae	Cu^{2+}	Ahmady-Asbchin et al. (2008)
Turbinaria turbinata	Brown algae	Cu^{2+} and Zn^{2+}	Madkour and dar (2021)
Ulva lactuca	Green algae	Pb^{2} Cd^{2+} Ni^{2+}	Jalali et al. (2002) Sari and Tuzen (2008a) Zakhama et al. (2011)
Cladophora glomerata	Green algae	Pb^{2+}	Jalali et al. (2002)
Ulva sp.	Green algae	Pb^{2+}, Cu^{2+}, Cd^{2+}, Zn^{2+}, and Ni^{2+}	Sheng et al. (2004)
Codium vermilara	Green algae	Pb^{2+}, Cu^{2+}, Cd^{2+}, Zn^{2+}, and Ni^{2+}	Romera et al. (2007)
Spirogyr ainsignis	Green algae	Pb^{2+}, Cu^{2+}, Cd^{2+}, Zn^{2+}, and Ni^{2+}	Romera et al. (2007)
Spirogyr neglecta	Green algae	Pb^{2+} Cu^{2+}	Romera et al. (2007) Singh et al. (2007)
Caulerpa lentillifera	Green algae	Pb^{2+}, Cu^{2+}, Cd^{2+}, and Zn^{2+}	Pavasant et al. (2006)
Spirogyra sp	Green algae	Pb^{2+}, Cu^{2+}, Cd^{2+}, and Zn^{2+}	Lee and Chang (2011) Rajfur et al. (2012)
Cladophora sp.	Green algae	Pb^{2+} and Cu^{2+}	Lee and Chang (2011)
Ulva fasciata	Green algae	Cu^{2+}	Karthikeyan et al. (2007)

(continued)

Table 11.1 (continued)

Seaweed	Type of algae	Metal ions	References
Chaetomorpha linum	Green algae	Cd^{2+}	Hashim (2004)
Oedogonium sp.	Green algae	Cd^{2+}	Gupta and Rastogi (2008)
Ulva rigida	Green algae	Sb^{3+}, Se^{4+}, and Se^{6+}	Filote et al. (2017)
Halimeda opuntia	Green algae	Cu^{2+} and Zn^{2+}	Madkour and dar (2021)
Ulva sp.	Green algae	Zn2+	Bădescu et al. (2017)
Gracilaria corticata	Red algae	Pb^{2+}	Jalali et al. (2002)
Gracilaria canaliculata	Red algae	Pb^{2+}	Jalali et al. (2002)
Polysiphonia violacea	Red algae	Pb^{2+}	Jalali et al. (2002)
Gracilaria sp.	Red algae	Pb^{2+}	Sheng et al. (2004)
		As, Fe, Ni, and Cd	Selvam et al. (2016)
Asparagopsis armata	Red algae	Pb^{2+}	Romera et al. (2007)
Jania rubens	Red algae	Pb^{2+}, Co^{2+}, and Cr	Ibrahim (2011)
Pterocladia capillacea	Red algae	Pb^{2+}, Co^{2+}, and Cr	Ibrahim (2011)
Corallina mediterranea	Red algae	Pb^{2+}, Co^{2+}, and Cr	Ibrahim (2011)
Galaxaura oblongata	Red algae	Pb^{2+}, Co^{2+}, and Cr	Ibrahim (2011)
Chondrus crispus	Red algae	Pb^{2+}	Romera et al. (2007)
Gelidium	Red algae	Pb^{2+}	Vilar et al. (2008)
Gracilaria changii	Red algae	Pb^{2+}	Hashim (2004)
Gracilaria edulis	Red algae	Pb^{2+}	Hashim (2004)
Gracilaria Salicornia	Red algae	Pb^{2+}	Hashim (2004)
Ceramium virgatum	Red algae	Pb^{2+} and Cd^{2+}	Sari and Tuzen (2008b)
Mastocarpus stellatus	Red algae	Pb^{2+}	Herrero et al. (2011)
Hypnea valentiae	Red algae	Pb^{2+}	Rathinam et al. (2010)
Palmaria palmate	Red algae	Cr	Murphy et al. (2008)
Polysiphonia lanosa	Red algae	Cr	Murphy et al. (2008)
Osmundea pinnatifida	Red algae	Cu^{2+} and Cd^{2+}	El Hassouni et al. (2014)
Chondracanthus chamissoi	Red algae	Pb^{2+} and Cd^{2+}	Yipmantin et al. (2011)

nutrients, seaweeds also compete with harmful algae for resources, inhibiting algal blooms and maintaining a healthier balance in the aquatic environment (Kim et al. 2020). Madkour et al. (2020) conducted a study on *Dictyota dichotoma* and found that this seaweed species efficiently assimilated ammonium and phosphorus, accounting for approximately 69.2–99.9% of available ammonium and 90.7–100% of available phosphate. The concentration of nutrients in the growth media significantly influenced the seaweed's growth rate. Notably, the nutritional composition, including lipids, carbohydrates, and proteins, increased with higher ammonium and phosphate levels. However, the researchers observed a decline in biomass at a concentration of 8/80 μM/L PO_4^{-3}/NH_4^+, as excessive nutrient levels proved toxic and hindered growth, leading to potential macroalgae mortality. This finding highlights the delicate balance required for optimal seaweed growth and underscores the importance of understanding nutrient dynamics in seaweed cultivation.

In addition to their role in nutrient uptake, seaweeds are often used in constructed wetlands and floating bioreactors to act as biofilters for wastewater treatment. They help remove pollutants, such as nitrogen and phosphorus, from effluent waters, contributing to the purification of polluted water bodies (Safferman et al. 2021). Tsagkamilis et al. (2010) conducted a comprehensive investigation on the potential application of seaweed for phosphate absorption in sewage treatment to combat eutrophication and improve water quality. The study involved data collection from both laboratory experiments and field studies, focusing on the sewage treatment system. Based on their findings, they proposed a continuous flow system with a water turnover rate of ¼ volume per hour. This system utilized a combination of 60% wastewater effluent and 40% seawater, with an initial biomass of *Ulva lactuca* at a concentration of 30 $g.L^{-1}$, replaced every 10 days. The results indicated that *U. lactuca* showed promising potential as a biofilter agent for efficient phosphate removal from wastewater effluent released by treatment plants.

Furthermore, the cultivation of seaweeds in aquaculture farms can be strategically designed to absorb excess nutrients from fish farming operations, reducing the environmental impact of aquaculture on surrounding ecosystems (Bhat and Mattoo 2018). The inclusion of seaweeds in integrated multitrophic aquaculture (IMTA) has been proposed as an environmentally sustainable approach for aquaculture production, providing a potential source of food and also aiding in aquatic bioremediation by efficiently eliminating inorganic nutrients from wastewater (Fleurence et al. 2012). Particularly, the integration of shrimp and green seaweed (*Ulva clathrate*) aquaculture has demonstrated a high efficiency in removing inorganic nutrients from water effluents (Copertino et al. 2009).

11.3.4 Radioactive Contaminant Remediation

The amount of radiopollutants transported into the sea through surface water leaching from contaminated soil or via river systems has increased due to a number of factors, including the deposition of pollutants from atmospheric contamination plumes, the deliberate and direct discharges of highly contaminated waters into the sea, and the transportation of radiopollutants into the sea. These sources contribute to the worsening saltwater pollution, which raises worries about potential effects on human health and marine ecosystems (Kanda 2013).

Radioactive contaminant remediation by seaweeds is an emerging field of study that explores the potential of seaweeds to remove radioactive substances from the environment. Seaweeds possess unique properties such as high surface area, complex cell wall structure, and natural ability to concentrate elements that allow them to efficiently accumulate and sequester various elements, including radionuclides, from water sources. Kawai et al. (2014) examined several macroalgal taxa, including *Saccharina* (*Laminaria*), *Eisenia*, *Pyropia* (*Porphyra*), *Monostroma*, and Gloiopeltis, as target species. However, the number of species included in the study was limited, and there were biases in terms of habitat, phylogenetic diversity, and life history, such as perennial, intertidal, subtidal, and annual species. Additionally, the measurement protocols used in the investigations varied, and the detection levels occasionally reached high values depending on the available facilities. As a result, comparing the data and understanding the overall trends of radioactive contamination in marine macroalgae and the variations in concentration factors (CFs) for radionuclides among different taxa posed a challenge.

Seaweeds play a crucial role as primary producers in coastal ecosystems, serving as a food source for various benthic animals and fish. Consequently, if these algae become contaminated with radioactive substances, the contaminants can be transferred to consumer animals through the food chain. This process of biological accumulation may lead to the concentration of radioactive contamination in higher trophic levels, posing potential risks to marine life and ecosystems (Kawai et al. 2014).

11.4 Factors Influencing Seaweeds' Phytoremediation Potential

The effectiveness of seaweeds in phytoremediation can be influenced by several factors that affect their ability to remove contaminants from the environment.

11.4.1 Seaweed Species

Different species of seaweeds have varying capabilities to accumulate and detoxify pollutants. Studies have shown variations in the phytoremediation potential of different seaweed taxa. According to recent studies, certain seaweed species have demonstrated remarkable capacity for absorbing light rare earth elements (REEs) from aquatic environments. For instance, *Ulva lactuca* showed a high absorption rate of approximately 60% for REEs, including yttrium (Y), lanthanum (La), cerium (Ce), praseodymium (Pr), neodymium (Nd), europium (Eu), gadolinium (Gd), terbium (Tb), and dysprosium (Dy) (Pinto et al. 2020, 2021). *Gracilaria gracilis* exhibited absorption capabilities for La, Ce, Pr, Gd, and Nd, while *Fucus vesiculosus* absorbed La, Ce, and Eu. Additionally, other seaweed species like *Ulva intestinalis*, *Osmundea pinnatifida*, and *Fucus spiralis* have also shown high potential for REE absorption from aqueous solutions, with bioconcentration factors of 2790, 1742, and 841, respectively (Fabre et al. 2021). Moreover, the red seaweed *Gracilaria gracilis* effectively removed 70% of 500 μg/L of various light REEs, including Y, Ce, Nd, Eu, and La, and further optimization led to the complete recovery of all these elements (Jacinto et al. 2018). Similar findings have been reported in various studies involving the difference in the efficiency of diverse species of seaweeds for phytoremediation of varied contaminants, such as heavy metals, nutrients, and organic pollutants (Baghour 2017; Nisha et al. 2019; Zokm et al. 2022).

Kawai et al. (2014) observed variations in the phytoremediation efficiency among seaweed species, which can be attributed to factors such as the structure of the cell wall, habitat preferences, and life history characteristics. Furthermore, variability in the genetic makeup of seaweed populations can influence their ability to tolerate and remediate pollutants. Additionally, epigenetic factors can play a role in how seaweeds respond to environmental stressors (Ferrari et al. 2023).

These unique traits of different seaweed species contribute to their diverse abilities in absorbing and removing contaminants from the environment. Understanding these differences is crucial for identifying the most suitable seaweed species for specific phytoremediation applications and optimizing their potential in environmental cleanup endeavors.

11.4.2 Environmental Factors

The growth of seaweeds is profoundly influenced by environmental factors. These factors include light availability, water temperature, nutrient concentration, salinity, and pH (El-Manaway and Rashedy 2022). Their influence on seaweed growth and metabolic activity can have a significant impact on their ability to effectively remediate pollutants. Furthermore, the presence of other contaminants in the environment can interact with the targeted pollutants and influence their uptake and accumulation by seaweeds. Some contaminants may compete for binding sites or interfere with cellular processes, affecting the overall phytoremediation efficiency (Farias et al. 2018).

11.4.3 Contaminant Type

The nature and concentration of pollutants in the environment can influence the uptake and accumulation ability of seaweeds. Some seaweeds may be more efficient in removing specific contaminants, such as heavy metals or organic pollutants (Arumugam et al. 2018) than others.

11.4.4 Seasonal Variations

Seasonal variations can have a significant impact on the efficiency of seaweeds in bioaccumulation and biosorption, as well as their overall growth and biomass production (Rashedy et al. 2022). In a study conducted by Madkour et al. (2019), the differential abilities of various macroalgae to accumulate specific heavy metals were investigated under local conditions. The results revealed clear seasonal fluctuations in the heavy metal contents of tested macroalgae, including *Caulerpa racemosa, Cystoseira myrica, Digenea simplex, Hormophysa cuneiformis, Padina pavonica,* and *Sargassum cinereum*. The highest levels of these metals in algal tissues were observed in spring and summer, while the lowest levels were recorded in winter. This variation may be attributed to the growth rates of different species, with the studied seaweeds flourishing the most during the summer months at the investigated sites.

11.4.5 Co-existing Species

The presence of other marine organisms, such as bacteria and other algae, can interact with seaweeds and influence their remediation capacity (Green-Gavrielidis et al. 2018).

Understanding these factors and their interplay is crucial for harnessing the full phytoremediation potential of seaweeds and optimizing their use in environmental cleanup strategies.

11.5 Conclusions and Future Perspectives

Seaweeds offer a promising and eco-friendly solution for phytoremediation, effectively removing various environmental pollutants such as heavy metals, nutrients, organic pollutants, and radioactive contaminants. Their rapid growth and adaptability to diverse marine environments make them valuable candidates for sustainable remediation strategies. Future advancements may involve exploring new seaweed species with enhanced phytoremediation capabilities and scaling up cultivation for large-scale projects. Addressing logistical challenges and establishing regulatory guidelines will be crucial for successful implementation. Seaweeds are poised to play a key role in combating pollution and creating a cleaner, more sustainable environment.

References

Ahmady-Asbchin S, Andrès Y, Gérente C, Cloirec PL (2008) Biosorption of Cu (II) from aqueous solution by *Fucus serratus*: surface characterization and sorption mechanisms. Bioresour Technol 99(14):6150–6155. https://doi.org/10.1016/j.biortech.2007.12.040

Arumugam N, Chelliapan S, Kamyab H, Thirugnana S, Othman N, Nasri NS (2018) Treatment of wastewater using seaweed: a review. Int J Environ Res Public Health 15(12):2851. https://doi.org/10.3390/ijerph15122851

Bădescu IS, Bulgariu D, Bulgariu L (2017) Alternative utilization of algal biomass (*Ulva* sp.) loaded with Zn(II) ions for improving of soil quality. J Appl Phycol 29(2):1069–1079. https://doi.org/10.1007/s10811-016-0997-y

Baghour M (2017) Effect of seaweeds in phyto-remediation. In: Biotechnological applications of seaweeds. Nova Publishers Sciences, Hauppauge, pp 47–83. ISBN 978-1-5361-0983-2

Banach JL, Hoek-van den Hil EF, van der Fels-Klerx HL (2020) Food safety hazards in the European seaweed chain. Compr Rev Food Sci Food Saf 19:332–364. https://doi.org/10.1111/1541-4337.12523

Barquilha CER, Cossich ES, Tavares CRG, Silva EA (2017) Biosorption of nickel(II) and copper(II) ions in batch and fixed-bed columns by free and immobilized marine algae *Sargassum* sp. J Clean Prod 150:58–64. https://doi.org/10.1016/j.jclepro.2017.02.199

Bermúdez YG, Rico ILR, Guibal E, de Hoces MC, Martín-Lara MÁ (2012) Biosorption of hexavalent chromium from aqueous solution by *Sargassum muticum* brown alga. Application of statistical design for process optimization. Chem Eng J 183:68–76. https://doi.org/10.1016/j.cej.2011.12.022

Bhat MA, Mattoo FA (2018) Seaweed culture: an eco-friendly approach for nutrient remediation from marine ecosystem. J Algal Biomass Util 9(2):218–224

Bulgariu L, Bulgariu D, Rusu C (2015) Springer handbook of marine biotechnology. Springer, Berlin, Heidelberg, pp 611–648

Cardoso SL, Costa CSD, Nishikawa E, da Silva MGC, Vieira MGA (2017) Biosorption of toxic metals using the alginate extraction residue from the brown algae *Sargassum filipendula* as a natural ion-exchanger. J Clean Prod 165:491–499. https://doi.org/10.1016/j.jclepro.2017.07.114

Copertino M, Tormena T, Seeliger U (2009) Biofiltering efficiency, uptake and assimilation rates of *Ulva clathrata* (Roth) J. Agardh (Clorophyceae) cultivated in shrimp aquaculture wastewater. J Appl Phycol 21:31–45. https://doi.org/10.1007/s10811-008-9357-x

Davis TA, Volesky B, Vieira RHS (2000) *Sargassum* seaweed as biosorbent for heavy metals. Water Res 34(17):4270–4278. https://doi.org/10.1016/s0043-1354(00)00177-9

El Hassouni H, Abdellaoui D, El Hani S, Bengueddour R (2014) Biosorption of cadmium(II) and copper(II) from aqueous solution using red alga (*Osmundea pinnatifida*) biomass. J Mater Environ Sci 5(4):967–974. https://www.jmaterenvironsci.com/Journal/vol5-4.html

El-Manaway IM, Rashedy SH (2022) The ecology and physiology of seaweeds: an overview. In: Ranga Rao A, Ravishankar GA (eds) Sustainable global resources of seaweeds, vol 1. Springer, Cham. https://doi.org/10.1007/978-3-030-91955-9_1

Eriksson BK, Bergström L, Johansson G (2018) Eutrophication-induced shifts in a Baltic Sea macroalgal community. Aquat Bot 145:21–28

Fabre E, Henriques B, Viana T, Pinto J, Costa M, Ferreira N, Tavares D, Vale C, Pinheiro-Torres J, Pereira E (2021) Optimization of Nd (III) removal from water by *Ulva* sp. and *Gracilaria* sp. through response surface methodology. J Environ Chem Eng 9:105946

Farias DR, Hurd CL, Eriksen RS, Macleod CK (2018) Macrophytes as bioindicators of heavy metal pollution in estuarine and coastal environments. Mar Pollut Bull 128:175–184. https://doi.org/10.1016/j.marpolbul.2018.01.023

Ferrari M, Muto A, Bruno L, Cozza R (2023) DNA methylation in algae and its impact on abiotic stress responses. Plants (Basel) 12(2):241. https://doi.org/10.3390/plants12020241

Filote C, Ungureanu G, Boaventura R, Santos S, Volf I, Botelho C (2017) Green macroalgae from the Romanian coast of Black Sea: physico-chemical characterization and future perspectives on their use as metal anions biosorbents. Process Saf Environ Prot: Trans Inst Chem Eng Part B 108:34–43. https://doi.org/10.1016/j.psep.2016.06.002

Fleurence M, Morançais M, Dumay J, Decottignies P, Turpin V, Munier M, Garcia-Bueno N, Jaouen P (2012) What are the prospects for using seaweed in human nutrition and for marine animals raised through aquaculture? J Trends in Food Science & Technology 27:57–61. https://doi.org/10.1016/j.tifs.2012.03.004

Ghasemi FF, Dobaradaran S, Keshtkar M, Mohammadi MJ, Ghaedi H, Soleimani F (2016) Biosorption of MN (II) from aqueous solution by *Sargassum hystrix* algae obtained from the

Persian Gulf: biosorption isotherm and kinetic. Int J Pharm Technol 8(3):18227–18238. http://eprints.bpums.ac.ir/7307

Green-Gavrielidis LA, MacKechnie F, Thornber CS, Gomez-Chiarri M (2018) Bloom-forming macroalgae (*Ulva* spp.) inhibit the growth of co-occurring macroalgae and decrease eastern oyster larval survival. Mar Ecol Prog Ser 595:27–37. https://doi.org/10.3354/meps12556

Gupta VK, Rastogi A (2008) Equilibrium and kinetic modelling of cadmium (II) biosorption by nonliving algal biomass *Oedogonium sp.* from aqueous phase. J Hazard Mater 153(1–2):759–766. https://doi.org/10.1016/j.jhazmat.2007.09.021

Hashim M (2004) Biosorption of cadmium by brown, green, and red seaweeds. Chem Eng J 97:249–255. https://doi.org/10.1016/s1385-8947(03)00216-x

Henriques B, Rocha LS, Lopes CB, Figueira P, Monteiro RJR, Duarte AC, Pardal MA, Pereira E (2015) Study on bioaccumulation and biosorption of mercury by living marine macroalgae: prospecting for a new remediation biotechnology applied to saline waters. Chem Eng J 281:759–770

Herrero R, Lodeiro P, García-Casal LJ, Vilariño T, Rey-Castro C, David C, Rodríguez P (2011) Full description of copper uptake by algal biomass combining an equilibrium NICA model with a kinetic intraparticle diffusion driving force approach. Bioresour Technol 102(3):2990–2997. https://doi.org/10.1016/j.biortech.2010.10.007

Holan ZR, Volesky B (1994) Biosorption of lead and nickel by biomass of marine algae. Biotechnol Bioeng 43(11):1001–1009. https://doi.org/10.1002/bit.260431102

Ibrahim WM (2011) Biosorption of heavy metal ions from aqueous solution by red macroalgae. J Hazard Mater 192(3):1827–1835. https://doi.org/10.1016/j.jhazmat.2011.07.019

Jacinto J, Henriques B, Duarte AC, Vale C, Pereira E (2018) Removal and recovery of Critical Rare Elements from contaminated waters by living *Gracilaria gracilis*. J Hazard Mater 344:531–538

Jalali R, Ghafourian H, Asef Y, Davarpanah SJ, Sepehr S (2002) Removal and recovery of lead using nonliving biomass of marine algae. J Hazard Mater 92(3):253–262. https://doi.org/10.1016/s0304-3894(02)00021-3

Kafle A, Timilsina A, Gautam A, Adhikari K, Bhattarai K, Aryal N (2022) Phytoremediation: mechanisms, plant selection and enhancement by natural and synthetic agents. Environ Adv 8:100–203. https://doi.org/10.1016/j.envadv.2022.100203

Kanda J (2013) Continuing ^{137}Cs release to the sea from the Fukushima Dai-ichi Nuclear Power Plant through 2012. Biogeosci Discuss 10:3577–3595. https://doi.org/10.5194/bgd-10-3577-2013

Karthikeyan S, Balasubramanian R, Iyer CSP (2007) Evaluation of the marine algae *Ulva fasciata* and *Sargassum sp.* for the biosorption of Cu(II) from aqueous solutions. Bioresour Technol 98(2):452–455. https://doi.org/10.1016/j.biortech.2006.01.010

Kawai H, Kitamura A, Mimura M, Mimura T, Tahara T, Aida D, Sato K, Sasaki H (2014) Radioactive cesium accumulation in seaweeds by the Fukushima 1 Nuclear Power Plant accident-two years' monitoring at Iwaki and its vicinity. J Plant Res 127(1):23–42. https://doi.org/10.1007/s10265-013-0603-1

Kim JK, Kim MJ, Lee SJ (2020) Potential utilization of seaweeds for nutrient reduction in fish farming effluents: a laboratory-scale study. Ocean Sci J 55(2):299–307

Kleinübing SJ, da Silva EA, da Silva MGC, Guibal E (2011) Equilibrium of Cu(II) and Ni(II) biosorption by marine alga *Sargassum filipendula* in a dynamic system: competitiveness and selectivity. Bioresour Technol 102(7):4610–4617. https://doi.org/10.1016/j.biortech.2010.12.049

Lee YC, Chang SP (2011) The biosorption of heavy metals from aqueous solution by *Spirogyra* and *Cladophora filamentous* macroalgae. Bioresour Technol 102(9):5297–5304. https://doi.org/10.1016/j.biortech.2010.12.103

Leitenmaier B, Küpper H (2013) Compartmentation and complexation of metals in hyperaccumulator plants. Front Plant Sci 4:374

Lodeiro P, Cordero B, Barriada JL, Herrero R, Sastre de Vicente ME (2005) Biosorption of cadmium by biomass of brown marine macroalgae. Bioresour Technol 96(16):1796–1803. https://doi.org/10.1016/j.biortech.2005.01.002

Luna AS, Costa ALH, da Costa ACA, Henriques CA (2010) Competitive biosorption of cadmium(II) and zinc(II) ions from binary systems by *Sargassum filipendula*. Bioresour Technol 101(14):5104–5111. https://doi.org/10.1016/j.biortech.2010.01.138

Madkour AG, Dar MA (2021) Biosorption of Cu and Zn in a batch system via dried macroalgae *Halimeda opuntia* and *Turbinaria turbinata*. Environ Res Eng Manag 77(1):76–84. https://doi.org/10.5755/j01.erem.77.1.27952

Madkour AG, Rashedey SH, Dar MA (2019) Spatial and temporal variations of heavy metals accumulation in some macroalgal flora of the Red Sea. Egypt J Aquat Biol Fish 23(4):539–549. https://doi.org/10.21608/ejabf.2019.60548

Madkour A, Rashedy S, Touliabah H (2020) Bioremediation of ammonium and phosphates and their effect on the macroalgae *Dictyota dichotoma* (Hudson) lamouroux native to the Red Sea Coast, Egypt. Egypt J Bot 60(3):671–680. https://doi.org/10.21608/ejbo.2020.21997.1430

Mahmood Z, Zahra S, Iqbal M, Raza MA, Nasir S (2017) Comparative study of natural and modified biomass of *Sargassum* sp. for removal of Cd^{2+} and Zn^{2+} from wastewater. Appl Water Sci 7(7):3469–3481. https://doi.org/10.1007/s13201-017-0624-3

Mansour AT, Alprol AE, Abualnaja KM, El-Beltagi HS, Ramadan KMA, Ashour M (2022) Dried brown seaweed's phytoremediation potential for methylene blue dye removal from aquatic environments. Polymers 14:1375

Mata YN, Blázquez ML, Ballester A, González F, Muñoz JA (2008) Characterization of the biosorption of cadmium, lead and copper with the brown alga *Fucus vesiculosus*. J Hazard Mater 158(2–3):316–323. https://doi.org/10.1016/j.jhazmat.2008.01.084

Murphy V, Hughes H, McLoughlin P (2008) Comparative study of chromium biosorption by red, green and brown seaweed biomass. Chemosphere 70(6):1128–1134. https://doi.org/10.1016/j.chemosphere.2007.08.015

Neveux N, Bolton JJ, Bruhn A, Roberts DA, Ras M (2018) The bioremediation potential of seaweeds: recycling nitrogen, phosphorus, and other waste products. Blue Biotechnol Prod Use Mar Mol 1:217–241

Nisha LL, Arthanareeswaran G, Poonguzhali TV, Mohan TA, Valentina J (2019) Phycoremediation of hydrocarbon using two marine seaweeds from the Bay of Bengal coast of India. Desalin Water Treat 156:378–386. https://doi.org/10.5004/dwt.2019.24284

Omar WM (2010) Perspectives on the use of algae as biological indicators for monitoring and protecting aquatic environments, with special reference to Malaysian freshwater ecosystems. Trop Life Sci Res 21(2):51–67

Pahlavanzadeh H, Keshtkar AR, Safdari J, Abadi Z (2010) Biosorption of nickel (II) from aqueous solution by brown algae: equilibrium, dynamic and thermodynamic studies. J Hazard Mater 175(13):304–310. https://doi.org/10.1016/j.jhazmat.2009.10.004

Pavasant P, Apiratikul R, Sungkhum V, Suthiparinyanont P, Wattanachira S, Marhaba TF (2006) Biosorption of Cu^{2+}, Cd^{2+}, Pb^{2+}, and Zn^{2+} using dried marine green macroalga *Caulerpa lentillifera*. Bioresour Technol 97(18):2321–2329. https://doi.org/10.1016/j.biortech.2005.10.032

Pinto J, Henriques B, Soares J, Costa M, Dias M, Fabre E, Lopes CB, Vale C, Pinheiro-Torres J, Pereira EA (2020) Green method based on living macroalgae for the removal of rare-earth elements from contaminated waters. J Environ Manag 263:110376

Pinto J, Costa M, Henriques B, Soares J, Dias M, Viana T, Ferreira N, Vale C, Pinheiro-Torres J, Pereira E (2021) Competition among rare earth elements on sorption onto six seaweeds. J Rare Earths 39:734–741

Rahhou A, Layachi M, Akodad M, El Ouamari N, Rezzoum NE, Skalli A, Oudra B, El Bakali M, Kolar M, Imperl J (2023) The bioremediation potential of *Ulva lactuca* (Chlorophyta) causing green tide in Marchica lagoon (NE Morocco, Mediterranean Sea): biomass, heavy metals, and health risk assessment. Water 15(7):1310. https://doi.org/10.3390/w15071310

Rajfur K, Kłos A, Wacławek M (2012) Sorption of copper (II) ions in the biomass of alga *Spirogyra* sp. Bioelectrochemistry 87:65–70. https://doi.org/10.1016/j.bioelechem.2011.12.007

Rashedy S, El-Mahdy S, El-Manawy I, Pereira L (2022) Spatial and temporal variations of macroalgal vegetation in the north-western Red Sea. Bot Mar 65(6):405–418. https://doi.org/10.1515/bot-2022-0046

Rathinam A, Maharshi B, Janardhanan SK, Jonnalagadda RR, Nair BU (2010) Biosorption of cadmium metal ion from simulated wastewaters using *Hypnea valentiae* biomass: a kinetic and thermodynamic study. Bioresour Technol 101(5):1466–1470. https://doi.org/10.1016/j.biortech.2009.08.008

Rathod M, Mody K, Basha S (2014) Efficient removal of phosphate from aqueous solutions by red seaweed, *Kappaphycus alverezii*. J Clean Prod 84:484–493

Romera E, González F, Ballester A, Blázquez ML, Muñoz JA (2007) Comparative study of biosorption of heavy metals using different types of algae. Bioresour Technol 98(17):3344–3353. https://doi.org/10.1016/j.biortech.2006.09.026

Rosa J, Leston S, Crespo D, Freitas A, Pouca ASV, Barbosa J, Lemos MFL, Pardal M, Ramos F (2020) Uptake of enrofloxacin from seawater to the macroalgae *Ulva* and its use in IMTA systems. Aquaculture 516:734609

Sadhukhan J, Gadkari S, Martinez-Hernandez E, Siew D, Shemfe M, Torres-Garciac E, Lyncha J (2019) Novel macroalgae (seaweed) biorefinery systems for integrated chemical, protein, salt, nutrient and mineral extractions and environmental protection by green synthesis and life cycle sustainability assessments. Green Chem 21:2635–2655

Safferman SI, Ferris MJ, Ivanov V (2021) Nutrient removal from industrial wastewater using seaweed systems: a review. J Environ Manag 292:112841

Sari A, Tuzen M (2008a) Biosorption of Pb (II) and Cd(II) from aqueous solution using green alga (*Ulva lactuca*) biomass. J Hazard Mater 152(1):302–308. https://doi.org/10.1016/j.jhazmat.2007.06.097

Sari A, Tuzen M (2008b) Biosorption of cadmium (II) from aqueous solution by red algae (*Ceramium virgatum*): equilibrium, kinetic and thermodynamic studies. J Hazard Mater 157(2):448–454. https://doi.org/10.1016/j.jhazmat.2008.01.008

Selvam SB, Chelliapan S, Din MFM, Shahperi R, Aris MAM (2016) Adsorption of heavy metals from matured leachate by *Gracilaria* Sp. extract. Res J Pharm Biol Chem Sci 7(3):15–20. https://www.rjpbcs.com/2016_7.3.html

Sheng PX, Ting YP, Chen JP, Hong L (2004) Sorption of lead, copper, cadmium, zinc, and nickel by marine algal biomass: characterization of biosorptive capacity and investigation of mechanisms. J Colloid Interface Sci 275(1):131–141. https://doi.org/10.1016/j.jcis.2004.01.036

Singh RK, Chavan SL, Sapkale PH (2007) Heavy metal concentrations in water, sediments and body tissues of red worm (*Tubifex* spp.) collected from natural habitats in Mumbai, India. Environ Monit Assess 129(3):471–481. https://doi.org/10.1007/s10661-006-9377-4

Stengel DB, Macken A, Morrison L, Morley N (2004) Zinc concentrations in marine macroalgae and a lichen from western Ireland in relation to phylogenetic grouping, habitat and morphology. Mar Pollut Bull 48:902–909

Sundhar S, JeyaShakila R, Jeyasekaran G (2019) Safety assessment of edible red seaweeds *Gracilaria* and *Gelidiella* of Gulf of Mannar in terms of OCP residual contamination. Environ Nanotechnol Monit Manag 12:100245

Tabaraki R, Nateghi A, Ahmady-Asbchin S (2014) Biosorption of lead (II) ions on *Sargassum ilicifolium*: application of response surface methodology. Int Biodeterior Biodegradation 93:145–152. https://doi.org/10.1016/j.ibiod.2014.03.022

Tsagkamilis P, Danielidis D, Dring MJ, Katsaros C (2010) Removal of phosphate by the green seaweed *Ulva lactuca* in a small-scale sewage treatment plant (Ios Island, Aegean Sea, Greece). J Appl Phycol 22(3):331–339

Ugurlu M, Karaoglu MH (2011) Adsorption of ammonium from an aqueous solution by fly ash and sepiolite: isotherm, kinetic and thermodynamic analysis. Microporous Mesoporous Mater 139:173–178

Uku JN, Mohamed CAR (2019) Seaweed community structure and eutrophication level in a tropical environment: a case study of Diani Beach, Kenya. West Indian Ocean J Mar Sci 17(2):115–124

Ungureanu G, Santos S, Boaventura R, Botelho C (2015) Biosorption of antimony by brown algae *S. muticum* and *A. nodosum*. Environ Eng Manag J 14(2):455–463. http://www.eemj.icpm.tuiasi.ro/issues/vol14/vol14no2.htm

Velásquez L, Dussan J (2009) Biosorption and bioaccumulation of heavy metals on dead and living biomass of *Bacillus sphaericus*. J Hazard Mater 167:713–716

Verma A, Kumar S, Kumar S (2016) Biosorption of lead ions from the aqueous solution by *Sargassum filipendula*: equilibrium and kinetic studies. J Environ Chem Eng 4(4):4587–4599. https://doi.org/10.1016/j.jece.2016.10.026

Vilar VJP, Botelho CMS, Boaventura RAR (2008) Copper removal by algae *Gelidium*, agar extraction algal waste and granulated algal waste: kinetics and equilibrium. Bioresour Technol 99(4):750–762. https://doi.org/10.1016/j.biortech.2007.01.042

Wan Maznah WO (2010) Perspectives on the use of algae as biological indicators for monitoring and protecting aquatic environments, with special reference to Malaysian freshwater ecosystems. Trop Life Sci Res 21:51–67

Yang L, Chen JP (2008) Biosorption of hexavalent chromium onto raw and chemically modified *Sargassum* sp. Bioresour Technol 99(2):297–307. https://doi.org/10.1016/j.biortech.2006.12.021

Yipmantin A, Maldonado HJ, Taulemesse JM, Guibal E (2011) Pb(II) and Cd(II) biosorption on *Chondracanthus chamissoi* (a red alga). J Hazard Mater 185(23):922–929. https://doi.org/10.1016/j.jhazmat.2010.09.108

Zabochnicka-Świątek M, Krzywonos M (2014) Potentials of biosorption and bioaccumulation processes for heavy metal removal. Pol J Environ Stud 23:551–561

Zakhama S, Dhaouadi H, Henni FM (2011) Nonlinear modelisation of heavy metal removal from aqueous solution using *Ulva lactuca* algae. Bioresour Technol 102(2):786–796. https://doi.org/10.1016/j.biortech.2010.08.107

Zhang C, Lu J, Wu J (2019) Adsorptive removal of polycyclic aromatic hydrocarbons by detritus of green tide algae deposited in coastal sediment. Sci Total Environ 670:320–327

Zhou Q, Zhang J, Fu J, Shi J, Jiang G (2008) Biomonitoring: an appealing tool for assessment of metal pollution in the aquatic ecosystem. Anal Chim Acta 606:135–150

Zokm GME, Ismail MM, Okbah MAE (2022) Seaweed as bioindicators of organic micropollutants polycyclic aromatic hydrocarbons (PAHs) and organochlorine pesticides (OCPs). Environ Sci Pollut Res Int 29(23):34738–34748. https://doi.org/10.1007/s11356-022-18634-z

Chapter 12
Brown Seaweed Biomass as Potential Raw Material for Biorefinery

Anurag A. K. Sharma and Ravi S. Baghel

Abstract More than 1800 brown seaweed species that come under Phaeophyta are known to have importance in the seaweed industry as well as from an ecological point of view. Traditionally, brown seaweed biomass was used for a long time to produce alginate. However, in the past few years, it has gained recognition as an important raw material to produce several other products, such as plant biostimulants, minerals, PUFAs, cellulose, and proteins. This chapter presents a brief account of brown seaweeds, starting with their introduction, composition, extraction process, products, and applications. The biorefinery approaches developed for the sustainable utilization of brown seaweed biomass were also discussed in detail.

Keywords Alginate · Fucoidans · Seaweed · Process · Products

12.1 Introduction

Brown seaweeds, taxonomically classified under Phaeophyceae, are a diverse group of marine algae that have garnered considerable importance due to their potential commercial and ecological values. With over 1800 recognized species and many more yet to be discovered, brown seaweeds offer a wide array of opportunities for research and utilization in cosmetics, nutraceuticals, pharmaceuticals, and even as functional foods (Lutfia et al. 2020). Brown seaweeds belong to the phylum Ochrophyta and are primarily found in cold-temperate to temperate marine waters on shallow rocky shores around the world (Peteiro 2018). They encompass four main classes: Fucales, Laminariales, Ectocarpales, and Dictyotales, each characterized by unique morphological and ecological features. The Fucales, commonly known as the "rockweeds," are often found in intertidal zones, while the Laminariales,

A. A. K. Sharma · R. S. Baghel (✉)
Biological Oceanography Division, CSIR-National Institute of Oceanography, Dona Paula, Panaji, Goa, India
e-mail: rsbaghel@nio.org

N. Trivedi et al. (eds.), *Recent Advances in Seaweed Biotechnology*,
https://doi.org/10.1007/978-981-96-0519-4_12

or kelps, form vast underwater forests in subtidal areas. The Dictyotales, on the other hand, are smaller and typically found in warm tropical waters. The Ectocarpales contains filamentous algae and is widely distributed in estuarine and marine habitats of temperate regions in both hemispheres (Bringloe et al. 2020; Montecinos et al. 2016). The diverse morphology of brown seaweeds includes macroscopic thalli with various growth forms such as erect, branched, or sheet-like structures. The pigmentation of brown seaweeds is due to the presence of the pigment fucoxanthin, which gives them their characteristic brown color. This pigment also plays a role in light harvesting and protection against harmful UV radiation. Brown seaweeds play vital ecological roles in marine ecosystems. They are primary producers, utilizing photosynthesis to convert sunlight into organic matter. Through this process, they contribute to the transfer of energy and nutrients within the marine food web and even act as a shelter for marine organisms (de Mesquita et al. 2019). Additionally, brown seaweeds form extensive underwater forests, providing habitat, shelter, and nursery areas for a wide range of marine organisms. These forests support high biodiversity and promote ecosystem stability (de Mesquita et al. 2019).

Brown seaweeds also have a significant impact on coastal ecosystems by acting as ecosystem engineers. Their dense structures attenuate wave energy, reduce coastal erosion, and stabilize sediments. Furthermore, brown seaweeds can influence local biogeochemical cycles by absorbing and releasing nutrients, such as nitrogen and phosphorus, thus influencing nutrient availability for other organisms (Krause-Jensen and Duarte 2014). The chemical composition of brown seaweeds is highly diverse and varies among species, locations, and environmental conditions. The major constituents of brown seaweed biomass include carbohydrates, proteins, lipids, and minerals. Carbohydrates, particularly laminarin and alginate, are the most abundant macromolecules in brown seaweeds. Laminarin, a storage polysaccharide, can constitute 15–35% of the dry weight of brown seaweeds, while alginate, a complex cell wall polysaccharide, can represent up to 30%. These carbohydrates have shown potential in various applications, including food additives, pharmaceuticals, and as a source of fermentable sugars for biofuel production (Li et al. 2021). Proteins are yet another significant component of brown seaweed biomass, typically comprising 10–25% of the dry weight. These proteins possess diverse functional properties and can be hydrolyzed to release bioactive peptides with potential health benefits. These bioactive peptides have shown antioxidant, antimicrobial, and anti-inflammatory activities, making them of interest in the development of functional foods and nutraceuticals (Begum et al. 2021). Lipids, are present in lower quantities (0.3–4.5% of the dry weight) and it is also been reported that cold-water species have higher lipid content than tropical species have and are of interest due to their potential as a source of valuable fatty acids and for biodiesel production. Brown seaweeds accumulate various minerals, including iodine, potassium, calcium, and magnesium, which contribute to their nutritional value and functional properties (Miyashita et al. 2013).

The diverse chemical composition of brown seaweeds has led to their exploration and utilization in various industries. In the food industry, brown seaweeds are used as ingredients in the production of functional foods, dietary supplements, and food additives due to their nutritional value and bioactive compounds (Gupta and Abu-Ghannam 2011). In medicine and healthcare, brown seaweeds have demonstrated promising bioactivities, including antioxidant, antimicrobial, antiviral, and anticancer properties (Baghel et al. 2023). They have been used traditionally in Asian cultures for their health-promoting effects and are now being investigated for their potential as a source of novel pharmaceutical compounds (Wijesekara et al. 2011). Further, brown seaweeds have gained attention as a renewable and sustainable resource for the production of biofuels and biochemical (Baghel 2023). The carbohydrates present in brown seaweed biomass can be converted into bioethanol, biogas, and platform chemicals through various fermentation and conversion processes. The lipids can be extracted and processed into biodiesel, offering a renewable alternative to fossil fuels (Alvarado-Morales et al. 2013; Hwangbo and Chu 2020). In agriculture, brown seaweeds are used as organic fertilizers and soil conditioners due to their rich mineral content and ability to enhance soil structure, water retention, and nutrient availability. Their application in agriculture can contribute to sustainable and eco-friendly farming practices (Khan et al. 2009). Furthermore, using seaweed as a feed source for livestock has been shown to have positive impacts on animal health, performance, and product quality. Seaweeds are also considered a sustainable alternative to conventional feed sources, as they contain a variety of nutrients and bioactive compounds and can be grown without the use of freshwater or arable land (Baghel et al. 2020). Additionally, seaweed-based feed additives have been shown to reduce methane emissions from ruminants, which could have significant environmental benefits (Ahmed et al. 2022). The impact of brown seaweeds on rumen fermentation is species-dependent and can be influenced by factors such as the inclusion level of macroalgae in the diet and the content of phlorotannins. Some studies have found that the inclusion of brown seaweeds in ruminant diets can reduce the production of total VFA, while others have observed a decrease in fiber degradability due to the presence of phlorotannins (Targett and Arnold 1998; Wang et al. 2009).

Looking ahead, there are exciting prospects for the utilization of brown seaweed biomass. Advances in biotechnology, metabolic engineering, and process optimization will contribute to enhancing the efficiency and sustainability of brown seaweed-based industries. Moreover, the integration of seaweed cultivation with other industries, such as aquaculture or wastewater treatment, can provide additional gains of economic and ecological importance. Brown seaweeds represent a valuable marine resource with diverse applications in various industries. Their taxonomy, ecological roles, chemical composition, and potential applications in food, medicine, and biorefinery processes highlight the significance of further research enhancing the gainful utilization of this resource. Understanding the biology, ecology, and chemistry of brown seaweeds is essential for sustainable management and harnessing their social, environmental, and economic potential in a rapidly changing world.

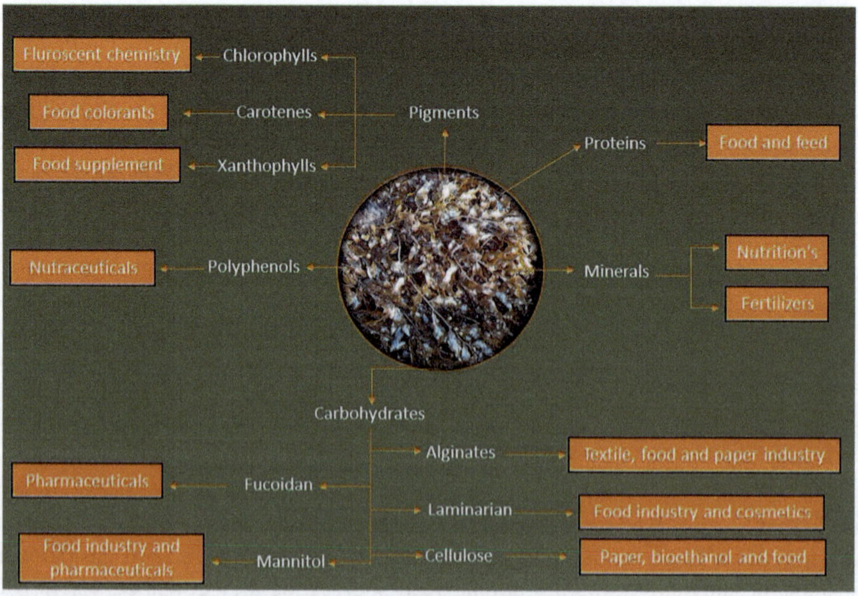

Fig. 12.1 Brown seaweed components and their applications

12.2 Components of Brown Seaweeds

Brown seaweeds are rich in a chemical composition having various components other than alginate and fucoidans such as pigments, proteins, cellulose, mannitol, lipids, and minerals which could be used in various sectors such as food, feed, nutraceuticals, agriculture, and pharmaceuticals (Fig. 12.1).

12.2.1 Pigments

Brown seaweeds feature a vast diversity of biosynthetic pigments, which gives them their distinctive coloring. (a) chlorophylls a and c; (b) xanthophylls, which is considered one of the most prevalent families of the pigments in these species, including violaxanthin and fucoxanthin as the major compounds; and (c) carotenes, are mainly represented by β-carotene and α-carotene. Various combinations of these pigments therefore cause the natural coloring of algae. For instance, it is crucial to note that brown algae, which are part of the Phaeophyceae family, are the main source of fucoxanthin and chlorophyll c (Garcia-Perez et al. 2022).

12.2.1.1 Chlorophylls

Among the chlorophylls, chlorophyll a, chlorophyll a isomer 1, and chlorophyll a isomer 2 are three different forms of chlorophyll a, another type of chlorophyll pigment. Chlorophyll a is the most common type of pigment widely distributed in land plants and algae. It absorbs light most efficiently at wavelengths around 430 and 662 nm. The three different forms of chlorophyll a that were found in brown algae have slightly different chemical structures, which may affect their function. This bioactive compound is widely used in the pharmaceutical, cosmetic, and food industries (da Silva Ferreira and Sant'Anna 2017).

Chlorophyll c is further divided into two isomers chlorophyll c2 and chlorophyll c1 that are the chlorophylls responsible for capturing light energy during photosynthesis. Chlorophyll c2 is a green pigment that absorbs light most efficiently at wavelengths around 470 nm, while chlorophyll c1 absorbs light most efficiently at wavelengths around 625 nm. As a nutraceutical agent having antioxidant, anti-inflammatory, antimutagenic, and antibacterial effects, chlorophyll has been taken for health reasons (da Silva Ferreira and Sant'Anna 2017).

12.2.1.2 Fucoxanthin

Fucoxanthin is the most abundant type of xanthophyll pigment that gives its characteristic brown color and represents 10% of the total carotenoids (Mohibbullah et al. 2022). Fucoxanthin has been shown to have various health benefits, including anti-inflammatory and anticancer properties. It also plays an important role in protecting the alga from excess light energy by transferring absorbed energy to other pigments (Lourenço-Lopes et al. 2021). Violaxanthin and auroxanthin are two other types of xanthophyll pigments found in brown algae. Xanthophylls play an important role in protecting the plant or alga from excess light energy by absorbing it and transferring it to other pigments for safe dissipation. Violaxanthin has absorption peaks at 418, 440, and 470 nm, while auroxanthin with absorption peaks at 380, 400, and 425 nm (Garcia-Perez et al. 2022). Fucoxanthin derivative is a derivative of fucoxanthin, which means it has a slightly different chemical structure than the original pigment. Its exact function in brown algae is not well understood, but it may play a role in protecting the algae from excess light energy. It absorbs light most efficiently at the wavelength 442 nm and a slight signal at 658 nm, a similar compound is reported also in microalgae *Tisochrysis lutea* (Gallego et al. 2020). Dihydrolutein and Zeaxanthin are two additional types of xanthophyll pigments found in brown algae. Like other xanthophylls, they help protect the plant or algae from excess light energy by absorbing it and transferring it to other pigments for safe dissipation.

12.2.1.3 β-Carotene

β-Carotene is a type of carotenoid pigment found exclusively in brown algae. It gives two major peaks at 450 and 476 nm. Carotenes is used as a coloring agent, antioxidant, and vitamin-A supplement. By absorbing light energy, carotenes function as photoreceptors in Phaeophyceae photosystems, driving photosynthetic electron transport. Additionally, these carotenoid pigments shield chloroplasts from too much light as well as reactive by-products produced during oxygenic photosynthesis.

12.2.2 Minerals

Seaweed-derived minerals market is valued at USD 2814.6 million in 2022 and is anticipated to reach USD 4455.3 million by 2032 (https://www.futuremarketinsights.com/reports/seaweed-derived-minerals-market, assessed on September 26, 2023). Brown seaweeds typically accumulate minerals from seawater, and their mineral content and quantity vary with species, location, environmental conditions, and their life span. Seaweeds are the richest source of vitamins like B-12, C, D, E, and K and some seaweed's calcium content is even more than that of milk (Vijay et al. 2017). K was the most abundant macroelement, present maximum in the *cystoseira indica* with 90.7 mg/g, and the lowest K was present in *Sargassum asperifolium* with 12 mg/g DW. Potassium was followed by sodium (Na) in most of the seaweeds, but it can be more than potassium, for example, *Spatoglossum asperum* Na content is 88.76 mg/g DW, whereas its K content is just 29.4 mg/g DW. Among the microelements present in seaweeds, Fe was the most abundant trace element and was the highest in *S. polycystum* with 277.60 mg/100 g DW and the lowest in *Sargassum tenerrimum* with 19.88 mg/100 g DW, followed by Zn and microelements like cadmium, cobalt, and molybdenum were present in the least amounts of <0.001 mg/g DW (Fouda et al. 2019; Kumar et al. 2011; Nazarudin et al. 2021). The species, wave exposure, physiological and environmental effects, as well as the processing and mineralization techniques, all affect the mineral fraction (Salehi et al. 2019). The mineral content also shows variation depending upon the seasons, for example, *Undaria pinnatifida* shows the highest mineral content in autumn with 14–54.8 g kg^{-1} of potassium, 2.7–7.0 g kg^{-1} of phosphate, 361–569 mg kg^{-1} of strontium, 62–140 mg kg^{-1} of iron, 8–103 mg kg^{-1} of zinc, 45–94 mg kg^{-1} of bromine, and 4–24 mg kg^{-1} of rubidium (Salomone and Riera 2020).

Brown seaweeds are known for their abundance of other minerals such as calcium, magnesium, potassium, iron, aluminum, and copper. These minerals play vital roles in various important physiological processes, including bone health, muscle function, nerve transmission, and oxygen transport. Calcium and potassium are the most prominent minerals found in brown seaweeds and species like *Padina gymnospora* and *Sargassum asperifolium* have been found to contain substantial amounts of calcium, contributing to their potential role in promoting bone health (Fouda et al. 2019; Tabarsa et al. 2012). Brown seaweeds are also rich in trace elements, including zinc, selenium, copper, and manganese. These trace elements are essential for enzyme function, immune system support, and antioxidant defense. Studies have indicated that brown seaweeds, such as *Undaria pinnatifida* and *Sargassum fusiforme*, exhibit significant trace element content, highlighting their potential as dietary sources for these essential micronutrients (Salomone and Riera 2020; Zhu et al. 2022). Minerals extracted from brown seaweeds exhibit a diverse range of applications in various industries due to their rich composition and bioactive properties. In agriculture, seaweed biostimulant-containing minerals serve as effective biofertilizers and soil conditioners, enhancing crop growth, nutrient uptake, and abiotic stress tolerance (Khan et al. 2009). Biostimulant market was about $ 2.5 billion in 2022 and from 2023 to 2030, the global seaweed extracts biostimulant market is predicted to expand rapidly at a CAGR of 10% (global seaweed new and emerging markets report 2023). The nutraceutical and food sectors utilize seaweed-derived minerals, such as iodine, calcium, magnesium, and iron, to enhance the nutritional value and sensory attributes of functional foods and dietary supplements (Gupta and Abu-Ghannam 2011). Additionally, these minerals find application in the cosmetics and personal care industry, where they contribute to the formulation of products with antioxidant, moisturizing, and antiaging properties, benefiting skin health and protection against UV radiation (Salehi et al. 2019).

12.2.3 Polysaccharides

12.2.3.1 Alginate

Alginate is a naturally occurring high molecular weight polysaccharide found abundantly in Phaeophyceae's cell wall. It is composed of guluronic acid (G) and mannuronic acid (M) residues that are (1→4)-linked linear co-polymer of β D-mannuronate and α-L-guluronate residues (Morris et al. 1980). Out of the total carbohydrate content, brown seaweeds contain 18.9–66.7% (Li et al. 2021) but this percentage varies according to the season and the depth at which they grow. Due to its adaptable qualities and numerous uses in the food, pharmaceutical, and biomedical industries, alginate has attracted a lot of attention. One of the alginate's key characteristics is its ability to form gels in the presence of divalent cations like calcium. Alginate is a fantastic thickening, stabilizing, and gelling agent in food items because of its gel-forming ability. It is widely used in the food industry for applications such as the encapsulation of flavors and nutrients, as well as in the formulation of gels, dressings, and confectionery (Qin et al. 2018). The global alginate market size was valued at USD 728.4 million in 2020 and is expected to grow at a compound annual growth rate (CAGR) of 5.0% from 2021 to 2028 (Chudasama et al. 2021). The market is expected to witness significant growth owing to the increasing product demand in the food applications as a thickening agent, emulsifier, and gelling agent. Increased acceptance from the food industry and higher demand for natural ingredients is the key drivers for product demand. Moreover, alginate exhibits biocompatibility and biodegradability, making it suitable for various pharmaceutical and biomedical applications. It has been utilized in wound dressings, drug delivery systems, tissue engineering scaffolds, and dental materials. The unique properties of alginate, including its ability to form hydrogels, provide controlled release of drugs and promote tissue regeneration (Puscaselu et al. 2020). Studies have also shown that alginate possesses potential health benefits, such as prebiotic effects, cholesterol-lowering properties, and anti-inflammatory activity. The consumption of alginate-rich brown seaweeds has been associated with improved gut health, reduced cholesterol levels, and modulation of the immune system (Liu et al. 2019). The applications and properties of alginate are highly dependent on factors such as the source of seaweed, extraction methods, and purification techniques. Further research and development are ongoing to explore the full potential of alginate and its derivatives for a wide range of industrial and biomedical applications.

12.2.3.2 Fucoidan

Brown seaweeds contain sulfated polysaccharides called fucoidans, which are renowned for their potential health advantages. Fucoidans are classified into three types based on their molecular weight, that is, low molecular weight fucoidan (LMWF) which is less than 10 kDa, middle molecular weight fucoidan (MMWF) that ranges between 10 and 10,000 kDa, and high molecular weight fucoidan (HMWF) which are more than 10,000 kDa (Senthilkumar et al. 2013). These complex molecules are composed of a backbone of (1→3)-linked -L-fucopyranose residues or alternating (1→4)-linked -L-fucopyranose and (1→3)-linked -L-fucopyranose residues and makes up to 3.94–8% of the total carbohydrate content (Li et al. 2021; van Weelden et al. 2019). Fucoidans also contain monosaccharides such glucose, galactose, xylose, mannose, arabinose, and fucose. In addition, it might also contain uronic acid, acetyl groups, and even proteins (Sanjeewa et al. 2017). Extensive examination has yielded conclusive findings that the fucose content of fucoidans falls within a broad spectrum of 4.45–84%. Additionally, it was observed that galactose constitutes 1.44–49% of the composition, glucose ranges from 0.2% to 45%, and xylose and mannose account for 0.3–16% of the total content, for example, the fucoidan found in *Ascophyllum nodosum* is composed of 52.1% fucose, glucose 21.3%, 16.5% xylose, 19% sulfate, and 6.1% galactose content. *Undaria pinnatifida* fucoidan is composed of 44.6% galactose, 50.9% fucose, 4.2% xylose, 0.3% mannose, and uronic acids which were not detected (Reyes et al. 2020).

Due to the significant differences in fucoidan structure because of its sugar composition, monosaccharide linkage, and sulfate position, they are diverse, and their structures may depend on the season, location, and growth stage and even the species of the seaweed (Fletcher et al. 2017). Fucoidans have been found to exhibit a variety of biological actions, including anti-inflammatory, antitumor, immunomodulatory, and antioxidant characteristics. They have shown promise in various areas of research, such as cancer therapy, cardiovascular health, immune system regulation, and wound healing (Wang et al. 2019). One of the significant aspects of fucoidans is their potential anticancer activity. Research has demonstrated that fucoidans can inhibit tumor growth, induce apoptosis in cancer cells, and suppress angiogenesis. Fucoidans, derived from brown seaweeds such as *Undaria pinnatifida*, have shown promising results in preclinical and clinical studies (Garg et al. 2019; van Weelden et al. 2019). Moreover, fucoidans have also exhibited immunomodulatory effects by regulating the activity of immune cells and modulating the production of immune mediators. They can enhance the immune response against pathogens and exert anti-inflammatory effects, making them potential candidates for the development of immunotherapeutic agents (Apostolova et al. 2020; Sanjeewa et al. 2017). It is important to note that the bioactivity and effectiveness of fucoidans can also vary depending on factors such as the extraction process, purification methods, and structural characteristics.

12.2.3.3 Laminarian

Laminaran is a type of storage polysaccharide found in brown seaweeds (Phaeophyceae) that serves as a reserve of energy for the seaweed. Laminaran exists in two forms, one soluble in cold water and the other insoluble form. It is a small size polysaccharide containing 20–30 glucosyl residues and is a linear β-1,3-glucan with occasional β-1,6-glucosidic linkages and it represents a total of 5.39–35% of algal dry weight (Li et al. 2021). The percentage of laminarian depends on Phaeophyceae species, extraction methods, and harvesting season, as they are found to be absent in spring, but in autumn and winter they are found to be maximum. Particularly in *L. digitata* and *L. saccharina*, the *Laminariales* genus is known to produce the highest amounts of laminarins, with contents reaching up to 35% of total dry weight. Other reported values of laminarin contents include those of *U. pinnatifida, A. nodosum, A. esculenta, and F. serratus*, with 3%, 4.5%, 11.1%, and up to 19% of total dry weight, respectively (Fletcher et al. 2017; Li et al. 2021). One of the main characteristics of laminaran is its solubility in cold form and it forms a viscous solution, and its solubility also increases when the degree of branching increases, and its functionality and properties make it suitable for various applications in the pharmaceutical and food industries. Laminaran has been studied for its potential health benefits and bioactive properties. By binding to receptors including complement receptor 3 and the glucan receptor, as well as to dectin-1 upon macrophages and white blood cells, it has been found to exhibit antibacterial and anticancer activity. This information sheds new light on how the innate immune system recognizes glucans (Karuppusamy et al. 2022; Sanjeewa et al. 2017). Additionally, laminarin is said to strengthen immunity, promote health, and prevent and treat diseases by selectively boosting the growth of advantageous gut bacteria through its impact on intestinal pH and short-chain fatty acid synthesis (Karuppusamy et al. 2022). Due to its unique properties, laminaran has been utilized as a functional ingredient in food products, including as a thickening agent, emulsifier, and stabilizer. The global laminarin market was USD two million in 2021 and market is projected to touch USD four million by 2028, exhibiting a compound annual growth rate (CAGR) of 5.0% from 2022 to 2030 (https://www.businessresearchinsights.com/ assessed on September 26, 2023). It has also been investigated for its possible use in immunomodulation by encouraging human skin fibroblast cells to adhere to one another for the purpose of healing wounds and by boosting the proliferation of human osteoblast cells for the purpose of forming bones (Kalasariya et al. 2021).

12.2.3.4 Cellulose

Cellulose is a natural organic polymer extracted mainly from lignocellulosic biomass of terrestrial origin. The annual production of cellulose is estimated at 180 billion tons, and the market demand for cellulose and its derivatives is constantly increasing (Baghel et al. 2021). The only crystalline polysaccharide identified in the walls of Phaeophyceae is cellulose and accounts for 1–8% on dry weight basis with an average molecular weight of 100,000 and a degree of polymerization (DP) of about 10,000. Cellulose is made up of a linear chain of several hundred to many thousands of β (1→4) connected D-glucose units. Strong alkali cannot harm it, but acid can quickly hydrolyze it into water-soluble sugars.

Cellulose contents show seasonal variations with a difference of 1–1.5% depending upon the season (Li et al. 2021). Brown seaweeds contain varying amounts of cellulose, which contributes to the structural integrity of the cell walls. However, the cellulose content in brown seaweeds is generally lower compared to land plants. The presence of cellulose in brown seaweeds provides mechanical strength and plasticity to the thallus, allowing it to withstand environmental stressors, such as wave action and tidal currents (Popper et al. 2011). Although cellulose in brown seaweeds has been less extensively studied compared to other polysaccharides, it holds potential for various applications. Cellulose has garnered significant interest from researchers over an extended period and finds widespread utilization across diverse applications owing to its plentiful accessibility, cost-effectiveness, and simplified processability (Motaung and Linganiso 2018). Researchers have explored the extraction and utilization of cellulose from brown seaweeds for the development of sustainable materials, including bioplastics, films, composites paper preparation, and nanocelluloses. The unique properties of cellulose, such as its biodegradability and biocompatibility, make it an attractive alternative to synthetic materials (Baghel et al. 2021).

One significant drawback in utilizing cellulose derived from seaweed is its comparatively lower cellulose content compared to lignocellulosic biomass. Although seaweed cellulose exhibits similar properties to that of cellulose obtained from lignocellulosic sources, the absence or minimal presence of lignin in seaweed biomass offers advantages. Consequently, there exists a vast opportunity to extract cellulose from seaweed and explore for value chain development, particularly when considering the increasing global production of seaweeds (Baghel et al. 2021).

12.2.4 Polyphenols

Polyphenols are water-soluble bioactive compounds known for their diverse health-promoting properties. Brown seaweeds are rich sources of polyphenols, including phlorotannins, a unique class of polyphenols specific to marine algae. Polyphenols are present in the cell vacuoles of algae, and their contents show seasonal variations and can vary within the parts of the thalli, such as the frond or basal part and even the old versus new thallus. These polymerized phloroglucinol units are used to create the chains of 1,3,5-trihydroxybenzene that make up these phlorotannins, are produced by the polyketide pathway, also known as the acetate-malonate pathway. Among the brown seaweeds highest content of phenolics was detected *Ecklonia bicyclis* that is 87 mg PCE/100 g, and the lowest content of phenols was detected in *Turbinaria conoides* that is 0.5 mg GAE/g. These are extremely hydrophilic and molecular size ranges from molecular sizes between 126 and 650 kDa (Charoensiddhi et al. 2020; Mekinić et al. 2019). Polyphenols have high biological activity, which makes brown algae a popular ingredient in nutraceutical and cosmeceutical products (Sanjeewa et al. 2016). Polyphenols have been proven to have antioxidative, antibacterial, anticancer, antiallergic, antidiabetes, antiaging, anti-inflammatory, antigrazing, and anti-HIV characteristics (Hakim and Patel 2020). They have been found to scavenge free radicals, protect against oxidative stress, and inhibit the growth of cancer cells (Wijesekara et al. 2011). Studies have shown that polyphenols from brown seaweeds can have positive impacts on human health. They have been investigated for their potential benefits in the prevention and management of various diseases, such as cardiovascular diseases, diabetes, obesity, and neurodegenerative disorders (Baldrick et al. 2018).

12.2.5 Proteins

Rising demand for plant-based foods results in an increase in global seaweed protein market size. The global seaweed protein industry was valued at $585 million in 2022 and is projected to reach $1.51 billion by 2030, growing at a CAGR of 14% from 2022 to 2030 (https://www.marketsandmarkets.com/Market-Reports/seaweed-protein-market-194056009.html#:~:text=The%20seaweed%20protein%20market%20sector,of%20%24585%20million%20in%202022. assessed on September 28, 2023). Brown seaweeds have protein contents as low as 2.81% of dry weight in *Cystoseira myrica* and as high as 24% of dry weight in *Sargassum* sp. and *Undaria* sp. (Fouda et al. 2019; O' Brien et al. 2022). Brown seaweeds are a trustworthy supply of necessary amino acids, but it has been claimed that essential amino acid levels are low to non-existent like methionine, and lysine amino acids found in phaeophyceae (O' Connor et al. 2020). The protein content shows variation depending upon the season and environmental factors such as light, temperature, and (Nazarudin et al. 2021). It has been documented that typically, the protein content within brown seaweeds exhibits higher levels during winter compared to summer due to the developmental phase of the seaweed plant, leading to enhanced photosynthetic activity. However, it is noteworthy to mention that protein content in *Sargassum wightii* was recorded highest in the summer, when compared to protein content in spring and winters (Kumar et al. 2015). The difference in the growth and protein production of seaweeds can be attributed to the assimilation of essential nutrients, particularly nitrogen, which predominantly takes place in the winter season where it has been proposed that this accumulation of nitrogen reserves is to support the rapid growth rates of seaweeds during summertime and where a strong link between nitrogen and protein content has also been found (Schiener et al. 2015). The protein composition of Phaeophyceae is diverse and complex, encompassing a wide range of proteins with varying functions and structures. These proteins play essential roles in different biological processes within the seaweeds, including growth, development, reproduction, and defense against environmental stressors. Brown seaweeds extracted proteins are involved in various biological processes, such as antioxidant, signal-modulating effects, and stress responses (Begum et al. 2021). A protein-rich red seaweed *Himanthalia elongata* was utilized to increase the phytochemical content of white wheat flour and even as a functional component in meat products to extend shelf life (O' Brien et al. 2022).

12.2.6 Mannitol

Mannitol is a sugar alcohol found abundantly in brown seaweeds, playing a crucial role in their physiology and biochemistry (Saravana et al. 2016). It serves as a major osmolyte, helping brown seaweeds to regulate water balance and cope with changes in salinity. Mannitol also acts as a carbon and energy storage compound, enabling brown seaweeds to endure periods of stress or unfavorable conditions (Dittami et al. 2011). The presence of mannitol in brown seaweeds has attracted interest due to its potential applications in various industries, including food, pharmaceuticals, and cosmetics. The versatile properties of mannitol make it valuable in various applications. In the food industry, mannitol is utilized as a low-calorie sweetener, texturizing agent, and bulking agent. In pharmaceuticals, it serves as an excipient in drug formulations and as a stabilizer in vaccines (Grembecka 2015; Tajima et al. 2018).

12.3 Tradition Extraction and Characterization Methods for Commercial Products

Commercially alginates and fucoidans are extracted from the brown seaweed biomass using different extraction methods:

12.3.1 Alginate Extraction and Characterization

Alginates were traditionally isolated from the biomass of dried alginophytic seaweed. Dry samples are first depigmented with formaldehyde and then washed. Then, dilute acid was used to treat the material before hydrolyzing it with Na_2CO_3 solution. Following the Na_2CO_3 hydrolysis, hydrolyzed material was filtered through muslin cloth to separate filtrate and residues. Alginates were precipitation from the filtrate using either alcohol, Cacl2 based on requirements and then dried to get the alginates. The Fourier transform infrared (FT-IR) spectroscopy is the basic method to characterize the alginates. One of the key characteristics of alginates is the presence of mannuronic acid and glucuronic acid residues, which can be identified using FT-IR analysis.

12.3.2 Fucoidan Extraction and Characterization

Fucoidan is commonly extracted from raw algal material by heating it in hot aqueous or acidic solutions for many hours at temperatures between 70 and 100 °C. After heating, the aqueous and residual parts are separated. To further isolate the fucoidan, a precipitation process using organic solvents or surfactants that specifically interact with sulfated polysaccharides is often performed immediately after extraction (Khairy and El-Shafay 2013; Rodríguez-Montesinos et al. 2013). The target polysaccharide undergoes the elimination of salts along with other small molecules using ethanol precipitation of fucoidans. Further anion-exchange chromatography is used to purify the fucoidans. The positions of the sulfates in the fucoidan structure can be determined using FT-IR spectroscopy. In the wavenumber 1500–700 cm^{-1} ranges, the majority of the information pertaining to the position of sulfate groups may be found. The overall amount of sulfate esters in polysaccharides is said to be represented by the wide signal at 1250–1260 cm^{-1} (–S–O antisymmetric stretching vibration of the sulfate group). The equatorial position of the sulfate ester's binding to the fucose to generate sulfate fucose is characterized by the signal at 820–825 cm^{-1}.

12.4 Biorefinery Approaches

In order to maximize the usage of biomass, intensive research has recently been carried out to develop integrated biorefining processes. First, approaches have been created that preserve alginate, one of the most crucial commercial products, while allowing for the extraction of additional useful by-products without compromising on the functional integrity of each product. The second category of approaches involves turning the entire biomass into many compounds without caring about alginate.

12.4.1 Extraction of Alginate with Other Co-Products

In recent years, there are several articles reporting the extraction of alginate along with the different co-products using different extraction methods and techniques (Table 12.1). As a median attempt, *Ascophyllum nodosum* biomass was valorized by preparing 4 products 14% of fucoidan, 11% of sugars, 21% of biochar along with 18% of alginate (Yuan and Macquarrie 2015) (Fig. 12.2a). In this process, about 65% of biomass was converted into products. In the first step, biomass subjected to ethanol extraction that removes the salts, pigments, mannitol, proteins present in the biomass. Residues leftover from ethanolic extraction were added with 0.1 M HCl and given thermal treatment at 90 °C for 15 min using microwave to extract fucoidan. Alginate was extracted from residues left from fucoidan extraction using 0.1 M Na_2CO_3 treatment following the precipitation with ethanol. At last, residues remained after alginate extraction was hydrolyzed using H_2SO_4 to get sugar. Baghel et al. (2020) demonstrated an integrated process for the valorization of brown seaweed, *Sargassum tenerrimum* biomass (Fig. 12.2b). In the first step, fresh biomass was crushed and separated into liquid fractions, sap, and solid parts. The solid part was bleached using $NaClO_2$ and then hydrolyzed using Na_2CO_3 to recover alginate. The residues left after alginate recovery were treated with NaOH to get protein and finally at last the cellulose. Also, effluents generated during processing were neutralized to make salt. By processing 1 kg of fresh biomass, 541 ml of sap, 32 g of alginic acid, 3.8 g of protein, 10 g of cellulose, and 115 g of salt were prepared. Overall, about 93% of the biomass was converted into products. Also, using the same *A. nodosum* biomass six different products (laminarin, mannitol, alginate, acetic acid, formic acid, and levulinic acids) were prepared with a set of treatments such as solvent extraction, acid extraction, microwave-assisted non-catalyzed (water only), acid-catalyzed (H_2SO_4), and metal cation-catalyzed (Pb II and Al III) (Tedesco et al. 2021). *Sargassum muticum* biomass was used to make alginate, and fucoidan/phloroglucinol, and 65% of the biomass was converted into the products (Flórez-Fernández et al. 2019). The biomass was subjected to non-isothermal autohydrolysis (150–170 °C) with water and then centrifuged to separate solid and liquid phases. Further, the liquid fraction was precipitated with $CaCl_2$ to separate the alginate and fucoidan/phloroglucinol-rich liquid fraction. The same biomass was valorized by producing four products that are sap, calcium alginate, biogas, and mineral amendment with mechanical and chemical treatments, autohydrolysis, and anaerobic digestion (Flórez-Fernández et al. 2021). The 100 gm of defrosted biomass was pressed to remove sap containing 20 gm of dry solids contained alginate, phlorotannins, minerals, and growth factors. Press solids obtained from pressing were subjected to autohydrolysis and filtered to get filtrate and solid residues. The calcium alginate was precipitated from filtrate, while the aqueous extract was found to be rich in sulfated fucoidan and phlorotannins. The solid residues were used to prepare mineral amendment and biogas by anaerobic digestion. Using abovementioned process, 4–5 additional products could be produced along with major commercial product alginate. Also, a high conversion rate (up to 93%) of biomass to products

Table 12.1 Integrated biorefinery processes for the preparation of alginates along with co-products

S. no.	Species	Techniques/treatments	Products	Reference
1	*Ascophyllum nodosum*	Thermal treatment, alcoholic extraction, acid, and alkali treatments	Sugars, fucoidan, alginate, and biochar	Yuan and Macquarrie (2015)
2	*Sargassum muticum*	Non-isothermal autohydrolysis	Alginate, fucoidan/ phloroglucinol	Flórez-Fernández et al. (2019)
3	*Durvillaea potatorum*	Acid treatment, alkali extraction	Alginate, laminarin, and fucoidan	Abraham et al. (2019)
4	*Sargassum tenerrimum*	Mechanical treatment, alkali treatment, acidic treatment	Sap, alginate, cellulose, and protein	Baghel et al. (2020)
5	*Carpophyllum flexuosum, Undaria pinnatifida, Ecklonia radiata, C. plumosum*	Ethanolic extraction, aqueous extraction, alkali extraction	Pigments, carbohydrate, mannitol, phlorotannins, and alginate	Zhang et al. (2020)
6	*Nizimuddinia zanardini*	Chemical treatments, hydrolysis, fermentation	Mannitol, protein, minerals, alginate, bioethanol, electricity	Nazemi et al. (2021)
7	*Durvillaea Antarctica, Laminaria digitata*	Mechanical treatment, chemical treatments, enzymatic hydrolysis, fermentation	Sodium alginate, L-lysine	Hoffmann et al. (2021)
8	*Sargassum muticum*	Mechanical treatment, autohydrolysis, chemical treatments, anaerobic digestion	Sap, calcium, mineral amendment, alginate, and biogas	Flórez-Fernández et al. (2021)
9	*Ascophyllum nodosum*	Ethanol extraction, acid extraction, microwave-assisted non-catalyzed (water only), acid catalyzed (H2SO4), and metal cation catalyzed (PbII and AlIII)	Laminarin, acetic acid, mannitol, alginate, formic acid, and levulinic acid	Tedesco et al. (2021)
10	*Saccharina latissima*	Enzyme-assisted extraction and hydrolysis	Alginate and laminarian	Herrera Barragán et al. (2022)
11	*Fucus spiralis*	Ethanolic extraction, acid extraction, and alkali treatment	Polyphenols, fucoidan, and alginate	Filote et al. (2022)
12	*Saccharina latissima* and *Alaria esculenta*	Sequential extraction and fractionation	Fucoidan, laminarian, cellulose, and alginate	Birgersson et al. (2023)

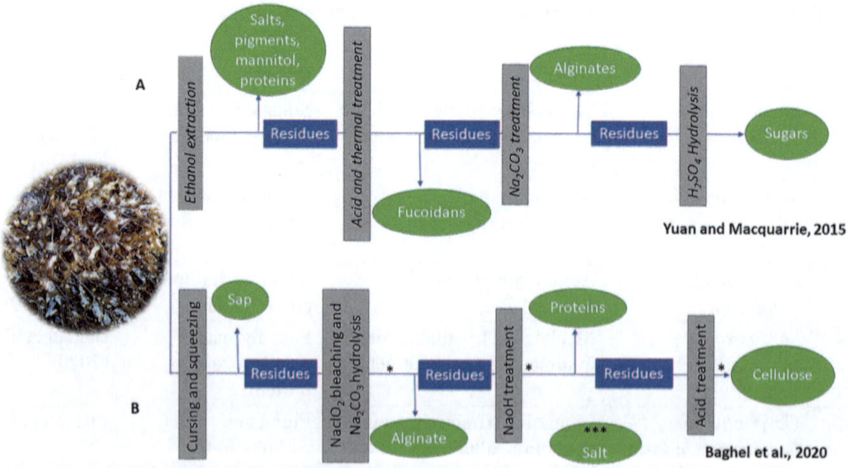

Fig. 12.2 Integrated biorefinery approaches used to convert whole seaweed biomass into variety of products along with alginates preparations

could be achieved. These processes are proof-of-concept stages and yet to be commercialized to better gains.

12.4.2 Conversion of Biomass into Several Products Without Alginates

In the second approach of processing, brown seaweed biomass is converted into a variety of chemicals without caring about alginates (Table 12.2). For example, three products ethanol, 2,3-butanediol, and lycopene were produced from Kelp (Lim et al. 2019), as well as methane, succinic acid, and feed was prepared from a brown seaweed *Laminaria digitata* by fermentation (Alvarado-Morales et al. 2015). *Sargassum* biomass was converted into biogas and fertilizer using hydrothermal pretreatment and anaerobic digestion (Thompson et al. 2021) and also bioethanol and biogas using acid hydrolysis, fermentation, and anaerobic digestion (Abomohra et al. 2021). Hydrothermal liquefaction is another important technique widely used for brown seaweed biomass biorefining. For example, the biomass of six different species such as *Laminaria hyperborean*, *L. digitate*, *Fucus vesiculosus*, *F. ceranoides*, *S. muticum*, *A. nodosum*, *H. elongate*, and *Pelvetia canaliculata* was subjected to hydrothermal liquification to prepare four different products gas, biocrude oil, biochar, and aqueous phase products (Raikova et al. 2017) (Fig. 12.3a). The HTL temp ranged from 250 to 350 °C and pressure of 50 to 180 bar. Similarly, bio-oil and biochar are produced from *Padina tetrastromatica* (Vimali et al. 2023). The low temperatures (160, 180, and 200 °C) HTL of *L. digitata* was done with three different concentrations of $H2SO_4$ (1%, 2%, and 4%) in a microwave. The highest yields

Table 12.2 Integrated biorefinery processes that convert biomass into variety of products without caring alginates

S. no.	Species	Techniques/treatments	Products	Reference
1	*Laminaria digitata*	Enzymatic hydrolysis, fermentation,	Methane, succinic acid, and feed	Alvarado-Morales et al. (2015)
2	*Sargassum muticum*	Mechanical treatment, alcoholic extraction, autohydrolysis	Fucoxanthin, phlorotannin, fucoidan, ethanol, salt, extracts, and saccharidic	Balboa et al. (2015)
3	*Laminaria hyperborea, Laminaria digitata, Ascophyllum nodosum,Fucus vesiculosus, F. ceranoides, Pelvetia canaliculata, Himanthalia elongata, Sargassum muticum*	Hydrothermal liquefaction (HTL)	Aqueousphase products, biochar, biocrude oil, and gas	Raikova et al. (2017)
4	*Saccharina japonica*	Pretreatment, enzymatic hydrolysis, and fermentation	Bioethanol and protein	Dickson et al. (2018)
5	*Ascophyllum nodosum, Fucus vesiculosus, Fucus serratus, Laminaria saccharina, L. digitata*	Ethanolic extraction, alkali extraction, fermentation	Biogas and bioproducts	Tedesco and Daniels (2018)
6	Kelp	Genetically engineered microbe, hydrolysis, and fermentation	Bioethanol, 2,3-butanediol and lycopene	Lim et al. (2019)
7	*Sargassum vulgare*	Acidic hydrolysis, simultaneous saccharification and fermentation (SSF), microbial fuel cells (MFCs)	Bioethanol and bioelectricity	Offei et al. (2019)
8	*Sargassum* sp.	Hydrothermal pretreatment (HTP) and anaerobic digestion (AD)	Biogas, fertilizer	Thompson et al. (2021)
9	*Sargassum muticum*	Microwave hydrothermal treatment, acid hydrolysis, and fermentation	Liquor, bioethanol	del Río et al. (2021)

(continued)

Table 12.2 (continued)

S. no.	Species	Techniques/treatments	Products	Reference
10	*Laminaria digitata, Saccharina latissimi, Fucus vesiculosus*	Water extraction, filtration, acid extraction	Laminarin, fucoidan, feed supplement	Zhang and Thomsen (2021)
11	*Sargassum* spp.	Acid hydrolysis, fermentation, anaerobic digestion	Bioethanol and biogas	Abomohra et al. (2021)
12	*Sargassum* spp.	Water extraction, ethanoic precipitation, solvent extraction, and acid-catalyzed transesterification	Polysaccharide and biodiesel	Ruangrit et al. (2023)
13	*Undaria pinnatifda waste*	Microwave-assisted extraction, hot water extraction	Proteins, fucoidan	Sasaki et al. (2022)
13	*Laminaria digitata*	Microwave-assisted hydrothermal carbonization	Hydrochar, levulinic acid, 5-hydroxymethylfurfural	Deng et al. (2022)
15	*Padina tetrastromatica*	Hydrothermal liquefaction	Bio-oil and biochar	Vimali et al. (2023)
16	*Sargassum swartzii, Padina tetrastromatica*	Acid and enzyme hydrolysis, solvent extraction, transesterification	Biodiesel and bioethanol	Jeyakumar et al. (2022)
17	*Laminaria digitata*	Enzymatic hydrolysis, fermentation	Acetone, butanol, and ethanol	Moenaert et al. (2023)

Fig. 12.3 Integrated biorefinery approaches used to convert whole seaweed biomass into a variety of products without carrying alginates

for levulinic acid (12%) and 5-hydroxymethylfurfural (5.8%) were recorded with 4% H_2SO_4 at 200 °C temperature, and also 15.9% hydrochar was obtained (Deng et al. 2022) (Fig. 12.3b). In this type of approaches, mostly products such as ethanol, acids, bio-oils, biogas, and biochars are produced. Such process allows to prepare at least 2–6 products from biomass. However, scale up for these approaches are highly needed to get data on technoeconomic assessments to ensure the viability at commercial scales.

12.5 *Sargassum* Tides and Their Valorization

In recent years, the biomass of washed-up seaweeds has reached record levels. Pelagic *Sargassum*, also known as golden tides, has been inundating beaches in the Caribbean, Gulf of Mexico, and West Africa in recent years, severely harming the local economy and ecosystem (Milledge and Harvey 2016). A variety of explanations have been advanced to explain such episodes, including an increase in nutrient loads, a shift in trade currents, and extremely high sea surface temperatures (Djakouré et al. 2017). At modest densities, wrack *Sargassum* biomass may be helpful to the ecosystem because it provides food and shelter to various species, aids in coastal erosion, and gives nutrients to beach environments. When biomass is extremely high, however, it can have a detrimental impact as the concentration of algae on the water's surface prevents light penetration and harms corals and benthic macroalgae (Gavio and Santos-Martínez 2018). The use of such washed biomass commercially for food, fuel, pharmaceuticals could help pay for clean-up and lessen the financial burden of these beautiful tides. Among the uses, preparation of bioplastic/packaging and biocomposite material is the latest application of *Sargassum* biomass utilization. The usage of green bioplastics made from algae as an alternative to traditional plastics has been made possible by a pressing concern with producing environmentally acceptable plastics. Using a multi-response statistical model, Lim et al. (2018) synthesized the bioplastic from alginate extracted from *Sargassum siliquosum* J. Agardh. Based on its yield and molecular weight, native alginate isolated from *S. siliquosum* without formaldehyde pretreatment was discovered to be the most acceptable option for the manufacture of bioplastic film. Kachaanun et al. (2022) used *Sargassum plagiophyllum* powder as reinforcement material for the preparation of wheat gluten biocomposites. The study suggests that the use of 1% seaweed powder as reinforcement material increases the tensile strength by twofold compared to a neat wheat gluten composite. Another study demonstrates the preparation and characterization of biodegradable algae-based bioplastics utilizing alginate derived from brown *Sargassum* sp. seaweeds, which are widely distributed in Sabah, Malaysia's coastal waters (Kanagesan et al. 2022). More recently, Mohammed et al. (2023), prepared alginate bioplastic composites for food packaging using *Sargassum natans* alginate. It was found that composite films entirely degraded in 14 days and had similar material qualities to HDPE, PET, and PLA in terms of OP and WVP. Although few studies have explored the use of *Sargassum* for the preparation of bioplastic or packaging material, extensive research is still needed to utilize this potential material for commercial-scale bioplastic production.

12.6 Challenges and Future Prospects

While brown seaweeds offer great potential as a feedstock for various applications, several challenges need to be addressed for their successful implementation on a larger scale. Harvesting practices should be carefully managed to ensure the conservation of natural seaweed populations and minimize environmental impacts. Efficient and cost-effective pretreatment methods need to be developed to optimize the extraction of valuable components from brown seaweed biomass. Furthermore, the scaling up of processes utilizing brown seaweeds requires the development of robust and scalable technologies. Process optimization, including the integration of different conversion steps and the selection of appropriate catalysts and enzymes, is crucial to maximize product yields and minimize energy consumption. In terms of commercialization, market demand and economic viability play significant roles. The development of high-value products from brown seaweeds should consider market trends, consumer acceptance, and cost competitiveness. Market studies and economic assessments are essential to evaluate the feasibility and profitability of ventures utilizing brown seaweed biomass.

12.7 Conclusion

The brown seaweeds are very rich in composition and could be a potential raw material for preparing variety of food, feed, and chemicals and fuels using integrated biorefining processes. Due to the richness of brown seaweed biomass, extensive research has been carried out in the last decade to develop advanced processes for the sustainable utilization of biomass. These processes utilized various advanced and green techniques, such as thermal treatment, autohydrolysis, microwave-assisted extraction, and enzymatic extraction, and provide the opportunity to extract a variety of chemicals without compromising the main product alginate. Such processes are yet to be commercialized for industrial gains. Further extensive research is needed in order to develop the purification techniques for compounds extracted from these seaweeds.

Acknowledgments The financial support received from projects GAP-3499 and OLP-2005 is gratefully acknowledged. The authors are grateful to the Director, CSIR-NIO (CSIR-National Institute of Oceanography), for providing the facilities and encouragement.

References

Abomohra AEF, El-Hefnawy ME, Wang Q, Huang J, Li L, Tang J, Mohammed S (2021) Sequential bioethanol and biogas production coupled with heavy metal removal using dry seaweeds: towards enhanced economic feasibility. J Clean Prod 316. https://doi.org/10.1016/j.jclepro.2021.128341

Abraham RE, Su P, Puri M, Raston CL, Zhang W (2019) Optimisation of biorefinery production of alginate, fucoidan and laminarin from brown seaweed *Durvillaea potatorum*. Algal Res 38. https://doi.org/10.1016/j.algal.2018.101389

Ahmed E, Batbekh B, Fukuma N, Hanada M, Nishida T (2022) Evaluation of different brown seaweeds as feed and feed additives regarding rumen fermentation and methane mitigation. Fermentation 8(10). https://doi.org/10.3390/fermentation8100504

Alvarado-Morales M, Boldrin A, Karakashev DB, Holdt SL, Angelidaki I, Astrup T (2013) Life cycle assessment of biofuel production from brown seaweed in Nordic conditions. Bioresour Technol 129:92–99. https://doi.org/10.1016/j.biortech.2012.11.029

Alvarado-Morales M, Gunnarsson IB, Fotidis IA, Vasilakou E, Lyberatos G, Angelidaki I (2015) *Laminaria digitata* as a potential carbon source for succinic acid and bioenergy production in a biorefinery perspective. Algal Res 9:126–132. https://doi.org/10.1016/j.algal.2015.03.008

Apostolova E, Lukova P, Baldzhieva A, Katsarov P, Nikolova M, Iliev I, Peychev L, Trica B, Oancea F, Delattre C, Kokova V (2020) Immunomodulatory and anti-inflammatory effects of fucoidan: a review. Polymers 12(10):1–25. MDPI AG. https://doi.org/10.3390/polym12102338

Baghel RS (2023) Developments in seaweed biorefinery research: a comprehensive review. Chem Eng J 454:140177. https://doi.org/10.1016/j.cej.2022.140177

Baghel RS, Suthar P, Gajaria TK, Bhattacharya S, Anil A, Reddy CRK (2020) Seaweed biorefinery: a sustainable process for valorising the biomass of brown seaweed. J Clean Prod 263. https://doi.org/10.1016/j.jclepro.2020.121359

Baghel RS, Reddy CRK, Singh RP (2021) Seaweed-based cellulose: applications, and future perspectives. Carbohydr Polym 267. Elsevier Ltd. https://doi.org/10.1016/j.carbpol.2021.118241

Baghel RS, Choudhary B, Pandey S, Pathak PK, Patel MK, Mishra A (2023) Rehashing our insight of seaweeds as a potential source of foods, nutraceuticals, and pharmaceuticals. Food Secur 12(19):3642. https://doi.org/10.3390/foods12193642

Balboa EM, Moure A, Domínguez H (2015) Valorization of *Sargassum muticum* biomass according to the biorefinery concept. Mar Drugs 13(6):3745–3760. https://doi.org/10.3390/md13063745

Baldrick FR, McFadden K, Ibars M, Sung C, Moffatt T, Megarry K, Thomas K, Mitchell P, Wallace JMW, Pourshahidi LK, Ternan NG, Corona G, Spencer J, Yaqoob P, Hotchkiss S, Campbell R, Moreno-Rojas JM, Cuevas FJ, Pereira-Caro G, Gill CIR (2018) Impact of a (poly)phenol-rich extract from the brown algae *Ascophyllum nodosum* on DNA damage and antioxidant activity in an overweight or obese population: a randomized controlled trial. Am J Clin Nutr 108(4):688–700. https://doi.org/10.1093/ajcn/nqy147

Begum R, Howlader S, Mamun-Or-Rashid ANM, Rafiquzzaman SM, Ashraf GM, Albadrani GM, Sayed AA, Peluso I, Abdel-Daim MM, Uddin MS (2021) Antioxidant and signal-modulating effects of brown seaweed-derived compounds against oxidative stress-associated pathology. Oxid Med Cell Longev 2021. Hindawi Limited. https://doi.org/10.1155/2021/9974890

Birgersson PS, Oftebro M, Strand WI, Aarstad OA, Sætrom GI, Sletta H, Arlov Ø, Aachmann FL (2023) Sequential extraction and fractionation of four polysaccharides from cultivated brown algae *Saccharina latissima* and *Alaria esculenta*. Algal Res 69:102928. https://doi.org/10.1016/j.algal.2022.102928

Bringloe TT, Starko S, Wade RM, Vieira C, Kawai H, De Clerck O, Cock JM, Coelho SM, Destombe C, Valero M, Neiva J, Pearson GA, Faugeron S, Serrão EA, Verbruggen H (2020) Phylogeny and evolution of the brown algae. Crit Rev Plant Sci 39(4):281–321. https://doi.org/10.1080/07352689.2020.1787679

Charoensiddhi S, Abraham RE, Su P, Zhang W (2020) Seaweed and seaweed-derived metabolites as prebiotics. Adv Food Nutr Res 91:97–156. Academic Press Inc. https://doi.org/10.1016/bs.afnr.2019.10.001

Chudasama NA, Sequeira RA, Moradiya K, Prasad K (2021) Seaweed polysaccharide based products and materials: an assessment on their production from a sustainability point of view. Molecules 26(9) MDPI AG. https://doi.org/10.3390/molecules26092608

da Silva Ferreira V, Sant'Anna C (2017) Impact of culture conditions on the chlorophyll content of microalgae for biotechnological applications. World J Microbiol Biotechnol 33(1) Springer Netherlands. https://doi.org/10.1007/s11274-016-2181-6

de Mesquita MMF, Crapez MAC, Teixeira VL, Cavalcanti DN (2019) Potential interactions bacteria-brown algae. J Appl Phycol 31(2):867–883. Springer Netherlands. https://doi.org/10.1007/s10811-018-1573-4

del Río PG, Gullón B, Pérez-Pérez A, Romaní A, Garrote G (2021) Microwave hydrothermal processing of the invasive macroalgae *Sargassum muticum* within a green biorefinery scheme. Bioresour Technol 340:125733. https://doi.org/10.1016/j.biortech.2021.125733

Deng C, Lin R, Kang X, Wu B, Ning X, Wall D, Murphy JD (2022) Co-production of hydrochar, levulinic acid and value-added chemicals by microwave-assisted hydrothermal carbonization of seaweed. Chem Eng J 441:135915. https://doi.org/10.1016/j.cej.2022.135915

Dickson R, Ryu J-H, Liu JJ (2018) Optimal plant design for integrated biorefinery producing bioethanol and protein from *Saccharina japonica*: a superstructure-based approach. Energy 164:1257–1270. https://doi.org/10.1016/j.energy.2018.09.007

Dittami SM, Gravot A, Renault D, Goulitquer S, Eggert A, Bouchereau A, Boyen C, Tonon T (2011) Integrative analysis of metabolite and transcript abundance during the short-term response to saline and oxidative stress in the brown alga *Ectocarpus siliculosus*. Plant Cell Environ 34(4):629–642. https://doi.org/10.1111/j.1365-3040.2010.02268.x

Djakouré S, Araujo M, Hounsou-Gbo A, Noriega C, Bourlès B (2017) On the potential causes of the recent Pelagic *Sargassum* blooms events in the tropical North Atlantic Ocean. Biogeosciences Discuss. https://doi.org/10.5194/bg-2017-346

Filote C, Lanez E, Popa VI, Lanez T, Volf I (2022) Characterization and bioactivity of polysaccharides separated through a (sequential) biorefinery process from *Fucus spiralis* brown macroalgae. Polymers 14(19). https://doi.org/10.3390/polym14194106

Fletcher HR, Biller P, Ross AB, Adams JMM (2017) The seasonal variation of fucoidan within three species of brown macroalgae. Algal Res 22:79–86. https://doi.org/10.1016/j.algal.2016.10.015

Flórez-Fernández N, Domínguez H, Torres MD (2019) Advances in the biorefinery of *Sargassum muticum*: valorisation of the alginate fractions. Ind Crop Prod 138. https://doi.org/10.1016/j.indcrop.2019.111483

Flórez-Fernández N, Illera M, Sánchez M, Lodeiro P, Torres MD, Elvira López-Mosquera M, Soto M, Sastre De Vicente M, Domínguez H (2021) Integrated valorization of *Sargassum muticum*. Biorefineries. https://doi.org/10.1016/j.cej.2020.125635

Fouda WA, Ibrahim WM, Ellamie AM, Ramadan G (2019) Biochemical and mineral compositions of six brown seaweeds collected from Red Sea at Hurghada Coast. Indian J Geo-Mar Sci 48(4):484–491

Gallego R, Tardif C, Parreira C, Guerra T, Alves MJ, Ibáñez E, Herrero M (2020) Simultaneous extraction and purification of fucoxanthin from *Tisochrysis lutea* microalgae using compressed fluids. J Sep Sci 43(9–10):1967–1977. https://doi.org/10.1002/jssc.202000021

Garcia-Perez P, Lourenço-Lopes C, Silva A, Pereira AG, Fraga-Corral M, Zhao C, Xiao J, Simal-Gandara J, Prieto MA (2022) Pigment composition of nine brown algae from the Iberian Northwestern coastline: influence of the extraction solvent. Mar Drugs 20(2). https://doi.org/10.3390/md20020113

Garg S, Afzal S, Elwakeel A, Sharma D, Radhakrishnan N, Dhanjal JK, Sundar D, Kaul SC, Wadhwa R (2019) Marine carotenoid fucoxanthin possesses anti-metastasis activity: molecular evidence. Mar Drugs 17(6). https://doi.org/10.3390/md17060338

Gavio B, Santos-Martínez A (2018) Floating *sargassum* in Serranilla bank, Caribbean Colombia, may jeopardize the race to the ocean of baby sea turtles. Acta Biol Colomb 23(3):311–314. https://doi.org/10.15446/abc.v23n3.68113

Grembecka M (2015) Sugar alcohols—their role in the modern world of sweeteners: a review. Eur Food Res Technol 241(1) Springer Verlag. https://doi.org/10.1007/s00217-015-2437-7

Gupta S, Abu-Ghannam N (2011) Recent developments in the application of seaweeds or seaweed extracts as a means for enhancing the safety and quality attributes of foods. Innov Food Sci Emerg Technol 12(4):600–609. https://doi.org/10.1016/j.ifset.2011.07.004

Hakim MM, Patel IC (2020) A review on phytoconstituents of marine brown algae. Futur J Pharm Sci 6(1). https://doi.org/10.1186/s43094-020-00147-6

Herrera Barragán JA, Olivieri G, Boboescu I, Eppink M, Wijffels R, Kazbar A (2022) Enzyme assisted extraction for seaweed multiproduct biorefinery: a techno-economic analysis. Front Mar Sci 9. https://doi.org/10.3389/fmars.2022.948086

Hoffmann SL, Kohlstedt M, Jungmann L, Hutter M, Poblete-Castro I, Becker J, Wittmann C (2021) Cascaded valorization of brown seaweed to produce l-lysine and value-added products using *Corynebacterium glutamicum* streamlined by systems metabolic engineering. Metab Eng 67:293–307. https://doi.org/10.1016/j.ymben.2021.07.010

Hwangbo M, Chu KH (2020) Recent advances in production and extraction of bacterial lipids for biofuel production. Sci Total Environ 734. Elsevier B.V. https://doi.org/10.1016/j.scitotenv.2020.139420

Jeyakumar N, Hoang AT, Nižetić S, Balasubramanian D, Kamaraj S, Lakshmana Pandian P, Sirohi R, Nguyen PQP, Nguyen XP (2022) Experimental investigation on simultaneous production of bioethanol and biodiesel from macro-algae. Fuel 329:125362. https://doi.org/10.1016/j.fuel.2022.125362

Kachaanun P, Numnuam A, Mayakun J, Kaewtatip K (2022) Utilization of brown alga (*Sargassum plagiophyllum*) as an efficient reinforcement material for application in wheat gluten biocomposites. J Appl Polym Sci 139(18):52080. https://doi.org/10.1002/app.52080

Kalasariya HS, Yadav VK, Yadav KK, Tirth V, Algahtani A, Islam S, Gupta N, Jeon BH (2021) Seaweed-based molecules and their potential biological activities: an eco-sustainable cosmetic. Molecules 26(17) MDPI. https://doi.org/10.3390/molecules26175313

Kanagesan K, Abdulla R, Derman E, Sabullah MK, Govindan N, Gansau JA (2022) A sustainable approach to green algal bioplastics production from brown seaweeds of Sabah, Malaysia. J King Saud Univ Sci 34(7):102268. https://doi.org/10.1016/j.jksus.2022.102268

Karuppusamy S, Rajauria G, Fitzpatrick S, Lyons H, McMahon H, Curtin J, Tiwari BK, O'Donnell C (2022) Biological properties and health-promoting functions of *laminarin*: a comprehensive review of preclinical and clinical studies. Mar Drugs 20(12) MDPI. https://doi.org/10.3390/md20120772

Khairy HM, El-Shafay SM (2013) Seasonal variations in the biochemical composition of some common seaweed species from the coast of Abu Qir Bay, Alexandria, Egypt. Oceanologia 55(2):435–452. https://doi.org/10.5697/oc.55-2.435

Khan W, Rayirath UP, Subramanian S, Jithesh MN, Rayorath P, Hodges DM, Critchley AT, Craigie JS, Norrie J, Prithiviraj B (2009) Seaweed extracts as biostimulants of plant growth and development. J Plant Growth Regul 28(4):386–399. https://doi.org/10.1007/s00344-009-9103-x

Krause-Jensen D, Duarte CM (2014) Expansion of vegetated coastal ecosystems in the future Arctic. Front Mar Sci 1(DEC). https://doi.org/10.3389/fmars.2014.00077

Kumar M, Kumari P, Trivedi N, Shukla MK, Gupta V, Reddy CRK, Jha B (2011) Minerals, PUFAs and antioxidant properties of some tropical seaweeds from Saurashtra coast of India. J Appl Phycol 23(5):797–810. https://doi.org/10.1007/s10811-010-9578-7

Kumar S, Sahoo D, Levine I (2015) Assessment of nutritional value in a brown seaweed *Sargassum wightii* and their seasonal variations. Algal Res 9:117–125. https://doi.org/10.1016/j.algal.2015.02.024

Li Y, Zheng Y, Zhang Y, Yang Y, Wang P, Imre B, Wong ACY, Hsieh YSY, Wang D (2021) Brown algae carbohydrates: structures, pharmaceutical properties, and research challenges. Mar Drugs 19(11) MDPI. https://doi.org/10.3390/md19110620

Lim J-Y, Siew Ling H, Chee S-Y, Wong C-L (2018) Sargassum siliquosum J. Agardh extract as potential material for synthesis of bioplastic film. J Appl Phycol 30. https://doi.org/10.1007/s10811-018-1603-2

Lim HG, Kwak DH, Park S, Woo S, Yang J-S, Kang CW, Kim B, Noh MH, Seo SW, Jung GY (2019) *Vibrio* sp. dhg as a platform for the biorefinery of brown macroalgae. Nat Commun 10(1):2486. https://doi.org/10.1038/s41467-019-10371-1

Liu J, Yang S, Li X, Yan Q, Reaney MJT, Jiang Z (2019) Alginate oligosaccharides: production, biological activities, and potential applications. Compr Rev Food Sci Food Saf 18(6):1859–1881. Blackwell Publishing Inc. https://doi.org/10.1111/1541-4337.12494

Lourenço-Lopes C, Fraga-Corral M, Jimenez-Lopez C, Carpena M, Pereira AG, Garcia-Oliveira P, Prieto MA, Simal-Gandara J (2021) Biological action mechanisms of fucoxanthin extracted from algae for application in food and cosmetic industries. Trends Food Sci Technol 117:163–181. Elsevier Ltd. https://doi.org/10.1016/j.tifs.2021.03.012

Lutfia FN, Isnansetyo A, Susidarti RA, Nursid M (2020) Chemical composition diversity of fucoidans isolated from three tropical brown seaweeds (Phaeophyceae) species. Biodiversitas 21(7):3170–3177. https://doi.org/10.13057/biodiv/d210739

Mekinić IG, Skroza D, Šimat V, Hamed I, Čagalj M, Perković ZP (2019) Phenolic content of brown algae (Phaeophyceae) species: extraction, identification, and quantification. Biomolecules 9(6) MDPI AG. https://doi.org/10.3390/biom9060244

Milledge JJ, Harvey PJ (2016) Golden tides: problem or golden opportunity? The valorisation of *Sargassum* from beach inundations. J Mar Sci Eng 4(3) MDPI AG. https://doi.org/10.3390/jmse4030060

Miyashita K, Mikami N, Hosokawa M (2013) Chemical and nutritional characteristics of brown seaweed lipids: a review. J Funct Foods 5(4):1507–1517. https://doi.org/10.1016/j.jff.2013.09.019

Moenaert A, López-Contreras AM, Budde M, Allahgholi L, Hou X, Bjerre AB, Örlygsson J, Karlsson EN, Friðjónsson ÓH, Hreggviðsson GÓ (2023) Evaluation of *Laminaria Digitata* hydrolysate for the production of bioethanol and butanol by fermentation. Fermentation 9(1). https://doi.org/10.3390/fermentation9010059

Mohammed A, Gaduan A, Chaitram P, Pooran A, Lee K-Y, Ward K (2023) *Sargassum* inspired, optimized calcium alginate bioplastic composites for food packaging. Food Hydrocoll 135:108192. https://doi.org/10.1016/j.foodhyd.2022.108192

Mohibbullah M, Haque MN, Sohag AAM, Hossain MT, Zahan MS, Uddin MJ, Hannan MA, Moon IS, Choi JS (2022) A systematic review on marine algae-derived fucoxanthin: an update of pharmacological insights. Mar Drugs 20(5) MDPI. https://doi.org/10.3390/md20050279

Montecinos AE, Peters AF, Rosko B, Desrut A, Valero M (2016) Species delimitation and phylogeographic analyses in the Ectocarpus subgroup siliculosi. J Phycol 53(1):17–31. https://doi.org/10.1111/jpy.12452-15-242

Morris ER, Rees DA, Thom D (1980) Characterisation of alginate composition and block-structure by circular dichroism. Carbohydr Res 81. https://doi.org/10.1016/S0008-6215(00)85661-X

Motaung TE, Linganiso LZ (2018) Critical review on agrowaste cellulose applications for biopolymers. Int J Plast Technol 22(2):185–216. Springer. https://doi.org/10.1007/s12588-018-9219-6

Nazarudin MF, Alias NH, Balakrishnan S, Wan Hasnan WNI, Noor Mazli NAI, Ahmad MI, Md Yasin IS, Isha A, Aliyu-Paiko M (2021) Chemical, nutrient and physicochemical properties of brown seaweed, S*argassum polycystum c. Agardh* (phaeophyceae) collected from Port Dickson, Peninsular Malaysia. Molecules 26(17). https://doi.org/10.3390/molecules26175216

Nazemi F, Karimi K, Denayer JFM, Shafiei M (2021) Techno-economic aspects of different process approaches based on brown macroalgae feedstock: a step toward commercialization of seaweed-based biorefineries. Algal Res 58:102366. https://doi.org/10.1016/j.algal.2021.102366

O' Brien R, Hayes M, Sheldrake G, Tiwari B, Walsh P (2022) Macroalgal proteins: a review. Food Secur 11(4) MDPI. https://doi.org/10.3390/foods11040571

O' Connor J, Meaney S, Williams GA, Hayes M (2020) Extraction of protein from four different seaweeds using three different physical pre-treatment strategies. Molecules 25(8). https://doi.org/10.3390/molecules25082005

Offei F, Mensah M, Kemausuor F, Thygesen A (2019) A biorefinery approach to bioethanol and bioelectricity co-production from tropical seaweeds. J Appl Phycol 31(6):3899–3913. https://doi.org/10.1007/s10811-019-01887-6

Peteiro C (2018) Alginate production from marine macroalgae, with emphasis on kelp farming, Springer series in biomaterials science and engineering, vol 11. Springer Science and Business Media, LLC, pp 27–66. https://doi.org/10.1007/978-981-10-6910-9_2

Popper ZA, Michel G, Hervé C, Domozych DS, Willats WGT, Tuohy MG, Kloareg B, Stengel DB (2011) Evolution and diversity of plant cell walls: from algae to flowering plants. Annu Rev Plant Biol 62:567–590. https://doi.org/10.1146/annurev-arplant-042110-103809

Puscaselu RG, Lobiuc A, Dimian M, Covasa M (2020) Alginate: from food industry to biomedical applications and management of metabolic disorders. Polymers 12(10):1–30. MDPI AG. https://doi.org/10.3390/polym12102417

Qin Y, Jiang J, Zhao L, Zhang J, Wang F (2018) Applications of alginate as a functional food ingredient. In: Biopolymers for food design. Elsevier Inc, pp 409–429. https://doi.org/10.1016/B978-0-12-811449-0.00013-X

Raikova S, Le CD, Beacham TA, Jenkins RW, Allen MJ, Chuck CJ (2017) Towards a marine biorefinery through the hydrothermal liquefaction of macroalgae native to the United Kingdom. Biomass Bioenergy 107:244–253. https://doi.org/10.1016/j.biombioe.2017.10.010

Reyes ME, Riquelme I, Salvo T, Zanella L, Letelier P, Brebi P (2020) Brown seaweed fucoidan in cancer: implications in metastasis and drug resistance. Mar Drugs 18(5) MDPI AG. https://doi.org/10.3390/md18050232

Rodríguez-Montesinos YE, Arvizu-Higuera DL, Hernández-Carmona G, Muñoz-Ochoa M, Murillo-Álvarez JI (2013) Seasonal variation of the agar quality and chemical composition of *Gracilaria veleroae* and *Gracilaria vermiculophylla* (Rhodophyceae, Gracilariaceae) from Baja California Sur, Mexico. Phycol Res 61(2):116–123. https://doi.org/10.1111/pre.12008

Ruangrit K, Chaipoot S, Phongphisutthinant R, Duangjan K, Phinyo K, Jeerapan I, Pekkoh J, Srinuanpan S (2023) A successful biorefinery approach of macroalgal biomass as a promising sustainable source to produce bioactive nutraceutical and biodiesel. Biomass Convers Biorefinery 13(2):1089–1099. https://doi.org/10.1007/s13399-021-01310-6

Sasaki C, Tamura S, Suzuki M, Etomi K, Nii N, Hayashi J, Kanemaru K (2022) Continuous microwave-assisted step-by-step extraction of bioactive water-soluble materials and fucoidan from brown seaweed *Undaria pinnatifida* waste. Biomass Convers Biorefin 14(6):7673–7682. https://doi.org/10.1007/s13399-022-03035-6

Salehi B, Sharifi-Rad J, Seca AML, Pinto DCGA, Michalak I, Trincone A, Mishra AP, Nigam M, Zam W, Martins N (2019) Current trends on seaweeds: looking at chemical composition, phytopharmacology, and cosmetic applications. Molecules 24(22) MDPI AG. https://doi.org/10.3390/molecules24224182

Salomone VN, Riera M (2020) Proximal composition of *Undaria pinnatifida* from San Jorge Gulf (Patagonia, Argentina). Biol Trace Elem Res 196(1):252–261. https://doi.org/10.1007/s12011-019-01905-1

Sanjeewa KKA, Kim EA, Son KT, Jeon YJ (2016) Bioactive properties and potentials cosmeceutical applications of phlorotannins isolated from brown seaweeds: a review. J Photochem Photobiol B: Biol 162:100–105. Elsevier B.V. https://doi.org/10.1016/j.jphotobiol.2016.06.027

Sanjeewa KKA, Lee JS, Kim WS, Jeon YJ (2017) The potential of brown-algae polysaccharides for the development of anticancer agents: an update on anticancer effects reported for fucoidan and laminaran. Carbohydr Polym 177:451–459. Elsevier Ltd. https://doi.org/10.1016/j.carbpol.2017.09.005

Saravana PS, Choi JH, Park YB, Woo HC, Chun BS (2016) Evaluation of the chemical composition of brown seaweed (*Saccharina japonica*) hydrolysate by pressurized hot water extraction. Algal Res 13:246–254. https://doi.org/10.1016/j.algal.2015.12.004

Schiener P, Black KD, Stanley MS, Green DH (2015) The seasonal variation in the chemical composition of the kelp species *Laminaria digitata, Laminaria hyperborea, Saccharina latissima* and *Alaria esculenta*. J Appl Phycol 27(1):363–373. https://doi.org/10.1007/s10811-014-0327-1

Senthilkumar K, Manivasagan P, Venkatesan J, Kim SK (2013) Brown seaweed fucoidan: biological activity and apoptosis, growth signaling mechanism in cancer. Int J Biol Macromol 60:366–374. Elsevier B.V. https://doi.org/10.1016/j.ijbiomac.2013.06.030

Tabarsa M, Rezaei M, Ramezanpour Z, Robert Waaland J, Rabiei R (2012) Fatty acids, amino acids, mineral contents, and proximate composition of some brown seaweeds. J Phycol 48(2):285–292. https://doi.org/10.1111/j.1529-8817.2012.01122.x

Tajima T, Tomita K, Miyahara H, Watanabe K, Aki T, Okamura Y, Matsumura Y, Nakashimada Y, Kato J (2018) Efficient conversion of mannitol derived from brown seaweed to fructose for fermentation with a thraustochytrid. J Biosci Bioeng 125(2):180–184. https://doi.org/10.1016/j.jbiosc.2017.09.002

Targett NM, Arnold TM (1998) Minireview predicting the effects of brown algal phlorotannins on marine herbivores in tropical and temperate oceans. J Phycol 34:195–205

Tedesco S, Daniels S (2018) Optimisation of biogas generation from brown seaweed residues: compositional and geographical parameters affecting the viability of a biorefinery concept. Appl Energy 228. https://doi.org/10.1016/j.apenergy.2018.06.120

Tedesco S, Hurst G, Randviir E, Francavilla M (2021) A comparative investigation of non-catalysed versus catalysed microwave-assisted hydrolysis of common North and South European seaweeds to produce biochemicals. Algal Res 60:102489. https://doi.org/10.1016/j.algal.2021.102489

Thompson TM, Ramin P, Udugama I, Young BR, Gernaey KV, Baroutian S (2021) Techno-economic and environmental impact assessment of biogas production and fertiliser recovery from pelagic Sargassum: a biorefinery concept for Barbados. Energy Convers Manag 245:114605. https://doi.org/10.1016/j.enconman.2021.114605

van Weelden G, Bobi M, Okła K, van Weelden WJ, Romano A, Pijnenborg JMA (2019) Fucoidan structure and activity in relation to anti-cancer mechanisms. Mar Drugs 17(1). https://doi.org/10.3390/md17010032

Vijay K, Velayathum P, Reshma R, Balasundari S, Jeyashakila R, Masilan K (2017) Proximate and mineral composition of brown seaweed from Gulf of Mannar. Int J Fish Aquat Stud 5(5):106–112. www.fisheriesjournal.com

Vimali E, Gunaseelan S, Devi VC, Mothil S, Arumugam M, Ashokkumar B, Moorthy IMG, Pugazhendhi A, Varalakshmi P (2023) Comparative study of different catalysts mediated FAME conversion from macroalga *Padina tetrastromatica* biomass and hydrothermal liquefaction facilitated bio-oil production. Chemosphere 292:133485. https://doi.org/10.1016/j.chemosphere.2021.133485

Wang Y, Alexander TW, Mcallister TA (2009) In vitro effects of phlorotannins from *Ascophyllum nodosum* (brown seaweed) on rumen bacterial populations and fermentation. J Sci Food Agric 89(13):2252–2260. https://doi.org/10.1002/jsfa.3717

Wang Y, Xing M, Cao Q, Ji A, Liang H, Song S (2019) Biological activities of fucoidan and the factors mediating its therapeutic effects: a review of recent studies. Mar Drugs 17(3). https://doi.org/10.3390/md17030183

Wijesekara I, Kim SK, Li Y, Li YX (2011) Phlorotannins as bioactive agents from brown algae. Process Biochem 46(12):2219–2224. https://doi.org/10.1016/j.procbio.2011.09.015

Yuan Y, Macquarrie DJ (2015) Microwave assisted step-by-step process for the production of fucoidan, alginate sodium, sugars and biochar from *Ascophyllum nodosum* through a biorefinery concept. Bioresour Technol 198:819–827. https://doi.org/10.1016/j.biortech.2015.09.090

Zhang X, Thomsen M (2021) Techno-economic and environmental assessment of novel biorefinery designs for sequential extraction of high-value biomolecules from brown macroalgae *Laminaria digitata, Fucus vesiculosus,* and *Saccharina latissima*. Algal Res 60:102499. https://doi.org/10.1016/j.algal.2021.102499

Zhang R, Yuen AKL, de Nys R, Masters AF, Maschmeyer T (2020) Step by step extraction of bio-actives from the brown seaweeds, *Carpophyllum flexuosum, Carpophyllum plumosum, Ecklonia radiata* and *Undaria pinnatifida*. Algal Res 52:102092. https://doi.org/10.1016/j.algal.2020.102092

Zhu Y, Christakos G, Wang H, Jin R, Wang Z, Li D, Liu Y, Xiao X, Wu J (2022) Distribution, accumulation and health risk assessment of trace elements in *Sargassum fusiforme*. Mar Pollut Bull 174:113155. https://doi.org/10.1016/j.marpolbul.2021.113155

Chapter 13
The Vibrant Spectra of the Oceans: Unravelling the Diversity, Functions, and Applications of Seaweed Pigments

Marta V. Freitas, Clélia Afonso, Leonel Pereira, and Teresa Mouga

Abstract Seaweeds are photosynthetic organisms that harvest sunlight and convert it into essential energy through a highly effective biochemical process powered by light, which in turn is captured and directed to the reactive centers by pigments. Seaweed pigments are fundamental to the seaweed photosynthetic process and are also responsible for their current taxonomic classification into three distinct divisions. The Chlorophyta comprises the green seaweeds, which hold high levels of both chlorophyll *a* and *b*, along with lesser amounts of β-carotene and xanthophylls. The Ochrophyta integrates the brown seaweeds (class Phaeophyceae), which are particularly rich in the xanthophyll fucoxanthin, while also showing lower levels of other xanthophylls, chlorophyll *a* and *c*, and *β*-carotene. The Rhodophyta houses the red seaweeds, which are characterized by holding high levels of phycoerythrin, as well as lower quantities of phycocyanin, chlorophyll *a*, *β*-carotene, and xanthophylls. However, regardless of their designation, seaweed species exhibit a significant range of colors and hues that go beyond "green," "brown," and "red," rendering them visually unique, due to the relative composition and abundance of each pigment within their living cells. Nowadays, science has unveiled not only the diversity of these pigments, but also their biological function. Targeted extraction methods

---

M. V. Freitas (✉)
MARE-Marine and Environmental Sciences Centre/ARNET-Aquatic Research Network, ESTM, Polytechnic of Leiria, Edifício Cetemares, Peniche, Portugal

MARE-Marine and Environmental Sciences Centre/ARNET-Aquatic Research Network, ESTM, Polytechnic of Leiria, Edifício Cetemares, Av. Porto de Pesca, Coimbra, Portugal
e-mail: marta.freitas@ipleiria.pt

C. Afonso · T. Mouga
MARE-Marine and Environmental Sciences Centre/ARNET-Aquatic Research Network, ESTM, Polytechnic of Leiria, Edifício Cetemares, Peniche, Portugal

L. Pereira
CFE - Centre for Functional Ecology: Science for People & Planet, Marine Resources, Conservation and Technology - Marine Algae Lab, University of Coimbra, Department of Life Sciences, Calçada Martim de Freitas, Coimbra, Portugal

N. Trivedi et al. (eds.), *Recent Advances in Seaweed Biotechnology*,
https://doi.org/10.1007/978-981-96-0519-4_13

and potential end-use applications for these pigments have been researched and developed, with the resulting product often designed to protect and promote human health. Today, humankind acknowledges the benefits of natural products obtained from the oceans, including those obtained from seaweed pigments. This chapter aims to supply insight on the diversity, functions, and applications of the pigments found in all the three divisions, thus contributing to a greater understanding of the benefits these functional molecules can provide to humanity.

Keywords Photosynthesis · Chlorophylls · Carotenoids · Phycobiliproteins

13.1 Introduction

13.1.1 The Role of Pigments in Photosynthesis

Photosynthesis, from the Greek *photo* (light) and *synthesis* (putting together) (Online Etymology Dictionary 2023), is the process where autotrophic organisms, such as plants, algae, and cyanobacteria, harvest solar energy and convert it into chemical energy to synthesize carbohydrates, essential for growth and development. With this process, the photoautotrophs incorporate carbon dioxide and water and release oxygen to the environment. Therefore, photosynthesis is the vital process that upkeeps the atmospheric oxygen and balances climatic processes, being thus considered the patron process of all life on Earth.

Seaweeds, also known as macroalgae, are photosynthetic aquatic organisms, which, unlike terrestrial plants or seagrass, lack true roots, stems, and leaves, and a vascular system (Pereira 2021). Being primary producers, they are fundamental in upkeeping marine trophic chains, in providing shelter and nursery grounds for marine life (Kasanah et al. 2022). Together with microalgae, macroalgae supply the greatest part of the photosynthetic ability of Earth, being therefore the dominant producers of the oxygen that we all depend on to live (Chapman 2013; Pereira 2021). In seaweeds, the metabolites that are tasked with capturing the light energy and convert it into chemical energy are known as pigments. These pigments are found in the living cells of photosynthetic organisms and are classified as photosynthetic pigments (e.g., chlorophylls) or accessory pigments (e.g., carotenoids and phycobiliproteins).

Seaweeds are found in seawater or brackish water, restrict to depths reachable by sunlight so they can perform photosynthesis, but can colonize a variety of ecological niches (Pereira 2021). Sunlight falls directly on seaweeds at specific times of the day, or it can first interact with sediment, macrophytic components, and the water itself as depth increases before reaching the seaweed tissues. The light spectral composition and irradiance that reaches the seaweed is thus highly diverse and ultimately determines its relative pigment composition, to ensure the photosynthetic process remains effective. Along with a wide range of other secondary metabolites (phenols, sterols, vitamins, among others), pigments are produced by seaweeds as a quick adaptation response to any abiotic changes in the environment (Talarico and Maranzana 2000). This is a fundamental ability that allows seaweeds to adapt to niches where abiotic changes accompany the tidal flux, such as tide pools (Pereira 2021).

Besides being directly involved in the photosynthetic process, pigments are also particularly effective photoprotective molecules able to shield the seaweed against ultraviolet radiation, which can reach biologically significant depths in aquatic ecosystems and can significantly damage the photosynthetic apparatus and hinder photosynthetic efficiency (Sen and Mallick 2021).

These seaweed pigments hold an amazing chromatic diversity, to which the organism owns the unique colors it displays. Seaweed pigments have also been settling the whole seaweed taxonomy since, at least, the last century (Pereira 2021). Green seaweeds, with pigmentation identical to that of vascular plants, are characterized by high amounts of chlorophyll *a* and *b* and small amounts of β-carotene and xanthophylls (lutein, neoxanthin, violaxanthin, and zeaxanthin) and thus belong to phylum Chlorophyta. Brown seaweeds owe their color to high concentrations of fucoxanthin, but also have low concentrations of chlorophyll *a* and *c*, violaxanthin, and *β*-carotene and therefore belong to phylum Ochrophyta, class Phaeophyceae. Finally, red seaweeds, which include species that are perfectly adapted to the low light levels, found at high ocean depths, hold high content in phycobilins (in particular R-phycoerythrin and R-phycocyanin), but low content in chlorophyll *a*, *β*-carotene, and xanthophylls (lutein and zeaxanthin), and thus belong to phylum Rhodophyta (Pereira and Neto 2015; Bonanno and Orlando-Bonaca 2018; Pereira 2021).

13.1.2 Diversity and Function of Pigments in Seaweeds

13.1.2.1 Chlorophylls

Chlorophylls, whose name comes from the Greek "khloros" (green) and "phullon" (leaf), are non-polar green pigments that are key players in the photosynthetic process, being specifically involved as light harvesters and in the energy transference and storage processes, functioning as natural solar cells (Mandal and Dutta 2020; Kato et al. 2020).

There are several kinds of chlorophyll, namely, chlorophyll *a*, chlorophyll *b*, chlorophyll c_1 and c_2, chlorophyll *d*, and chlorophyll *f*. All chlorophyll types are optimized to efficiently target and use the diverse spectral components of solar energy. Chlorophylls capture light from both ends of the electromagnetic spectrum, namely the blue and the red region, while scattering wavelengths from the green region. While all chlorophylls are able to absorb solar energy, not all are able to follow up, by converting it into chemical energy (Mandal and Dutta 2020). In seaweeds, there are only three types of chlorophylls: chlorophyll *a*, chlorophyll *b*, and chlorophyll *c*.

Chlorophyll *a* is present in all seaweed groups (Pereira 2021), as well as in all oxygenic photosynthetic organisms, being specifically found in the photosystem cores and in the light-harvesting antennas within chloroplasts (Kato et al. 2020). In fact, the reaction center of the photosystem I off all oxygenic photosynthetic organisms only contain this type of chlorophyll, which has its peak absorption at wavelengths 430 and 664 nm (Larkum et al. 2021). Regarding its chemical structure, chlorophyll *a* holds a porphyrin ring with a hetero polycyclic structure that surrounds a central Mg^{2+} ion. Around this central ion, nitrogen atoms are coordinately positioned, and an ester linkage attaches the polycyclic structure to a long phytol tail, being this a feature shared by all chlorophyll species (Mandal and Dutta 2020).

Chlorophylls *b* and *c* can only be found in green and brown seaweeds, respectively, with chlorophyll *b* being also found in plants. Both pigments are found only in the light-harvesting antennas (Kato et al. 2020). While chlorophyll *b* has its peak absorption at wavelengths 460 and 647 nm, chlorophyll *c* has its peak absorption at wavelengths 442 and 630 nm (c_1), and 444 and 630 nm (c_2) (Larkum et al. 2021). Chlorophyll *b* shares the same molecular structure or chlorophyll *a*, but while the former contains a methyl group ($-CH_3$) in the second ring, chlorophyll *b* contains an aldehydic group (-cho) instead (Mandal and Dutta 2020). Chlorophylls *d* and *f* are only found in Cyanobacteria. Chlorophyll derivatives such as purpurin-18 *a*, pheophytin *d*, and phytyl-purpurin-18 *a* have been studied within the scope of chlorophyll profile assessment in a number of commercially valuable seaweed species from all taxonomic groups (Chen et al. 2017a).

13.1.2.2 Carotenoids

In nature, carotenoids are the most widely distributed pigments, being found in a wide diversity of photosynthetic organisms, namely algae, plants, photosynthetic bacteria, and several archaea and fungi (Koizumi et al. 2018; Maoka 2020). The animal kingdom, on the other hand, lacks the capacity to produce these pigments, likely because of the evolutionary loss of the necessary genes encoding enzymes for carotenoid biosynthesis (Bohn et al. 2021). Carotenoids show warm colors such as yellow, orange, red, and purple. Structurally, carotenoids are tetraterpene pigments, where most consist of eight isoprene units with a 40-carbon skeleton and a general structure of a polyene chain with nine conjugated double bonds and an end group at both ends of the polyene chain (Maoka 2020).

Carotenoids form a crucial component of the light-harvesting antenna complex within the thylakoid membrane of chloroplasts (Koizumi et al. 2018). Carotenoids are essential compounds involved in photosynthesis and photoprotection. In photosynthesis, their primary role is to enhance light absorption in the blue spectrum, effectively extending the range of light absorption (Bohn et al. 2021), as they harvest and transfer light energy to chlorophylls through singlet-singlet transfer. In photoprotection, carotenoids stand as the first line of defense against oxidative damage, by scavenging free radicals through energy transfer and chemical reaction (Zhao et al. 2022). Whenever chlorophylls have any excess energy, carotenoid photoprotective mechanisms fall in place by absorbing this energy surplus through triplet-triplet transfer and subsequent release by polyene vibration. Carotenoids also have the ability to quench free radicals such as singlet oxygens that are produced by oxygen, light, and the photosynthetic process itself (Schubert et al. 2006; Viera et al. 2018; Maoka 2020; Swapnil et al. 2021).

Carotenoids are divided into two main groups, carotenes and xanthophylls. Carotenes are unsaturated hydrocarbons (Osório et al. 2020), existing about 50 kinds in nature (Britton et al. 2004), being α-carotene, β-carotene, and lycopene well-known examples (Poojary et al. 2016). Carotenes, essential photosynthetic pigments, play a crucial role in primary light absorption (Aldred et al. 2009) as well as in the photoprotective system of seaweeds (Asada 2006; Takahashi and Murata 2008). Although various carotenes can be detected in seaweeds, depending on the algal species, the presence of β-carotene stands out as one of the major carotenes commonly found (Murray and Capelli 2020). Xanthophylls are similar to carotenes but have, at least, one functional group with oxygen and have a highly diverse structure. These pigments may be present as protein complexes, fatty acid esters, glycosides, and sulfates and are found in a far greater number in nature, with 800 kinds described, until 2018 (Britton et al. 2004; Maoka 2009). Popular examples include lutein, fucoxanthin, astaxanthin, zeaxanthin, and β-cryptoxanthin (Poojary et al. 2016; Pereira et al. 2021). Xanthophylls are essential in photo-inhibition protection, by reducing and dissipating the energy reaching the photosynthetic reaction centers, through non-photochemical quenching (Maoka 2020).

Carotenoids are distinctly distributed across algae taxonomical groups, and their profile can be used for taxonomic classification (Poojary et al. 2016). Green algae have β-carotene, lutein, violaxanthin, neoxanthin, and zeaxanthin (Quitério et al. 2021), and their synthesis share common pathways with those observed in terrestrial plant species (Eismann et al. 2020). Brown seaweeds are especially rich in xanthophylls, with the major compounds being neoxanthins A and B, fucoxanthin, and violaxanthin, but they also have carotenes, especially α-carotene and β-carotene (Yuan et al. 2018; Park and Lee 2021; Quitério et al. 2021). Fucoxanthin is the main pigment responsible to convey brown algae their color and is one of the most abundant carotenoids in nature (about 10% of natural carotenoid production) (Osório et al. 2020). Red seaweeds usually exhibit a pattern composed of α-carotene and β-carotene and from one to three xanthophylls: lutein, antheraxanthin, and zeaxanthin (Esteban et al. 2009; Quitério et al. 2021), and differences in carotenoid composition are found between classes within this phylum (Takaichi et al. 2016).

13.1.2.3 Phycobiliproteins

Phycobiliproteins are soluble protein pigments, brilliant colored and fluorescent, that are found in the photosynthetic light-harvesting antenna complexes in Rhodophyta macro- and microalgae, and also in Cyanobacteria, Cryptomonads, and Glaucophytes (Chini Zittelli et al. 2023). These natural-soluble pigments are proteins, that exhibit fluorescence emission (Qiang et al. 2021). In most red algae, distinct types of phycobiliproteins are assembled into supramolecular complexes anchored to the thylakoid membrane of the algae cell, through peptide links. These complexes, known as phycobilisomes, are highly efficient in capturing solar energy and transferring it into photosystems and are also acknowledged for being effective photo-protectors against high radiances, and opportune nutrient reservoirs whenever nitrogen and phosphorus are scarce (Sonani et al. 2016). Phycobilisomes assume distinct forms, being usually bundled, double-barrel-shaped, semicircular, or semiellipsoidal (Qiang et al. 2021).

Phycobiliproteins are grouped into four categories according to their absorption properties, and the distinct color transmitted by the chromophores: phycoerythrin (λ max = 540–570 nm, red-pink), phycocyanin (λ max = 610–625 nm, blue), allophycocyanin (λ max = 650–660 nm, blue-green), and phycoerythrocyanin (λ max = 560–600 nm, magenta) (Sonani et al. 2016; Kannaujiya et al. 2017, 2021; Puzorjov and McCormick 2020). These pigments are further categorized according to their source, namely, C for cyanobacteria, R for Rhodophyta algae, and B for algae from the order Bangiales. With the exception of phycoerythrocyanin, which is exclusive to cyanobacteria (Dagnino-Leone et al. 2022), all abovementioned phycobiliproteins can be found in red algae.

Phycobiliproteins have, as main components, protein subunits α and β that form monomeric αβ units. These units self-assemble and organize themselves on the thylakoid membrane to form phycobilisomes, which are large protein complexes (Chini Zittelli et al. 2023). The phycobiliproteins that compose a phycobilisome are variable within and between biological species (Chini Zittelli et al. 2023), and while red seaweeds have substantial quantities of phycoerythrin, and fair amounts of phycocyanin and allophycocyanin, they do not have phycoerythrocyanin, which is a pigment characteristic of Cyanobacteria (Basheva et al. 2018).

In Rhodophyta, phycobiliproteins comprise up 50% of all water-soluble proteins (Dumay et al. 2014). These molecules are rearranged in an antenna or rod-like structure, with a central allophycocyanin core from where the subunits phycoerythrin, phycocyanin, and allophycocyanin loop outward in an orderly fashion: phycoerythrin (the pigment with the highest energy state) stands placed in the outermost part of the phycobilisome to capture the light energy and transmit it to an adjacent phycocyanin which, in turn, transfer this energy to the allophycocyanin (the pigment with the lowest energy state), placed next to the chlorophyll *a* photosystem reaction center (Qiang et al. 2021; Chini Zittelli et al. 2023). This unidirectional and highly effective energy capture and transport system allows red seaweeds to be remarkably effective in harvesting a wider range of light wavelength and optimizing the whole photosynthetic process, when compared to other seaweed groups (Hsieh-Lo et al. 2019). In the present era, it is recognized that the function of phycobiliproteins goes beyond just light absorption. Phycobilisomes also seemingly play a crucial role as photo-protectors against intense light levels and also serves as a nutrient source during periods of inadequate nitrogen and phosphorus supply (Jesús et al. 2016). In seaweeds, phycoerythrin and phycocyanin are found within the Rhodophyta phyllo (Hsieh-Lo et al. 2019).

Phycoerythrin, the crimson pigment, derives part of its name from the Greek word "erythros" meaning "red," and it holds the greatest variety of chromophores among phycobiliproteins (Dagnino-Leone et al. 2022). Phycoerythrin serves as a photosynthetic pigment found in red macroalgae (Rossano et al. 2003), red microalgae (Román et al. 2002), and cryptomonads (Van Der Weij-De Wit et al. 2006). As stated before for all phycobiliproteins, the classification of phycoerythrin is determined by its original source, leading to R-Phycoerythrin (phycoerythrin obtained from Rhodophyta, with a peak absorption wavelength of 545–565 nm), C-Phycoerythrin (phycoerythrin obtained from Cyanobacteria, with a peak absorption wavelength of 540–570 nm), and B-Phycoerythrin (phycoerythrin obtained from Bangiales, a specific family of filamentous red macroalgae, with a peak absorption wavelength of 546–565 nm) (Fleurence 2003; Kannaujiya et al. 2021). Additionally, these compounds exhibit slight variations in their absorption characteristics, besides their origin (Mysliwa-Kurdziel and Solymosi 2016). Under conditions of low light but high nutrient load, phycoerythrin stands as the primary soluble protein produced by the cell of red algae (Glazer 1994).

Phycocyanin, the blue pigment, gets part of its name from the word "cyan" in English, which has its origins in the Greek word "kyanos." Phycoerythrin is categorized based on its source; therefore, R-Phycocyanin is found in Rhodophyta, C-Phycocyanin in Cyanobacteria and Cryptophyta, and B-Phycocyanin in Bangiophyceae (Dumay and Morançais 2016). Besides being from distinct sources, these pigments also exhibit slight variations in their absorption characteristics (Fernández-Rojas et al. 2014; Mysliwa-Kurdziel and Solymosi 2016).

Allophycocyanin, the blue-green pigment, derives its name from the Greek words "allos" (other) and "kyanos" (blue). Within phycobilisomes, allophycocyanin is located at the core, where assembled trimers $(\alpha\beta)3$ and phycocyanobilin bind to an α or β phycobilisome subunit (Chen and Jiang 2018).

13.1.3 Prospects and Applications of Seaweed Pigments

Marine biotechnology, or blue biotechnology, pertains to the application of marine resources and tools for a diversity of applications, namely medical, cosmeceutical, environmental, and industrial. Blue biotechnology is a field with amazing potential and promise, as 70% of the planet is covered by ocean, which in turn harbors 90% of all Earth's organisms (Thompson et al. 2017), and yet a huge slice of this biodiversity remains unexplored. Therefore, there is great potential in unraveling novel natural marine products from the materials, extracts, and molecules the oceans provide (Kasanah et al. 2022), with algae, both micro and macro, being no exception to this. However, regarding the biotechnological application of algae in industry, research endeavors have been mainly focused on microalgae (Dufossé et al. 2005; Begum et al. 2016; Morocho-Jácome et al. 2020; Silva et al. 2020; Chini Zittelli et al. 2023; Sun et al. 2023; Borovkov et al. 2023), although efforts have recently targeted a number of seaweed species from all the three groups, outlined further below.

Seaweeds hold a variety of bioactive compounds with diverse structures, including natural pigments (Osório et al. 2020). Due to their natural properties, there is an increasing demand for these seaweed-derived compounds, as consumers today prefer natural and eco-friendly products and strive for healthier lifestyles. Consequently, markets have started investing in and shifting toward natural-origin products, to replace highly processed and synthetic alternatives. In terms of pigments, synthetic colorants have traditionally been used, due to their accessibility, low cost, high yield, and stable coloring properties (Muñoz-Miranda and Iñiguez-Moreno 2023). However, the development of synthetic colorants involves the inclusion of harmful substances, like heavy metals or dyes in high concentrations, which not only pose risks to the environment but also to human health, potentially leading to cancer occurrences (Nawaz et al. 2020).

On the other hand, natural pigments obtained from seaweeds are known to have biological activities with substantial relevance on a pharmaceutical and biomedical context, such as antioxidant and free-radical scavenging activities. These promote human health (Muñoz-Miranda and Iñiguez-Moreno 2023) and include antioxidant, antimicrobial (Gomes et al. 2022), antitumoral, antimutagenic, antidiabetic, anti-obesity, photoprotective, and neuroprotective activities (Lanfer-Marquez et al. 2005; Stahl and Sies 2012; Chini Zittelli et al. 2023). Due to their notable spectroscopic properties, a number of these natural pigments have also been regarded as a valuable tool in biomedical research, therapeutic science, and clinical diagnosis (Dashwood 1997; Pagels et al. 2019; Li et al. 2019a).

Moreover, these pigments can be applied as substitute of synthetic dies and colorants, which appeals to the food, cosmetic, and textile industry, among other applications. Examples include the inclusion of these pigments in food products such as bakery, dairy, functional foods, and deserts and in cosmetic products aimed to care and protect the skin against UV radiation and to prevent and treat hair loss (Chini Zittelli et al. 2023).

Techniques to extract pigments from natural sources are nowadays known to be fast and efficient, regarding yield, time, and costs (Muñoz-Miranda and Iñiguez-Moreno 2023). However, these advantages do not seem to be seen yet in protocols designed to obtain pigments from macroalgal biomass. Terrestrial plants, fungi, yeasts, bacteria, and microalgae are yet the prime providers of pigments (Dufossé 2016; Muñoz-Miranda and Iñiguez-Moreno 2023), especially carotenoids and phycocyanins, on an industrial scale (Dufossé 2016). These pigments have been applied mostly on the cosmetic and human nutrition sectors (World Intellectual Property Organization (WIPO) 2016), and patents are regularly developed that focus especially on chlorophylls and carotenoids (World International Property Organization 2021). Specifically, regarding seaweeds, the most valuable cultivated species from where pigments may be extracted are the green seaweeds *Ulva clathrata, Monostroma nitidum,* and *Caulerpa* sp., the brown seaweeds *Saccharina japonica, Undaria pinnatifida,* and *Sargassum fusiforme*, and the red seaweeds *Porphyra* sp., *Eucheuma* sp., *Kappaphycus alvarezii*, and *Gracilaria* sp. (Muñoz-Miranda and Iñiguez-Moreno 2023).

13.1.3.1 Prospects and Applications of Chlorophylls

The ability of chlorophyll to act as a solar power cell, by harvesting light energy and store it as chemical energy, has inspired scientists to study and harness the process to obtain chemical fuel from solar energy through chlorophyll centric technologies, namely for hydrogen production (Mandal and Dutta 2020). The chlorophyll used to explore this venue was chlorophyll *a*, although it is not obtained from seaweeds, but rather from microalgae of the genus *Spirulina* (Saiki and Amao 2002; Tomonou and Amao 2002; Himeshima and Amao 2005; Takeuchi and Amao 2005).

The activity of chlorophylls involves not only solar energy, but also electromagnetic radiation within the visible range (Mandal and Dutta 2020) and fluorescence properties (Padhi et al. 2021). This means this pigment has unique properties and is an excellent and versatile biomaterial that must be considered to the benefit of humankind.

The biological activities and health benefit effects of natural chlorophylls present in seaweeds have been recently studied. However, studies focusing on the qualitative and quantitative characterization of their chlorophyll profile are yet scarce (Chen et al. 2017a). Chlorophyll is known for its antioxidant and antibacterial activities (Lanfer-Marquez et al. 2005; Pérez-Gálvez et al. 2020). Additionally, chlorophylls have been clinically applied for years in photodynamic therapy of tumors and their chemo-preventive properties are well acknowledged. Chlorophylls also show antimutagenic activity in genotoxicity assays and are able to protect against cancer biomarkers (Dashwood 1997).

Chlorophyll has been tested as a biomarker to screen plant health and physiological condition and detect diseases, mostly on crops relevant to agriculture (Chaerle et al. 2004; Tung et al. 2013; Guo and Tan 2015), but also to measure coral reef health (Febbo et al. 2012) and survey macroalgal blooms (Lin et al. 2011). Chlorophyll markers were also used to test the effects of environmental stress upon plants (Peguero-Pina et al. 2008; Wang et al. 2018; Li et al. 2019b; van der Westhuizen et al. 2020; Stefanov et al. 2023), to monitor nutrient levels in crops (Schächtl et al. 2005; Long et al. 2013; Liu et al. 2022; Chen et al. 2023) and herbicide levels both in soil (Huang et al. 2012) and water samples (Bi Fai et al. 2007).

Due to their significant fluorescent ability, chlorophylls have been considered in the development of optical and electrochemical biosensors. Devised applications include chlorophyll-based biosensors to assess microalgal cell viability in ballast water (Wang et al. 2013; Li et al. 2022), to detect harmful agents (Merz et al. 1996; Frense et al. 1998; Védrine et al. 2003; Husu et al. 2013), to measure dissolved oxygen in water (Brasil Silva et al. 2017), to detect biological warfare agents in gases (Sanders et al. 2001), and to detect chlorophyll levels in fecal matter from livestock (Pemberton et al. 2007). Assessing herbicide levels in the environment is a very popular application for chlorophyll-based optical sensors (Mandal and Dutta 2020).

Additionally, chlorophylls have been clinically applied for years in photodynamic therapy of tumors, and their chemo-preventive properties are acknowledged. Chlorophylls also show antimutagenic activity in genotoxicity assays and show promise as anticancer agents (Dashwood 1997).

Chlorophylls are additionally being considered as a pigment in the textile, cosmetic, and animal feed industries (Solymosi and Mysliwa-Kurdziel 2016; Janarthanan and Senthil Kumar 2017; Jesumani et al. 2019; Morais et al. 2021). A synthetic formulation of chlorophyll, registered as the food additive E140, is extensively used in the food industry to color foods and beverages.

Typically, chlorophyll has predominantly been sourced from microalgae and occasionally from higher plants, in the various applications explored thus far. However, despite not extracting chlorophyll from macroalgae, the abovementioned examples are still referenced to illustrate the diverse array of potential applications for chlorophylls. In seaweeds, chlorophylls and chlorophyll derivatives have been extracted essentially for research purposes, namely from the green seaweeds *Boodleopsis verticillata*, *Cladophora graminea* (Urrea-Victoria et al. 2023), *Halimeda macroloba* (Nazarudin et al. 2020), *Rhizoclonium riparium* (Urrea-Victoria et al. 2023), *Ulva* spp. (formerly *Enteromorpha* spp.) (Chen et al. 2017a), and *Ulva* spp. (Lin et al. 2011; Zhu et al. 2017; Chen et al. 2017a; Nazarudin et al. 2020), the brown seaweeds *Cladosiphon okamuranus* (Fujii et al. 2012), *Sargassum ilicifolium* (Nazarudin et al. 2020), *Laminaria ochroleuca,* and *Undaria pinnatifida* (Chen et al. 2017a), and the red seaweeds *Bostrychia* spp., *Catenella impudica* (Urrea-Victoria et al. 2023), *Chondrus crispus* (Robertson et al. 2015), *Gracilaria salicornia* (Lumbessy et al. 2021), *Gracilaria tenuistipitata* (Torres et al. 2014), *Grateloupia elliptica* (Lee et al. 2021a), *Kappaphycus alvarezii* (Brotosudarmo et al. 2018), *Palmaria palmata* (Robertson et al. 2015), and *Porphyra* spp. (Robertson et al. 2015; Chen et al. 2017a), to cite a few examples. Chlorophyll *a* from red seaweeds, identified chemically as CAS 479-61-8, has received the endorsement of the European Food Safety Authority (EFSA) for inclusion in food processing. Concurrently, it remains under ongoing scrutiny as a potential pigment for various sectors including textiles, animal feed, and cosmetics (Ferreira et al. 2021; Muñoz-Miranda and Iñiguez-Moreno 2023).

There are two primary methods commonly used to extract chlorophylls from seaweeds. The first method involves the use of organic solvents, such as alcoholic or acetone-based solutions, while the second method utilizes cell disruption techniques, such as mechanical or ultrasonication methods. There is, as well, a growing interest in adopting innovative and environmentally friendly approaches to extract chlorophyll from seaweeds. Some of these promising "green" techniques include microwave-assisted extraction and green solvent extraction methods (Samarasinghe et al. 2012; Zhu et al. 2017; Chen and Roca 2019; Martins et al. 2023).

13.1.3.2 Prospects and Applications of Carotenoids

Carotenoids have been the focus of investigation since the early nineteenth century (Takaichi 2011). Today, we know these pigments are crucial for our body, but must be acquired from dietary sources as, since as mentioned, the animal kingdom cannot synthetize them (Gammone et al. 2015). Fruits, vegetables, and their processed variants are the most popular sources to obtain this compound (Darvin et al. 2011;

Gammone et al. 2015). Carotenoids also enrich the nutritional content of various other natural food sources, such as eggs, fish, algae, fungi, and yeasts (Meléndez-Martínez et al. 2020; Rapoport et al. 2021). In the human diet, approximately 50 different carotenoids can be identified (Gammone et al. 2015).

Carotenoids have undergone extensive research and are recognized for their significant biological activities, particularly their potent antioxidant properties in scavenging free radicals as their main acting mechanism (Krinsky and Johnson 2005; Pangestuti and Kim 2011a). Consequently, carotenoids have been extensively used as nutraceuticals and cosmeceuticals. Carotenoids hold significant recognition in the cosmetic industry, contributing to improved skin health by bolstering dermal defense against UV radiation (Swapnil et al. 2021), and recent research suggests that carotenoids also influence skin moisture, texture, and elasticity as well (Zerres and Stahl 2020). As antioxidant agents, they work synergistically with other reducing agents like polyphenols and vitamins (C and E) (Chan et al. 2015). Carotenoids also play a crucial role in enhancing immune defenses and reducing the risk of chronic diseases and age-related ailments, including inflammatory diseases, cardiovascular disorders, bone/skin/eye conditions, diabetes, and cancer (Fiedor and Burda 2014; Gammone et al. 2015; Kulczyński et al. 2017; Eggersdorfer and Wyss 2018; Viera et al. 2018; Ojulari et al. 2020; Dias et al. 2021; Swapnil et al. 2021; Pereira et al. 2021). Moreover, they promote mental and metabolic well-being during pregnancy and early life (Dias et al. 2021). However, the most noteworthy biological role carotenoids have is their innate role as a provitamin A, rendering them effective in preventing vitamin A deficiencies (Maeda et al. 2018; Meléndez-Martínez et al. 2020; Quitério et al. 2021). Additionally, carotenoids serve as sources of color, odor, and taste in various contexts (Yabuzaki 2017).

Also, carotenoids have been used as feed supplements in aquaculture (Vílchez et al. 2011; Lim et al. 2018; Nakano and Wiegertjes 2020; Elbahnaswy and Elshopakey 2023), and vastly considered as a natural alternative pigment as food colorant (Poojary et al. 2016). However, the current major commercial carotenoids are chemically synthetized (Rodriguez-Amaya 2015), instead of being obtained from natural sources, which have raised concerns regarding public health, safety, and environment burden, among consumers. Consequently, current market trend is to grow toward the commercialization of natural carotenoid-based products (Poojary et al. 2016).

Terrestrial plants are one of the primary sources of natural-based carotenoids; however, the possibility of obtained carotenoids from algae instead has garnered attention from industry due to the advantages they offer regarding cost, scale, time, and yield (Poojary et al. 2016). Extensive research has already been published, which focus either on macroalgae carotenoids in general (Schubert et al. 2006), or focus in specific *taxa* or species, such as the green seaweeds *Caulerpa prolifera* (Hegazi et al. 1998), *Caulerpa lentillifera* (Balasubramaniam et al. 2020), the brown seaweeds *Padina australis* (Brotosudarmo et al. 2018), *Padina pavonica* (Hegazi et al. 1998), *Saccharina latissima* (Gevaert et al. 2002), and *Sargassum polycystum* (Balasubramaniam et al. 2020), and the red seaweeds *Euchema denticulatum* (Balasubramaniam et al. 2020), *Gracilaria tenuistipitata* (Torres et al. 2014), *Jania rubens* (Hegazi et al. 1998), *Kappaphycus alvarezii* (Brotosudarmo et al. 2018), and *Neopyropia yezoensis* (Koizumi et al. 2018; Xie et al. 2020). It has also been suggested to cultivate *Ulva* sp., a green macroalga with a wide array of commercial purposes, to obtain raw material for carotenoid enrichment and production, to strategically improve its nutritional value (Eismann et al. 2020).

Various methods have been employed for carotenoid extraction, with liquid-liquid extraction being the most common approach, followed recently by alternative extraction procedures such as ultrasound-assisted extraction, microwave-assisted extraction, enzymatically assisted extraction, pressurized liquid extraction, extraction methods based on electrotechnology and supercritical/subcritical fluid extraction (Singh et al. 2015; Poojary et al. 2016; Xu et al. 2017; Saini and Keum 2018). By having a high degree of hydrophobicity, carotenoids are effectively extracted with non-polar solvents such as *n*-hexane, diethyl ether, dimethyl ether, dichloromethane, or mixtures of organic solvents such as acetone and octane, or green/food-grade solvents such as ethanol, limonene, and solutions of water and organic solvents (Poojary et al. 2016). The analysis of carotenoids in seaweeds is conducted using high-performance liquid chromatography (HPLC), in which identification is achieved by comparing their retention times and absorption spectra with authentic standards (Hegazi et al. 1998; Koizumi et al. 2018). Carotenoids that are extensively researched and are worth mentioning further below are the well-known β-carotene for its ubiquitous presence in seaweeds, and the xanthophylls fucoxanthin, for its high presence in brown seaweeds, and astaxanthin, for its reportedly strong activity, particularly when sourced from algae. Lutein and zeaxanthin will also be briefly mentioned.

β-Carotene is possibly the most well-known carotene, holding several well-documented benefits for human health, such as antioxidant, anti-inflammatory, anticancer, and antiaging (Mayne 1996; Sies and Stahl 2004; Stahl and Sies 2012; Carpena et al. 2021). This carotene has, therefore, an outstanding potential and applicability in the food, cosmetic, and pharmaceutical industries (Phan et al. 2018; Carpena et al. 2021). In seaweeds, research has shown that it is possible to manipulate β-carotene yields in the red seaweed *Neopyropia yezoensis* (Xie et al. 2020). However, to date, the most efficient source to obtain β-carotene is the green microalgae *Dunaliella salina* (Barbosa et al. 2023), provided its production is properly induced by manipulation of culture conditions (Lamers et al. 2008). Nevertheless, the majority of its industrial production results from chemical synthesis (Ye et al. 2008).

Fucoxanthin, a brown pigment, is one of the most abundant marine carotenoids, is effortlessly found in brown seaweeds (Pereira et al. 2021) and has a wide range of benefits for human health. Fucoxanthin is renowned for its ability to modulate processes in the central nervous system. Additionally, it exhibits antioxidant, antibacterial (Karpiński and Adamczak 2019; Liu et al. 2019), antiphotoaging (Noviendri et al. 2022), antiallergic, anticancer (Pádua et al. 2015; Méresse et al. 2020; Ming et al. 2021), anti-inflammatory (Yang et al. 2013), antidiabetic (Park et al. 2011), antiobese (Ojulari et al. 2020), antiangiogenic (Pereira 2016), and antiobesity activities, while also showing potential for treating neurodegenerative diseases (Schepers et al. 2020; Yang et al. 2021), which makes this pigment particularly sought out by the pharmaceutical, cosmetic, and food industries (Peng et al. 2011; Aziz et al. 2020; Ojulari et al. 2020). The main algal provider of fucoxanthin is the brown seaweed *Undaria pinnatifida*, that which produce this pigment in high quantities (Liu et al. 2019; Garcia-Perez et al. 2022; Lourenço-Lopes et al. 2023), but other brown species have also been looked at, such as *Fucus* spp. (Garcia-Perez et al. 2022; Resende et al. 2023), *Laminaria* spp. (Ming et al. 2021; Garcia-Perez et al. 2022), *Ascophyllum nodosum*, *Bifurcaria bifurcata*, *Himanthalia elongata*, *Pelvetia canaliculata* (Garcia-Perez et al. 2022), *Sargassum* spp. (Yang et al. 2013; Nomura et al. 2013; Conde et al. 2015; Garcia-Perez et al. 2022), and *Sphaerotrichia divaricata* (Maeda et al. 2018). Additionally, brown algae from the genus *Alaria*, *Ascophyllum*, *Carpophyllum*, *Cladosiphon*, *Cystophora*, *Cystoseira*, *Dictyota*, *Ecklonia*, *Ectocarpus*, *Eisenia*, *Himanthalia*, *Ishige*, *Kjellmaniella*, *Myagropsis*, *Padina*, *Pelvetia*, *Petalonia*, *Saccharina*, *Sargassum*, *Scytosiphon*, *Sporochnus,* and *Turbinaria* have been targeted as well (Muradian et al. 2015), and their commercial exploration could be an hypothesis to put forward, provided such can be done sustainably by, for example, cultivating the required biomass if possible.

Astaxanthin, a red pigment, is worth mention as it has an antioxidant activity 10 times stronger than a number of other carotenoids such as β-carotene, lutein, and zeaxanthin (O'Connor and O'Brien 1998). This pigment is known for being able to suppress free radical production and inhibit the cellular toxicity induced by reactive oxygen species, which thus labels it as being a super vitamin E (Miki 1991). Astaxanthin has also great potential as a novel geroneuroprotector (compound that slow the rate of biological aging) due to its chemical-physical properties (Schubert et al. 2018; Sorrenti et al. 2020). Additionally, it shares many of the health benefits previously mentioned for other carotenoid classes, namely as a promoter of cardiovascular health, immune response, cancer prevention, and skin health (Davinelli et al. 2018; Fakhri et al. 2018) This renders this pigment particularly interesting, as astaxanthin, when extracted from algae, reportedly has the strongest biological activity when compared to that obtained from other biological sources (Ye et al. 2019). Such has been the key factor triggering the global demand for this pigment on an aquaculture context, as astaxanthin targets a wide range of aquatic animal health conditions, improves their survival, growth, fitness, and reproductivity, and dramatically improves their coloration, thus boosting their market value (Lim et al. 2018; Elbahnaswy and Elshopakey 2023). In seaweeds, research has shown that the content of astaxanthin changes according to season in the red seaweed *Catenella caespitosa* (Banerjee et al. 2009). Brown seaweeds, such as *Undaria pinnatifida*, *Saccharina japonica* (formerly *Laminaria japonica*), and *Sargassum fusiforme* (formerly *Hizikia fusiformis*), are rich in astaxanthin and fucoxanthin; these seaweeds feature in the traditional Okinawan diet, known for its healthy aging properties (Willcox et al. 2014). However, the main source of astaxanthin among algae, besides the commercially produced chemical counterpart, has been the green microalgae *Haematococcus lacustris* (formerly *Haematococcus pluvialis*) and *Chromochloris zofingiensis* (formerly *Chlorella zofingiensis*) (Han et al. 2013).

Lutein and zeaxanthin are carotenoids whose biotechnological potential has been studied. Being stored in the eye macula (Murillo et al. 2019; Gomes et al. 2022), both carotenoids have proven their usefulness as antitumoral compounds (Firdous et al. 2015), as photoprotective agents (González et al. 2003), and as eye health promoters (Stahl and Sies 2005; Ma and Lin 2010; Murillo et al. 2019; Saha et al. 2020; Becerra et al. 2020), and therefore are explored in food, cosmetics, and pharmaceutical industries (Shi et al. 2002; Fernández-Sevilla et al. 2010; Saha et al. 2020; Becerra et al. 2020). Either one of these carotenoids has been studied from macroalgae such as the red seaweeds *Corallina officinalis* (Sajilata et al. 2008; Koo et al. 2012), *Gracilaria corticata*, *Grateloupia filicina* (Bhat et al. 2021), and *Neophyropia yezoensis* (Xie et al. 2020). However, and again similarly to other carotenoids, the production of these pigments, namely zeaxanthin, using microalgae instead remains an attractive venue (Becerra et al. 2020; Bourdon et al. 2021).

13.1.3.3 Prospects and Applications of Phycobiliproteins

Phycobiliproteins exhibit remarkable spectroscopic properties, including high absorption coefficient, excitation and emission spectra, quantum yield, low interference, quenching stability, and water solubility (Niu et al. 2013; Kannaujiya et al. 2017, 2021). These exceptional properties have led to their widespread application and documentation in various fields, such as biomedical research, clinical diagnostics, therapeutic science, and the cosmeceutical and pharmaceutical industries (Pagels et al. 2019; Li et al. 2019a; Dagnino-Leone et al. 2020; Chini Zittelli et al. 2023). Notably, these pigments have found extensive use as photosynthesizers in cancer therapy (Bei et al. 2002) and fluorescent sensors in flow cytometry and immunofluorescence microscopy (Leney et al. 2018).

The primary reason for the significant commercial interest in phycobiliproteins lies in the health benefits they offer, acting as antioxidants and free-radical scavengers (Yabuta et al. 2010; Pangestuti and Kim 2011b; Nagaraj et al. 2012; Hemlata et al. 2018), with therapeutic and nutraceutical effects (Fernández-Rojas et al. 2014; Manirafasha et al. 2020; Dagnino-Leone et al. 2022). Moreover, they have demonstrated effectiveness in neuroprotection, antibacterial (Hemlata et al. 2018), antiviral (Shih et al. 2003), anti-inflammatory (Lee et al. 2017), antiallergic (Lee et al. 2021c), antitumoral (Pattarayan et al. 2017; Hao et al. 2018; Hemlata et al. 2018), antiaging (Lee et al. 2021c) and have shown promise as agents for combating Alzheimer's disease. Additionally, they have been found to provide hepatoprotective and immunomodulatory benefits (Nagaraj et al. 2012; Cian et al. 2012) and are considered as potential hypocholesterolemia agents (Sonani et al. 2016; Kannaujiya et al. 2021).

Phycobiliproteins are considered the primary food colorants and preservatives obtained from algae and highly regarded as an excellent green alternative to their chemical counterparts (Chen et al. 2017b; Nowruzi et al. 2022; Dagnino-Leone et al. 2022). Phycoerythrin has extensive applications as a natural dye in the food and cosmetic sectors (Mysliwa-Kurdziel and Solymosi 2016; Kannaujiya et al.

2021; Ma et al. 2022; Chen et al. 2022). For example, phycoerythrin extracted from the microalgae *Porphyridium* sp. has garnered significant attention and research due to its ability to emit a vibrant yellow fluorescence. This unique characteristic has led to its thorough exploration and experimentation in the creation of food products with singular visual effects. These include visually captivating cake decorations, soft drinks, and alcoholic beverages that exhibit fluorescence under ultraviolet light or when subjected to specific pH levels (Dufossé et al. 2005).

The potential application and market value of phycobiliproteins is determined by their purity level, specifically the ratio of A_{max} PBP/A_{280}, and consequently, the potential application from all the possibilities listed above. When this ratio is greater than or equal to 0.7, it is considered food grade, for ratios between 0.7 and 3.9, it is categorized as reagent grade, and lastly a ratio of Amax PBP/A280 equal to or greater than 4.0 indicates analytical grade. Both reagent and analytical grade phycobiliproteins are priced higher than food-grade phycobiliproteins (Hsieh-Lo et al. 2019; Qiang et al. 2021).

Within macroalgae, phycoerythrin has been studied not only from the highly acknowledged red seaweeds *Gelidium* sp. (Mittal and Raghavarao 2018; Mittal et al. 2019), *Gracilaria* spp. (Nguyen et al. 2020; Pereira et al. 2020; Zhao et al. 2020), *Grateloupia turuturu* (Munier et al. 2014, 2015; Le Guillard et al. 2015), *Palmaria palmata* (Dumay et al. 2013; Lee et al. 2017), and *Porphyra*/*Pyropia* sp. (Wang et al. 2020; Sano et al. 2020; Huang et al. 2021), but also from several other non-commercial or otherwise less-known species (Ismail and Osman 2016; Malairaj et al. 2016; Sfriso et al. 2018; Saluri et al. 2019; Freitas et al. 2022). However, considering the amazing diversity of red seaweeds worldwide, and the vibrant red hues some Rhodophyta hold, this probably leaves out a great number of less known, but potentially equally promising species. Being phycoerythrin a protein, red seaweeds with reportedly high protein content may look particularly promising to explore as well. Examples of such seaweeds include *Chondrus crispus*, *Osmundea pinnatifida*, *Pterocladiella capillacea*, *Pyropia tenera* (formerly *Neopyropia tenera*) and *Pyropia yezoensis*, besides the abovementioned *Palmaria palmata* and *Porphyra* sp. (Pereira 2011, 2016, 2023).

The primary interest in phycobiliproteins lies not only in their high protein yield but also in their easy-to-perform extraction methods. Typically, in macroalgae, phycoerythrin is studied and processed using various methods, which include grinding/maceration (Francavilla et al. 2015; Pereira et al. 2020), often combined with freeze-thaw techniques (Sfriso et al. 2018; Zhao et al. 2020) or liquid nitrogen (Francavilla et al. 2013; Saluri et al. 2020). The main solvents that are often chosen to complement these techniques are phosphate buffer (Ismail and Osman 2016; Nguyen et al. 2017; Pereira et al. 2020; Sano et al. 2020) and distilled water (Lee et al. 2017; Zhao et al. 2020). Less commonly applied extraction methods include ultrasounds (Le Guillard et al. 2015; Mittal et al. 2017) and enzymatic hydrolysis (Fleurence 2003). Separation and purification of the extracted phycoerythrin are usually performed through chromatography techniques (Niu et al. 2010; Gu et al. 2018; Saluri et al. 2020; Zhao et al. 2020) or gel electrophoresis (MacColl et al. 1996; Galland-Irmouli et al. 2000; Malairaj et al. 2016).

Phycocyanin is the phycobiliprotein that holds the largest market and shares the same properties and applications of phycoerythrin. This brilliant blue colored pigment presents remarkable fluorescent abilities, and it can be processed into fluorescent reagents, probes, and tracers. These are then applied in the context of biomedical research and biological engineering, namely in medical diagnosis and immunology trials (Liu et al. 2016). Phycocyanin also stands out as the most used natural blue pigment in the food industry, being employed to color various products, such as desserts, with a hardly comparable color, since interestingly, there are limited alternatives approved as natural blue food colorants (Brauch 2016). One key advantage and distinctive feature of phycocyanin, apart from its role as a food colorant, is its potential to promote health, being therefore labeled as a pharmaceutically active phyco-pigment (Vinothkanna and Sekar 2020). Phycobiliprotein has an attested nutritional value (Farag et al. 2016), nutraceutical, therapeutical (Piniella-Matamoros et al. 2021) (113), and bioactive potential (Liu et al. 2016), especially as anticancer agent (Jiang et al. 2017; Braune et al. 2021), and shows, as well, remote-sensing abilities (Ogashawara 2020).

Phycocyanin is primarily obtained from the Cyanobacteria *Arthrospira platensis*, which is by far the main source for this pigment, probably in part due to being a highly researched and known organism (Sahin 2018; Wollina et al. 2018; Marková et al. 2020; Araujo et al. 2021; Ragusa et al. 2021), but also due to its cosmopolitan distribution on water and earth (Kannaujiya et al. 2021), thus being easy to obtain. However, the microalgae *Cyanidium caldarium* (Eisele et al. 2000) and *Galdieria sulphuraria* (Chini Zittelli et al. 2023) have also been recently considered for the same purpose. There is an outstanding number of published research and reviews regarding this pigment in the literature (de Morais et al. 2018; Prado et al. 2018; Pagels et al. 2019; Morocho-Jácome et al. 2020; Braune et al. 2021; Cottas et al. 2021; Jaeschke et al. 2021; Patel et al. 2022).

In macroalgae, the Bangiophyceae species such as *Pyropia haitanensis* (as *Neoporphyra haitanensis*) (Liu et al. 2015) and *Bangia atropurpurea* (Chang et al. 2011) have provided B-phycocyanin that demonstrates promising capabilities as an antiallergic agent. Additionally, B-phycocyanin isolated from *Porphyra* sp. exhibits potential as an antioxidant agent (Huang et al. 2021). As part of an effort to enhance our understanding of R-phycocyanin's structure and function in red macroalgae, R-phycocyanin from *Polysiphonia stricta* was isolated, purified, and thoroughly studied (Wang et al. 2014). A mix of several well-known and less-known red seaweed species was also recently evaluated for phycocyanin content (Freitas et al. 2022).

Extraction and purification methods that target phycocyanin are common to those applied to phycoerythrin. However, they are seldom applied or optimized as seaweeds are outpaced by Cyanobacteria, which remains the predominant source of phycocyanin for the industry. Consequently, dedicated studies focused on phycocyanin derived from macroalgae may face challenges in gaining recognition, traction, or support in the coming times. There are few studies use apply phosphate buffer (Lüder et al. 2001; Wang et al. 2014; Ismail and Osman 2016) to extract phycocyanin from red macroalgae, that can be purified by chromatography (Kuddus et al. 2013; Wang et al. 2014; Huang et al. 2021) and electrophoresis (Huang et al. 2021).

Compared to either phycoerythrin or phycocyanin, allophycocyanin is not extensively studied, regardless of the organism. However, its distinctive blue-green color is almost unparalleled by any other natural pigment (Mysliwa-Kurdziel and Solymosi 2016), which would likely enhance its value as a truly unique compound. Again, cyanobacteria are the prime source of this pigment in academic settings (Chen et al. 2016; Soulier and Bryant 2021), and focus is given to understand structural mechanisms (Dagnino-Leone et al. 2017) and synthesis (Dagnino-Leone et al. 2020; Guo et al. 2020). In seaweeds, allophycocyanin been successfully extracted from various Rhodophyta species using phosphate buffer (Lüder et al. 2001; Ismail and Osman 2016). Efforts have also been made to produce high-quality recombinant allophycocyanin in *Escherichia coli* from *Gracilaria chilensis*, intended for biotechnological and biomedical applications (Dagnino-Leone et al. 2020).

Examples of phycobiliprotein commercialization can be found at a commercial level. Some companies profit from phycobiliprotein-derived natural dyes, with microalgae being prominently featured as the source for their products. Several companies promote their microalgae-based natural dyes for use in cosmetic products. Notably, phycocyanin from *Arthrospira* sp. and phycoerythrin from *Porphyridium* sp. are used in this context. Moreover, R-Phycoerythrin is distributed as a laboratory consumable, sourced from *Pyropia tenera* (as *Neopyropia tenera*) or *Neogastroclonium subarticulatum*, specifically for biotechnological and biomedical research purposes. Research on these pigments, obtained from macroalgae, is still scarce. Recent studies focused on the optimization of culture conditions to manipulate phycobiliprotein content and production of *Colaconema* sp. (Lee et al. 2021b), *Halymenia floresii* (Godínez-Ortega et al. 2008), *Halymenia durvillei* (Rula et al. 2021), *Gracilaria caudata,* and *Laurencia catarinensis* (Souza et al. 2019). Such endeavors have, however, progressed little beyond the academic context, which may be partly due to the challenges inherent in extraction procedures, discussed below.

13.1.3.4 Current Challenges and Perspectives

One of the primary challenges in exploring pigments derived from seaweed sources lies in the differing cultural perspectives on seaweeds themselves. Throughout history, seaweeds have held a significant place in the cultures and traditions of Eastern countries, serving various purposes beyond human consumption and health, as extensively documented in the literature (Pereira 2011, 2016, 2018, 2023; Abreu et al. 2015; Carvalho and Pereira 2015; Wells et al. 2017; Leandro et al. 2020; Pereira and Valado 2021; Morais et al. 2021; Kalasariya et al. 2022). On the other hand, the Western part of the world still tends to embrace a lifestyle that emphasizes the abundance and diversity of available food options (Pereira 2023). However, this diversity often includes highly unhealthy choices, such as the widely prevalent "fast food," which has become a response to the fast-paced lifestyle in the West. These processed foods are high in calories and unhealthy fats, artificially flavored with synthetic additives, while lacking nutritional value and appeal. Consequently, in a considerable portion of the Western world, food is primarily viewed and consumed as a necessity for survival, with a number of exceptions such as countries following a Mediterranean diet, for example. In the East, on the other hand, food consumption and preparation are integral to the traditions of many cultures, and the act of eating is often considered a revered and ceremonious activity.

It is no surprise that seaweeds have gained widespread recognition in Eastern cultures, while the Western world still holds reservations about them. In many shoreside Western countries, seaweeds were once consumed as a last resort in times of strife, such as warfare and poverty, so they may now be held as a gloomy reminder of such difficult times back then. Only a few Western countries incorporate seaweeds into their cultural practices (Pérez-Lloréns et al. 2020), and in most places, they are considered a relatively new and unexplored addition to a healthy diet. Nonetheless, extensive research has already been conducted, particularly in academic settings, but the full transition to integrating seaweeds into the mainstream industry and making them commonly available in grocery stores is yet to be accomplished.

Seaweeds, as a food source, have not yet captured the interest of the typical Western society, which often prefers to stick with familiar options rather than explore new alternatives, coupled with an inherent lack of skill, dare, or imagination, in incorporating seaweeds to their daily diet. The lack of interest leads to minimal demand, making industries hesitant to invest in seaweed products when it seems people do care little about them yet. However, recently, both Eastern and Western societies have started to reflect on their lifestyle and the potential consequences for their health and the well-being of future generations. People have become more conscious of the quality of their food, clothing, and other mundane choices and are now making smarter decisions that consider not only personal health but also the environment. There is a growing demand for higher-quality products in the market, and individuals are actively seeking alternatives for both food and materials, reflecting a shift toward a more mindful and sustainable way of living.

Seaweeds exemplify this kind of resource. In contemporary times, although encountering skepticism, particularly from individuals entrenched in traditional customs, seaweeds are conveniently accessible, at the very least, in large retail outlets and commercial spaces. Naturally, there exists a potential downside to this situation, centered around the apprehension of endangering the excessive use of native seaweed populations. However, this issue can be mitigated by implementing effective resource management strategies and practicing sustainable cultivation of the desired species. Besides all their benefits, numerous seaweed species have the added advantage that can be sustainably cultivated, allowing the preservation of natural populations, either in controlled, indoor settings or in outdoor, shored settings. Naturally, challenges remain regarding process sustainability, cost-effectiveness, and economic viability. The diversity of the natural pigments obtained from seaweeds determines their stability against a range of external conditions, such as environmental or those applied in technological settings (Cserháti 2006).

There are challenges also inherent to the recovery of pigments, or any other compound, from a biological matrix as challenging as that of a seaweed. Their extracellular material hinders the full recovery and yield of chlorophylls, whereas microalgae pose a significantly easier endeavor (Chen et al. 2017a). Then, there are the challenges pertaining the pigments themselves. Natural pigments are highly unstable, being particularly sensitive to temperature and light, and the biomass quickly adjusts its pigment composition to the external conditions. While this may present an advantage regarding the manipulation of culture conditions to obtain the desired pigment, a hardly overlooked disadvantage lies in the fact that it is quite challenging to preserve the quality and stability of the metabolites as soon as they are extracted from the biomass. Processes have always taken this into consideration, but process scale-up efforts are challenging.

The isolation of carotenoids can be performed by conventional processing technologies, such as solvent extraction. However, such methods present several limitations, such as high solvent consumption and time-consuming steps, which also result in poor efficiency and selectivity of the end product (Poojary et al. 2016). Certain precautions need to be taken during carotene extraction to minimize degradation and isomerization. Factors such as exposure to light, heat, and oxygen can considerably reduce the efficiency of extraction. Therefore, it is essential to handle and process the samples promptly (Rodriguez-Amaya and Kimura 2004; Rosso and de Rosso and Mercadante 2007). Certain methodologies also have proven to be quite costly, such as the supercritical fluid extraction methodology, or cause carotene degradation, such as ultrasound-assisted extraction and microwave-assisted extraction, that involves the use of high temperatures during the process (Poojary et al. 2016). Alternatives to such sensitive techniques are also highly desired, and to isolate carotenoids from micro- and macroalgae novel techniques such as microwaves, ultrasounds, high-pressure homogenization, pulsed electric fields, liquid pressurization, supercritical fluids, and subcritical fluids, have been recently considered (Poojary et al. 2016). Chemical synthesis of carotenoids is currently commercially made for astaxanthin, and possible to be performed for fucoxanthin although for the latter pigment, such process is quite expensive, and therefore, the standardization of the extraction and purification methods targeting algae-sourced fucoxanthin is a fairly appealing endeavor (Pereira et al. 2021).

Phycobiliproteins are applied in the industry in its water-soluble protein form, and its commercial uses heavily rely on comprehensive knowledge and advanced technology, which are still relatively limited (Mysliwa-Kurdziel and Solymosi 2016). To begin with, efforts have been made to achieve success in culture settings in order to obtain not only macroalgae, but also cyanobacteria, as the source material to extract pigments, particularly regarding achieving a balance between biomass growth rates and phycobiliprotein productivity (Dagnino-Leone et al. 2022). Then, there is also the fact that the majority of innovative applications involving phycobiliproteins are currently only disclosed in patents (Sekar and Chandramohan 2008; Mysliwa-Kurdziel and Solymosi 2016). Currently, the commercialization of phycobiliproteins is hindered not only by their chemical instability, which reduces their shelf life, but also by the high level of purity that is required for pharmaceutical applications, which involves several steps. These steps increase both production time and costs (Chini Zittelli et al. 2023) and may compromise the pigment quality due to manipulation alone. The low yields often obtained during production also limit implementation and setting in the food and cosmetic industries. Additionally, these compounds have limited chemical stability, particularly under heat and light conditions, making them susceptible to denaturation.

To address these issues, researchers have compared and reviewed strategies to improve phycobiliprotein yields and chemical stability (Hsieh-Lo et al. 2019), although said research has been more focused on phycocyanin and phycoerythrin, and such leaving out allophycocyanin. Most research under this topic also focus on phycobiliproteins obtained from *Spirulina platensis* (Nowruzi et al. 2022), leaving out a wide number of likely promising species, either micro or macroalgae. In the food sector, recent progress reviews the stability of phycobiliproteins under a wide range of the acidic, light, and temperature conditions in food storage. Phycobiliproteins are significantly more stable when coupled with other natural preservatives or encapsulated by using algal materials such as alginates and carrageenans, through nanotechnology (Nowruzi et al. 2022). Although this is still a procedure seldom applied, research focusing on this methodology has been recently published (Braga et al. 2016; Seyed Yagoubi et al. 2017; Patel et al. 2020).

Currently, the purification of phycobiliproteins remains expensive, urging the consideration of more effective and cost-efficient extraction and purification methods, especially on a larger scale (Pereira et al. 2020). Most laboratory protocols share common points, such as meticulous biomass preparation and the use of specialized equipment. However, the subtle variations between extraction protocols make it challenging to establish a universally efficient procedure capable of sustainably process large quantities of algae at a minimal cost. While researchers achieve impressive yields in the laboratory, these results are not directly applicable from a commercial perspective, as there is no industrial-scale viability. To bridge this gap, research must shift focus to larger-scale settings and incorporate industrial parameters to develop a commercial protocol that delivers high yield and purity of the pigment (Beattie et al. 2018; Carlos et al. 2021). This emphasis on industrial-scale viability remains relevant and critical in the current context.

13.2 Conclusion

Natural pigments derived from seaweeds exhibit structural diversity and a wide range of vivid colors, while also serving various functions and applications. Chlorophylls, carotenoids, and phycobiliproteins are pivotal components of the seaweed photosynthesis process. They are finely tuned to capture solar energy within specific wavelengths of the light spectrum and convert it into chemical energy, facilitating the synthesis of essential carbohydrates for the organism. Moreover, many of these pigments play a crucial photoprotective role, shielding the seaweed from excessive radiation.

Seaweed-derived pigments are increasingly preferred over synthetic alternatives, reflecting the contemporary society's heightened awareness and concern for health, lifestyle choices, and the long-term impact on both the environment and personal well-being. The manifold benefits offered by chlorophylls, carotenoids, and phycobiliproteins for human health are well established, encompassing a diverse array of bioactivities. Within the realm of biotechnology, these pigments are in high demand across the food, cosmetic, and pharmaceutical industries due to their unique properties.

Nonetheless, challenges persist in the acquisition of these pigments from macroalgae. The ongoing overexploitation of natural seaweed resources underscores the urgency in developing species-specific cultivation techniques. The inherent physiological variations across different seaweed taxa necessitate the refinement of extraction and purification protocols, which are still undergoing optimization to achieve maximum yields at the species level. This optimization is crucial as these processes transition to an industrial scale, while maintaining sustainability and eco-friendliness. These complexities explain why, within the realm of algae, chlorophylls, carotenoids, and phycobiliproteins are primarily sourced from microalgae. These microalgae benefit from well-mastered culture techniques and optimized extraction methods. On the other hand, similar endeavors with macroalgae are mostly confined to research purposes. Nonetheless, ongoing research highlights the remarkable potential of macroalgae as abundant sources of high-quality natural pigments, warranting further thorough exploration.

Acknowledgments This work was funded by national funds through FCT-Fundação para a Ciência e a Tecnologia, I.P., within the scope of the project MARE—Centro de Ciências do Mar e do Ambiente, UIDB/04292/2020, UIDP/04292/2020 (DOI:10.54499/UIDB/04292/2020, DOI:10.54499/UIDP/04292/2020) and ARNET – Aquatic Research Network, LA/P/0069/2020 (DOI: 10.54499/LA/P/0069/2020), and the doctoral grant awarded to Marta V. Freitas, UI/BD/150957/2021 (https://doi.org/10.54499/UI/BD/150957/2021).

References

Abreu MH, Pereira R, Sassi J-F (2015) Marine algae and the global food industry. In: Pereira L, Neto JM (eds) Marine algae: biodiversity, taxonomy, environmental assessment, and biotechnology, 1st edn. CRC Press, Boca Raton, pp 300–319

Aldred EM, Buck C, Vall K (2009) Terpenes. In: Aldred EM (ed) Pharmacology: a handbook for complementary healthcare professionals. Elsevier, Amsterdam, p 362

Araujo GS, Santiago CS, Moreira RT et al (2021) Nutrient removal by *Arthrospira platensis* cyanobacteria in cassava processing wastewater. J Water Process Eng 40:101826. https://doi.org/10.1016/j.jwpe.2020.101826

Asada K (2006) Production and scavenging of reactive oxygen species in chloroplasts and their functions. Plant Physiol 141:391–396. https://doi.org/10.1104/pp.106.082040

Aziz E, Batool R, Akhtar W et al (2020) Xanthophyll: health benefits and therapeutic insights. Life Sci 240:117104. https://doi.org/10.1016/j.lfs.2019.117104

Balasubramaniam V, June Chelyn L, Vimala S et al (2020) Carotenoid composition and antioxidant potential of *Eucheuma denticulatum*, *Sargassum polycystum* and *Caulerpa lentillifera*. Heliyon 6:e04654. https://doi.org/10.1016/j.heliyon.2020.e04654

Banerjee K, Ghosh R, Homechaudhuri S, Mitra A (2009) Seasonal variation in the biochemical composition of red seaweed (*Catenella repens*) from Gangetic delta, northeast coast of India. J Earth Syst Sci 118:497–505. https://doi.org/10.1007/S12040-009-0045-2

Barbosa M, Inácio LG, Afonso C, Maranhão P (2023) The microalga *Dunaliella* and its applications: a review. Appl Phycol 4:99–120. https://doi.org/10.1080/26388081.2023.2222318

Basheva D, Moten D, Stoyanov P et al (2018) Content of phycoerythrin, phycocyanin, allophycocyanin and phycoerythrocyanin in some cyanobacterial strains: applications. Eng Life Sci 18:861–866. https://doi.org/10.1002/elsc.201800035

Beattie SW, Morançais M, Déléris P et al (2018) Extraction of phycocyanin and phycoerythrin pigments. In: Charrier B, Wichard T, Reddy CRK (eds) Protocols for macroalgae research. CRC Press, Boca Raton, pp 249–265

Becerra MO, Contreras LM, Lo MH et al (2020) Lutein as a functional food ingredient: stability and bioavailability. J Funct Foods 66:103771. https://doi.org/10.1016/j.jff.2019.103771

Begum H, Yusoff FMD, Banerjee S et al (2016) Availability and utilization of pigments from microalgae. Crit Rev Food Sci Nutr 56:2209–2222. https://doi.org/10.1080/10408398.2013.764841

Bei H, Guang-Ce W, Chen-Kui Z, Zhen-Gang L (2002) The experimental research of R-phycoerythrin subunits on cancer treatment: a new photosensitizer in PDT. Cancer Biother Radiopharm 17:35–42. https://doi.org/10.1089/10849780252824055

Bhat I, Haripriya G, Jogi N, Mamatha BS (2021) Carotenoid composition of locally found seaweeds of Dakshina Kannada district in India. Algal Res 53:102154. https://doi.org/10.1016/j.algal.2020.102154

Bi Fai P, Grant A, Reid B (2007) Chlorophyll a fluorescence as a biomarker for rapid toxicity assessment. Environ Toxicol Chem 26:1520–1531. https://doi.org/10.1897/06-394R1.1

Bohn T, Bonet ML, Borel P et al (2021) Mechanistic aspects of carotenoid health benefits-where are we now? Nutr Res Rev 34:276–302. https://doi.org/10.1017/S0954422421000147

Bonanno G, Orlando-Bonaca M (2018) Chemical elements in Mediterranean macroalgae. A review. Ecotoxicol Environ Saf 148:44–71. https://doi.org/10.1016/j.ecoenv.2017.10.013

Borovkov AB, Gudvilovich IN, Lelekov AS, Avsiyan AL (2023) Effect of specific irradiance on productivity and pigment and protein production of *Porphyridium purpureum* (Rhodophyta) semi-continuous culture. Bioresour Technol 374:128771. https://doi.org/10.1016/J.BIORTECH.2023.128771

Bourdon L, Jensen AA, Kavanagh JM, McClure DD (2021) Microalgal production of zeaxanthin. Algal Res 55:102266. https://doi.org/10.1016/j.algal.2021.102266

Braga ARC, Figueira F, da Silveira JT et al (2016) Improvement of thermal stability of C-phycocyanin by nanofiber and preservative agents. J Food Process Preserv 40:1264–1269. https://doi.org/10.1111/jfpp.12711

Brasil Silva E, Ferreira Pinto PV, Bamutsha Chretien J et al (2017) Green optical dissolved oxygen sensor based on a chlorophyll–zinc complex extracted from the plant *Brassica oleracea* L. Appl Opt 56:9951. https://doi.org/10.1364/AO.56.009951

Brauch JE (2016) Underutilized fruits and vegetables as potential novel pigment sources. In: Carle R, Schweiggert R (eds) Handbook on natural pigments in food and beverages: industrial applications for improving food color. Elsevier, Duxford, pp 305–335

Braune S, Krüger-Genge A, Kammerer S et al (2021) Phycocyanin from *Arthrospira platensis* as potential anti-cancer drug: review of in vitro and in vivo studies. Life 11:91. https://doi.org/10.3390/life11020091

Britton G, Liaaen-Jensen S, Pfander H (2004) Carotenoids. Birkhäuser Basel, Basel

Brotosudarmo T, Heriyanto, Shioi Y et al (2018) Composition of the main dominant pigments from potential two edible seaweeds. Philipp J Sci 147:47–55

Carlos TAV, dos Santos Pires Cavalcante KM, de Cássia Evangelista Oliveira F et al (2021) Pressurized extraction of phycobiliproteins from *Arthrospira platensis* and evaluation of its effect on antioxidant and anticancer activities of these biomolecules. J Appl Phycol 33:929–938. https://doi.org/10.1007/s10811-020-02358-z

Carpena M, Caleja C, Pereira E et al (2021) Red seaweeds as a source of nutrients and bioactive compounds: optimization of the extraction. Chemosensors 9:132. https://doi.org/10.3390/chemosensors9060132

Carvalho LG, Pereira L (2015) Review of marine algae as source of bioactive metabolites: a marine biotechnological approach. In: Pereira L, Neto JM (eds) Marine algae: biodiversity, taxonomy, environmental assessment, and biotechnology. CRC Press, Boca Raton, pp 195–227

Chaerle L, Hagenbeek D, De Bruyne E et al (2004) Thermal and chlorophyll-fluorescence imaging distinguish plant-pathogen interactions at an early stage. Plant Cell Physiol 45:887–896. https://doi.org/10.1093/PCP/PCH097

Chan PT, Matanjun P, Yasir SM, Tan TS (2015) Antioxidant activities and polyphenolics of various solvent extracts of red seaweed, *Gracilaria changii*. J Appl Phycol 27:2377–2386. https://doi.org/10.1007/s10811-014-0493-1

Chang CJ, Yang YH, Liang YC et al (2011) A novel phycobiliprotein alleviates allergic airway inflammation by modulating immune responses. Am J Respir Crit Care Med 183:15–25. https://doi.org/10.1164/rccm.201001-0009OC

Chapman RL (2013) Algae: the world's most important "plants"—an introduction. Mitig Adapt Strateg Glob Chang 18:5–12. https://doi.org/10.1007/s11027-010-9255-9

Chen H, Jiang P (2018) Combinational biosynthesis and characterization of fusion proteins with tandem repeats of allophycocyanin holo-α subunits, and their application as bright fluorescent labels for immunofluorescence assay. J Biosci Bioeng 126:778–782. https://doi.org/10.1016/j.jbiosc.2018.06.004

Chen K, Roca M (2019) Cooking effects on bioaccessibility of chlorophyll pigments of the main edible seaweeds. Food Chem 295:101–109. https://doi.org/10.1016/J.FOODCHEM.2019.05.092

Chen H, Liu Q, Zhao J, Jiang P (2016) Biosynthesis, spectral properties and thermostability of cyanobacterial allophycocyanin holo-α subunits. Int J Biol Macromol 88:88–92. https://doi.org/10.1016/j.ijbiomac.2016.03.050

Chen K, Ríos JJ, Pérez-Gálvez A, Roca M (2017a) Comprehensive chlorophyll composition in the main edible seaweeds. Food Chem 228:625–633. https://doi.org/10.1016/J.FOODCHEM.2017.02.036

Chen X, Wu M, Yang Q, Wang S (2017b) Preparation, characterization of food grade phycobiliproteins from *Porphyra haitanensis* and the application in liposome-meat system. LWT 77:468–474. https://doi.org/10.1016/J.LWT.2016.12.005

Chen H, Qi H, Xiong P (2022) Phycobiliproteins—a family of algae-derived biliproteins: productions, characterization and pharmaceutical potentials. Mar Drugs 20:450. https://doi.org/10.3390/md20070450

Chen K, Ma T, Ding J et al (2023) Effects of straw return with nitrogen fertilizer reduction on rice (*Oryza sativa* L.) morphology, photosynthetic capacity, yield and water–nitrogen use efficiency traits under different water regimes. Agronomy 13:133. https://doi.org/10.3390/AGRONOMY13010133

Chini Zittelli G, Lauceri R, Faraloni C et al (2023) Valuable pigments from microalgae: phycobiliproteins, primary carotenoids, and fucoxanthin. Photochem Photobiol Sci 22:1733–1789. https://doi.org/10.1007/S43630-023-00407-3

Cian RE, López-Posadas R, Drago SR et al (2012) Immunomodulatory properties of the protein fraction from *Phorphyra columbina*. J Agric Food Chem 60:8146–8154. https://doi.org/10.1021/jf300928j

Conde E, Moure A, Domínguez H (2015) Supercritical CO2 extraction of fatty acids, phenolics and fucoxanthin from freeze-dried *Sargassum muticum*. J Appl Phycol 27:957–964. https://doi.org/10.1007/s10811-014-0389-0

Cottas AG, Ribeiro EJ, Watanabe EO, Ferreira JDS (2021) Evaluation of extraction methods and purification by aqueous two-phase systems of phycocyanin from *Anabaena variabilis* and *Nostoc* sp. Braz J Chem Eng 38:617–627. https://doi.org/10.1007/s43153-021-00131-4

Cserháti T (2006) Liquid chromatography of natural pigments and synthetic dyes, 1st edn. Elsevier Science, Amsterdam

Dagnino-Leone J, Figueroa M, Mella C et al (2017) Structural models of the different trimers present in the core of phycobilisomes from *Gracilaria chilensis* based on crystal structures and sequences. PLoS One 12:e0177540. https://doi.org/10.1371/journal.pone.0177540
Dagnino-Leone J, Figueroa M, Uribe E et al (2020) Biosynthesis and characterization of a recombinant eukaryotic allophycocyanin using prokaryotic accessory enzymes. Microbiology 9:e989. https://doi.org/10.1002/mbo3.989
Dagnino-Leone J, Figueroa CP, Castañeda ML et al (2022) Phycobiliproteins: structural aspects, functional characteristics, and biotechnological perspectives. Comput Struct Biotechnol J 20:1506–1527. https://doi.org/10.1016/J.CSBJ.2022.02.016
Darvin ME, Sterry W, Lademann J, Vergou T (2011) The role of carotenoids in human skin. Molecules 16:10491. https://doi.org/10.3390/MOLECULES161210491
Dashwood RH (1997) Chlorophylls as anticarcinogens (Review). Int J Oncol 10:721–727. https://doi.org/10.3892/ijo.10.4.721
Davinelli S, Nielsen ME, Scapagnini G (2018) Astaxanthin in skin health, repair, and disease: a comprehensive review. Nutrients 10:522. https://doi.org/10.3390/nu10040522
de Morais MG, da Fontoura Prates D, Moreira JB et al (2018) Phycocyanin from microalgae: properties, extraction and purification, with some recent applications. Ind Biotechnol 14:30–37. https://doi.org/10.1089/ind.2017.0009
de Rosso VV, Mercadante AZ (2007) Identification and quantification of carotenoids, by HPLC-PDA-MS/MS, from Amazonian fruits. J Agric Food Chem 55:5062–5072. https://doi.org/10.1021/jf0705421
Dias MG, Borge GIA, Kljak K et al (2021) European database of carotenoid levels in foods. Factors affecting carotenoid content. Food Secur 10:912. https://doi.org/10.3390/foods10050912
Dufossé L (2016) Current and potential natural pigments from microorganisms (bacteria, yeasts, fungi, microalgae). In: Carle R, Schweiggert RM (eds) Handbook on natural pigments in food and beverages: industrial applications for improving food color. Woodhead Publishing, Amsterdam, pp 337–354
Dufossé L, Galaup P, Yaron A et al (2005) Microorganisms and microalgae as sources of pigments for food use: a scientific oddity or an industrial reality? Trends Food Sci Technol 16:389–406. https://doi.org/10.1016/j.tifs.2005.02.006
Dumay J, Morançais M (2016) Proteins and pigments. In: Fleurence J, Levine IA (eds) Seaweed in health and disease prevention. Elsevier Academic Press, London, pp 275–318
Dumay J, Clément N, Morançais M, Fleurence J (2013) Optimization of hydrolysis conditions of *Palmaria palmata* to enhance R-phycoerythrin extraction. Bioresour Technol 131:21–27. https://doi.org/10.1016/j.biortech.2012.12.146
Dumay J, Morançais M, Munier M et al (2014) Phycoerythrins: valuable proteinic pigments in red seaweeds. Adv Bot Res 71:321–343. https://doi.org/10.1016/B978-0-12-408062-1.00011-1
Eggersdorfer M, Wyss A (2018) Carotenoids in human nutrition and health. Arch Biochem Biophys 652:18–26. https://doi.org/10.1016/j.abb.2018.06.001
Eisele LE, Bakhru SH, Liu X et al (2000) Studies on C-phycocyanin from *Cyanidium caldarium*, a eukaryote at the extremes of habitat. BBA-Bioenergetics 1456:99–107. https://doi.org/10.1016/S0005-2728(99)00110-3
Eismann AI, Perpetuo Reis R, Ferreira da Silva A, Negrão Cavalcanti D (2020) *Ulva* spp. carotenoids: responses to environmental conditions. Algal Res 48:101916. https://doi.org/10.1016/j.algal.2020.101916
Elbahnaswy S, Elshopakey GE (2023) Recent progress in practical applications of a potential carotenoid astaxanthin in aquaculture industry: a review. Fish Physiol Biochem 2023:1–30. https://doi.org/10.1007/S10695-022-01167-0
Esteban R, Martínez B, Fernández-Marín B et al (2009) Carotenoid composition in Rhodophyta: insights into xanthophyll regulation in *Corallina elongata*. Eur J Phycol 44:221–230. https://doi.org/10.1080/09670260802439109
Fakhri S, Abbaszadeh F, Dargahi L, Jorjani M (2018) Astaxanthin: a mechanistic review on its biological activities and health benefits. Pharmacol Res 136:1–20. https://doi.org/10.1016/j.phrs.2018.08.012

Farag MR, Alagawany M, El-Hack MEA, Dhama K (2016) Nutritional and healthical aspects of *Spirulina* (*Arthrospira*) for poultry, animals and human. Int J Pharmacol 12:36–51. https://doi.org/10.3923/ijp.2016.36.51

Febbo E, Richard C, Horlin E et al (2012) PAM chlorophyll a fluorometry for monitoring health of corals along the coast of Qatar. Society of Petroleum Engineers – SPE/APPEA Int conference on health, safety and environment in oil and gas exploration and production 2012: protecting people and the environment – evolving challenges 2:1159–1172. https://doi.org/10.2118/156815-MS

Fernández-Rojas B, Hernández-Juárez J, Pedraza-Chaverri J (2014) Nutraceutical properties of Phycocyanin. J Funct Foods 11:375–392. https://doi.org/10.1016/j.jff.2014.10.011

Fernández-Sevilla JM, Acién Fernández FG, Molina Grima E (2010) Biotechnological production of lutein and its applications. Appl Microbiol Biotechnol 86:27–40. https://doi.org/10.1007/s00253-009-2420-y

Ferreira MS, Resende DISP, Sousa Lobo JM et al (2021) Marine ingredients for sensitive skin: market overview. Mar Drugs 19:464. https://doi.org/10.3390/MD19080464

Fiedor J, Burda K (2014) Potential role of carotenoids as antioxidants in human health and disease. Nutrients 6:466–488. https://doi.org/10.3390/nu6020466

Firdous AP, Kuttan G, Kuttan R (2015) Anti-inflammatory potential of carotenoid meso-zeaxanthin and its mode of action. Pharm Biol 53:961–967. https://doi.org/10.3109/13880209.2014.950673

Fleurence J (2003) R-phycoerythrin from red macroalgae: strategies for extraction and potential application in biotechnology. Appl Biotechnol Food Sci Policy 1:1–6

Francavilla M, Franchi M, Monteleone M, Caroppo C (2013) The red seaweed *Gracilaria gracilis* as a multi products source. Mar Drugs 11:3754–3776. https://doi.org/10.3390/md11103754

Francavilla M, Manara P, Kamaterou P et al (2015) Cascade approach of red macroalgae *Gracilaria gracilis* sustainable valorization by extraction of phycobiliproteins and pyrolysis of residue. Bioresour Technol 184:305–313. https://doi.org/10.1016/j.biortech.2014.10.147

Freitas MV, Inácio LG, Martins M et al (2022) Primary composition and pigments of 11 red seaweed species from the Center of Portugal. J Mar Sci Eng 10:1168. https://doi.org/10.3390/jmse10091168

Frense D, Müller A, Beckmann D (1998) Detection of environmental pollutants using optical biosensor with immobilized algae cells. Sens Actuators B Chem 51:256–260. https://doi.org/10.1016/S0925-4005(98)00203-2

Fujii R, Kita M, Doe M et al (2012) The pigment stoichiometry in a chlorophyll a/c type photosynthetic antenna. Photosynth Res 111:165–172. https://doi.org/10.1007/S11120-011-9698-1/TABLES/2

Galland-Irmouli AV, Pons L, Luçon M et al (2000) One-step purification of R-phycoerythrin from the red macroalga *Palmaria palmata* using preparative polyacrylamide gel electrophoresis. J Chromatogr B: Biomed Sci Appl 739:117–123

Gammone MA, Riccioni G, D'Orazio N (2015) Carotenoids: potential allies of cardiovascular health? Food Nutr Res 59:26762. https://doi.org/10.3402/fnr.v59.26762

Garcia-Perez P, Lourenço-Lopes C, Silva A et al (2022) Pigment composition of nine brown algae from the Iberian Northwestern coastline: influence of the extraction solvent. Mar Drugs 20:113. https://doi.org/10.3390/MD20020113

Gevaert F, Creach A, Davoult D et al (2002) Photo-inhibition and seasonal photosynthetic performance of the seaweed *Laminaria saccharina* during a simulated tidal cycle: chlorophyll fluorescence measurements and pigment analysis. Plant Cell Environ 25:859–872. https://doi.org/10.1046/J.1365-3040.2002.00869.X

Glazer AN (1994) Phycobiliproteins – a family of valuable, widely used fluorophores. J Appl Phycol 6:105–112. https://doi.org/10.1007/BF02186064

Godínez-Ortega JL, Snoeijs P, Robledo D et al (2008) Growth and pigment composition in the red alga *Halymenia floresii* cultured under different light qualities. J Appl Phycol 20:253–260. https://doi.org/10.1007/S10811-007-9241-0/FIGURES/5

Gomes L, Monteiro P, Cotas J et al (2022) Seaweeds' pigments and phenolic compounds with antimicrobial potential. Biomol Concepts 13:89–102. https://doi.org/10.1515/BMC-2022-0003

González S, Astner S, An W et al (2003) Dietary lutein/zeaxanthin decreases ultraviolet B-induced epidermal hyperproliferation and acute inflammation in hairless mice. J Invest Dermatol 121:399–405. https://doi.org/10.1046/j.1523-1747.2003.12355.x

Gu D, Lazo-Portugal R, Fang C et al (2018) Purification of R-phycoerythrin from *Gracilaria lemaneiformis* by centrifugal precipitation chromatography. J Chromatogr B Analyt Technol Biomed Life Sci 1087–1088:138–141. https://doi.org/10.1016/j.jchromb.2018.04.047

Guo Y, Tan J (2015) Recent advances in the application of chlorophyll a fluorescence from photosystem II. Photochem Photobiol 91:1–14. https://doi.org/10.1111/PHP.12362

Guo Y, Zang X, Cao X et al (2020) Cloning and expression of Allophycocyanin gene from *Gracilariopsis lemaneiformis* and studying the binding sites of phycocyanobilin on its α and β subunits. J Appl Phycol 32:2657–2671. https://doi.org/10.1007/s10811-020-02102-7

Han D, Li Y, Hu Q (2013) Astaxanthin in microalgae: pathways, functions and biotechnological implications. Algae 28:131–147. https://doi.org/10.4490/ALGAE.2013.28.2.131

Hao S, Li S, Wang J et al (2018) Phycocyanin reduces proliferation of melanoma cells through downregulating GRB2/ERK signaling. J Agric Food Chem 66:10921–10929. https://doi.org/10.1021/acs.jafc.8b03495

Hegazi MM, Pérez-Ruzafa A, Almela L, Candela ME (1998) Separation and identification of chlorophylls and carotenoids from *Caulerpa prolifera*, *Jania rubens* and *Padina pavonica* by reversed-phase high-performance liquid chromatography. J Chromatogr A 829:153–159. https://doi.org/10.1016/S0021-9673(98)00803-6

Hemlata, Afreen S, Fatma T (2018) Extraction, purification and characterization of phycoerythrin from *Michrochaete* and its biological activities. Biocatal Agric Biotechnol 13:84–89. https://doi.org/10.1016/j.bcab.2017.11.012

Himeshima N, Amao Y (2005) Green process for hydrogen production from cellulose derivative using visible light-harvesting function of Mg chlorophyll-a. Green Chem 7:742–746. https://doi.org/10.1039/B504377H

Hsieh-Lo M, Castillo G, Ochoa-Becerra MA, Mojica L (2019) Phycocyanin and phycoerythrin: strategies to improve production yield and chemical stability. Algal Res 42:101600. https://doi.org/10.1016/j.algal.2019.101600

Huang Y, Thomson SJ, Molin WT et al (2012) Early detection of soybean plant injury from glyphosate by measuring chlorophyll reflectance and fluorescence. J Agric Sci 4:p117. https://doi.org/10.5539/JAS.V4N5P117

Huang C-H, Chen W-C, Gao Y-H et al (2021) Enzyme-assisted method for phycobiliproteins extraction from *Porphyra* and evaluation of their bioactivity. Processes 9:560. https://doi.org/10.3390/pr9030560

Husu I, Rodio G, Touloupakis E et al (2013) Insights into photo-electrochemical sensing of herbicides driven by *Chlamydomonas reinhardtii* cells. Sens Actuators B Chem 185:321–330. https://doi.org/10.1016/J.SNB.2013.05.013

Ismail MM, Osman MEH (2016) Seasonal fluctuation of photosynthetic pigments of most common red seaweeds species collected from Abu Qir, Alexandria, Egypt. Rev Biol Mar Oceanogr 51:515–525. https://doi.org/10.4067/S0718-19572016000300004

Jaeschke DP, Teixeira IR, Marczak LDF, Mercali GD (2021) Phycocyanin from Spirulina: a review of extraction methods and stability. Food Res Int 143:110314. https://doi.org/10.1016/j.foodres.2021.110314

Janarthanan M, Senthil Kumar M (2017) The properties of bioactive substances obtained from seaweeds and their applications in textile industries. J Ind Text 48:361–401. https://doi.org/10.1177/1528083717692596

Jesumani V, Du H, Aslam M et al (2019) Potential use of seaweed bioactive compounds in skincare – a review. Mar Drugs 17:688. https://doi.org/10.3390/MD17120688

Jesús VC, Gutiérrez-Rebolledo GA, Hernández-Ortega M et al (2016) Methods for extraction, isolation and purification of C-phycocyanin: 50 years of research in review. Int J Food Nutr Sci 3:275–284. https://doi.org/10.15436/2377-0619.16.946

Jiang L, Wang Y, Yin Q et al (2017) Phycocyanin: a potential drug for cancer treatment. J Cancer 8:3416–3429. https://doi.org/10.7150/jca.21058

Kalasariya HS, Pereira L, Patel NB (2022) Pioneering role of marine macroalgae in cosmeceuticals. Phycology 2:172–203. https://doi.org/10.3390/PHYCOLOGY2010010

Kannaujiya VK, Sundaram S, Sinha RP (2017) Phycobiliproteins: recent developments and future applications, 1st edn. Springer Singapore, Singapore

Kannaujiya VK, Kumar D, Singh V, Sinha RP (2021) Advances in phycobiliproteins research: innovations and commercialization. In: Sinha RP, Häder D-P (eds) Natural bioactive compounds: technological advancements. Academic Press, London, pp 57–81

Karpiński TM, Adamczak A (2019) Fucoxanthin—an antibacterial carotenoid. Antioxidants 8. https://doi.org/10.3390/ANTIOX8080239/ANTIOXIDANTS_08_00239_PDF.PDF

Kasanah N, Ulfah M, Imania O et al (2022) Rhodophyta as potential sources of photoprotectants, antiphotoaging compounds, and hydrogels for cosmeceutical application. Molecules 27:7788. https://doi.org/10.3390/MOLECULES27227788

Kato K, Shinoda T, Nagao R et al (2020) Structural basis for the adaptation and function of chlorophyll *f* in photosystem I. Nat Commun 11:238. https://doi.org/10.1038/S41467-019-13898-5

Koizumi J, Takatani N, Kobayashi N et al (2018) Carotenoid profiling of a red seaweed *Pyropia yezoensis*: insights into biosynthetic pathways in the order Bangiales. Mar Drugs 16:1–14. https://doi.org/10.3390/md16110426

Koo SY, Cha KH, Song DG et al (2012) Optimization of pressurized liquid extraction of zeaxanthin from *Chlorella ellipsoidea*. J Appl Phycol 24:725–730. https://doi.org/10.1007/s10811-011-9691-2

Krinsky NI, Johnson EJ (2005) Carotenoid actions and their relation to health and disease. Mol Asp Med 26:459–516. https://doi.org/10.1016/j.mam.2005.10.001

Kuddus M, Singh P, Thomas G, Al-Hazimi A (2013) Recent developments in production and biotechnological applications of C-phycocyanin. Biomed Res Int 2013:742859. https://doi.org/10.1155/2013/742859

Kulczyński B, Gramza-Michałowska A, Kobus-Cisowska J, Kmiecik D (2017) The role of carotenoids in the prevention and treatment of cardiovascular disease – current state of knowledge. J Funct Foods 38:45–65. https://doi.org/10.1016/j.jff.2017.09.001

Lamers PP, Janssen M, De Vos RCH et al (2008) Exploring and exploiting carotenoid accumulation in *Dunaliella salina* for cell-factory applications. Trends Biotechnol 26:631–638

Lanfer-Marquez UM, Barros RMC, Sinnecker P (2005) Antioxidant activity of chlorophylls and their derivatives. Food Res Int 38:885–891. https://doi.org/10.1016/j.foodres.2005.02.012

Larkum AWD, Grossman AR, Raven JA (2021) Photosynthesis in algae: biochemical and physiological mechanisms. Springer International Publishing, New York

Le Guillard C, Dumay J, Donnay-Moreno C et al (2015) Ultrasound-assisted extraction of R-phycoerythrin from *Grateloupia turuturu* with and without enzyme addition. Algal Res 12:522–528. https://doi.org/10.1016/j.algal.2015.11.002

Leandro A, Pereira L, Gonçalves AMM (2020) Diverse applications of marine macroalgae. Mar Drugs 18:1–15. https://doi.org/10.3390/md18010017

Lee D, Nishizawa M, Shimizu Y, Saeki H (2017) Anti-inflammatory effects of dulse (*Palmaria palmata*) resulting from the simultaneous water-extraction of phycobiliproteins and chlorophyll a. Food Res Int 100:514–521. https://doi.org/10.1016/J.FOODRES.2017.06.040

Lee HG, Lu YA, Je JG et al (2021a) Effects of ethanol extracts from *Grateloupia elliptica*, a red seaweed, and its chlorophyll derivative on 3T3-L1 adipocytes: suppression of lipid accumulation through downregulation of adipogenic protein expression. Mar Drugs 19:91. https://doi.org/10.3390/MD19020091

Lee MC, Yeh HY, Jhang FJ et al (2021b) Enhancing growth, phycoerythrin production, and pigment composition in the red alga *Colaconema* sp. through optimal environmental conditions in an indoor system. Bioresour Technol 333:125199. https://doi.org/10.1016/J.BIORTECH.2021.125199

Lee PT, Yeh HY, Lung WQC et al (2021c) R-phycoerythrin from *Colaconema formosanum* (Rhodophyta), an anti-allergic and collagen promoting material for cosmeceuticals. Appl Sci (Switzerland) 11:9425. https://doi.org/10.3390/APP11209425

Leney AC, Tschanz A, Heck AJR (2018) Connecting color with assembly in the fluorescent B-phycoerythrin protein complex. FEBS J 285:178–187. https://doi.org/10.1111/febs.14331

Li W, Su HN, Pu Y et al (2019a) Phycobiliproteins: molecular structure, production, applications, and prospects. Biotechnol Adv 37:340–353. https://doi.org/10.1016/j.biotechadv.2019.01.008

Li Y, Song H, Zhou L et al (2019b) Tracking chlorophyll fluorescence as an indicator of drought and rewatering across the entire leaf lifespan in a maize field. Agric Water Manag 211:190–201. https://doi.org/10.1016/j.agwat.2018.09.050

Li N, Liu Z, Wang P et al (2022) Effects of sodium hypochlorite treatment on the chlorophyll fluorescence in photosystem II of microalgae. Sci Total Environ 833:155192. https://doi.org/10.1016/j.scitotenv.2022.155192

Lim KC, Yusoff FM, Shariff M, Kamarudin MS (2018) Astaxanthin as feed supplement in aquatic animals. Rev Aquac 10:738–773. https://doi.org/10.1111/RAQ.12200

Lin A, Shen S, Wang G et al (2011) Comparison of chlorophyll and photosynthesis parameters of floating and attached *Ulva prolifera*. J Integr Plant Biol 53:25–34. https://doi.org/10.1111/j.1744-7909.2010.01002.x

Liu Q, Wang Y, Cao M et al (2015) Anti-allergic activity of R-phycocyanin from *Porphyra haitanensis* in antigen-sensitized mice and mast cells. Int Immunopharmacol 25:465–473. https://doi.org/10.1016/j.intimp.2015.02.032

Liu Q, Huang Y, Zhang R et al (2016) Medical application of *Spirulina platensis* derived C-phycocyanin. Evid Based Complement Alternat Med 2016. https://doi.org/10.1155/2016/7803846

Liu Z, Sun X, Sun X et al (2019) Fucoxanthin isolated from *Undaria pinnatifida* can interact with *Escherichia coli* and *lactobacilli* in the intestine and inhibit the growth of pathogenic bacteria. J Ocean Univ China 18:926–932. https://doi.org/10.1007/s11802-019-4019-y

Liu Z, Li W, Xu Z et al (2022) Effects of different nitrogen forms and concentrations on seedling growth traits and physiological characteristics of *Populus simonii* × *P. nigra*. J For Res (Harbin) 33:1593–1606. https://doi.org/10.1007/S11676-021-01447-0

Long JR, Ma GH, Wan YZ et al (2013) Effects of nitrogen fertilizer level on chlorophyll fluorescence characteristics in flag leaf of super hybrid rice at late growth stage. Rice Sci 20:220–228. https://doi.org/10.1016/S1672-6308(13)60138-9

Lourenço-Lopes C, Carreira-Casais A, Carperna M et al (2023) Emerging technologies to extract fucoxanthin from *Undaria pinnatifida*: microwave vs. ultrasound assisted extractions. Mar Drugs 21:282. https://doi.org/10.3390/md21050282

Lüder UH, Knoetzel J, Wiencke C (2001) Acclimation of photosynthesis and pigments to seasonally changing light conditions in the endemic antarctic red macroalga *Palmaria decipiens*. Polar Biol 24:231–236. https://doi.org/10.1007/s003000100260

Lumbessy SY, Junaidi M, Diniarti N et al (2021) Identification of chlorophyll pigment on *Gracilaria salicornia* seaweed. IOP Conf Ser Earth Environ Sci 681:12017. https://doi.org/10.1088/1755-1315/681/1/012017

Ma L, Lin XM (2010) Effects of lutein and zeaxanthin on aspects of eye health. J Sci Food Agric 90:2–12. https://doi.org/10.1002/jsfa.3785

Ma J, Hu J, Sha X et al (2022) Phycobiliproteins, the pigment-protein complex form of natural food colorants and bioactive ingredients. Crit Rev Food Sci Nutr 4:1–19. https://doi.org/10.1080/10408398.2022.2128714

MacColl R, Eisele LE, Williams EC, Bowser SS (1996) The discovery of a novel R-phycoerythrin from an Antarctic red alga. J Biol Chem 271:17157–17160. https://doi.org/10.1074/jbc.271.29.17157

Maeda H, Fukuda S, Izumi H, Saga N (2018) Anti-oxidant and fucoxanthin contents of brown alga Ishimozuku (*Sphaerotrichia divaricata*) from the West Coast of Aomori, Japan. Mar Drugs 16:0255. https://doi.org/10.3390/md16080255

Malairaj S, Muthu S, Gopal VB et al (2016) Qualitative and quantitative determination of R-phycoerythrin from *Halymenia floresia* (Clemente) C. Agardh by polyacrylamide gel using electrophoretic elution technique. J Chromatogr A 1454:120–126. https://doi.org/10.1016/j.chroma.2016.05.063

Mandal R, Dutta G (2020) From photosynthesis to biosensing: chlorophyll proves to be a versatile molecule. Sens Int 1:100058. https://doi.org/10.1016/j.sintl.2020.100058

Manirafasha E, Guo L, Jing K (2020) Nutraceutical and pharmaceutical applications of phycobiliproteins. In: Jacob-Lopes E, Queiroz MI, Zepka LQ (eds) Pigments from microalgae handbook. Springer, Cham, pp 577–654

Maoka T (2009) Recent progress in structural studies of carotenoids in animals and plants. Arch Biochem Biophys 483:191–195. https://doi.org/10.1016/J.ABB.2008.10.019

Maoka T (2020) Carotenoids as natural functional pigments. J Nat Med 74:1–16. https://doi.org/10.1007/s11418-019-01364-x

Marková I, Koníčková R, Vaňková K et al (2020) Anti-angiogenic effects of the blue-green alga *Arthrospira platensis* on pancreatic cancer. J Cell Mol Med 24:2402–2415. https://doi.org/10.1111/jcmm.14922

Martins M, Fernandes APM, Torres-Acosta MA et al (2023) Extraction of chlorophyll from wild and farmed *Ulva* spp. using aqueous solutions of ionic liquids. Sep Purif Technol 254:117589. https://doi.org/10.1016/J.SEPPUR.2020.117589

Mayne ST (1996) Beta-carotene, carotenoids, and disease prevention in humans. FASEB J 10:690–701. https://doi.org/10.1096/fasebj.10.7.8635686

Meléndez-Martínez AJ, Mandić AI, Bantis F et al (2020) A comprehensive review on carotenoids in foods and feeds: status quo, applications, patents, and research needs. Crit Rev Food Sci Nutr 5:1–51. https://doi.org/10.1080/10408398.2020.1867959

Méresse S, Fodil M, Fleury F, Chénais B (2020) Fucoxanthin, a marine-derived carotenoid from brown seaweeds and microalgae: a promising bioactive compound for cancer therapy. Int J Mol Sci 21:1–27. https://doi.org/10.3390/ijms21239273

Merz D, Geyer M, Moss DA, Ache HJ (1996) Chlorophyll fluorescence biosensor for the detection of herbicides. Anal Bioanal Chem 354:299–305. https://doi.org/10.1007/S0021663540299

Miki W (1991) Biological functions and activities of animal carotenoids. Pure Appl Chem 63:141–146. https://doi.org/10.1351/PAC199163010141/MACHINEREADABLECITATION/RIS

Ming JX, Wang ZC, Huang Y et al (2021) Fucoxanthin extracted from *Laminaria japonica* inhibits metastasis and enhances the sensitivity of lung cancer to Gefitinib. J Ethnopharmacol 265:113302. https://doi.org/10.1016/j.jep.2020.113302

Mittal R, Raghavarao KSMS (2018) Extraction of R-Phycoerythrin from marine macro-algae, *Gelidium pusillum*, employing consortia of enzymes. Algal Res 34:1–11. https://doi.org/10.1016/j.algal.2018.07.002

Mittal R, Tavanandi HA, Mantri VA, Raghavarao KSMS (2017) Ultrasound assisted methods for enhanced extraction of phycobiliproteins from marine macro-algae, *Gelidium pusillum* (Rhodophyta). Ultrason Sonochem 38:92–103. https://doi.org/10.1016/j.ultsonch.2017.02.030

Mittal R, Sharma R, Raghavarao K (2019) Aqueous two-phase extraction of R-Phycoerythrin from marine macro-algae, *Gelidium pusillum*. Bioresour Technol 280:277–286. https://doi.org/10.1016/j.biortech.2019.02.044

Morais T, Cotas J, Pacheco D, Pereira L (2021) Seaweeds compounds: an ecosustainable source of cosmetic ingredients? Cosmetics 8:8. https://doi.org/10.3390/COSMETICS8010008

Morocho-Jácome AL, Ruscinc N, Martinez RM et al (2020) (Bio)Technological aspects of microalgae pigments for cosmetics. Appl Microbiol Biotechnol 104:9513–9522. https://doi.org/10.1007/s00253-020-10936-x

Munier M, Jubeau S, Wijaya A et al (2014) Physicochemical factors affecting the stability of two pigments: R-phycoerythrin of *Grateloupia turuturu* and B-phycoerythrin of *Porphyridium cruentum*. Food Chem 150:400–407. https://doi.org/10.1016/j.foodchem.2013.10.113

Munier M, Morançais M, Dumay J et al (2015) One-step purification of R-phycoerythrin from the red edible seaweed *Grateloupia turuturu*. J Chromatogr B Analyt Technol Biomed Life Sci 992:23–29. https://doi.org/10.1016/j.jchromb.2015.04.012

Muñoz-Miranda LA, Iñiguez-Moreno M (2023) An extensive review of marine pigments: sources, biotechnological applications, and sustainability. Aquat Sci 85:68. https://doi.org/10.1007/s00027-023-00966-8

Muradian Kh, Vaiserman A, Min K-J, Fraifeld VE (2015) Fucoxanthin and lipid metabolism: A minireview. Nutr Metab Cardiovasc Dis 25:891–897. https://doi.org/10.1016/j.numecd.2015.05.010

Murillo AG, Hu S, Fernandez ML (2019) Zeaxanthin: metabolism, properties, and antioxidant protection of eyes, heart, liver, and skin. Antioxidants 8:390. https://doi.org/10.3390/antiox8090390

Murray MT, Capelli B (2020) β-carotene and other carotenoids. In: Textbook of natural medicine, pp 443–450.e2. https://doi.org/10.1016/B978-0-323-43044-9.00057-1

Mysliwa-Kurdziel B, Solymosi K (2016) Phycobilins and phycobiliproteins used in food industry and medicine. Mini-Rev Med Chem 17:1173–1193. https://doi.org/10.2174/1389557516666160912180155

Nagaraj S, Arulmurugan P, Rajaram MG et al (2012) Hepatoprotective and antioxidative effects of C-phycocyanin from *Arthrospira maxima* SAG 25780 in CCl4-induced hepatic damage rats. Biomed Prev Nutr 2:81–85. https://doi.org/10.1016/j.bionut.2011.12.001

Nakano T, Wiegertjes G (2020) Properties of carotenoids in fish fitness: a review. Mar Drugs 18. https://doi.org/10.3390/MD18110568

Nawaz A, Chaudhary R, Shah Z et al (2020) An overview on industrial and medical applications of bio-pigments synthesized by marine bacteria. Microorganisms 9:1–24. https://doi.org/10.3390/MICROORGANISMS9010011

Nazarudin MF, Isha A, Mastuki SN et al (2020) Chemical composition and evaluation of the α-glucosidase inhibitory and cytotoxic properties of marine algae *Ulva intestinalis*, *Halimeda macroloba*, and *Sargassum ilicifolium*. Evid Based Complement Alternat Med 2020:2753945. https://doi.org/10.1155/2020/2753945

Nguyen HPT, Morançais M, Fleurence J, Dumay J (2017) *Mastocarpus stellatus* as a source of R-phycoerythrin: optimization of enzyme assisted extraction using response surface methodology. J Appl Phycol 29:1563–1570. https://doi.org/10.1007/s10811-016-1024-z

Nguyen HPT, Morançais M, Déléris P et al (2020) Purification of R-phycoerythrin from a marine macroalga *Gracilaria gracilis* by anion-exchange chromatography. J Appl Phycol 32:553–561. https://doi.org/10.1007/s10811-019-01947-x

Niu JF, Chen ZF, Wang GC, Zhou BC (2010) Purification of phycoerythrin from *Porphyra yezoensis* Ueda (Bangiales, Rhodophyta) using expanded bed absorption. J Appl Phycol 22:25–31. https://doi.org/10.1007/s10811-009-9420-2

Niu J, Xu M, Wang G et al (2013) Comprehensive extraction of agar and R-phycoerythrin from *Gracilaria lemaneiformis* (Bangiales, Rhodophyta). Indian J Mar Sci 42:21–28

Nomura M, Kamogawa H, Susanto E et al (2013) Seasonal variations of total lipids, fatty acid composition, and fucoxanthin contents of *Sargassum horneri* (Turner) and *Cystoseira hakodatensis* (Yendo) from the northern seashore of Japan. J Appl Phycol 25:1159–1169. https://doi.org/10.1007/s10811-012-9934-x

Noviendri D, Hasrini RF, Subaryono, Marraskuranto E (2022) Biopigments (phycoerythrin, fucoxanthin and siphonaxanthin) from seaweeds and their potential applications as ingredients in cosmeceutical industries: a review. IOP Conf Ser Earth Environ Sci 1118:012028. https://doi.org/10.1088/1755-1315/1118/1/012028

Nowruzi B, Konur O, Amir S, Anvar A (2022) The stability of the phycobiliproteins in the adverse environmental conditions relevant to the food storage. Food Bioprocess Technol 15:2646–2663. https://doi.org/10.1007/s11947-022-02855-8

O'Connor I, O'Brien N (1998) Modulation of UVA light-induced oxidative stress by beta-carotene, lutein and astaxanthin in cultured fibroblasts. J Dermatol Sci 16:226–230. https://doi.org/10.1016/s0923-1811(97)00058-3

Ogashawara I (2020) Determination of phycocyanin from space—a bibliometric analysis. Remote Sens 12:567. https://doi.org/10.3390/rs12030567

Ojulari OV, Gi Lee S, Nam JO (2020) Therapeutic effect of seaweed derived xanthophyl carotenoid on obesity management; overview of the last decade. Int J Mol Sci 21:2502. https://doi.org/10.3390/ijms21072502

Online Etymology Dictionary Photosynthesis | Etymology, origin and meaning of photosynthesis by etymonline. https://www.etymonline.com/word/photosynthesis. Accessed 13 Apr 2023

Osório C, Machado S, Peixoto J et al (2020) Pigments content (Chlorophylls, fucoxanthin and phycobiliproteins) of different commercial dried algae. Separations 7:1–14. https://doi.org/10.3390/separations7020033

Padhi B, Chauhan G, Kandoi D et al (2021) A comparison of chlorophyll fluorescence transient measurements, using handy pea and fluorpen fluorometers. Photosynthetica 59:399–408. https://doi.org/10.32615/PS.2021.026

Pádua D, Rocha E, Gargiulo D, Ramos AA (2015) Bioactive compounds from brown seaweeds: phloroglucinol, fucoxanthin and fucoidan as promising therapeutic agents against breast cancer. Phytochem Lett 14:91–98. https://doi.org/10.1016/j.phytol.2015.09.007

Pagels F, Guedes AC, Amaro HM et al (2019) Phycobiliproteins from cyanobacteria: chemistry and biotechnological applications. Biotechnol Adv 37:422–443. https://doi.org/10.1016/j.biotechadv.2019.02.010

Pangestuti R, Kim SK (2011a) Neuroprotective effects of marine algae. Mar Drugs 9:803–818. https://doi.org/10.3390/MD9050803

Pangestuti R, Kim SK (2011b) Biological activities and health benefit effects of natural pigments derived from marine algae. J Funct Foods 3:255–266. https://doi.org/10.1016/j.jff.2011.07.001

Park JJ, Lee WY (2021) Anti-glycation effects of brown algae extracts and its phenolic compounds. Food Biosci 41:101042. https://doi.org/10.1016/J.FBIO.2021.101042

Park HJ, Lee MK, Park YB et al (2011) Beneficial effects of *Undaria pinnatifida* ethanol extract on diet-induced-insulin resistance in C57BL/6J mice. Food Chem Toxicol 49:727–733. https://doi.org/10.1016/j.fct.2010.11.032

Patel AS, Lakshmibalasubramaniam SP, Nayak B (2020) Steric stabilization of phycobiliprotein loaded liposome through polyethylene glycol adsorbed cellulose nanocrystals and their impact on the gastrointestinal tract. Food Hydrocoll 98:105252. https://doi.org/10.1016/j.foodhyd.2019.105252

Patel SN, Sonani RR, Roy D et al (2022) Exploring the structural aspects and therapeutic perspectives of cyanobacterial phycobiliproteins. 3 Biotech 12(9):1–15. https://doi.org/10.1007/S13205-022-03284-2

Pattarayan D, Rajarajan D, Ayyanar S et al (2017) C-phycocyanin suppresses transforming growth factor-β1-induced epithelial mesenchymal transition in human epithelial cells. Pharmacol Rep 69:426–431. https://doi.org/10.1016/j.pharep.2016.12.013

Peguero-Pina JJ, Morales F, Gil-Pelegrín E (2008) Frost damage in *Pinus sylvestris* L. stems assessed by chlorophyll fluorescence in cortical bark chlorenchyma. Ann For Sci 65:813–813. https://doi.org/10.1051/FOREST:2008068/METRICS

Pemberton RM, Amine A, Hart JP (2007) Voltammetric behavior of chlorophyll a at a screen-printed carbon electrode and its potential role as a biomarker for monitoring fecal contamination. Anal Lett 37:1625–1643. https://doi.org/10.1081/AL-120037592

Peng J, Yuan JP, Wu CF, Wang JH (2011) Fucoxanthin, a marine carotenoid present in brown seaweeds and diatoms: metabolism and bioactivities relevant to human health. Mar Drugs 9:1806–1828. https://doi.org/10.3390/md9101806

Pereira L (2011) A review of the nutrient composition of selected edible seaweeds. In: Pomin VH (ed) Seaweed: ecology, nutrient composition and medicinal uses, 1st edn. Nova Science Publishers, Inc., Coimbra, pp 15–47

Pereira L (2016) Edible seaweeds of the world, 1st edn. CRC Press, Boca Raton

Pereira L (2018) Nutritional composition of the main edible algae. In: Pereira L (ed) Therapeutic and nutritional uses of algae. CRC Press, Boca Raton, pp 65–127

Pereira L (2021) Macroalgae. Encyclopedia 1:177–188. https://doi.org/10.3390/ENCYCLOPEDIA1010017

Pereira L (2023) Atlantic algae as food and their extracts. Explor Foods Foodomics 1:15–31. https://doi.org/10.37349/EFF.2023.00003

Pereira L, Neto JM (2015) Marine algae – biodiversity, taxonomy, environmental assessment, and biotechnology, 1st edn. CRC Press, Boca Raton

Pereira L, Valado A (2021) The seaweed diet in prevention and treatment of the neurodegenerative diseases. Mar Drugs 19:1–25. https://doi.org/10.3390/md19030128

Pereira T, Barroso S, Mendes S et al (2020) Optimization of phycobiliprotein pigments extraction from red algae *Gracilaria gracilis* for substitution of synthetic food colorants. Food Chem 321:126688. https://doi.org/10.1016/j.foodchem.2020.126688

Pereira AG, Otero P, Echave J et al (2021) Xanthophylls from the sea: algae as source of bioactive carotenoids. Mar Drugs 19:188. https://doi.org/10.3390/MD19040188

Pérez-Gálvez A, Viera I, Roca M (2020) Carotenoids and chlorophylls as antioxidants. Antioxidants 9:505. https://doi.org/10.3390/antiox9060505

Pérez-Lloréns JL, Mouritsen OG, Rhatigan P et al (2020) Seaweeds in mythology, folklore, poetry, and life. J Appl Phycol 32:3157–3182. https://doi.org/10.1007/s10811-020-02133-0

Phan MAT, Bucknall M, Arcot J (2018) Interactive effects of β-carotene and anthocyanins on cellular uptake, antioxidant activity and anti-inflammatory activity in vitro and ex vivo. J Funct Foods 45:129–137. https://doi.org/10.1016/j.jff.2018.03.021

Piniella-Matamoros B, Marín-Prida J, Pentón-Rol G (2021) Nutraceutical and therapeutic potential of Phycocyanobilin for treating Alzheimer's disease. J Biosci 46:1–16. https://doi.org/10.1007/s12038-021-00161-7

Poojary MM, Barba FJ, Aliakbarian B et al (2016) Innovative alternative technologies to extract carotenoids from microalgae and seaweeds. Mar Drugs 14:214. https://doi.org/10.3390/md14110214

Prado JM, Veggi PC, Náthia-Neves G, Meireles MAA (2018) Extraction methods for obtaining natural blue colorants. Curr Anal Chem 16:504–532. https://doi.org/10.2174/1573411014666181115125740

Puzorjov A, McCormick AJ (2020) Phycobiliproteins from extreme environments and their potential applications. J Exp Bot 71:3827–3842. https://doi.org/10.1093/JXB/ERAA139

Qiang X, Wang L, Niu J et al (2021) Phycobiliprotein as fluorescent probe and photosensitizer: a systematic review. Int J Biol Macromol 193:1910–1917. https://doi.org/10.1016/J.IJBIOMAC.2021.11.022

Quitério E, Soares C, Ferraz R et al (2021) Marine health-promoting compounds: recent trends for their characterization and human applications. Food Secur 10:3100. https://doi.org/10.3390/FOODS10123100

Ragusa I, Nardone GN, Zanatta S et al (2021) Spirulina for skin care: a bright blue future. Cosmetics 8:7. https://doi.org/10.3390/cosmetics8010007

Rapoport A, Guzhova I, Bernetti L et al (2021) Carotenoids and some other pigments from fungi and yeasts. Metabolites 11:92. https://doi.org/10.3390/metabo11020092

Resende ACR, Pereira R, Nunes C et al (2023) Photosynthetic pigment and carbohydrate profiling of *Fucus vesiculosus* from an Iberian Coastal Lagoon. Plants 12:1324. https://doi.org/10.3390/plants12061324

Robertson RC, Guihéneuf F, Bahar B et al (2015) The anti-inflammatory effect of algae-derived lipid extracts on lipopolysaccharide (LPS)-stimulated human THP-1 macrophages. Mar Drugs 13:5402–5424. https://doi.org/10.3390/md13085402

Rodriguez-Amaya DB (2015) Food carotenoids: chemistry, biology and technology. Wiley-Blackwell, Chichester

Rodriguez-Amaya D, Kimura M (2004) HarvestPlus handbook for carotenoid analysis. HarvestPlus, Washington DC

Román RB, Alvárez-Pez JM, Fernández FGA, Grima EM (2002) Recovery of pure B-phycoerythrin from the microalga *Porphyridium cruentum*. J Biotechnol 93:73–85. https://doi.org/10.1016/S0168-1656(01)00385-6

Rossano R, Ungaro N, D'Ambrosio A et al (2003) Extracting and purifying R-phycoerythrin from Mediterranean red algae *Corallina elongata* Ellis & Solander. J Biotechnol 101:289–293. https://doi.org/10.1016/S0168-1656(03)00002-6

Rula NAM, Ganzon-Fortes ET, Pante MJR, Trono GC (2021) Influence of light, water motion, and stocking density on the growth and pigment content of *Halymenia durvillei* (Rhodophyceae) under laboratory conditions. J Appl Phycol 33:2367–2377. https://doi.org/10.1007/S10811-021-02474-4/FIGURES/6

Saha SK, Ermis H, Murray P (2020) Marine microalgae for potential lutein production. Appl Sci (Switzerland) 10:6457. https://doi.org/10.3390/APP10186457

Sahin SC (2018) The potential of *Arthrospira platensis* extract as a tyrosinase inhibitor for pharmaceutical or cosmetic applications. S Afr J Bot 119:236–243. https://doi.org/10.1016/j.sajb.2018.09.004

Saiki Y, Amao Y (2002) Visible light induced biohydrogen production from sucrose using the photosensitization of Mg chlorophyll-a. Bioconjug Chem 13:898–901. https://doi.org/10.1021/BC025506G/ASSET/IMAGES/LARGE/BC025506GF00004.JPEG

Saini RK, Keum YS (2018) Carotenoid extraction methods: a review of recent developments. Food Chem 240:90–103. https://doi.org/10.1016/j.foodchem.2017.07.099

Sajilata MG, Singhal RS, Kamat MY (2008) The carotenoid pigment zeaxanthin – a review. Compr Rev Food Sci Food Saf 7:29–49. https://doi.org/10.1111/j.1541-4337.2007.00028.x

Saluri M, Kaldmäe M, Tuvikene R (2019) Extraction and quantification of phycobiliproteins from the red alga *Furcellaria lumbricalis*. Algal Res 37:115–123. https://doi.org/10.1016/j.algal.2018.11.013

Saluri M, Kaldmäe M, Tuvikene R (2020) Reliable quantification of R-phycoerythrin from red algal crude extracts. J Appl Phycol 32:1421–1428. https://doi.org/10.1007/s10811-019-01968-6

Samarasinghe N, Fernando S, Lacey R, Faulkner WB (2012) Algal cell rupture using high pressure homogenization as a prelude to oil extraction. Renew Energy 48:300–308. https://doi.org/10.1016/J.RENENE.2012.04.039

Sanders CA, Rodriguez M, Greenbaum E (2001) Stand-off tissue-based biosensors for the detection of chemical warfare agents using photosynthetic fluorescence induction. Biosens Bioelectron 16:439–446. https://doi.org/10.1016/S0956-5663(01)00158-0

Sano F, Murata K, Niwa K (2020) Identification, growth, and pigment content of a spontaneous green mutant of *Pyropia kinositae* (Bangiales, Rhodophyta). J Appl Phycol 32:1983–1994. https://doi.org/10.1007/S10811-020-02085-5/TABLES/1

Schächtl J, Huber G, Maidl FX et al (2005) Laser-induced chlorophyll fluorescence measurements for detecting the nitrogen status of wheat (*Triticum aestivum* L.) canopies. Precis Agric 6:143–156. https://doi.org/10.1007/S11119-004-1031-Y/METRICS

Schepers M, Martens N, Tiane A et al (2020) Edible seaweed-derived constituents: an undisclosed source of neuroprotective compounds. Neural Regen Res 15:790–795. https://doi.org/10.4103/1673-5374.268894

Schubert N, García-Mendoza E, Pacheco-Ruiz I (2006) Carotenoid composition of marine red algae. J Phycol 42:1208–1216. https://doi.org/10.1111/j.1529-8817.2006.00274.x

Schubert D, Currais A, Goldberg J et al (2018) Geroneuroprotectors: effective geroprotectors for the brain. Trends Pharmacol Sci 39:1004–1007

Sekar S, Chandramohan M (2008) Phycobiliproteins as a commodity: trends in applied research, patents and commercialization. J Appl Phycol 20:113–136. https://doi.org/10.1007/s10811-007-9188-1

Sen S, Mallick N (2021) Mycosporine-like amino acids: algal metabolites shaping the safety and sustainability profiles of commercial sunscreens. Algal Res 58:102425. https://doi.org/10.1016/J.ALGAL.2021.102425

Seyed Yagoubi A, Shahidi F, Mohebbi M et al (2017) Preparation, characterization and evaluation of physicochemical properties of phycocyanin-loaded solid lipid nanoparticles and nanostructured lipid carriers. J Food Meas Charact 1:378–385. https://doi.org/10.1007/S11694-017-9650-Y

Sfriso AA, Gallo M, Baldi F (2018) Phycoerythrin productivity and diversity from five red macroalgae. J Appl Phycol 30:2523–2531. https://doi.org/10.1007/s10811-018-1440-3

Shi XM, Jiang Y, Chen F (2002) High-yield production of lutein by the green microalga *Chlorella protothecoides* in heterotrophic fed-batch culture. Biotechnol Prog 18:723–727. https://doi.org/10.1021/bp0101987

Shih SR, Tsai KN, Li YS et al (2003) Inhibition of enterovirus 71-induced apoptosis by allophycocyanin isolated from a blue-green alga *Spirulina platensis*. J Med Virol 70:119–125. https://doi.org/10.1002/jmv.10363

Sies H, Stahl W (2004) Nutritional protection against skin damage from sunlight. Annu Rev Nutr 24:173–200. https://doi.org/10.1146/annurev.nutr.24.012003.132320

Silva SC, Ferreira ICFR, Dias MM, Filomena Barreiro M (2020) Microalgae-derived pigments: a 10-year bibliometric review and industry and market trend analysis. Molecules 25:3406. https://doi.org/10.3390/molecules25153406

Singh A, Ahmad S, Ahmad A (2015) Green extraction methods and environmental applications of carotenoids-a review. RSC Adv 5. https://doi.org/10.1039/c5ra10243j

Solymosi K, Mysliwa-Kurdziel B (2016) Chlorophylls and their derivatives used in food industry and medicine. Mini-Rev Med Chem 17:1194–1222. https://doi.org/10.2174/1389557516666161004161411

Sonani RR, Rastogi RP, Patel R, Madamwar D (2016) Recent advances in production, purification and applications of phycobiliproteins. World J Biol Chem 7:100. https://doi.org/10.4331/WJBC.V7.I1.100

Sorrenti V, Davinelli S, Scapagnini G et al (2020) Astaxanthin as a putative geroprotector: molecular basis and focus on brain aging. Mar Drugs 18:351

Soulier N, Bryant DA (2021) The structural basis of far-red light absorbance by allophycocyanins. Photosynth Res 147:11–26. https://doi.org/10.1007/s11120-020-00787-y

Souza JMC, Castro JZ, Critchley AT, Yokoya NS (2019) Physiological responses of the red algae *Gracilaria caudata* (Gracilariales) and *Laurencia catarinensis* (Ceramiales) following treatment with a commercial extract of the brown alga Ascophyllum nodosum (AMPEP). J Appl Phycol 31:1883–1888. https://doi.org/10.1007/S10811-018-1683-Z/FIGURES/4

Stahl W, Sies H (2005) Bioactivity and protective effects of natural carotenoids. Biochim Biophys Acta Mol basis Dis 1740:101–107. https://doi.org/10.1016/j.bbadis.2004.12.006

Stahl W, Sies H (2012) β-Carotene and other carotenoids in protection from sunlight. Am J Clin Nutr 96:1179S–1184S. https://doi.org/10.3945/AJCN.112.034819

Stefanov M, Rashkov G, Borisova P, Apostolova E (2023) Sensitivity of the photosynthetic apparatus in maize and sorghum under different drought levels. Plants 12:1863. https://doi.org/10.3390/plants12091863

Sun H, Wang Y, He Y et al (2023) Microalgae-derived pigments for the food industry. Mar Drugs 21:82. https://doi.org/10.3390/MD21020082

Swapnil P, Meena M, Singh SK et al (2021) Vital roles of carotenoids in plants and humans to deteriorate stress with its structure, biosynthesis, metabolic engineering and functional aspects. Curr Plant Biol 26:100203. https://doi.org/10.1016/j.cpb.2021.100203

Takahashi S, Murata N (2008) How do environmental stresses accelerate photoinhibition? Trends Plant Sci 13:178–182. https://doi.org/10.1016/j.tplants.2008.01.005

Takaichi S (2011) Carotenoids in algae: distributions, biosyntheses and functions. Mar Drugs 9:1101–1118. https://doi.org/10.3390/md9061101

Takaichi S, Yokoyama A, Mochimaru M et al (2016) Carotenogenesis diversification in phylogenetic lineages of Rhodophyta. J Phycol 52:329–338. https://doi.org/10.1111/jpy.12411

Takeuchi Y, Amao Y (2005) Biohydrogen production from sucrose using the light-harvesting function of zinc chlorophyll-a. Bull Chem Soc Jpn 78:622–625. https://doi.org/10.1246/BCSJ.78.622

Talarico L, Maranzana G (2000) Light and adaptive responses in red macroalgae: an overview. J Photochem Photobiol B 56:1–11. https://doi.org/10.1016/S1011-1344(00)00046-4

Thompson CC, Kruger RH, Thompson FL (2017) Unlocking marine biotechnology in the developing world. Trends Biotechnol 35:1119–1121. https://doi.org/10.1016/J.TIBTECH.2017.08.005

Tomonou Y, Amao Y (2002) Visible light induced hydrogen production with Mg chlorophyll-a from spirulina and colloidal platinum. Biometals 15:391–395. https://doi.org/10.1023/A:1020213122186

Torres PB, Chow F, Furlan CM et al (2014) Standardization of a protocol to extract and analyze chlorophyll a and carotenoids in *Gracilaria tenuistipitata* var. *liui*. Zhang and Xia (Rhodophyta). Braz J Oceanogr 62:57–63. https://doi.org/10.1590/S1679-87592014068106201

Tung J, Goodwin PH, Hsiang T (2013) Chlorophyll fluorescence for quantification of fungal foliar infection and assessment of the effectiveness of an induced systemic resistance activator. Eur J Plant Pathol 136:301–315. https://doi.org/10.1007/S10658-012-0164-5/FIGURES/8

Urrea-Victoria V, Geraldes V, Pinto E, Castellanos L (2023) Photosynthetic pigments and photoprotective metabolites of Colombian pacific marine macroalgae in response to contrasting ultraviolet-index periods. J Exp Mar Biol Ecol 564. https://doi.org/10.1016/J.JEMBE.2023.151908

Van Der Weij-De Wit CD, Doust AB, Van Stokkum IHM et al (2006) How energy funnels from the phycoerythrin antenna complex to photosystem I and photosystem II in cryptophyte *Rhodomonas* CS24 cells. J Phys Chem B 110:25066–25073. https://doi.org/10.1021/jp061546w

van der Westhuizen MM, Oosterhuis DM, Berner JM, Boogaers N (2020) Chlorophyll a fluorescence as an indicator of heat stress in cotton (*Gossypium hirsutum* L.). S Afr J Plant Soil 37:116–119. https://doi.org/10.1080/02571862.2019.1665721

Védrine C, Leclerc JC, Durrieu C, Tran-Minh C (2003) Optical whole-cell biosensor using Chlorella vulgaris designed for monitoring herbicides. Biosens Bioelectron 18:457–463. https://doi.org/10.1016/S0956-5663(02)00157-4

Viera I, Pérez-Gálvez A, Roca M (2018) Bioaccessibility of marine carotenoids. Mar Drugs 16:397. https://doi.org/10.3390/md16100397

Vílchez C, Forján E, Cuaresma M et al (2011) Marine carotenoids: biological functions and commercial applications. Mar Drugs 9:319–333. https://doi.org/10.3390/md9030319

Vinothkanna A, Sekar S (2020) Diagnostic applications of phycobiliproteins. In: Pigments from microalgae handbook, pp 585–610. https://doi.org/10.1007/978-3-030-50971-2_24

Wang J, Sun J, Song Y et al (2013) A label-free microfluidic biosensor for activity detection of single microalgae cells based on chlorophyll fluorescence. Sensors 13:16075–16089. https://doi.org/10.3390/S131216075

Wang L, Qu Y, Fu X et al (2014) Isolation, purification and properties of an R-phycocyanin from the phycobilisomes of a marine red macroalga *Polysiphonia urceolata*. PLoS One 9:e101724. https://doi.org/10.1371/journal.pone.0087833

Wang Z, Li G, Sun H et al (2018) Effects of drought stress on photosynthesis and photosynthetic electron transport chain in young apple tree leaves. Biol Open 7:bio035279. https://doi.org/10.1242/BIO.035279

Wang C, Shen Z, Cui X et al (2020) Response surface optimization of enzyme-assisted extraction of R-phycoerythrin from dry *Pyropia yezoensis*. J Appl Phycol 32:1429–1440. https://doi.org/10.1007/s10811-019-01963-x

Wells ML, Potin P, Craigie JS et al (2017) Algae as nutritional and functional food sources: revisiting our understanding. J Appl Phycol 29:949–982. https://doi.org/10.1007/s10811-016-0974-5

Willcox DC, Scapagnini G, Willcox BJ (2014) Healthy aging diets other than the Mediterranean: a focus on the Okinawan diet. Mech Ageing Dev 136–137:148–162. https://doi.org/10.1016/j.mad.2014.01.002

Wollina U, Voicu C, Gianfaldoni S et al (2018) *Arthrospira platensis* – potential in dermatology and beyond. Open Access Maced J Med Sci 6:176–180. https://doi.org/10.3889/oamjms.2018.033

World Intellectual Property Organization (WIPO) (2016) Patent landscape report: microalgae-related technologies. Patent Landscape Reports

World International Property Organization (2021) PATENTSCOPE. In: Patentscope. https://www.wipo.int/patentscope/en/. Accessed 7 Dec 2021

Xie X, Lu X, Wang L et al (2020) High light intensity increases the concentrations of β-carotene and zeaxanthin in marine red macroalgae. Algal Res 47:101852. https://doi.org/10.1016/j.algal.2020.101852

Xu DP, Li Y, Meng X et al (2017) Natural antioxidants in foods and medicinal plants: extraction, assessment and resources. Int J Mol Sci 18:96. https://doi.org/10.3390/ijms18010096

Yabuta Y, Fujimura H, Kwak CS et al (2010) Antioxidant activity of the phycoerythrobilin compound formed from a dried Korean purple laver (*Porphyra* sp.) during *in vitro* digestion. Food Sci Technol Res 16:347–352. https://doi.org/10.3136/fstr.16.347

Yabuzaki J (2017) Carotenoids database: structures, chemical fingerprints and distribution among organisms. Database 2017:bax004. https://doi.org/10.1093/database/bax004

Yang EJ, Ham YM, Lee WJ et al (2013) Anti-inflammatory effects of apo-9′-fucoxanthinone from the brown alga, *Sargassum muticum*. DARU, J Pharm Sci 21:62. https://doi.org/10.1186/2008-2231-21-62

Yang M, Xuan Z, Wang Q et al (2021) Fucoxanthin has potential for therapeutic efficacy in neurodegenerative disorders by acting on multiple targets. Nutr Neurosci:1–14. https://doi.org/10.1080/1028415X.2021.1926140

Ye ZW, Jiang JG, Wu GH (2008) Biosynthesis and regulation of carotenoids in *Dunaliella*: progresses and prospects. Biotechnol Adv 26:352–360. https://doi.org/10.1016/j.biotechadv.2008.03.004

Ye KX, Fan TT, Keen LJ, Han BN (2019) A review of pigments derived from marine natural products. Isr J Chem 59:327–338. https://doi.org/10.1002/ijch.201800154

Yuan Y, Zhang J, Fan J et al (2018) Microwave assisted extraction of phenolic compounds from four economic brown macroalgae species and evaluation of their antioxidant activities and inhibitory effects on α-amylase, α-glucosidase, pancreatic lipase and tyrosinase. Food Res Int 113:288–297. https://doi.org/10.1016/J.FOODRES.2018.07.021

Zerres S, Stahl W (2020) Carotenoids in human skin. Biochim Biophys Acta Mol Cell Biol Lipids 1865:158588. https://doi.org/10.1016/J.BBALIP.2019.158588

Zhao P, Wang X, Niu J et al (2020) Agar extraction and purification of R-phycoerythrin from *Gracilaria tenuistipitata*, and subsequent wastewater treatment by *Ulva prolifera*. Algal Res 47:101862. https://doi.org/10.1016/j.algal.2020.101862

Zhao W, Hu CM, Zhou W et al (2022) Carotenoids participate in adaptation/resistance of daily desiccation in the intertidal red alga *Neopyropia yezoensis* (Bangiales, Rhodophyta). Algal Res 61:102606. https://doi.org/10.1016/j.algal.2021.102606

Zhu Z, Wu Q, Di X et al (2017) Multistage recovery process of seaweed pigments: investigation of ultrasound assisted extraction and ultra-filtration performances. Food Bioprod Process 104:40–47. https://doi.org/10.1016/J.FBP.2017.04.008

Chapter 14
Role of Seaweed as a Functional Ingredients in Nutraceuticals, Pharmaceuticals, Cosmetics, and Edible Salts

Haresh S. Kalasariya and Leonel Pereira

Abstract Seaweed, a marine botanical treasure, emerges as an exceptional source of versatile functional ingredients poised to redefine multiple industries, encompassing nutraceuticals, pharmaceuticals, cosmetics, and culinary applications. This exploration begins with seaweed's role as a sustainable biomass source, unveiling its diverse attributes that span across the biochemical spectrum, including polysaccharides, proteins, and lipids, influenced by environmental factors and harvesting timing. The chapter provides a comprehensive overview of extraction techniques, shedding light on their efficiencies and the quality of obtained compounds, encompassing solvent extraction, enzymatic hydrolysis, and supercritical fluid extraction. Nutraceutical applications spotlight the extraction, utilization, and health benefits of seaweed-derived bioactive compounds, complemented by case studies showcasing commercial viability. In the pharmaceutical domain, seaweed-derived compounds exhibit valuable properties, including anticancer, anti-inflammatory, and antimicrobial potential, with meticulous attention to challenges, opportunities, and regulatory considerations. The cosmetic and skincare sectors integrate seaweed-derived ingredients, harnessing their unique properties and catering to market trends, while edible salts and seasonings infused with seaweed's essence introduce innovation to culinary creations. The chapter scrutinizes bioprocessing technologies specific to seaweed-based functional ingredient production, emphasizing sustainability and waste reduction. Emerging trends encompass personalized nutrition, clean and green beauty, pharmaceutical innovations, and seaweed's role as a sustainable protein source, reflecting evolving consumer preferences, scientific advancements, and environmental sustainability. In summary, this chapter encapsulates the richness of seaweed-derived functional ingredients, promising transformative impacts across

H. S. Kalasariya
Centre for Natural Products Discovery, School of Pharmacy and Biomolecular Sciences, Liverpool John Moores University, Liverpool, UK

L. Pereira (✉)
Department of Life Sciences, University of Coimbra, Coimbra, Portugal
e-mail: leonel.pereira@uc.pt

N. Trivedi et al. (eds.), *Recent Advances in Seaweed Biotechnology*,
https://doi.org/10.1007/978-981-96-0519-4_14

industries. It epitomizes collaborative efforts, technological strides, and environmental consciousness driving seaweed to the forefront of global challenges, spanning nutrition, health, sustainability, and environmental preservation. Seaweed's journey from ocean depths to scientific and industrial prominence underscores nature's enduring potential.

Keywords Seaweed · Functional ingredients · Nutraceuticals · Pharmaceuticals · Cosmetics · Extraction techniques · Bioprocessing · Sustainability · Bioactive compounds · Marine resources

14.1 Introduction

Seaweeds, a diverse assemblage of marine macroalgae, have garnered considerable attention as a valuable and sustainable biomass reservoir, exhibiting versatile applications across multiple industries (Sudhakar et al. 2018). Seaweeds, often hailed as "nature's superfood," have long been valued for their ecological and nutritional significance. Their functional components are increasingly being explored for their potential applications in nutraceuticals, pharmaceuticals, cosmetics, and edible salts (Acquah et al. 2020). These remarkable marine macrophytes possess a unique capacity to harness solar energy, converting it into valuable biomass through the process of photosynthesis. Their rapid growth rates and minimal resource demands make seaweeds a promising, sustainable, and renewable resource. A noteworthy characteristic of seaweeds lies in their ability to absorb and accumulate essential minerals and trace elements from the surrounding seawater. These attributes endow seaweeds not only with exceptional nutritional qualities but also underline their potential as a source of multifunctional components with diverse industrial applications (Choudhary et al. 2021). Functional ingredients, defined as substances offering distinct health or cosmetic benefits beyond their fundamental nutritional attributes, have garnered escalating importance across diverse industries. This surge in demand for such ingredients is primarily driven by consumer preferences for products promoting well-being and augmenting the overall quality of life (Helkar et al. 2016). Within the domain of nutraceuticals, functional components derived from seaweed are highly esteemed for their abundance of vitamins, minerals, antioxidants, and bioactive compounds (Peñalver et al. 2020). These constituents have the potential to bestow an array of health advantages, encompassing enhancements in immune function, cardiovascular well-being, and digestive health. Furthermore, these nutraceuticals derived from seaweed exhibit promise in both the prevention and management of chronic diseases (Lordan et al. 2011). Seaweed-based compounds have garnered substantial interest as prospective pharmaceutical agents (Lomartire and Gonçalves 2022a, b). Their distinct biochemical profiles present promising avenues for drug discovery, particularly within the domains of cancer research, anti-inflammatory therapies, and antimicrobial treatments. Seaweed-based pharmaceuticals hold the capacity to catalyse innovation in healthcare by offering novel therapeutic options. Seaweed's impact extends to the cosmetics sector, where it serves as a wellspring of novel ingredients for skincare and beauty products (Polat et al. 2023). Its inherent richness in polysaccharides, vitamins, and minerals positions it as a highly sought-after constituent in formulations designed to enhance skin health and appearance (Kim 2014). Seaweed-based cosmetics provide benefits encompassing hydration, antiageing, and skin rejuvenation. Edible salts infused with seaweed not only impart unique flavours but also introduce a novel dimension to culinary experiences. These salts, frequently enriched with sea minerals and the subtle umami of seaweed, are cherished by culinary professionals and gastronomy enthusiasts alike. Seaweed-infused salts contribute to the burgeoning trend of sustainable and palatable cuisine (Kumar et al. 2023).

The present chapter is structured to provide a comprehensive exploration of seaweed-derived functional ingredients and their diverse applications. It commences by delving into the multifarious seaweed species and their distinctive characteristics, thereby establishing the groundwork for the extraction and utilization of functional compounds. A brief view on a variety of extraction techniques will be scrutinized to underscore their efficiency and efficacy in acquiring high-quality ingredients. In addition, this chapter will subsequently transition to a detailed discussion of specific applications in nutraceuticals, pharmaceuticals, cosmetics, and culinary contexts, spotlighting successful case studies and expounding upon the health and economic advantages associated with seaweed-derived ingredients.

Seaweeds, also known as macroalgae, encompass a diverse range of marine photosynthetic multicellular eukaryotic organisms thriving in aquatic ecosystems worldwide. These organisms have gained increasing attention due to their potential as a sustainable source of functional ingredients across various industries (Balina et al. 2017). The exploration of the various seaweed species commonly employed for the extraction of functional constituents is conducted, with an examination of the fundamental biochemical elements that render them valuable and scrutiny of the pivotal factors influencing their composition (Kim 2012). It can be broadly categorized into three primary taxonomic groups predicated on their pigmentation and photosynthetic pigments: brown algae (Ochrophyta and Phaeophyceae), red algae (Rhodophyta), and green algae (Chlorophyta). Each of these categories encompasses a plethora of distinct species, each possessing unique characteristics and biochemical profiles (Kılınç et al. 2013). The selection of a specific seaweed species for functional ingredient extraction hinges upon the intended application and the specific bioactive compounds of interest.

Marine macroalgae represent a diverse group of eukaryotic, photosynthetic, and multicellular organisms inhabiting aqueous environments, commonly referred to as seaweeds. These ubiquitous organisms populate saltwater realms and coastal regions, and taxonomically (Kalasariya and Pereira 2022; Kalasariya et al. 2021, 2022b). Each of these taxonomic groups exhibits distinctive characteristics and profiles of bioactive compounds. For example, brown algae such as *Ascophyllum nodosum* and *Laminaria digitata* (Fig. 14.1) are frequently harnessed for the production of functional ingredients. Noteworthy for their rich content of laminarin, fucoidan, and alginates, these species have garnered attention (Belanche et al. 2016). Laminarin is a β-glucan with potential immunomodulatory properties, while fucoidan exhibits diverse bioactivities, including antioxidant and anti-inflammatory effects. Alginates, due to their gelling and emulsifying properties, find application in the food and pharmaceutical industries (Hou et al. 2020).

Red algae species like *Porphyra* sp. (nori), *Palmaria palmata* (dulse), and *Gracilaria* sp. (Fig. 14.2) are renowned for their use in food, cosmetics, and nutraceuticals (Holdt and Kraan 2011). These species boast an abundance of phycoerythrin, agar, carrageenan, and other bioactive compounds. Phycoerythrin is a red pigment with potential antioxidant properties, while agar and carrageenan serve as widely used hydrocolloids, functioning as gelling agents and stabilizers in the food industry (Shafie et al. 2022).

Green algae species such as *Ulva lactuca* (macroalga) and *Chlorella vulgaris* (microalga) (Fig. 14.3) are highly regarded for their elevated protein content, chlorophyll, and an array of other pigments (Suganya et al. 2016).

Seaweed's utilization for the extraction of functional ingredients is predicated upon its intricate biochemical composition. Among the principal constituents of interest, seaweeds are distinguished by their abundant polysaccharide content, encompassing alginate, agar, carrageenan, laminarin, and fucoidan (Wijesekara and Karunarathna 2017). These polysaccharides exhibit distinct structural and

Fig. 14.1 *Ascophyllum nodosum* (left) and *Laminaria digitata* (right) (Phaeophyceae). Scale bar = 1 cm

Fig. 14.2 *Porphyra* sp. (left), *Palmaria palmata* (centre), and *Gracilaria* (right) (Rhodophyta). Scale bar = 1 cm

Fig. 14.3 *Ulva lactuca* (scale bar = 1 cm) (left) and *Chlorella vulgaris* (scale bar = 10 μm) (right - © ACOI - Coimbra Collection of Algae) (Chlorophyta)

functional attributes, rendering them invaluable across a spectrum of industries. Notably, alginate finds utility in the food industry as a thickening and gelling agent, while fucoidan holds promise for health-related benefits, including anticoagulant and anti-inflammatory properties (Tanna and Mishra 2019). Seaweeds harbour proteins that serve as a wellspring of essential amino acids, along with the presence of bioactive peptides with potential health-enhancing properties. These proteins are of particular interest in the development of functional foods and nutraceuticals (Li-Chan 2015). Furthermore, seaweeds encompass lipids, including indispensable fatty acids such as Omega-3 and Omega-6. These lipids find applications in the production of dietary supplements, notably owing to their potential advantages for cardiovascular health and mitigating inflammation (Wu et al. 2016). In addition to these constituents, seaweeds are endowed with essential minerals such as iodine, calcium, magnesium, and iron. The mineral composition is subject to variation among species and can be influenced by environmental factors. Iodine, in particular, stands as a crucial component of thyroid hormones and plays a pivotal role in thyroid function (Soetan et al. 2010; Cornish et al. 2017).

14.1.1 Factors Affecting Seaweed Composition

The biochemical composition of seaweed is a dynamic parameter, subject to variability contingent upon several influential factors. Environmental conditions play a crucial role, as seaweed composition is notably responsive to prevailing conditions within its habitat. Parameters such as water temperature, salinity, light intensity, and nutrient availability exert considerable influence over the abundance of bioactive constituents. Certain species, for instance, exhibit heightened production of secondary metabolites as a response to environmental stressors (Hafting et al. 2015). The timing of seaweed harvest holds paramount significance in the procurement of the desired functional ingredients. Variables including the growth stage, seasonal variation, and reproductive cycle of the seaweed can have discernible effects on the concentration and quality of targeted compounds. Harvesting at the opportune moment is imperative to optimize the yield of functional ingredients (Chauton et al. 2021). Additionally, seaweed species may manifest regional disparities in their biochemical makeup, a consequence of disparities in localized environmental factors. This geographic diversity provides opportunities for tailored ingredient sourcing predicated on specific chemical profiles (Lange et al. 2020).

14.1.2 Extraction of Functional Ingredients from Seaweed

The extraction of functional ingredients from seaweed constitutes a pivotal process in unlocking the bioactive potential of these marine organisms (Wijesinghe and Jeon 2012). Various extraction techniques have been developed to isolate and concentrate specific compounds with valuable properties. This section provides an in-depth exploration of the diverse methodologies employed in the extraction process, encompassing solvent extraction, enzymatic hydrolysis, and supercritical fluid extraction, with a comparative analysis of their efficiency and the quality of the obtained compounds (Liaqat et al. 2023). For instance, solvent extraction, a conventional method extensively employed for the isolation of functional ingredients from seaweed, involves the utilization of organic solvents, such as ethanol, methanol, or water, to dissolve and extract target compounds from the seaweed biomass (Rodriguez Garcia and Raghavan 2022). The choice of solvent hinges upon the nature of the desired functional ingredient and its solubility properties (da Silva et al. 2016a). Renowned for its versatility, solvent extraction permits the extraction of a wide array of compounds, including polysaccharides, phenolic compounds, and lipids. It is favoured for its high extraction efficiency and relatively modest equipment costs (Thakur et al. 2022). Nevertheless, challenges may arise in terms of solvent toxicity and the potential presence of residual solvent in the final product, necessitating meticulous optimization and purification steps (Darvishmanesh et al. 2011).

Second, enzymatic hydrolysis emerges as an eco-friendly and selective extraction technique relying on enzymes to disassemble complex macromolecules in seaweed, such as polysaccharides and proteins, into smaller, bioactive fragments. This process may be tailored to target specific compounds of interest, delivering several merits, including heightened specificity and selectivity, reduced environmental impact, and the potential for generating bioactive peptides and oligosaccharides with augmented functionality (Vollet Marson et al. 2020). The quality of the extracted compounds typically excels in terms of purity and bioactivity. However, enzymatic hydrolysis may entail prolonged processing times and the use of specialized enzymes, necessitating a careful consideration of the trade-offs between extraction efficiency and production costs (Caldeira et al. 2020).

Supercritical fluid extraction (SFE), an advanced technique, leverages supercritical fluids, often carbon dioxide (CO_2), as a solvent to extract functional ingredients from seaweed. Under specific temperature and pressure conditions, CO_2 transitions into a supercritical state, manifesting the properties of both a gas and a liquid (da Silva et al. 2016b; Herrero et al. 2015). This unique state empowers it to effectively permeate solid matrices and selectively extract target compounds. SFE offers multiple advantages, including increased selectivity, minimal solvent residues, and a reduced environmental footprint. It is particularly suitable for extracting volatile compounds and thermally sensitive bioactives (Zia et al. 2022). The quality of the extracted compounds often excels, as the gentle extraction conditions minimize degradation. Nevertheless, the implementation of SFE necessitates specialized equipment and expertise, rendering it a more costly option than some other extraction methods.

Choosing the appropriate extraction method is a critical step in the process, as there are numerous extraction techniques available today. Traditional extraction methods include heat-assisted extraction (maceration), percolation, and Soxhlet extraction, as documented by Padilla et al. (2005). Maceration, a classical method, involves immersing marine algae in suitable solvent or solvent combinations (Kim et al. 2013a, b, c). At an industrial scale, ethanol is often preferred due to its economic advantages (Stengel and Connan 2015). Methanol, ethanol, acetone, water, and ethyl ethanoate, in varying proportions, are commonly employed solvents for extraction, with selection based on polarity. Hydroalcoholic solvents are particularly effective due to the hydrophilic nature of the compounds involved. Some studies have suggested improved extraction of phenolic compounds through the addition of acids such as citric acid, tartaric acid, or HCl (Santos-Buelga et al. 2012; Vieira et al. 2018). Moreover, Soxhlet extraction, a traditional method, offers high yields but has drawbacks, including the degradation of temperature-sensitive compounds. Phenolic acids, tannins, and anthocyanins necessitate a large quantity of solvents and extended processing times. Soxhlet extraction is also a continuous process, allowing for solvent recycling, which reduces time and solvent usage compared to maceration and percolation (Ospina et al. 2017). However, the constant heating at the solvent's boiling point may damage temperature-sensitive components, affecting subsequent analysis (Matulja et al. 2022). The Soxhlet method has been employed for extracting lipophilic compounds as well (Santos-Buelga et al. 2012). Lastly, heat-assisted extraction can be categorized into two phases: a fast phase and a slower one. The fast phase involves transferring compounds from the matrix surface to the solvent, while the slower phase encompasses diffusion from the matrix into the solvent. Key factors influencing extraction include sample type, solvent type, temperature, and time. Nevertheless, heat-assisted extraction necessitates filtration, decantation, or clarification to separate solid components, and it consumes significant time and solvents. Consequently, these classical methods are considered inefficient and environmentally unfriendly due to their heavy reliance on organic solvents (Dai and Mumper 2010).

Advancements in extraction techniques have led to the development of more efficient and environmentally friendly methods. Pressurized liquid extraction (PLE), also known as extraction with a pressurized solvent, operates at high pressure (1–15 MPa), short processing times, and temperatures ranging from 50 to 200 °C, using a minimal amount of nontoxic solvent. Otero et al. (2019) observed high extraction yields using PLE for brown alga *Laminaria ochroleuca*. Microwave-assisted extraction is primarily utilized for extracting polyphenols and polysaccharides. This technique can be performed under open or closed vessel conditions and relies on electromagnetic waves that alter cell structures. Two mechanisms, ionic conduction and dipole rotation, transform electromagnetic energy into heat (Kalil et al. 2017; Cikoš et al. 2018). Variables such as extraction type, microwave frequency, solid-to-solvent ratio, temperature, pressure, and time play significant roles. However, high microwave power and elevated temperature may lead to the degradation of phenolic compounds (Pinela et al. 2016). Ultrasound-assisted extraction (UAE) is another method for extracting phenolic compounds from algae. It offers advantages such as low temperature, shorter processing times, and minimal solvent usage. However, ultrasonication may elevate the temperature, potentially impacting the stability of phenolic compounds. Other factors like frequency, power, time, temperature, and solid-solvent ratio must be considered. UAE employs ultrasound waves with frequencies ranging from 20 to 100 kHz, generating bubbles due to pressure fluctuations. These bubbles collapse, leading to cavitation, which breaks down particles near the liquid-solid interface, releasing bioactive compounds (Heleno et al. 2016).

Subcritical water extraction (SWE) requires the application of water at higher temperatures (100–374 °C) than its boiling point, under high pressure (10–60 bar), maintaining its liquid state for 5–10 min. Parameters such as pressure, time, temperature, and solvent choice impact the extraction process (da Silva et al. 2016b; Herrero et al. 2015). The supercritical CO_2 (SC-CO_2) method utilizes CO_2 as a nontoxic gas, acting as both a liquid and a gas, making extraction more straightforward. Ethanol can be used to modulate CO_2 polarity, while low temperature and pressure are employed to minimize the degradation of phenolic compounds. Various seaweed species have been explored for a wide range of biochemicals and their potential benefits to the skin. The selection of an extraction method for seaweed-derived functional ingredients should consider factors such as the nature of the target compounds, production scale, environmental concerns, and cost-efficiency (Holgate and Ruddy 2023).

In order to select the appropriate extraction methods for seaweed-derived functional ingredients, a comprehensive comparative analysis is indispensable. Key considerations encompass the specific functional compounds sought, where solvent extraction's versatility must be balanced against potential solvent residues, enzymatic hydrolysis's promise of high purity tempered by longer processing times, and supercritical fluid extraction's suitability for thermally sensitive compounds. Environmental impact is pivotal, with enzymatic hydrolysis and supercritical fluid extraction emerging as more environmentally friendly options due to their reduction in toxic solvent use and waste generation. Furthermore, the selection process necessitates evaluating equipment and cost factors, with solvent extraction often favoured for its cost-effectiveness, while enzymatic hydrolysis and supercritical fluid extraction entail specialized equipment and potentially higher capital and operational costs. The dimension of selectivity is vital; enzymatic hydrolysis and supercritical fluid extraction excel in selectively isolating specific bioactive compounds, in contrast to solvent extraction, which may yield a broader spectrum of extracted compounds. Quality is another determinant, as enzymatic hydrolysis and supercritical fluid extraction often yield high-quality extracts with minimal degradation, necessitating fewer purification steps compared to solvent extraction. The choice of extraction method is influenced by a complex interplay of these factors, including the nature of the target compounds, environmental sustainability, equipment availability, and production costs. Researchers and industry professionals should meticulously weigh these considerations to optimize the extraction process, ensuring the quality and sustainability of the final functional ingredients. As the field continues to evolve, ongoing innovations in extraction techniques hold the potential to enhance efficiency and sustainability in seaweed-based ingredient production.

14.1.3 Nutraceuticals from Seaweed

The realm of nutraceuticals, a fusion of "nutrition" and "pharmaceuticals," is utilizing seaweed and seaweed-derived bioactive compounds that have potential nutritional value and health-enhancing attributes (Calado et al. 2018). In this section, the extraction and utilization of bioactive compounds from seaweed for nutraceutical applications briefly focused mainly on the potential health benefits and medical uses of these seaweed-derived nutraceutical actions and spotlight successful market products. Seaweeds stand as a bountiful source of bioactive compounds, encompassing vitamins, minerals, antioxidants, and other phytochemicals, rendering them a valuable resource for the nutraceutical industry (Pradhan et al. 2022). Seaweeds abound with essential vitamins like vitamin A, vitamin C, and various B vitamins, including B_1 (thiamine), B_2 (riboflavin), B_9 (folate), and B_{12} (cobalamin). These vitamins play pivotal roles in diverse physiological processes, including immune function, energy metabolism, and skin health (Beane et al. 2021). They also serve as exceptional resources of minerals like iodine, calcium, magnesium, and iron. Iodine, in particular, assumes a crucial role in thyroid health and hormone production, while calcium contributes to both bone health and muscle function (Gharibzahedi and Jafari 2017). They are abundant in natural antioxidants, comprising polyphenols, flavonoids, and carotenoids. These compounds combat oxidative stress, diminish inflammation, and shield cells from damage induced by free radicals. Antioxidants derived from seaweed hold the potential to promote overall health and longevity (Kumar et al. 2021b; Matanjun et al. 2008). In addition, certain seaweed species incorporate omega-3 and omega-6 fatty acids, known for their benefits in heart health, brain function, and the mitigation of inflammation (Saini and Keum 2018).

The extraction of these bioactive compounds from seaweed can be achieved through a variety of techniques, including solvent extraction, enzymatic hydrolysis, and supercritical fluid extraction, as previously discussed. Once extracted, these compounds are incorporated into nutraceutical products in diverse forms such as capsules, tablets, powders, or liquid supplements. Seaweed-derived nutraceuticals proffer a wide spectrum of health advantages and potential medical applications owing to their rich biochemical profiles: certain seaweeds, such as kelp, with their high iodine content, effectively support thyroid function and guard against iodine deficiency disorders like goitre. Seaweed's natural antioxidants combat oxidative stress and inflammation, diminishing the risk of chronic ailments, such as cardiovascular disease and cancer (Bhatt et al. 2022). Some seaweed compounds, like fucoxanthin, have been subject to research for their potential to aid in weight management by stimulating fat metabolism and reducing fat accumulation. Seaweed-derived prebiotics and fibres may bolster a healthy gut microbiome, potentially ameliorating digestive disorders and enhancing nutrient absorption (Praveen et al. 2019; Shannon et al. 2021). Seaweeds' vitamin and mineral content, along with their immune-modulating polysaccharides, have the potential to enhance immune function and resilience against infections. Compounds from seaweed, such as fucoidan, exhibit anti-inflammatory properties, suggesting potential applications in managing chronic inflammatory conditions (Khalid et al. 2018).

The market for nutraceuticals derived from seaweed has experienced remarkable growth, propelled by the increasing awareness among consumers of the health benefits associated with these marine resources. A series of successful case studies serve as compelling evidence of the commercial viability and effectiveness of such products. Fucoidan, a sulphated polysaccharide found in brown seaweeds, has been subject to extensive research for its potential health advantages. Certain supplements based on fucoidan claim to bolster immune function, mitigate inflammation, and enhance cardiovascular health (Jayawardena et al. 2022). Kelp-based supplements, enriched with iodine, are positioned as natural thyroid support products, addressing concerns about iodine deficiency and the maintenance of optimal thyroid function (Andersson et al. 2007). Algal oil supplements, sourced from microalgae like *Schizochytrium* and *Ulkenia*, present a plant-based alternative to fish oil for obtaining Omega-3 fatty acids (Katiyar and Arora 2020). These supplements cater to consumers seeking heart and brain health benefits without relying on fish-derived sources. Carrageenan, derived from red seaweed species such as *Eucheuma* sp. and *Kappaphycus* sp. (Fig. 14.4), serves various purposes, including as a food additive and pharmaceutical ingredient (Shafie et al. 2022). It functions as a thickening and gelling agent in a range of products, including nutraceuticals (Imeson 2011). Carrageenan-based supplements are valued for their capacity to enhance digestive health by forming a protective layer within the gastrointestinal tract, potentially reducing symptoms of diarrhoea and indigestion (Manzoor et al. 2020).

Astaxanthin, a potent carotenoid with antioxidant and anti-inflammatory properties, is obtained from the microalga *Haematococcus lacustris* (formerly *Haematococcus pluvialis*) (Fig. 14.5) of the Chlorophyceae class. Astaxanthin supplements have gained attention for their potential benefits in skin health, eye health, and the reduction of exercise-induced muscle damage (Mularczyk et al. 2020). Some studies suggest that astaxanthin may enhance endurance and skin elasticity.

Fucoxanthin, a carotenoid found in brown seaweed species such as *Undaria pinnatifida* (wakame) (Fig. 14.6) and *Saccharina japonica* (formerly *Laminaria japonica*), of the Phaeophyceae class (kombu), is marketed for its potential in weight management by promoting fat oxidation and inhibiting fat cell proliferation (Din et al. 2022). Research has indicated that fucoxanthin may aid in weight reduction and improve metabolic parameters.

Algal oil supplements sourced from microalgae like *Schizochytrium* and *Ulkenia* offer a plant-based alternative to fish oil for obtaining Omega-3 fatty acids (Saini et al. 2021; Patel et al. 2021). These supplements are rich in docosahexaenoic acid (DHA) and eicosapentaenoic acid (EPA), essential for heart and brain health. They are especially popular among vegetarians and vegans and have been demonstrated to confer similar cardiovascular benefits to fish oil (Santos et al. 2020). With the ongoing advancement of our scientific understanding of marine-derived

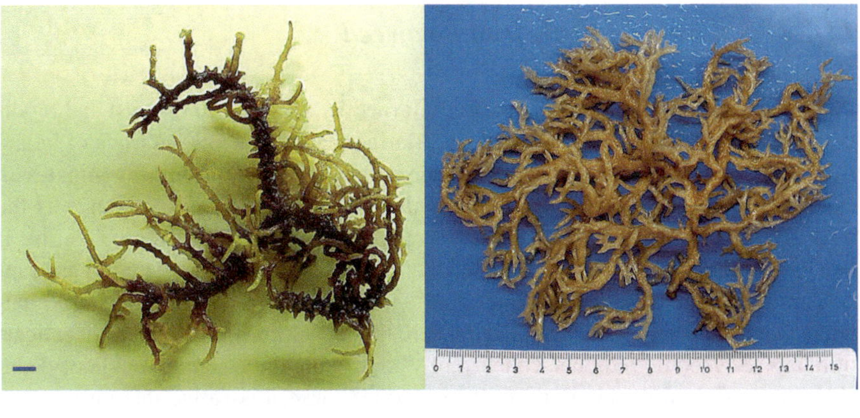

Fig. 14.4 *Eucheuma* (left) and *Kappaphycus* (right). Scale bar = 1 cm

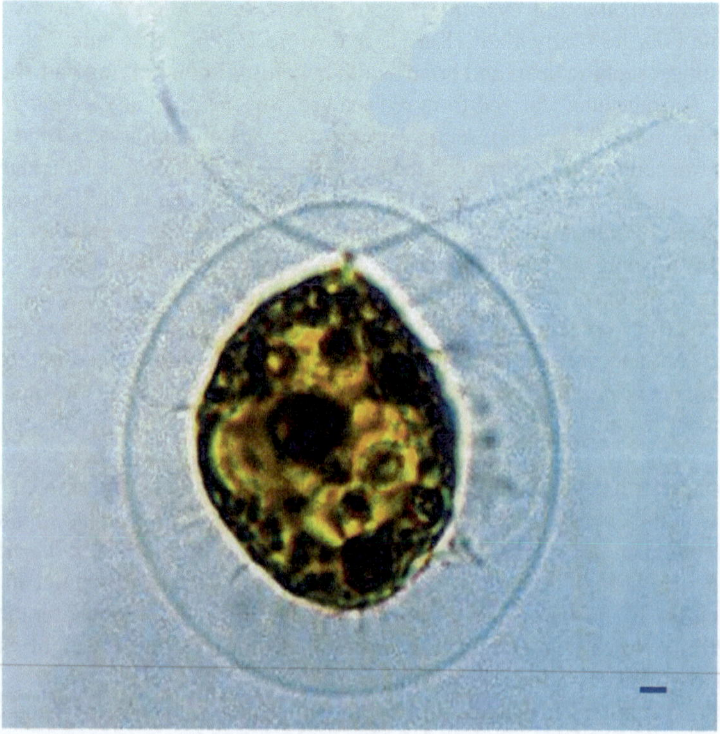

Fig. 14.5 *Haematococcus lacustris* (scale bar = 10 μm) (© ACOI - Coimbra Collection of Algae)

compounds, the anticipation arises for the creation of increasingly innovative seaweed-based nutraceuticals. These advancements are expected to play a crucial role in enhancing overall well-being and promoting sustainability.

14.1.4 Pharmaceuticals from Seaweed

Pharmaceutical research has increasingly turned its attention to the vast and biodiverse world beneath the ocean's surface. Marine macroalgae offer a trove of bioactive compounds with remarkable pharmaceutical potential. This section initiates an exploration of seaweed-derived compounds that are positioned to advance as the next generation of pharmaceutical agents (Lomartire and Gonçalves 2022a, b). It delves into their versatile applications, encompassing properties related to anticancer, anti-inflammatory, and antimicrobial uses. Additionally, it uncovers the challenges and opportunities associated with the development of pharmaceuticals utilizing seaweed ingredients, concurrently addressing the essential regulatory considerations and approvals required to introduce these innovative therapies to the market. Some macroalgae are rich sources of bioactive molecules with

Fig. 14.6 *Undaria pinnatifida* (wakame) (Phaeophyceae). Scale bar = 1 cm

demonstrated anticancer properties. Compounds like fucoidan, derived from brown algae, have exhibited potential in inhibiting cancer cell proliferation, inducing apoptosis (programmed cell death) and suppressing tumour angiogenesis. Preclinical studies have shown promise in the development of fucoidan-based anticancer drugs (Do-Amaral et al. 2020; Lu et al. 2021). Inflammation is at the root of many chronic diseases. Seaweed compounds, particularly polysaccharides and polyphenols, possess anti-inflammatory properties. For instance, carrageenan, extracted from red seaweed, has been studied for its potential to reduce inflammation and may find applications in treating inflammatory conditions (Campo et al. 2009).

Seaweeds are known to produce secondary metabolites with antimicrobial activity. These compounds have shown efficacy against a wide range of pathogens, including bacteria, fungi, and viruses. For example, sulphated polysaccharides from brown algae have demonstrated antiviral properties and are being explored as potential treatments for viral infections (Hans et al. 2021). One of the primary challenges in seaweed-based pharmaceutical development is the extraction and purification of bioactive compounds. Seaweed extracts can be complex mixtures, requiring sophisticated separation techniques to isolate the target molecules efficiently (Quitério et al. 2022). Achieving consistent product quality is vital for pharmaceutical applications. Standardizing the composition and bioactivity of seaweed-derived compounds is a critical challenge, especially when working with natural sources subject to environmental variations (Leandro et al. 2020). Ensuring that the active ingredients are absorbed effectively within the human body is essential. Improving the bioavailability of seaweed-derived compounds is a focus of ongoing research (Martins et al. 2022). Comprehensive safety assessments and toxicological studies are necessary to establish the safety profiles of seaweed-based pharmaceuticals. This includes evaluating potential side effects, interactions, and long-term effects (Singh et al. 2012). Navigating regulatory pathways for novel pharmaceuticals can be complex and time-consuming. Obtaining regulatory approvals from agencies like the Food and Drug Administration (FDA) or the European Medicines Agency (EMA) requires rigorous preclinical and clinical trials, as well as compliance with good manufacturing practices (GMP) (Souto et al. 2020). Protecting intellectual property and securing funding for research and development are critical aspects of bringing seaweed-based pharmaceuticals to market. Collaborations between academia and industry play a pivotal role in advancing these innovations (Duarte et al. 2022). Regulatory considerations and approvals are pivotal steps in the pharmaceutical development process, and seaweed-based therapies are no exception: Before entering clinical trials, extensive preclinical studies are conducted to evaluate the safety and efficacy of seaweed-derived compounds (Kalasariya et al. 2022a). These studies provide crucial data for regulatory submissions. Clinical trials in humans follow a well-defined protocol and proceed through phases I to III to assess safety, dosage, and efficacy. The data generated from these trials form the basis for regulatory submissions (Andrade et al. 2016). Pharmaceutical developers must compile comprehensive dossiers containing data from preclinical and clinical studies, chemistry, manufacturing, and controls (CMC) data, and information on product quality and safety which are assessed for the benefit-risk profiles of seaweed-based pharmaceuticals by regulatory agencies (Afonso et al. 2019; Gegg and Wells 2019).

The exploration of seaweed-derived compounds with pharmaceutical potential is a promising frontier in drug discovery and development. These compounds offer exciting possibilities for addressing unmet medical needs in areas like cancer, inflammation, and infectious diseases (Bouley et al. 2023). However, the path to pharmaceutical success is fraught with challenges, including extraction complexities, bioavailability concerns, and stringent regulatory requirements. As research in this field advances and regulatory pathways are navigated, seaweed-based pharmaceuticals hold the potential to revolutionize healthcare and expand the repertoire of therapeutic options available to patients worldwide (Acharya et al. 2018).

By looking at successful studies, fucoidan has garnered significant attention for its potential as an anticancer therapy. It exhibits various mechanisms of action, including apoptosis induction, inhibition of angiogenesis, and modulation of immune responses, making it a promising candidate for cancer treatment (De Silva et al. 2012). However, extracting and purifying fucoidan can be challenging due to variations in seaweed sources and environmental conditions, necessitating optimization of extraction methods for consistent quality (García-Vaquero et al. 2017). Addressing the limited bioavailability of fucoidan requires the development of novel delivery systems to enhance absorption, and clinical trials are underway to evaluate the safety and efficacy of fucoidan-based therapies in various cancer types (Luthuli et al. 2019). Preclinical studies demonstrating fucoidan's potential in inhibiting tumour growth in animal models have led to ongoing clinical trials assessing its effectiveness in human cancer patients, with initial results showing promise in cancer treatment and management (Hsu and Hwang 2019). In addition, Carrageenan's antiviral properties have been explored for their potential in preventing viral infections by forming a protective barrier on mucosal surfaces, preventing viruses from binding to host cells. Achieving consistent quality and standardization of carrageenan products is essential, as variations in structure can affect antiviral efficacy (Rahman et al. 2022). Extensive safety and efficacy studies are required to demonstrate the compound's effectiveness against various viruses without adverse effects (Pergolizzi et al. 2010). Developing carrageenan-based antiviral products, such as topical gels and nasal sprays, requires careful formulation and testing. Some carrageenan-based products have been approved for use in preventing the spread of certain viral infections, offering a new approach to antiviral protection (Leibbrandt et al. 2010). Alginate's exceptional wound-healing properties stem from its ability to create a moist environment that promotes tissue repair, making it valuable for wound dressings and tissue engineering applications. Ensuring that alginate-based materials are biocompatible and do not elicit adverse reactions in the wound environment is crucial (Sun and Tan 2013). Developing alginate-based products with appropriate mechanical strength for different wound types and depths is a key consideration (Tan et al. 2023). Clinical trials are necessary to demonstrate the effectiveness of alginate-based wound dressings in promoting wound healing and tissue regeneration (Kurakula et al. 2020). In clinical settings, alginate-based wound dressings and tissue scaffolds have been developed and widely adopted. Studies have confirmed their ability to enhance wound healing, reduce infection risk, and support tissue regeneration in various applications, including diabetic foot ulcers and burn injuries (Moura et al. 2013). Moreover, seaweed-derived compounds, particularly polyphenols and peptides, exhibit antimicrobial properties that can be harnessed for the treatment of skin infections, such as bacterial and fungal infections (Jiménez-González et al. 2023; Ruocco et al. 2016). Developing effective formulations for topical application that deliver the active compounds to the site of infection while minimizing skin irritation is a critical aspect of product development. Preclinical and clinical studies are necessary to assess the safety and efficacy of seaweed-derived antimicrobial agents (Karuppusamy et al. 2022). Navigating regulatory pathways for the approval of seaweed-derived antimicrobial products for skin infections is a crucial step. Research and development efforts have culminated in the creation of topical antimicrobial products containing seaweed-derived compounds (Lomartire and Gonçalves 2022a, b), and these products have shown promise in treating a range of skin infections, including bacterial and fungal conditions. Ongoing clinical trials aim to validate their safety and efficacy.

14.1.5 Seaweed in Cosmetics and Skincare

Natural and sustainable ingredients have become a prominent trend in cosmetics and skincare, and among these coveted treasures from the sea, seaweed-derived components (see Table 14.1) have gained significant attention in the beauty industry. In this section, the exploration focuses on seaweed-derived ingredients integrated into cosmetic formulations, including products like alginate masks and seaweed extracts (Tan et al. 2013). The manifold skincare benefits and cosmetic properties offered by these marine wonders are unveiled. Additionally, an analysis of the market trends and consumer demand propelling the surge in seaweed-based cosmetic products is presented (Sarker and Mahmudul Islam 2022; Dolbeth and Arenas 2022). Seaweed-derived bioactive constituents are rich in potential biological activities, making them highly suitable for cosmeceutical products. Compounds derived from macroalgae have gained considerable importance in the preparation of cosmeceutical products (Brunt and Burgess 2018). Polysaccharides play a pivotal role in cosmetics, serving as moisturizers, emulsifiers, wound healing agents, and thickening agents (Percival 1979). For instance, fucoidan, sourced from *Chnoospora minima* (Phaeophyceae), exhibits anti-inflammatory properties by inhibiting lipopolysaccharides-induced nitric oxide production, inducible nitric oxide production, cyclooxygenase-2, and prostaglandin E2 levels (Fernando et al. 2017). Similarly, polysaccharides from *Fucus vesiculosus* (Phaeophyceae) (Fig. 14.7) have shown antiageing, anti-melanogenic, anticancer, and antioxidant activities by stimulating collagen production, inhibiting tyrosinase, reducing melanoma growth, and preventing oxidation (Ariede et al. 2017; Wang et al. 2017a, b; Teas and Irhimeh 2017) (Fig. 14.7).

In addition, a sulphated polysaccharide from *Padina tetrastromatica* (Phaeophyceae) has exhibited anti-inflammatory activity through COX-2 and iNOS inhibitions in an experimental model of paw oedema in rats (Mansouri et al. 2015). Moreover, polyunsaturated fatty acids from *Undaria pinnatifida* (Phaeophyceae) have demonstrated anti-inflammatory activity on mouse ear edema and erythema (Khan et al. 2009). Various seaweed species, including *Osmundaria obtusilo*, *Palisada flagellifera* (Rhodophyta), and *Sargassum horneri* (Phaeophyceae), have displayed antioxidant and antiageing activities (Vasconcelos et al. 2019; Santos et al. 2019; Kim et al. 2013a, b, c). These activities encompass collagenase inhibition, tyrosinase inhibition, and matrix metalloproteinase (MMP) expression modulation. Moreover, seaweed-derived mycosporine amino acids from *Porphyra* sp. (Rhodophyta) have shown collagenase inhibition and control over MMP expression (Wang et al. 2017a, b; Sakai et al. 2011). Mycosporine amino acids from *Palmaria palmata* (Rhodophyta) have also exhibited antiageing activity through collagenase inhibition (Hartmann et al. 2015). Several macroalgae species, including *Laurencia pacifica*, *Palisada rigida* (formerly *Laurencia rigida*), *Wilsonosiphonia howei* (formerly *Polysiphonia howei*), *Rhodomela confervoides*, and *Schizymenia dubyi* (Rhodophyta), have been identified as rich sources of phenolic compounds (Veluchamy and Palaniswamy 2020; Azam et al. 2017; Li et al. 2012). Hyaluronic

Table 14.1 Seaweed-derived biologically active compounds and their biological potentials

No.	Name of macroalgae	Cosmetic benefits	Compounds	References
1	*Scytosiphon lomentaria* (P)	Antioxidant, radical scavengers, chelators	Amino acids	Heo et al. (2005) Gupta et al. (2019) Salehi et al. (2019)
2	*Gracilaria vermiculophylla* (R)	Antioxidant, UV protector	Porphyra-334, palythine, asterina-330, shinorine	Barceló-Villalobos et al. (2017)
3	*Ulva lactuca* (C), *Asparagopsis armata* (R)	Antiageing, antiwrinkles, improves collagen formation	MAAs, amino acids	Leandro et al. (2020)
4	*Pelvetia canaliculata* (P)	Antioxidant, collagen formation, proteoglycan synthesis	Amino acids	Hupel et al. (2011)
5	*Gracilaria chilensis*, *Pyropia plicata*, *Champia novae-zelandiae* (R)	Anti UV, antioxidant	MAAs	Orfanoudaki et al. (2019)
6	*Ulva lactuca* (C)	Enhance collagen and elastin synthesis	Arginine, aspartic acid, glycine	Orfanoudaki et al. (2019)
7	*Porphyra umbilicalis* (R)	Antiageing	MAAs (2:1 ratio of porphyra-334 and shinorine)	Guglielmo and Montanari (2008)
8	*Neopyropia yezoensis* (R)	Anti-inflammatory	Peptide PPy1	Lee et al. (2015)
9	*Palmaria palmata* (R)	UV protector	MAAs	Yuan et al. (2009)
10	*Sargassum polycystum* (P)	Anti-melanogenic or skin whitening effect	Amino acids and amines	Song et al. (2009) Chan et al. (2011)
11	*Porphyra umbilicalis* (R)	Moisturization, skin protector, antiwrinkle, protection against roughness	Porphyra-334, shinorine	Gianeti and Maia Campos (2014)
12	*Porphyra yezoensis* f. *coreana* (R)	Enhance elastin and collagen formation, reduce MMP expression	Peptides, PYP1-5, porphyra-334	Ryu et al. (2014)

(continued)

Table 14.1 (continued)

No.	Name of macroalgae	Cosmetic benefits	Compounds	References
13	*Macrocystis pyrifera* (P)	Antioxidant, antidiabetic, antiageing	Phlorotannins, phloroeckol, tetrameric phloroglucinol	Leyton et al. (2016)
14	*Ascophyllum nodosum* (P)	MMP inhibition	Ascophyllan	Abu et al. (2015)
15	*Cystoseira foeniculacea* (P)	Antioxidant	Polyphenol	Messina et al. (2019)
16	*Stephanocystis hakodatensis* (P)	Antioxidant	Phenol	Airanthi et al. (2011)
17	*Ecklonia cava* subsp. *Stolonifera* (P)	Protection against UVB radiation	Fucofuroeckol-A	Vo et al. (2018)
18	*Corallina pilulifera* (R)	Antiageing, anti-inflammation, antioxidants, antiallergic, UV screens	Phlorotannins	Thomas and Kim (2013)
19	*Ishige foliacea* (P)	Anti-melanogenic, inhibition of tyrosinase and melanin synthesis	Phlorotannin	Kim et al. (2013a, b, c, 2015)
20	*Laminaria ochroleuca* (P)	Antioxidant	Polyphenol	del Olmo et al. (2019)
21	*Caulerpa racemosa* (C)	Tyrosinase inhibitor	Flavonoids, hydroquinone, saponins	Yang et al. (2014)
22	*Ecklonia cava* (P)	UV B protective	Dioxinodehydroeckol	Ryu et al. (2015)
23	*Ecklonia cava*, *Ecklonia cava* subsp. *stolonifera* (P)	Antiallergic	Eckol, 6,6′-bieckol, doeckol, phlorofucofuroeckol-A, 8,8′-bieckol	Sugiura et al. (2021)
24	*Eisenia bicyclis*, *Ecklonia cava* subsp. *stolonifera*	Hepatoprotective, anti-tyrosinase	Phlorofucofuroeckol A	Kim et al. (2011) Catarino et al. (2021)
25	*Eisenia arborea*, *Ecklonia bicyclis* (P)	Anti-inflammation, hyaluronidase inhibitor, antiwrinkle	Phlorotannins	Sugiura et al. (2006a, b)
26	*Eisenia arborea* (P)	Anti-allergic	Phlorofucofuroeckol A	Sugiura et al. (2006a, b)

(continued)

Table 14.1 (continued)

No.	Name of macroalgae	Cosmetic benefits	Compounds	References
27	*Ascophyllum nodosum*, *Fucus serratus*, *Himanthalia elongata*, *Sargassum muticum* (P)	Antioxidant, antibacterial, antiviral, photoprotection, anti-inflammatory	Phlorotannins	Casas et al. (2016) Eom et al. (2017) Gager et al. (2020)
28	*Undaria pinnatifida* (P)	Anti-inflammatory	PUFA	Sanghvi and Martin Lo (2010)
29	*Ulva lactuca* (P)	In-vitro and in-vivo Nrf2-ARE activation, cell-protective, antioxidant	Fatty acids such as C18 and C16 type	Wang et al. (2013)
30	Phaeophyceae (brown algae) (P)	Antioxidant	Unsaturated fatty acids	Susanto et al. (2016)
31	*Ulva lactuca* (P)	Inhibition of elastase, enhance collagen synthesis	Lipopeptides	Castejón et al. (2021)
32	*Himanthalia elongata* (P)	Antioxidant, antimicrobial	Fatty acids and volatile compounds	Plaza et al. (2010)
33	*Porphyridium purpureum* (R)	Antioxidant, anti-inflammatory, anti-photoaging	Eicosapentaenoic acid, docosahexaenoic acid, eicosatetraenoic acid, polyunsaturated omega-3 fatty acids	Li et al. (2019)
34	*Ulva rigida* (C), *Gracilaria* sp. (R), *Fucus vesiculosus*, *Saccharina latissima* (P)	Antioxidant	Lipidic profile	Neto et al. (2018)
35	*Sargassum fusiforme* (P)	Protection against photodamage, UVB protector, MMP inhibition, enhanced procollagen formation, anti-inflammatory	Fucosterol	Sun et al. (2015) Hwang et al. (2014)
36	*Gracilariopsis longissima* (R), *Saccharina japonica* (P)	Anti-inflammatory	(8E)-10-oxo-8-octadecenoic acid, (E)-9-oxo-10-octadecenoic acid, myristic acid, palmitic acid	Patra et al. (2015)

(continued)

Table 14.1 (continued)

No.	Name of macroalgae	Cosmetic benefits	Compounds	References
37	*Silvetia siliquosa* (P)	Antioxidants, stimulate antioxidant enzymes such as catalase, glutathione peroxidase	Fucosterol	Lee et al. (2003, 2004)
38	*Fucus vesiculosus* (P)	Anticoagulant antioxidant, enhancer of skin fibroblast formation	Fucoidan	Rupérez et al. (2002)
39	*Neoporphyra haitanensis* (R)	Antioxidant	Porphyran	Zhang et al. (2003, 2004)
40	*Saccharina longicruris* (P)	Anti-inflammation, antioxidant, reconstruction of dermis	Laminaran	Ayoub et al. (2015) Ozanne et al. (2020)
41	*Saccharina longicruris* (P)	Enhance fibroblast formation, increase synthesis of matrix metalloproteinase (MMP) complex and collagen-1	Galactofucans	Rioux et al. (2013)
42	*Eucheuma denticulatum* (R)	Antioxidant, photoprotection	Carrageenan	Thevanayagam et al. (2014)
43	*Gelidium* sp. (R)	Thickener	Agar	Wells et al. (2017)
44	*Ascophyllum* sp., *Fucus* sp., *Sargassum* sp., *Undaria* sp. (P)	Anticellulite	Laminaran	Pereira et al. (2013)
45	*Saccharina cichorioides* (P)	Anti-atopic dermatitis	Fucoidan	Yang (2012)
46	*Corallina officinalis* (R)	Antioxidant	Sulphated polysaccharides	Yang et al. (2011)
47	*Macrocystis* sp., *Lessonia* sp., *Laminaria* sp. (P)	Gelling and stabilizing agent, moisturizer, chelator	Alginate	Malinowska (2011) Tønnesen and Karlsen (2002)
48	*Kjellmaniella crassifolia*	Antiageing, antiwrinkle	Fucoidan	
49	Brown algae (P)	Thickening agent gelling agent	Alginate	Morais et al. (2021)
50	*Sargassum vachellianum* (P)	Skin moisturizer and protectors	Polysaccharides	Jesumani et al. (2020)

(continued)

Table 14.1 (continued)

No.	Name of macroalgae	Cosmetic benefits	Compounds	References
51	*Fucus vesiculosus*, *Laminaria digitata*, *Undaria pinnatifida* (P)	Antioxidant, antiageing, anticoagulant, increased skin fibroblast formulation	Fucoidan	Rupérez et al. (2002) Je et al. (2009)
52	*Ascophyllum nodosum* (P)	Anti-elastase, gelatinase A inhibition, inhibition of interleukin-1 beta in fibroblast cells	Fucoidan	Pereira (2018)
53	*Ecklonia cava* (P)	Photoprotection against UV-B	Phlorotannins	Heo et al. (2009) Ko et al. (2011)
54	*Saccharina japonica* (P)	Inhibition of tyrosinase and melanogenesis in UVB irradiated	Fucoxanthin	Shimoda et al. (2010)
55	*Cladosiphon okamuranus* (P)	Antioxidant, DPPH inhibition	Fucoxanthin	Mise et al. (2011)
56	*Gracilaria gracilis*, *Porpyridium* sp. (R)	Antioxidant, skin whitening activity by anti-melanogenic activity	Phycobiliprotein pigment such as R-phycoerythrin, phycocyanin, allophycocyanins	
57	*Cladophora glomera®*(C)	Antibacterial, antioxidant, colourants, deodorizer	Chlorophyll a, chlorophyll b, chlorophyll c, chlorophyll d	
58	*Portieria* (R)	Antioxidants, anti-inflammatory, colourants, radical scavenger	Phycobiliproteins, phycoerythrin, phycocyanin	
59	*Cladophora glomerata* (C)	Tissue growth stimulators	Chlorophyll	
60	*Neopyropia yezoensis* (R)	Antioxidant, anti-inflammatory	Porphyran	
61	*Himanthalia elongata* (P)	Antioxidant	Fucoxanthin extract	
62	*Ascophyllum nodosum* (P)	Antiageing, antiwrinkle	Fucoxanthin	
63	*Fucus vesiculosus* (P)	Antioxidant	Fucoxanthin	
64	Phaeophyta	Anti-photoaging	Fucoxanthin	

(continued)

Table 14.1 (continued)

No.	Name of macroalgae	Cosmetic benefits	Compounds	References
65	*Sargassum siliquastrum* (P)	Anti-melanogenic (skin whitening effect), antioxidant, anti-inflammatory	Fucoxanthin	
66	*Gelidium crinale* (R)	Antioxidant	Carotenoids	

Fig. 14.7 *Fucus vesisculosus* (Phaeophyceae). Scale bar = 1 cm

acid (HA) is a well-recognized skincare ingredient celebrated for its remarkable capacity to hydrate and enhance skin plumpness. Naturally present in the skin and connective tissues, HA plays a vital role in maintaining skin moisture levels and elasticity (Hammon et al. 2019). In recent years, cosmetic manufacturers have turned to seaweed as a source of HA due to its unique advantages. Seaweed-derived hyaluronic acid is obtained from specific seaweed species, particularly red and brown algae (Jesumani et al. 2019). This natural source aligns with the growing consumer preference for clean and naturally derived skincare ingredients. Seaweed extracts are rich in vitamins, minerals, and antioxidants, offering an array of skin-nourishing benefits (Kalasariya et al. 2023). When combined with hyaluronic acid, these extracts create a potent skincare ingredient that not only hydrates but also provides essential nutrients to the skin (Juncan et al. 2021). Seaweed hyaluronic acid offers a synergy between the hydration properties of HA and the skin-loving components of seaweed extracts. This combination delivers a comprehensive skincare solution that goes beyond hydration to address various skin concerns (Pereira 2018). Seaweed hyaluronic acid acts as a moisture magnet, drawing and retaining water molecules within the skin. This deep hydration helps improve skin plumpness and elasticity, reducing the appearance of fine lines and wrinkles (Borumand and Sibilla 2015). By locking in moisture, seaweed hyaluronic acid contributes to a strengthened skin barrier. A robust skin barrier is essential for protecting the skin from environmental stressors and preventing moisture loss (Menon and Kligman 2009). The natural soothing properties of seaweed extracts, combined with the calming effect of hyaluronic acid, make this ingredient suitable for sensitive and irritated skin types. Seaweed hyaluronic acid's ability to hydrate and nourish the skin can promote a more youthful complexion by reducing the signs of premature ageing, such as dullness and dryness (Mistry 2017). Seaweed-derived ingredients have established themselves as highly sought-after components in cosmetic formulations, offering a broad spectrum of functions and adaptability.

Alginate mask which mainly contains alginate, a polysaccharide derived from brown seaweeds, plays a pivotal role in the formulation of alginate masks. Renowned for their ability to provide a cooling, hydrating, and soothing effect on the skin, these masks are made possible by the unique gel-like texture formed when alginate is combined with water. This texture adheres to facial contours, ensuring the delivery of active ingredients while creating a moisture-locking barrier (Cascone and Lamberti 2020). Cosmetic products benefit from a wide array of seaweed extracts sourced from various species of red, brown, and green algae. These extracts are abundant in vitamins, minerals, antioxidants, and bioactive compounds that revitalize and nourish the skin. Commonly found in serums, creams, toners, and masks, they contribute to enhanced skin health and appearance (Ashokkumar et al. 2022). Certain cosmetic products incorporate seaweed-derived hyaluronic acid, which combines the hydrating prowess of hyaluronic acid with the nutrient-rich advantages of seaweed extracts. This synergy elevates moisture retention, imparts a plumping effect to the skin, and promotes a more youthful complexion (Susano et al. 2022). Seaweed-derived components offer an array of skin benefits and cosmetic properties, making them highly valuable additions to skincare products, for example, their extracts function as natural humectants, attracting and retaining moisture in the skin, leading to improved hydration levels and a softer, smoother complexion. Enriched with antioxidants such as polyphenols, as well as vitamins C and E, seaweed combats free radicals, reducing oxidative stress on the skin, thereby fostering a more youthful appearance and offering protection against premature ageing (Mansuri et al. 2021). They also comprise anti-inflammatory compounds, including fucoidan and various polyphenols, which effectively soothe irritated skin, diminishing redness and puffiness (Obluchinskaya et al. 2021). They can also create a protective barrier on the skin's surface, thereby locking in moisture and acting as a shield against external pollutants and irritants (Graça et al. 2022). Moreover, it incorporates natural exfoliating agents such as alpha hydroxy acids (AHAs) that gently remove dead skin cells, promoting a brighter and more radiant complexion (Kadam Vaishali et al. 2013). The cosmetics and skincare industry is witnessing a substantial surge in the demand for seaweed-based products, driven by several market trends and consumer preferences: Consumers increasingly seek natural and eco-friendly cosmetic ingredients. Seaweed aligns with this trend, being a renewable marine resource with minimal environmental impact. Clean beauty products, free from harmful chemicals and additives, are gaining popularity (Asthana et al. 2021). Seaweed-based cosmetics often align with this clean beauty ethos due to their natural origin and minimal processing. As the global population ages, there is a growing demand for skincare products that offer antiageing benefits. Seaweed's antioxidant properties and skin-renewing capabilities make it an attractive option for addressing these concerns (Yarosh 2008). Skin hydration and overall skin health are perennial consumer priorities. Seaweed's moisturizing and nutrient-rich properties make it a preferred choice in products targeting these concerns (Lomartire and Gonçalves 2022a, b). Consumers are increasingly conscious of ethical and sustainable sourcing practices. Brands that emphasize responsible seaweed harvesting and eco-friendly production processes appeal to this segment of consumers (Elkhattat and Medhat 2022).

14.1.6 Edible Salts and Seasonings from Seaweed

Seaweed, often celebrated for its role as an essential marine resource, extends its influence into the culinary world through the production of edible salts and seasonings. This section explores the fascinating process of deriving edible salts from seaweed, such as sea salt and seaweed-infused salts, and delves into the culinary applications and flavour profiles of these seaweed-based salts. Moreover, it highlights the culinary innovation and transformative impact of seaweed in the food industry (Raja et al. 2022). Sea salt, also known as solar salt or sea crystal salt, is one of the most ancient forms of salt production, dating back to the days when salt was harvested by evaporating seawater. This method involves collecting seawater in salt pans and allowing it to evaporate naturally under the sun, leaving behind sea salt crystals. These crystals are then harvested, cleaned, and packaged (Lei et al. 2022). The result is a natural and minimally processed salt with a distinctive flavour profile that reflects the unique characteristics of the sea from which it was derived. Seaweed-infused salts are a culinary innovation that combines the savoury umami of seaweed with the mineral-rich qualities of sea salt (Magnusson et al. 2019; Leandro et al. 2019). To make these salts, dried seaweed, often varieties like nori, dulse, or kombu, is blended with sea salt. The seaweed imparts its flavours and aromas to the salt, creating a distinctive seasoning with a hint of the ocean (Mouritsen et al. 2012; Milne 2016). Seaweed-based salts are prized for their ability to enhance umami, the fifth basic taste. Umami is characterized by its savoury, rich, and satisfying flavour, and seaweed, with its natural glutamate content, contributes to this sensation. Chefs and home cooks alike use seaweed-based salts to elevate the savoury aspects of a dish, whether it is a classic pasta sauce or a simple bowl of popcorn (O'Connor 2017; De Rovira Sr 2017). Seaweed-based salts naturally complement seafood dishes. The oceanic notes in the salts harmonize with the flavours of fish, shellfish, and other aquatic delicacies, providing a well-rounded taste experience (Mouritsen 2009). In addition, seaweed-based salts are excellent for seasoning soups, stews, and broths. They infuse the dishes with depth and complexity, making them more satisfying and flavourful (Mouritsen et al. 2018). Seaweed-based salts are particularly popular in vegetarian and vegan cuisine. They bring a hint of the sea to plant-based dishes, adding a layer of complexity that mimics the flavours of seafood (Figueroa Hurtado 2023; Sarkar and Choudhury 2017). Its presence in the culinary world extends beyond the realm of salts and seasonings. It has sparked innovation and creativity in the food industry, leading to the development of novel culinary experiences: dried and ground seaweed is transformed into flakes and powders that can be used as garnishes, seasonings, or as a primary ingredient in dishes (Glicksman 2019). These products allow chefs to experiment with seaweed's flavours and textures in various culinary creations. Seaweed-based condiments like seaweed pesto, tapenade, and "salsa verde" have emerged as flavourful and versatile accompaniments to a wide range of dishes (O'Connor 2017; Kumar et al. 2021b). They introduce a burst of umami and marine nuances. The popularity of seaweed snacks, such as seaweed crisps and chips, has surged in recent years. These snacks offer a satisfying crunch and the addictive umami of seaweed, providing a healthier alternative to traditional snacks (Milne 2016; McLean 2015). Seaweed's cultivation and harvest are often more sustainable and environmentally friendly compared to land-based crops. As the culinary world embraces sustainability, seaweed's role as a responsible ingredient gains prominence (Fleurence and Levine 2016; Arnold 2008). The production of edible salts and seasonings from seaweed marks a culinary renaissance rooted in sustainability, flavour enhancement, and creativity. Seaweed-based salts and seasonings offer a unique and wholesome way to infuse dishes with the essence of the sea while promoting responsible sourcing practices. As chefs and food enthusiasts continue to explore the diverse applications of seaweed in culinary artistry, the future holds exciting gastronomic possibilities that honour both tradition and innovation. Seaweed's journey from the depths of the ocean to the world's dining tables represents a flavourful and sustainable evolution in the culinary landscape (Mouritsen et al. 2018; Cordeiro and van Hal 2022).

Examples of seaweed-based culinary innovations and applications in the food industry (Ebrahimzadeh et al. 2023; Yesuraj et al. 2022) include seaweed-infused pasta and noodles which have gained popularity as a nutritious and flavourful alternative to traditional wheat-based options. These products offer a subtle sea taste and a pleasing texture, making them a hit in various cuisines (Kumar et al. 2021b; Li et al. 2014). Second, nori, a type of red seaweed, is a staple in Japanese cuisine and is widely recognized as the outer wrapping for sushi rolls. Nori's delicate umami flavour and crisp texture play a fundamental role in sushi's taste and presentation (Pérez-Lloréns 2019). Chefs and food manufacturers have developed seaweed-infused oils, dressings, and vinaigrettes. These products add a layer of complexity to salads, grilled vegetables, and seafood dishes, combining the earthy notes of olive oil or vinegar with the subtle marine essence of seaweed (Houston and Milne 2012; Debbarma et al. 2022). Beyond traditional seaweed snacks, powdered seaweed seasonings are available for sprinkling on popcorn, roasted nuts, and other snacks. These seasonings provide a burst of umami and a unique twist to familiar treats (Jayakody and Vanniarachchy 2024). It is a key ingredient in various Asian soups and broths, such as miso soup and dashi. It imparts a delicate sea flavour and serves as a source of nutrients and minerals (Mouritsen 2013). Moreover, seaweed-based sauces, like seaweed soy sauce or seaweed-infused hot sauces, have made their way into contemporary culinary creations. They contribute an extra layer of depth and umami to a wide range of dishes (Tanna and Mishra 2019). Its savoury and mineral-rich qualities have even found their way into beverages. Some innovative products include seaweed-infused cocktails, mocktails, and health drinks that offer a unique flavour profile and added nutritional benefits (Wahyuni 2023). It is also increasingly recognized for its functional properties, such as its role as a natural thickening agent and a source of dietary fibre. It is used to enhance the texture and nutritional value of food products, including plant-based meat alternatives, dairy alternatives, and baked goods (Peñalver et al. 2020). These diverse culinary applications showcase seaweed's versatility and its ability to elevate the flavour and nutrition of a wide array of dishes and food products. As consumers seek out innovative and sustainable food options, seaweed continues to make waves in the culinary world, offering a unique blend of taste, nutrition, and environmental responsibility (Boyd et al. 2020).

14.1.7 Bioprocessing Techniques for Sustainable and Valuable Functional Ingredients from Seaweed

The bioprocessing of seaweed stands at the forefront of endeavours to attain sustainable and valuable functional ingredients. This section is dedicated to the exploration of bioprocessing techniques customized for the production of functional ingredients derived from seaweed. These techniques focus on fermentation, enzymatic conversion, and bioreactor systems and concurrently, investigate the sustainability considerations and strategies for waste reduction linked with these methods (Qin 2018; Farghali et al. 2023). Key methods employed in the bioprocessing of seaweed harness the intrinsic capabilities of microorganisms, enzymes, and bioreactors for the conversion of raw seaweed biomass into high-value functional ingredients. Fermentation, one of these key methods, entails the deliberate cultivation of microorganisms like bacteria or yeast on seaweed biomass to convert it into desired compounds. This technique is adept at yielding a diverse array of products, encompassing bioactive compounds, enzymes, and organic acids. Notably, enzymes, the biocatalysts present in nature, play a pivotal role in the bioprocessing of seaweed (Wijesinghe and Jeon 2012). Enzymatic conversion, another vital method, focuses on the precise application of specific enzymes to break down seaweed components into valuable functional ingredients. This method exhibits a high degree of precision and can selectively target particular compounds, such as polysaccharides or proteins. Furthermore, bioreactors, controlled environments that facilitate either microbial fermentation or enzymatic conversion of seaweed biomass, play an indispensable role in optimizing conditions. These conditions encompass variables like temperature, pH, oxygen levels, and agitation, which collectively maximize product yield and quality.

Fermentation stands as a versatile and sustainable method in the bioprocessing of seaweed for the production of functional ingredients. It is instrumental in producing various bioactive compounds, including polysaccharides, antioxidants, and organic acids (Zhao et al. 2021). For instance, fermentation can yield seaweed polysaccharides like alginate and fucoidan, which find applications in pharmaceuticals, cosmetics, and food. Moreover, fermentation has the potential to enhance the bioavailability of nutrients in seaweed, rendering them more accessible for absorption within the human body. In addition to these advantages, it can ameliorate the characteristic "seaweed" taste, thus expanding the palatability of seaweed-based ingredients for a wider range of culinary applications. Enzymatic conversion, in its precision-driven approach, plays a central role in breaking down the complex components of seaweed into functional ingredients (Moroney et al. 2015). It is the vehicle for extracting bioactive compounds, such as phlorotannins, peptides, and lipids from seaweed, which subsequently find applications in nutraceuticals and cosmetics. Furthermore, it serves as a tool for modifying the structure of seaweed polysaccharides, enhancing their functional properties for use in the food industry (Jayakody et al. 2023). In addition, bioreactor systems, the third crucial component in seaweed bioprocessing, provide a controlled environment that is indispensable for microbial fermentation or enzymatic conversion (Khan et al. 2018; Costa et al. 2018). These systems are pivotal in optimizing the conditions that significantly impact product yield and quality. Moreover, they can be scaled up to meet the demands of industrial-scale production, ensuring both consistency and efficiency in the bioprocessing of seaweed. Bioreactors offer the advantage of precise control over environmental factors, including temperature, pH, and oxygen levels, culminating in optimized yields and product quality. Moreover, sustainability is an underlying principle that permeates seaweed bioprocessing (Yusoff et al. 2019). Seaweed cultivation inherently plays a role in carbon dioxide (CO_2) capture and storage, mitigating climate change. It demands relatively less land and freshwater when compared to traditional crops. Additionally, it is characterized by the avoidance of chemical fertilizers and pesticides, thereby reducing its environmental impact (Zhang et al. 2022). Furthermore, the by-products arising from seaweed bioprocessing, such as residual biomass or wastewater, can be repurposed for various applications, effectively minimizing waste (Baghel et al. 2020).

The techniques used in seaweed bioprocessing, which encompass fermentation, enzymatic conversion, and bioreactor systems, hold immense promise for the sustainable production of functional ingredients. They enable the transformation of seaweed biomass into valuable bioactive compounds, offering a wealth of opportunities across a spectrum of sectors, including healthcare and food (Lange et al. 2020; Reboleira et al. 2021). Moreover, the environmentally responsible nature of seaweed bioprocessing, underscored by its intrinsic sustainability and potential for waste reduction, positions it as a pivotal player in the pursuit of sustainable and valuable functional ingredients (Lorbeer 2016). Incorporating these sustainability aspects and waste reduction strategies into seaweed bioprocessing not only aligns with environmentally responsible practices but also contributes to a holistic approach to seaweed cultivation and functional ingredient production. As the global demand for sustainable and eco-friendly solutions continues to rise, seaweed bioprocessing stands as a beacon of innovation in harnessing the potential of the sea while safeguarding its delicate ecosystems. Seaweed cultivation plays a significant role in sequestering carbon dioxide (CO_2) from the atmosphere. As seaweed grows, it undergoes photosynthesis, absorbing CO_2 and aiding in carbon capture and storage. This not only mitigates climate change but also benefits ocean health by reducing ocean acidification, a result of excess atmospheric CO_2 absorbed by the oceans (Yong et al. 2022; Duarte et al. 2017). Seaweed farming contributes to the mitigation of eutrophication, a process where excess nutrients, often from agricultural runoff, lead to harmful algal blooms and the creation of oxygen-depleted "dead zones" in water bodies. Seaweeds absorb nutrients like nitrogen and phosphorus, which, when cultivated near these areas, can effectively reduce nutrient levels in coastal waters. This, in turn, promotes water quality and marine biodiversity (Fei 2004). Seaweed bioprocessing generates valuable by-products, including residual biomass post-extraction or processing. These by-products can be repurposed for various applications, minimizing waste. For example, seaweed biomass can serve as animal feed, a soil conditioner, or a resource for bioenergy production through anaerobic digestion (White and White 2020; Desai and Reddy 2024). Some seaweed cultivation systems are designed as closed-loop systems, where waste and nutrient-rich effluents from one stage of production are redirected for use in subsequent stages. This practice significantly reduces the environmental footprint of seaweed farming while minimizing nutrient release into surrounding ecosystems (Kumar et al. 2021a). It also stands out for its significantly lower freshwater requirements compared to traditional agriculture. This water efficiency is especially advantageous in regions facing water scarcity or where agricultural runoff poses ecological challenges (Langton et al. 2019). Seaweed-based biopolymers and materials are emerging as sustainable alternatives to conventional plastics for packaging. These materials are biodegradable and have the potential to reduce plastic pollution and waste (Westlake et al. 2023). Moreover, it also acts as a natural buffer, stabilizes coastal sediments, and helps protect coastal communities from the adverse effects of erosion and storm surges (Sigren et al. 2014). And, particularly within integrated multi-trophic aquaculture (IMTA) systems, promotes nutrient cycling. In IMTA, seaweed cultivators often co-locate with other aquaculture operations, such as fish farms. Seaweeds efficiently absorb excess nutrients released by the fish, reducing environmental impacts and enhancing overall sustainability (Campbell et al. 2019). Certain seaweed species demonstrate the capacity to accumulate heavy metals and pollutants from the surrounding water, a process known as bioremediation. This capability can be harnessed for environmental cleanup efforts in polluted aquatic environments (Mani and Kumar 2014). The integration of these sustainability aspects and waste reduction strategies underscores the environmentally responsible nature of seaweed bioprocessing. It plays a pivotal role in the quest for sustainable and eco-friendly solutions, striking a balance between harnessing the potential of the sea and protecting its delicate ecosystems (Table 14.1).

14.1.8 Future Trends and Challenges in Seaweed-Derived Functional Ingredients

Personalized nutrition, driven by the recognition of significant individual dietary variations, stands as an evolving paradigm in the nutrition industry. This emerging trend seeks to tailor nutritional interventions according to an individual's unique health profile, lifestyle, and genetic makeup. Seaweed-derived ingredients hold great promise in contributing to this movement. Seaweed bioprocessing technologies are continuously advancing, enabling the targeted extraction and formulation of specific bioactive compounds from seaweed sources (Gallego et al. 2019; Rodríguez-González et al. 2022). These compounds have the potential to address diverse health concerns, providing antioxidant-rich extracts for those seeking oxidative stress management or mineral-rich derivatives for individuals with specific dietary deficiencies. Furthermore, seaweed-based functional ingredients can be tailored to optimize nutrient bioavailability. Techniques like enzymatic hydrolysis or nanotechnology allow the controlled release of bioactive compounds, ensuring alignment with an individual's digestive capabilities, and thereby maximizing nutrient absorption (Cassani et al. 2023). In the cosmetics industry, a profound shift towards clean and green beauty products is underway, reflecting consumer demand for sustainable and environmentally responsible formulations. Seaweed-based ingredients, celebrated for their natural origins and eco-friendly attributes, are poised to maintain their prominence in skincare and cosmetic formulations. The practice of seaweed cultivation itself aligns with these principles of clean beauty, as it is both sustainable and low-impact (Nilsson and Martin 2022; Bizzaro et al. 2022). Seaweed's rapid growth rates, minimal freshwater requirements, and carbon sequestration capabilities resonate with the eco-conscious values of clean beauty (Samanta et al. 2015). Moreover, seaweed's rich composition of vitamins, minerals, antioxidants, and polysaccharides makes it an ideal choice for skin care products designed to nourish, hydrate, and rejuvenate the skin. Alginate masks, seaweed extracts, and seaweed-derived hyaluronic acid exemplify the sustainable elegance that seaweed brings to cosmetics.

Seaweed-derived compounds with pharmaceutical potential are a subject of intense research and development. Among these compounds are anticancer agents and anti-inflammatory agents, which hold promise for addressing critical health challenges (Thun et al. 2002). Seaweed is a rich source of bioactive compounds such as phlorotannins and sulphated polysaccharides, both of which exhibit anticancer and anti-inflammatory properties. Research is ongoing to elucidate their mechanisms of action and optimize their pharmaceutical applications. Seaweed-derived materials, including micro- and nanostructures, are being explored for their potential as drug-delivery vehicles (Venkatesan et al. 2015). These systems enable the targeted and controlled release of pharmaceutical compounds, improving drug efficacy while reducing side effects (Mansour et al. 2010).

As the world seeks sustainable alternatives to traditional animal-based proteins, seaweed protein emerges as a contender. Seaweed protein offers a complete amino acid profile, making it an attractive option for plant-based diets, meat substitutes, and dietary supplements (Raja et al. 2022; Day et al. 2022). Rich in essential amino acids, vitamins, and minerals, seaweed protein's potential to serve as a protein source in plant-based diets positions it as a vital component in addressing global nutritional challenges. Furthermore, seaweed cultivation requires minimal land and freshwater resources compared to traditional protein sources like livestock. Additionally, it contributes to carbon sequestration and environmental preservation. The sustainability of seaweed sourcing and processing is of paramount importance for the long-term viability of this industry. This includes adhering to responsible farming practices to avoid monocultures, minimize ecosystem disruption, and prevent the spread of invasive species (Sugumaran et al. 2022). The development of energy-efficient processing techniques and transportation methods is crucial to minimize the carbon footprint associated with seaweed production and processing (Herrero and Ibanez 2015). Furthermore, strategies to reduce waste and repurpose by-products generated during seaweed bioprocessing contribute to a circular economy (Dahiya et al. 2022). The preservation of marine ecosystems and biodiversity remains imperative by mitigating any negative impacts associated with seaweed farming and harvesting (Duarte et al. 2022). While the field of seaweed-derived functional ingredients holds significant promise, there are notable challenges and research gaps that warrant attention and resolution. Standardization and quality control are crucial aspects requiring consideration. Developing standardized methods for the extraction and characterization of seaweed-derived compounds is imperative to guarantee consistent product quality (Monteiro et al. 2022). Furthermore, establishing clear and comprehensive regulatory frameworks is of utmost importance. These frameworks should facilitate the approval and appropriate labelling of seaweed-derived functional ingredients across various industries, including nutraceuticals and pharmaceuticals. This would provide both manufacturers and consumers with the necessary guidance and confidence in these products (Calado et al. 2018; Bradly et al. 2021). Bioprocessing optimization is another key area demanding continuous research efforts. Achieving higher yields, reducing production costs, and enhancing overall process efficiency are essential goals. Research in this domain is vital for unlocking the full potential of seaweed-derived functional ingredients (Potvin et al. 2012). Last but not least, comprehensive studies focusing on health and safety aspects are essential. These studies should investigate the safety and long-term effects of seaweed-derived functional ingredients in various applications.

Such research is paramount to ensure the responsible and widespread adoption of these innovative products (Farghali et al. 2023).

14.2 Conclusion

Seaweed emerges as an exceptionally versatile source of functional ingredients, holding the potential to revolutionize a multitude of industries, spanning from nutrition, cosmetics, and pharmaceuticals, to food. Throughout this chapter, the various dimensions of seaweed bioprocessing technologies have been meticulously explored, emphasizing their profound significance as a sustainable and abundant resource, destined to shape the future of functional ingredients. The array of compounds derived from seaweed offers a diverse range of advantages, including manifold health benefits, distinctive flavour profiles, and eco-friendly attributes. These attributes collectively establish seaweed-derived functional ingredients as invaluable assets in our relentless pursuit of sustainable and innovative solutions. As research continually progresses, consumer preferences evolve, and environmental sustainability takes centre stage in decision-making, the trajectory of seaweed-derived functional ingredients appears poised for sustained growth, firmly establishing themselves as transformative elements in our daily lives. The prospects for future developments in seaweed-based products are undeniably bright. Interdisciplinary collaboration, coupled with ever-advancing technologies and a steadfast commitment to environmental stewardship, fuels the progress of this field. Seaweed, often viewed as a humble oceanic resource, is now positioned to play a pivotal role in addressing the pressing global challenges pertaining to nutrition, health, sustainability, and environmental conservation. Its journey, from the depths of the sea to the forefront of scientific and industrial innovation, serves as a compelling testament to the enduring potential that nature's bounty offers to humanity.

References

Abu R, Jiang Z, Ueno M, Isaka S, Nakazono S, Okimura T, Cho K, Yamaguchi K, Kim D, Oda T (2015) Anti-metastatic effects of the sulfated polysaccharide ascophyllan isolated from *Ascophyllum nodosum* on B16 melanoma. Biochem Biophys Res Commun 458(4):727–732

Acharya PC, Shetty S, Fernandes C, Suares D, Maheshwari R, Tekade RK (2018) Preformulation in drug research and pharmaceutical product development. In: Dosage form design considerations. Academic Press, London, pp 1–55

Acquah C, Tibbetts SM, Pan S, Udenigwe C (2020) Nutritional quality and bioactive properties of proteins and peptides from microalgae. In: Handbook of microalgae-based processes and products. Academic Press, London, pp 493–531

Afonso NC, Catarino MD, Silva AM, Cardoso SM (2019) Brown macroalgae as valuable food ingredients. Antioxidants 8(9):365

Airanthi MWA, Hosokawa M, Miyashita K (2011) Comparative antioxidant activity of edible Japanese brown seaweeds. J Food Sci 76(1):C104–C111

Andersson M, De Benoist B, Darnton-Hill I, Delange F (eds) (2007) Iodine deficiency in Europe: a continuing public health problem. World Health Organization, Geneva, pp 1–86

Andrade EL, Bento AF, Cavalli J, Oliveira SK, Schwanke RC, Siqueira JM, Freitas CS, Marcon R, Calixto JB (2016) Non-clinical studies in the process of new drug development-Part II: Good laboratory practice, metabolism, pharmacokinetics, safety and dose translation to clinical studies. Braz J Med Biol Res 49:e5646

Ariede MB, Candido TM, Jacome ALM, Velasco MVR, de Carvalho JCM, Baby AR (2017) Cosmetic attributes of algae – a review. Algal Res 25:483–487

Arnold S (2008) Seaweed, power, and markets: a political ecology of the Caluya Islands, Philippines. MA thesis, York University

Ashokkumar V, Jayashree S, Kumar G, Sharmili SA, Gopal M, Dharmaraj S, Chen WH, Kothari R, Manasa I, Park JH, Shruthi S (2022) Recent developments in biorefining of macroalgae metabolites and their industrial applications – a circular economy approach. Bioresour Technol 359:127235

Asthana N, Pal K, Aljabali AA, Tambuwala MM, de Souza FG, Pandey K (2021) Polyvinyl alcohol (PVA) mixed green–clay and aloe vera based polymeric membrane optimization: peel-off mask formulation for skin care cosmeceuticals in green nanotechnology. J Mol Struct 1229:129592

Ayoub A, Pereira JM, Rioux LE, Turgeon SL, Beaulieu M, Moulin VJ (2015) Role of seaweed laminaran from *Saccharina longicruris* on matrix deposition during dermal tissue-engineered production. Int J Biol Macromol 75:13–20

Azam MS, Choi J, Lee MS, Kim HR (2017) Hypopigmenting effects of brown algae-derived phytochemicals: a review on molecular mechanisms. Mar Drugs 15(10):297

Baghel RS, Suthar P, Gajaria TK, Bhattacharya S, Anil A, Reddy CRK (2020) Seaweed biorefinery: a sustainable process for valorising the biomass of brown seaweed. J Clean Prod 263:121359

Balina K, Romagnoli F, Blumberga D (2017) Seaweed biorefinery concept for sustainable use of marine resources. Energy Procedia 128:504–511

Barceló-Villalobos M, Figueroa FL, Korbee N, Álvarez-Gómez F, Abreu MH (2017) Production of mycosporine-like amino acids from *Gracilaria vermiculophylla* (Rhodophyta) cultured through one year in an integrated multi-trophic aquaculture (IMTA) system. Mar Biotechnol 19:246–254

Beane KE, Redding MC, Wang X, Pan JH, Le B, Cicalo C, Jeon S, Kim YJ, Lee JH, Shin EC, Li Y (2021) Effects of dietary fibers, micronutrients, and phytonutrients on gut microbiome: a review. Appl Biol Chem 64(1):1–18

Belanche A, Jones E, Parveen I, Newbold CJ (2016) A metagenomics approach to evaluate the impact of dietary supplementation with *Ascophyllum nodosum* or Laminaria digitata on rumen function in rusitec fermenters. Front Microbiol 7:299

Bhatt N, Al-Subhi L, Waly M (2022) Seaweeds as functional food: a comprehensive review of its antioxidants and therapeutic merits against oxidative stress-mediated chronic diseases. In: Sustainable global resources of seaweeds. Volume 2: Food, pharmaceutical and health applications. Springer International Publishing, Cham, pp 77–91

Bizzaro G, Vatland AK, Pampanin DM (2022) The One-Health approach in seaweed food production. Environ Int 158:106948

Borumand M, Sibilla S (2015) Effects of a nutritional supplement containing collagen peptides on skin elasticity, hydration and wrinkles. J Med Nutr Nutraceuticals 4(1):47

Bouley TA, Machalaba C, Keast J, Gerwick WH, Fleming LE (2023) Marine biotechnology: a One Health approach to linking life on land to life underwater. In: Oceans and human health. Academic Press, London, pp 149–180

Boyd CE, D'Abramo LR, Glencross BD, Huyben DC, Juarez LM, Lockwood GS, McNevin AA, Tacon AG, Teletchea F, Tomasso JR Jr, Tucker CS (2020) Achieving sustainable aquaculture: historical and current perspectives and future needs and challenges. J World Aquacult Soc 51(3):578–633

Bradly N, Syddall V, Ingram C, Clarkson R, Elliot A, Major R, Adams S (2021) Stocktake and characterisation of Aotearoa New Zealand's seaweed sector: market and regulatory focus. Sustainable Seas, Ko ngā moana whakauka

Brunt EG, Burgess JG (2018) The promise of marine molecules as cosmetic active ingredients. Int J Cosmet Sci 40(1):1–15

Calado R, Leal MC, Gaspar H, Santos S, Marques A, Nunes ML, Vieira H (2018) How to succeed in marketing marine natural products for nutraceutical, pharmaceutical and cosmeceutical markets. In: Grand challenges in marine biotechnology. Springer, Cham, pp 317–403

Caldeira C, Vlysidis A, Fiore G, De Laurentiis V, Vignali G, Sala S (2020) Sustainability of food waste biorefinery: a review on valorisation pathways, techno-economic constraints, and environmental assessment. Bioresour Technol 312:123575

Campbell I, Macleod A, Sahlmann C, Neves L, Funderud J, Øverland M, Hughes AD, Stanley M (2019) The environmental risks associated with the development of seaweed farming in Europe-prioritizing key knowledge gaps. Front Mar Sci 6:107

Campo VL, Kawano DF, da Silva Jr DB, Carvalho I (2009) Carrageenans: biological properties, chemical modifications and structural analysis – a review. Carbohydr Polym 77(2):167–180

Casas MP, Rodríguez-Hermida V, Pérez-Larrán P, Conde E, Liveri MT, Ribeiro D, Fernandes E, Domínguez H (2016) In vitro bioactive properties of phlorotannins recovered from hydrothermal treatment of *Sargassum muticum*. Sep Purif Technol 167:117–126

Cascone S, Lamberti G (2020) Hydrogel-based commercial products for biomedical applications: a review. Int J Pharm 573:118803

Cassani L, Marcovich NE, Gomez-Zavaglia A (2023) Seaweed bioactive compounds: promising and safe inputs for the green synthesis of metal nanoparticles in the food industry. Crit Rev Food Sci Nutr 63(11):1527–1550

Castejón N, Thorarinsdottir KA, Einarsdóttir R, Kristbergsson K, Marteinsdóttir G (2021) Exploring the potential of Icelandic seaweeds extracts produced by aqueous pulsed electric fields-assisted extraction for cosmetic applications. Mar Drugs 19(12):662

Catarino MD, Amarante SJ, Mateus N, Silva AM, Cardoso SM (2021) Brown algae phlorotannins: a marine alternative to break the oxidative stress, inflammation and cancer network. Foods 10(7):1478

Chan YY, Kim KH, Cheah SH (2011) Inhibitory effects of *Sargassum polycystum* on tyrosinase activity and melanin formation in B16F10 murine melanoma cells. J Ethnopharmacol 137(3):1183–1188

Chauton MS, Forbord S, Mäkinen S, Sarno A, Slizyte R, Mozuraityte R, Standal IB, Skjermo J (2021) Sustainable resource production for manufacturing bioactives from micro-and macroalgae: examples from harvesting and cultivation in the Nordic region. Physiol Plant 173(2):495–506

Choudhary P, Khade M, Savant S, Musale A, Chelliah MS, Dasgupta S (2021) Empowering blue economy: from underrated ecosystem to sustainable industry. J Environ Manag 291:112697

Cikoš AM, Jokić S, Šubarić D, Jerković I (2018) Overview on the application of modern methods for the extraction of bioactive compounds from marine macroalgae. Mar Drugs 16(10):348

Cordeiro CM, van Hal J (2022) A systems integral approach in exploring creative innovation in culinary research: the example of seaweed in the context of the new nordic cuisine. Int J Food Stud 11(1):106–128

Cornish ML, Critchley AT, Mouritsen OG (2017) Consumption of seaweeds and the human brain. J Appl Phycol 29:2377–2398

Costa JA, Treichel H, Kumar V, Pandey A (2018) Advances in solid-state fermentation. In: Current developments in biotechnology and bioengineering. Elsevier, Amsterdam, pp 1–17

da Silva BV, Barreira JC, Oliveira MBP (2016a) Natural phytochemicals and probiotics as bioactive ingredients for functional foods: extraction, biochemistry and protected-delivery technologies. Trends Food Sci Technol 50:144–158

da Silva RP, Rocha-Santos TA, Duarte AC (2016b) Supercritical fluid extraction of bioactive compounds. TrAC Trends Anal Chem 76:40–51

Dahiya D, Sharma H, Rai AK, Nigam PS (2022) Application of biological systems and processes employing microbes and algae to Reduce, Recycle, Reuse (3Rs) for the sustainability of circular bioeconomy. AIMS Microbiol 8(1):83

Dai J, Mumper RJ (2010) Plant phenolics: extraction, analysis and their antioxidant and anticancer properties. Molecules 15(10):7313–7352

Darvishmanesh S, Firoozpour L, Vanneste J, Luis P, Degrève J, Van der Bruggen B (2011) Performance of solvent resistant nanofiltration membranes for purification of residual solvent in the pharmaceutical industry: experiments and simulation. Green Chem 13(12):3476–3483

Day L, Cakebread JA, Loveday SM (2022) Food proteins from animals and plants: differences in the nutritional and functional properties. Trends Food Sci Technol 119:428–442

De Rovira Sr D (2017) Dictionary of flavors. Wiley, Chichester

De Silva DD, Rapior S, Fons F, Bahkali AH, Hyde KD (2012) Medicinal mushrooms in supportive cancer therapies: an approach to anti-cancer effects and putative mechanisms of action. Fungal Divers 55:1–35

Debbarma J, Viji P, Rao BM, Ravishankar CN (2022) Seaweeds: potential applications of the aquatic vegetables to augment nutritional composition, texture, and health benefits of food and food products. In: Sustainable global resources of seaweeds. Volume 2: Food, pharmaceutical and health applications. Springer International Publishing, Cham, pp 3–54

del Olmo A, Picon A, Nuñez M (2019) High pressure processing for the extension of Laminaria ochroleuca (kombu) shelf-life: a comparative study with seaweed salting and freezing. Innovative Food Sci Emerg Technol 52:420–428

Desai AS, Reddy VK (2024) Seaweeds derived by-products. In: Applications of seaweeds in food and nutrition. Elsevier, Amsterdam, pp 225–234

Din NAS, Mohd Alayudin AS, Sofian-Seng NS, Rahman HA, Mohd Razali NS, Lim SJ, Wan Mustapha WA (2022) Brown algae as functional food source of fucoxanthin: a review. Foods 11(15):2235

Do-Amaral CCF, Pacheco BS, Seixas FK, Pereira CMP, Collares T (2020) Antitumoral effects of fucoidan on bladder cancer. Algal Res 47:101884

Dolbeth M, Arenas F (2022) Marine ecosystems: types, their importance, and main impacts. In: Life below water. Springer International Publishing, Cham, pp 591–607

Duarte CM, Wu J, Xiao X, Bruhn A, Krause-Jensen D (2017) Can seaweed farming play a role in climate change mitigation and adaptation? Front Mar Sci 4:100

Duarte CM, Bruhn A, Krause-Jensen D (2022) A seaweed aquaculture imperative to meet global sustainability targets. Nat Sustain 5(3):185–193

Ebrahimzadeh S, Biswas D, Roy S, McClements DJ (2023) Incorporation of essential oils in edible seaweed-based films: a comprehensive review. Trends Food Sci Technol 135:43–56

Elkhattat D, Medhat M (2022) Creativity in packaging design as a competitive promotional tool. Inf Sci Lett 11(1):135–145

Eom SH, Lee EH, Park K, Kwon JY, Kim PH, Jung WK, Kim YM (2017) Eckol from Eisenia bicyclis inhibits inflammation through the Akt/NF-κB signaling in Propionibacterium acnes-induced human keratinocyte HaCaT cells. J Food Biochem 41(2):e12312

Farghali M, Mohamed IM, Osman AI, Rooney DW (2023) Seaweed for climate mitigation, wastewater treatment, bioenergy, bioplastic, biochar, food, pharmaceuticals, and cosmetics: a review. Environ Chem Lett 21(1):97–152

Fei X (2004) Solving the coastal eutrophication problem by large scale seaweed cultivation. In: Asian Pacific phycology in the 21st century: prospects and challenges: proceeding of the second Asian Pacific Phycological forum, held in Hong Kong, China, 21–25 June 1999. Springer Netherlands, Dordrecht, pp 145–151

Fernando IS, Sanjeewa KA, Samarakoon KW, Lee WW, Kim HS, Kang N, Ranasinghe P, Lee HS, Jeon YJ (2017) A fucoidan fraction purified from *Chnoospora minima*; a potential inhibitor of LPS-induced inflammatory responses. Int J Biol Macromol 104:1185–1193

Figueroa Hurtado VN (2023) Characterization of texture and flavor compounds of selected Chilean seaweed. Doctoral thesis, Catholic University of Chile

Fleurence J, Levine IA (eds) (2016) Seaweed in health and disease prevention. Academic Press, Amsterdam

Gager L, Connan S, Molla M, Couteau C, Arbona JF, Coiffard L, Cérantola S, Stiger-Pouvreau V (2020) Active phlorotannins from seven brown seaweeds commercially harvested in Brittany (France) detected by 1 H NMR and in vitro assays: temporal variation and potential valorization in cosmetic applications. J Appl Phycol 32:2375–2386

Gallego R, Bueno M, Herrero M (2019) Sub-and supercritical fluid extraction of bioactive compounds from plants, food-by-products, seaweeds and microalgae – an update. TrAC Trends Anal Chem 116:198–213

García-Vaquero M, Rajauria G, O'Doherty JV, Sweeney T (2017) Polysaccharides from macroalgae: recent advances, innovative technologies and challenges in extraction and purification. Food Res Int 99:1011–1020

Gegg P, Wells V (2019) The development of seaweed-derived fuels in the UK: an analysis of stakeholder issues and public perceptions. Energy Policy 133:110924

Gharibzahedi SMT, Jafari SM (2017) The importance of minerals in human nutrition: bioavailability, food fortification, processing effects and nanoencapsulation. Trends Food Sci Technol 62:119–132

Gianeti MD, Maia Campos PM (2014) Efficacy evaluation of a multifunctional cosmetic formulation: the benefits of a combination of active antioxidant substances. Molecules 19(11):18268–18282

Glicksman M (2019) Red seaweed extracts (agar, carrageenans, furcellaran). In: Food hydrocolloids. CRC Press, Boca Raton, pp 73–113

Graça A, Martins AM, Ribeiro HM, Marques Marto J (2022) Indirect consequences of coronavirus disease 2019: skin lesions caused by the frequent hand sanitation and use of personal protective equipment and strategies for their prevention. J Dermatol 49(9):805–817

Guglielmo M, Montanari D (2008) Cosmetic composition with a lifting effect for sustaining relaxed tissues. Patent WO2008146116 A, 2, p 2008

Gupta PL, Rajput M, Oza T, Trivedi U, Sanghvi G (2019) Eminence of microbial products in cosmetic industry. Nat Prod Bioprospect 9:267–278

Hafting JT, Craigie JS, Stengel DB, Loureiro RR, Buschmann AH, Yarish C, Edwards MD, Critchley AT (2015) Prospects and challenges for industrial production of seaweed bioactives. J Phycol 51(5):821–837

Hammon KA, Amanda Dahl BS, Guttman N, Fong M (2019) Novel hydrolyzed chicken sternal cartilage extract improves facial epidermis and connective tissue in healthy adult females: a randomized, double-blind, placebo-controlled trial. Altern Ther Health Med 25(5):12–29

Hans N, Malik A, Naik S (2021) Antiviral activity of sulfated polysaccharides from marine algae and its application in combating COVID-19: mini review. Bioresour Technol Rep 13:100623

Hartmann A, Gostner J, Fuchs JE, Chaita E, Aligiannis N, Skaltsounis L, Ganzera M (2015) Inhibition of collagenase by mycosporine-like amino acids from marine sources. Planta Med 81(10):813–820

Heleno SA, Diz P, Prieto MA, Barros L, Rodrigues A, Barreiro MF, Ferreira IC (2016) Optimization of ultrasound-assisted extraction to obtain mycosterols from *Agaricus bisporus* L. by response surface methodology and comparison with conventional Soxhlet extraction. Food Chem 197:1054–1063

Helkar PB, Sahoo AK, Patil NJ (2016) Review: Food industry by-products used as a functional food ingredients. Int J Waste Resour 6(3):1–6

Heo SJ, Park EJ, Lee KW, Jeon YJ (2005) Antioxidant activities of enzymatic extracts from brown seaweeds. Bioresour Technol 96(14):1613–1623

Heo SJ, Ko SC, Cha SH, Kang DH, Park HS, Choi YU, Kim D, Jung WK, Jeon YJ (2009) Effect of phlorotannins isolated from *Ecklonia cava* on melanogenesis and their protective effect against photo-oxidative stress induced by UV-B radiation. Toxicol In Vitro 23(6):1123–1130

Herrero M, Ibanez E (2015) Green processes and sustainability: an overview on the extraction of high added-value products from seaweeds and microalgae. J Supercrit Fluids 96:211–216

Herrero M, del Pilar Sánchez-Camargo A, Cifuentes A, Ibáñez E (2015) Plants, seaweeds, microalgae and food by-products as natural sources of functional ingredients obtained using pressurized liquid extraction and supercritical fluid extraction. TrAC Trends Anal Chem 71:26–38

Holdt SL, Kraan S (2011) Bioactive compounds in seaweed: functional food applications and legislation. J Appl Phycol 23:543–597

Holgate O, Ruddy A (2023) Potential impacts of large-scale seaweed farming on global warming: a review. In: OCEANS 2023-Limerick, pp 1–6

Hou C, Chen L, Yang L, Ji X (2020) An insight into anti-inflammatory effects of natural polysaccharides. Int J Biol Macromol 153:248–255

Houston F, Milne X (2012) Seaweed and eat it: a family foraging and cooking adventure. Random House

Hsu HY, Hwang PA (2019) Clinical applications of fucoidan in translational medicine for adjuvant cancer therapy. Clin Transl Med 8(1):1–18

Hupel M, Lecointre C, Meudec A, Poupart N, Gall EA (2011) Comparison of photoprotective responses to UV radiation in the brown seaweed *Pelvetia canaliculata* and the marine angiosperm *Salicornia ramosissima*. J Exp Mar Biol Ecol 401(1–2):36–47

Hwang E, Park SY, Sun ZW, Shin HS, Lee DG, Yi TH (2014) The protective effects of fucosterol against skin damage in UVB-irradiated human dermal fibroblasts. Mar Biotechnol 16:361–370

Imeson A (ed) (2011) Food stabilisers, thickeners and gelling agents. Wiley, New York

Jayakody MM, Vanniarachchy MPG (2024) Seaweed-derived snack foods: products and processing. In: Applications of seaweeds in food and nutrition. Elsevier, Amsterdam, pp 215–224

Jayakody MM, Kaushani KG, Vanniarachchy MPG, Wijesekara I (2023) Hydrocolloid and water soluble polymers used in the food industry and their functional properties: a review. Polym Bull 80(4):3585–3610

Jayawardena TU, Nagahawatta DP, Fernando IPS, Kim YT, Kim JS, Kim WS, Lee JS, Jeon YJ (2022) A review on fucoidan structure, extraction techniques, and its role as an immunomodulatory agent. Mar Drugs 20(12):755

Je JY, Park PJ, Kim EK, Park JS, Yoon HD, Kim KR, Ahn CB (2009) Antioxidant activity of enzymatic extracts from the brown seaweed *Undaria pinnatifida* by electron spin resonance spectroscopy. LWT Food Sci Technol 42(4):874–878

Jesumani V, Du H, Aslam M, Pei P, Huang N (2019) Potential use of seaweed bioactive compounds in skincare—a review. Mar Drugs 17(12):688

Jesumani V, Du H, Pei P, Aslam M, Huang N (2020) Comparative study on skin protection activity of polyphenol-rich extract and polysaccharide-rich extract from *Sargassum vachellianum*. PLoS One 15(1):e0227308

Jiménez-González C, Agrasar AMT, Mallo F, Rúa ML, Fuciños C (2023) Red seaweed proteins: valuable marine-origin compounds with encouraging applications. Algal Res 75:103262

Juncan AM, Moisă DG, Santini A, Morgovan C, Rus LL, Vonica-Ţincu AL, Loghin F (2021) Advantages of hyaluronic acid and its combination with other bioactive ingredients in cosmeceuticals. Molecules 26(15):4429

Kadam Vaishali S, Chintale Ashwini GDKP, Deshmukh Kshitija P, Nalwad Digambar N (2013) Cosmeceuticals an emerging concept: a comprehensive review. Int J Res Pharm Chem 3(2):308–316

Kalasariya HS, Pereira L (2022) Dermo-cosmetic benefits of marine macroalgae-derived phenolic compounds. Appl Sci 12(23):11954

Kalasariya HS, Patel NB, Yadav A, Perveen K, Yadav VK, Munshi FM, Yadav KK, Alam S, Jung YK, Jeon BH (2021) Characterization of fatty acids, polysaccharides, amino acids, and minerals in marine macroalga *Chaetomorpha crassa* and evaluation of their potentials in skin cosmetics. Molecules 26(24):7515

Kalasariya HS, Patel NB, Gacem A, Alsufyani T, Reece LM, Yadav VK, Awwad NS, Ibrahium HA, Ahn Y, Yadav KK, Jeon BH (2022a) Marine Alga *Ulva fasciata*-derived molecules for the potential treatment of SARS-CoV-2: an in silico approach. Mar Drugs 20(9):586

Kalasariya HS, Pereira L, Patel NB (2022b) Pioneering role of marine macroalgae in cosmeceuticals. Phycology 2(1):172–203

Kalasariya HS, Pereira L, Patel NB (2023) Comprehensive phytochemical analysis and bioactivity evaluation of *Padina boergesenii*: unveiling its prospects as a promising cosmetic component. Mar Drugs 21(7):385

Kalil SJ, Moraes CC, Sala L, Burkert CA (2017) Bioproduct extraction from microbial cells by conventional and nonconventional techniques. In: Food bioconversion. Academic Press, London, pp 179–206

Karuppusamy S, Rajauria G, Fitzpatrick S, Lyons H, McMahon H, Curtin J, Tiwari BK, O'Donnell C (2022) Biological properties and health-promoting functions of laminarin: a comprehensive review of preclinical and clinical studies. Mar Drugs 20(12):772

Katiyar R, Arora A (2020) Health promoting functional lipids from microalgae pool: a review. Algal Res 46:101800

Khalid S, Abbas M, Saeed F, Bader-Ul-Ain H, Suleria HAR (2018) Therapeutic potential of seaweed bioactive compounds. IntechOpen, London

Khan MNA, Yoon SJ, Choi JS, Park NG, Lee HH, Cho JY, Hong YK (2009) Anti-edema effects of brown seaweed (*Undaria pinnatifida*) extract on phorbol 12-myristate 13-acetate-induced mouse ear inflammation. Am J Chin Med 37(02):373–381

Khan MI, Shin JH, Kim JD (2018) The promising future of microalgae: current status, challenges, and optimization of a sustainable and renewable industry for biofuels, feed, and other products. Microb Cell Fact 17(1):1–21

Kılınç B, Cirik S, Turan G, Tekogul H, Koru E (2013) Seaweeds for food and industrial applications. In: Food industry. IntechOpen

Kim KT (2012) Seasonal variation of seaweed components and novel biological function of fucoidan extracted from brown algae in Quebec. Doctoral dissertation, Université Laval

Kim SK (2014) Marine cosmeceuticals. J Cosmet Dermatol 13(1):56–67

Kim SM, Kang K, Jeon JS, Jho EH, Kim CY, Nho CW, Um BH (2011) Isolation of phlorotannins from *Eisenia bicyclis* and their hepatoprotective effect against oxidative stress induced by tert-butyl hyperoxide. Appl Biochem Biotechnol 165:1296–1307

Kim JA, Ahn BN, Kong CS, Kim SK (2013a) The chromene sargachromanol E inhibits ultraviolet A-induced ageing of skin in human dermal fibroblasts. Br J Dermatol 168(5):968–976

Kim KN, Yang HM, Kang SM, Kim D, Ahn G, Jeon YJ (2013b) Octaphlorethol A isolated from *Ishige foliacea* inhibits α-MSH-stimulated induced melanogenesis via ERK pathway in B16F10 melanoma cells. Food Chem Toxicol 59:521–526

Kim SM, Kang SW, Jeon JS, Jung YJ, Kim WR, Kim CY, Um BH (2013c) Determination of major phlorotannins in *Eisenia bicyclis* using hydrophilic interaction chromatography: seasonal variation and extraction characteristics. Food Chem 138(4):2399–2406

Kim KN, Yang HM, Kang SM, Ahn G, Roh SW, Lee W, Kim D, Jeon YJ (2015) Whitening effect of octaphlorethol A isolated from *Ishige foliacea* in an in vivo zebrafish model. J Microbiol Biotechnol 25(4):448–451

Ko SC, Cha SH, Heo SJ, Lee SH, Kang SM, Jeon YJ (2011) Protective effect of *Ecklonia cava* on UVB-induced oxidative stress: in vitro and in vivo zebrafish model. J Appl Phycol 23:697–708

Kumar AN, Dissanayake PD, Masek O, Priya A, Lin CSK, Ok YS, Kim SH (2021a) Recent trends in biochar integration with anaerobic fermentation: win-win strategies in a closed-loop. Renew Sust Energ Rev 149:111371

Kumar Y, Tarafdar A, Badgujar PC (2021b) Seaweed as a source of natural antioxidants: therapeutic activity and food applications. J Food Qual 2021:1–17

Kumar A, Hanjabam MD, Kishore P, Uchoi D, Panda SK, Mohan CO, Chatterjee NS, Zynudheen AA, Ravishankar CN (2023) Exploitation of seaweed functionality for the development of food products. Food Bioprocess Technol 16:1873–1903

Kurakula M, Rao GK, Kiran V, Hasnain MS, Nayak AK (2020) Alginate-based hydrogel systems for drug releasing in wound healing. In: Alginates in drug delivery. Academic Press, Cambridge, MA, pp 323–358

Lange L, Bak UG, Hansen SCB, Gregersen O, Harmsen P, Karlsson EN, Meyer A, Mikkelsen MD, Van Den Broek L, Hreggviðsson GÓ (2020) Opportunities for seaweed biorefinery. In: Sustainable seaweed technologies. Elsevier, Amsterdam, pp 3–31

Langton R, Augyte S, Price N, Forster J, Noji T, Grebe G, St Gelais A, Byron CJ (2019) An ecosystem approach to the culture of seaweed. NOAA Tech. Memo. NMFS-F/SPO-195, 24 p

Leandro A, Pereira L, Gonçalves AM (2019) Diverse applications of marine macroalgae. Mar Drugs 18(1):17

Leandro A, Pacheco D, Cotas J, Marques JC, Pereira L, Gonçalves AM (2020) Seaweed's bioactive candidate compounds to food industry and global food security. Life 10(8):140

Lee S, Lee YS, Jung SH, Kang SS, Shin KH (2003) Anti-oxidant activities of fucosterol from the marine algae *Pelvetia siliquosa*. Arch Pharm Res 26:719–722

Lee YS, Jung SH, Lee SH, Shin KH (2004) Effects of the extracts from the marine algae *Pelvetia siliquosa* on hyperlipidemia in rats. Korean J Pharmacogn 35(2):143–146

Lee HA, Kim IH, Nam TJ (2015) Bioactive peptide from *Pyropia yezoensis* and its anti-inflammatory activities. Int J Mol Med 36(6):1701–1706

Lei Z, Sun X, Zhu S, Dong K, Liu X, Wang L, Zhang X, Qu L, Zhang X (2022) Nature inspired MXene-decorated 3D honeycomb-fabric architectures toward efficient water desalination and salt harvesting. Nanomicro Lett 14(1):10

Leibbrandt A, Meier C, König-Schuster M, Weinmüllner R, Kalthoff D, Pflugfelder B, Graf P, Frank-Gehrke B, Beer M, Fazekas T, Unger H (2010) Iota-carrageenan is a potent inhibitor of influenza A virus infection. PLoS One 5(12):e14320

Leyton A, Pezoa-Conte R, Barriga A, Buschmann AH, Mäki-Arvela P, Mikkola JP, Lienqueo ME (2016) Identification and efficient extraction method of phlorotannins from the brown seaweed *Macrocystis pyrifera* using an orthogonal experimental design. Algal Res 16:201–208

Li K, Li XM, Gloer JB, Wang BG (2012) New nitrogen-containing bromophenols from the marine red alga *Rhodomela confervoides* and their radical scavenging activity. Food Chem 135(3):868–872

Li M, Zhu KX, Guo XN, Brijs K, Zhou HM (2014) Natural additives in wheat-based pasta and noodle products: opportunities for enhanced nutritional and functional properties. Compr Rev Food Sci Food Saf 13(4):347–357

Li T, Xu J, Wu H, Jiang P, Chen Z, Xiang W (2019) Growth and biochemical composition of *Porphyridium purpureum* SCS-02 under different nitrogen concentrations. Mar Drugs 17(2):124

Liaqat F, Xu L, Khazi MI, Ali S, Rahman MU, Zhu D (2023) Extraction, purification, and applications of vanillin: a review of recent advances and challenges. Ind Crop Prod 204:117372

Li-Chan EC (2015) Bioactive peptides and protein hydrolysates: research trends and challenges for application as nutraceuticals and functional food ingredients. Curr Opin Food Sci 1:28–37

Lomartire S, Gonçalves AM (2022a) An overview of potential seaweed-derived bioactive compounds for pharmaceutical applications. Mar Drugs 20(2):141

Lomartire S, Gonçalves AM (2022b) Novel technologies for seaweed polysaccharides extraction and their use in food with -therapeutically applications—a review. Foods 11(17):2654

Lorbeer A (2016) Advanced and integrated technologies for the valorization of South Australia's unique seaweed resources. Doctoral dissertation, Flinders University, School of Medicine

Lordan S, Ross RP, Stanton C (2011) Marine bioactives as functional food ingredients: potential to reduce the incidence of chronic diseases. Mar Drugs 9(6):1056–1100

Lu KY, Jheng PR, Lu LS, Rethi L, Mi FL, Chuang EY (2021) Enhanced anticancer effect of ROS-boosted photothermal therapy by using fucoidan-coated polypyrrole nanoparticles. Int J Biol Macromol 166:98–107

Luthuli S, Wu S, Cheng Y, Zheng X, Wu M, Tong H (2019) Therapeutic effects of fucoidan: a review on recent studies. Mar Drugs 17(9):487

Magnusson M, Glasson CR, Vucko MJ, Angell A, Neoh TL, de Nys R (2019) Enrichment processes for the production of high-protein feed from the green seaweed *Ulva ohnoi*. Algal Res 41:101555

Malinowska P (2011) Algae extracts as active cosmetic ingredients. Zesz Nauk/Uniw Ekon Poznaniu 212:123–129

Mani D, Kumar C (2014) Biotechnological advances in bioremediation of heavy metals contaminated ecosystems: an overview with special reference to phytoremediation. Int J Environ Sci Technol 11:843–872

Mansour HM, Sohn M, Al-Ghananeem A, DeLuca PP (2010) Materials for pharmaceutical dosage forms: molecular pharmaceutics and controlled release drug delivery aspects. Int J Mol Sci 11(9):3298–3322

Mansouri MT, Hemmati AA, Naghizadeh B, Mard SA, Rezaie A, Ghorbanzadeh B (2015) A study of the mechanisms underlying the anti-inflammatory effect of ellagic acid in carrageenan-induced paw edema in rats. Indian J Pharmacol 47(3):292

Mansuri R, Diwan A, Kumar H, Dangwal K, Yadav D (2021) Potential of natural compounds as sunscreen agents. Pharmacogn Rev 15(29):47

Manzoor M, Singh J, Bandral JD, Gani A, Shams R (2020) Food hydrocolloids: functional, nutraceutical and novel applications for delivery of bioactive compounds. Int J Biol Macromol 165:554–567

Martins B, Vieira M, Delerue-Matos C, Grosso C, Soares C (2022) Biological potential, gastrointestinal digestion, absorption, and bioavailability of algae-derived compounds with neuroprotective activity: a comprehensive review. Mar Drugs 20(6):362

Matanjun P, Mohamed S, Mustapha NM, Muhammad K, Ming CH (2008) Antioxidant activities and phenolics content of eight species of seaweeds from north Borneo. J Appl Phycol 20:367–373

Matulja D, Vranješević F, Kolympadi Markovic M, Pavelić SK, Marković D (2022) Anticancer activities of marine-derived phenolic compounds and their derivatives. Molecules 27(4):1449

McLean AL (2015) Asian American food culture. Bloomsbury Publishing, Santa Barbara

Menon GK, Kligman AM (2009) Barrier functions of human skin: a holistic view. Skin Pharmacol Physiol 22(4):178–189

Messina CM, Renda G, Laudicella VA, Trepos R, Fauchon M, Hellio C, Santulli A (2019) From ecology to biotechnology, study of the defense strategies of algae and halophytes (from Trapani Saltworks, NW Sicily) with a focus on antioxidants and antimicrobial properties. Int J Mol Sci 20(4):881

Milne X (2016) The seaweed cookbook. Penguin, London

Mise T, Ueda M, Yasumoto T (2011) Production of fucoxanthin-rich powder from *Cladosiphon okamuranus*. Adv J Food Sci Technol 3(1):73–76

Mistry N (2017) Guidelines for formulating anti-pollution products. Cosmetics 4(4):57

Monteiro P, Cotas J, Pacheco D, Figueirinha A, da Silva GJ, Pereira L, Gonçalves AMM (2022) Seaweed as food: how to guarantee their quality? In: Sustainable global resources of seaweeds. Volume 2: Food, pharmaceutical and health applications. Springer International Publishing, Cham, pp 309–321

Morais T, Cotas J, Pacheco D, Pereira L (2021) Seaweeds compounds: an ecosustainable source of cosmetic ingredients? Cosmetics 8(1):8

Moroney NC, O'Grady MN, Lordan S, Stanton C, Kerry JP (2015) Seaweed polysaccharides (laminarin and fucoidan) as functional ingredients in pork meat: an evaluation of anti-oxidative potential, thermal stability and bioaccessibility. Mar Drugs 13(4):2447–2464

Moura LI, Dias AM, Carvalho E, de Sousa HC (2013) Recent advances on the development of wound dressings for diabetic foot ulcer treatment—a review. Acta Biomater 9(7):7093–7114

Mouritsen OG (2009) Sushi: food for the eye, the body and the soul. Springer Science & Business Media, New York

Mouritsen OG (2013) Seaweeds: edible, available, and sustainable. University of Chicago Press, Chicago

Mouritsen OG, Williams L, Bjerregaard R, Duelund L (2012) Seaweeds for umami flavour in the New Nordic Cuisine. Flavour 1(1):1–12

Mouritsen OG, Rhatigan P, Pérez-Lloréns JL (2018) World cuisine of seaweeds: science meets gastronomy. Int J Gastron Food Sci 14:55–65

Mularczyk M, Michalak I, Marycz K (2020) Astaxanthin and other nutrients from *Haematococcus pluvialis*—multifunctional applications. Mar Drugs 18(9):459

Neto RT, Marçal C, Queirós AS, Abreu H, Silva AM, Cardoso SM (2018) Screening of *Ulva rigida*, *Gracilaria* sp., *Fucus vesiculosus* and *Saccharina latissima* as functional ingredients. Int J Mol Sci 19(10):2987

Nilsson J, Martin M (2022) Exploratory environmental assessment of large-scale cultivation of seaweed used to reduce enteric methane emissions. Sustain Prod Consum 30:413–423

O'Connor K (2017) Seaweed: a global history. Reaktion Books, London

Obluchinskaya ED, Pozharitskaya ON, Flisyuk EV, Shikov AN (2021) Formulation, optimization and in vivo evaluation of fucoidan-based cream with anti-inflammatory properties. Mar Drugs 19(11):643

Orfanoudaki M, Hartmann A, Karsten U, Ganzera M (2019) Chemical profiling of mycosporine-like amino acids in twenty-three red algal species. J Phycol 55(2):393–403

Ospina M, Castro-Vargas HI, Parada-Alfonso F (2017) Antioxidant capacity of Colombian seaweeds: 1. Extracts obtained from *Gracilaria mammillaris* by means of supercritical fluid extraction. J Supercrit Fluids 128:314–322

Otero P, López-Martínez MI, García-Risco MR (2019) Application of pressurized liquid extraction (PLE) to obtain bioactive fatty acids and phenols from *Laminaria ochroleuca* collected in Galicia (NW Spain). J Pharm Biomed Anal 164:86–92

Ozanne H, Toumi H, Roubinet B, Landemarre L, Lespessailles E, Daniellou R, Cesaro A (2020) Laminarin effects, a β-(1,3)-glucan, on skin cell inflammation and oxidation. Cosmetics 7(3):66

Padilla M, Palma M, Barroso CG (2005) Determination of phenolics in cosmetic creams and similar emulsions. J Chromatogr A 1091(1–2):83–88

Patel A, Karageorgou D, Katapodis P, Sharma A, Rova U, Christakopoulos P, Matsakas L (2021) Bioprospecting of thraustochytrids for omega-3 fatty acids: a sustainable approach to reduce dependency on animal sources. Trends Food Sci Technol 115:433–444

Patra JK, Das G, Baek KH (2015) Chemical composition and antioxidant and antibacterial activities of an essential oil extracted from an edible seaweed, *Laminaria japonica* L. Molecules 20(7):12093–12113

Peñalver R, Lorenzo JM, Ros G, Amarowicz R, Pateiro M, Nieto G (2020) Seaweeds as a functional ingredient for a healthy diet. Mar Drugs 18(6):301

Percival E (1979) The polysaccharides of green, red and brown seaweeds: their basic structure, biosynthesis and function. Br Phycol J 14(2):103–117

Pereira L (2018) Seaweeds as source of bioactive substances and skin care therapy—cosmeceuticals, algotheraphy, and thalassotherapy. Cosmetics 5(4):68

Pereira L, Gheda SF, Ribeiro-Claro PJ (2013) Analysis by vibrational spectroscopy of seaweed polysaccharides with potential use in food, pharmaceutical, and cosmetic industries. Int J Carbohydr Chem 2013:537202

Pérez-Lloréns JL (2019) Seaweed consumption in the Americas. Gastronomica 19(4):49–59

Pergolizzi J, Aloisi AM, Dahan A, Filitz J, Langford R, Likar R, Mercadante S, Morlion B, Raffa RB, Sabatowski R, Sacerdote P (2010) Current knowledge of buprenorphine and its unique pharmacological profile. Pain Pract 10(5):428–450

Pinela J, Prieto MA, Carvalho AM, Barreiro MF, Oliveira MBP, Barros L, Ferreira IC (2016) Microwave-assisted extraction of phenolic acids and flavonoids and production of antioxidant ingredients from tomato: a nutraceutical-oriented optimization study. Sep Purif Technol 164:114–124

Plaza M, Santoyo S, Jaime L, Reina GGB, Herrero M, Señoráns FJ, Ibáñez E (2010) Screening for bioactive compounds from algae. J Pharm Biomed Anal 51(2):450–455

Polat S, Trif M, Rusu A, Šimat V, Čagalj M, Alak G, Meral R, Özogul Y, Polat A, Özogul F (2023) Recent advances in industrial applications of seaweeds. Crit Rev Food Sci Nutr 63(21):4979–5008

Potvin G, Ahmad A, Zhang Z (2012) Bioprocess engineering aspects of heterologous protein production in *Pichia pastoris*: a review. Biochem Eng J 64:91–105

Pradhan B, Bhuyan PP, Patra S, Nayak R, Behera PK, Behera C, Behera AK, Ki JS, Jena M (2022) Beneficial effects of seaweeds and seaweed-derived bioactive compounds: current evidence and future prospective. Biocatal Agric Biotechnol 39:102242

Praveen MA, Parvathy KK, Balasubramanian P, Jayabalan R (2019) An overview of extraction and purification techniques of seaweed dietary fibers for immunomodulation on gut microbiota. Trends Food Sci Technol 92:46–64

Qin Y (2018) Applications of bioactive seaweed substances in functional food products. In: Bioactive seaweeds for food applications. Academic Press, San Diego, pp 111–134

Quitério E, Grosso C, Ferraz R, Delerue-Matos C, Soares C (2022) A critical comparison of the advanced extraction techniques applied to obtain health-promoting compounds from seaweeds. Mar Drugs 20(11):677

Rahman MM, Islam MR, Shohag S, Hossain ME, Shah M, Shuvo SK, Khan H, Chowdhury MAR, Bulbul IJ, Hossain MS, Sultana S (2022) Multifaceted role of natural sources for COVID-19 pandemic as marine drugs. Environ Sci Pollut Res 29(31):46527–46550

Raja K, Kadirvel V, Subramaniyan T (2022) Seaweeds, an aquatic plant-based protein for sustainable nutrition – a review. Future Foods 5:100142

Reboleira J, Silva S, Chatzifragkou A, Niranjan K, Lemos MF (2021) Seaweed fermentation within the fields of food and natural products. Trends Food Sci Technol 116:1056–1073

Rioux LE, Moulin V, Beaulieu M, Turgeon SL (2013) Human skin fibroblast response is differentially regulated by galactofucan and low molecular weight galactofucan. Bioact Carbohydr Diet Fibre 1(2):105–110

Rodriguez Garcia SL, Raghavan V (2022) Green extraction techniques from fruit and vegetable waste to obtain bioactive compounds—a review. Crit Rev Food Sci Nutr 62(23):6446–6466

Rodríguez-González I, Díaz-Reinoso B, Domínguez H (2022) Intensification strategies for the extraction of polyunsaturated fatty acids and other lipophilic fractions from seaweeds. Food Bioprocess Technol 15(5):978–997

Ruocco N, Costantini S, Guariniello S, Costantini M (2016) Polysaccharides from the marine environment with pharmacological, cosmeceutical and nutraceutical potential. Molecules 21(5):551

Rupérez P, Ahrazem O, Leal JA (2002) Potential antioxidant capacity of sulfated polysaccharides from the edible marine brown seaweed *Fucus vesiculosus*. J Agric Food Chem 50(4):840–845

Ryu J, Park SJ, Kim IH, Choi YH, Nam TJ (2014) Protective effect of porphyra-334 on UVA-induced photoaging in human skin fibroblasts. Int J Mol Med 34(3):796–803

Ryu B, Ahn BN, Kang KH, Kim YS, Li YX, Kong CS, Kim SK, Kim DG (2015) Dioxinodehydroeckol protects human keratinocyte cells from UVB-induced apoptosis modulated by related genes Bax/Bcl-2 and caspase pathway. J Photochem Photobiol B Biol 153:352–357

Saini RK, Keum YS (2018) Omega-3 and omega-6 polyunsaturated fatty acids: dietary sources, metabolism, and significance—a review. Life Sci 203:255–267

Saini RK, Prasad P, Sreedhar RV, Akhilender Naidu K, Shang X, Keum YS (2021) Omega-3 polyunsaturated fatty acids (PUFAs): emerging plant and microbial sources, oxidative stability, bioavailability, and health benefits—a review. Antioxidants 10(10):1627

Sakai S, Komura Y, Nishimura Y, Sugawara T, Hirata T (2011) Inhibition of mast cell degranulation by phycoerythrin and its pigment moiety phycoerythrobilin, prepared from *Porphyra yezoensis*. Food Sci Technol Res 17(2):171–177

Salehi B, Sharifi-Rad J, Seca AM, Pinto DC, Michalak I, Trincone A, Mishra AP, Nigam M, Zam W, Martins N (2019) Current trends on seaweeds: looking at chemical composition, phytopharmacology, and cosmetic applications. Molecules 24(22):4182

Samanta KK, Basak S, Chattopadhyay SK (2015) Specialty chemical finishes for sustainable luxurious textiles. In: Handbook of sustainable luxury textiles and fashion: volume 1. Springer, Singapore, pp 145–184

Sanghvi AM, Martin Lo Y (2010) Present and potential industrial applications of macro-and microalgae. Recent Pat Food Nutr Agric 2(3):187–194

Santos JP, Torres PB, dos Santos DY, Motta LB, Chow F (2019) Seasonal effects on antioxidant and anti-HIV activities of Brazilian seaweeds. J Appl Phycol 31:1333–1341

Santos HO, Price JC, Bueno AA (2020) Beyond fish oil supplementation: the effects of alternative plant sources of omega-3 polyunsaturated fatty acids upon lipid indexes and cardiometabolic biomarkers—an overview. Nutrients 12(10):3159

Santos-Buelga C, Gonzalez-Manzano S, Dueñas M, Gonzalez-Paramas AM (2012) Extraction and isolation of phenolic compounds. In: Natural products isolation. Humana Press, New York, pp 427–464

Sarkar NS, Choudhury S (2017) Algae as source of natural flavour enhancers – a mini review. Plant Sci Today 4(4):172–176

Sarker S, Mahmudul Islam M (2022) Marine protected area and biodiversity conservation. In: Life below water. Springer International Publishing, Cham, pp 629–644

Shafie MH, Kamal ML, Zulkiflee FF, Hasan S, Uyup NH, Abdullah S, Hussin NAM, Tan YC, Zafarina Z (2022) Application of Carrageenan extract from red seaweed (Rhodophyta) in cosmetic products: a review. J Indian Chem Soc 99(9):100613

Shannon E, Conlon M, Hayes M (2021) Seaweed components as potential modulators of the gut microbiota. Mar Drugs 19(7):358

Shimoda H, Tanaka J, Shan SJ, Maoka T (2010) Anti-pigmentary activity of fucoxanthin and its influence on skin mRNA expression of melanogenic molecules. J Pharm Pharmacol 62(9):1137–1145

Sigren JM, Figlus J, Armitage AR (2014) Coastal sand dunes and dune vegetation: restoration, erosion, and storm protection. Shore Beach 82(4):5–12

Singh D, Gupta R, Saraf SA (2012) Herbs–are they safe enough? An overview. Crit Rev Food Sci Nutr 52(10):876–898

Soetan KO, Olaiya CO, Oyewole OE (2010) The importance of mineral elements for humans, domestic animals and plants: a review. Afr J Food Sci 4(5):200–222

Song TY, Chen CH, Yang NC, Fu CS, Chang YT, Chen CL (2009) The correlation of in vitro mushroom tyrosinase activity with cellular tyrosinase activity and melanin formation in melanoma cells A2058. J Food Drug Anal 17(3):4

Souto EB, Silva GF, Dias-Ferreira J, Zielinska A, Ventura F, Durazzo A, Lucarini M, Novellino E, Santini A (2020) Nanopharmaceutics: Part I—Clinical trials legislation and good manufacturing practices (GMP) of nanotherapeutics in the EU. Pharmaceutics 12(2):146

Stengel DB, Connan S (2015) Natural products from marine algae. Methods Mol Biol 1308:1–37

Sudhakar K, Mamat R, Samykano M, Azmi WH, Ishak WFW, Yusaf T (2018) An overview of marine macroalgae as bioresource. Renew Sust Energ Rev 91:165–179

Suganya T, Varman M, Masjuki HH, Renganathan S (2016) Macroalgae and microalgae as a potential source for commercial applications along with biofuels production: a biorefinery approach. Renew Sust Energ Rev 55:909–941

Sugiura Y, Matsuda K, Yamada Y, Nishikawa M, Shioya K, Katsuzaki H, Imai K, Amano H (2006a) Isolation of a new anti-allergic phlorotannin, phlorofucofuroeckol-B, from an edible brown alga, *Eisenia arborea*. Biosci Biotechnol Biochem 70(11):2807–2811

Sugiura Y, Takeuchi Y, Kakinuma M, Amano H (2006b) Inhibitory effects of seaweeds on histamine release from rat basophile leukemia cells (RBL-2H3). Fish Sci 72:1286–1291

Sugiura Y, Kinoshita Y, Misumi S, Yamatani H, Katsuzaki H, Hayashi Y, Murase N (2021) Correlation between the seasonal variations in phlorotannin content and the antiallergic effects of the brown alga *Ecklonia cava* subsp. *stolonifera*. Algal Res 58:102398

Sugumaran R, Padam BS, Yong WTL, Saallah S, Ahmed K, Yusof NA (2022) A retrospective review of global commercial seaweed production—current challenges, biosecurity and mitigation measures and prospects. Int J Environ Res Public Health 19(12):7087

Sun J, Tan H (2013) Alginate-based biomaterials for regenerative medicine applications. Materials 6(4):1285–1309

Sun Z, Mohamed MAA, Park SY, Yi TH (2015) Fucosterol protects cobalt chloride induced inflammation by the inhibition of hypoxia-inducible factor through PI3K/Akt pathway. Int Immunopharmacol 29(2):642–647

Susano P, Silva J, Alves C, Martins A, Pinteus S, Gaspar H, Goettert MI, Pedrosa R (2022) Mitigating the negative impacts of marine invasive species – *Sargassum muticum* – a key seaweed for skincare products development. Algal Res 62:102634

Susanto E, Fahmi AS, Abe M, Hosokawa M, Miyashita K (2016) Lipids, fatty acids, and fucoxanthin content from temperate and tropical brown seaweeds. Aquat Procedia 7:66–75

Tan SP, O'Sullivan L, Luz Prieto M, McLoughlin P, Lawlor PG, Hughes H, Gardiner GE (2013) Seaweed antimicrobials: isolation, characterization, and potential use in functional foods. In: Bioactive compounds from marine foods: plant and animal sources. Wiley, Chichester, pp 269–312

Tan J, Luo Y, Guo Y, Zhou Y, Liao X, Li D, Lai X, Liu Y (2023) Development of alginate-based hydrogels: crosslinking strategies and biomedical applications. Int J Biol Macromol 239:124275

Tanna B, Mishra A (2019) Nutraceutical potential of seaweed polysaccharides: structure, bioactivity, safety, and toxicity. Compr Rev Food Sci Food Saf 18(3):817–831

Teas J, Irhimeh MR (2017) Melanoma and brown seaweed: an integrative hypothesis. J Appl Phycol 29(2):941–948

Thakur A, Kumar A, Kaya S, Vo DVN, Sharma A (2022) Suppressing inhibitory compounds by nanomaterials for highly efficient biofuel production: a review. Fuel 312:122934

Thevanayagam H, Mohamed SM, Chu WL (2014) Assessment of UVB-photoprotective and antioxidative activities of carrageenan in keratinocytes. J Appl Phycol 26:1813–1821

Thomas NV, Kim SK (2013) Beneficial effects of marine algal compounds in cosmeceuticals. Mar Drugs 11(1):146–164

Thun MJ, Henley SJ, Patrono C (2002) Nonsteroidal anti-inflammatory drugs as anticancer agents: mechanistic, pharmacologic, and clinical issues. J Natl Cancer Inst 94(4):252–266

Tønnesen HH, Karlsen J (2002) Alginate in drug delivery systems. Drug Dev Ind Pharm 28(6):621–630

Vasconcelos JB, de Vasconcelos ER, Urrea-Victoria V, Bezerra PS, Reis TN, Cocentino AL, Navarro DM, Chow F, Areces AJ, Fujii MT (2019) Antioxidant activity of three seaweeds from tropical reefs of Brazil: potential sources for bioprospecting. J Appl Phycol 31:835–846

Veluchamy C, Palaniswamy R (2020) A review on marine algae and its applications. Asian J Pharm Clin Res 13(3):21–27

Venkatesan J, Lowe B, Anil S, Manivasagan P, Kheraif AAA, Kang KH, Kim SK (2015) Seaweed polysaccharides and their potential biomedical applications. Starch-Stärke 67(5–6):381–390

Vieira V, Prieto MA, Barros L, Coutinho JA, Ferreira IC, Ferreira O (2018) Enhanced extraction of phenolic compounds using choline chloride based deep eutectic solvents from *Juglans regia* L. Ind Crop Prod 115:261–271

Vo TS, Kim SK, Ryu B, Ngo DH, Yoon NY, Bach LG, Hang NTN, Ngo DN (2018) The suppressive activity of fucofuroeckol-A derived from brown algal *Ecklonia stolonifera* Okamura on UVB-induced mast cell degranulation. Mar Drugs 16(1):1

Vollet Marson G, Belleville MP, Lacour S, Dupas Hubinger M (2020) Membrane fractionation of protein hydrolysates from by-products: recovery of valuable compounds from spent yeasts. Membranes 11(1):23

Wahyuni T (2023) Total carotenoid content of Padina sp. seaweed-infused oil with virgin coconut oil. In: IOP conference series: earth and environmental science, vol 1137, No. 1. IOP Publishing, p 012038

Wang R, Paul VJ, Luesch H (2013) Seaweed extracts and unsaturated fatty acid constituents from the green alga *Ulva lactuca* as activators of the cytoprotective Nrf2–ARE pathway. Free Radic Biol Med 57:141–153

Wang HMD, Li XC, Lee DJ, Chang JS (2017a) Potential biomedical applications of marine algae. Bioresour Technol 244:1407–1415

Wang ZJ, Xu W, Liang JW, Wang CS, Kang Y (2017b) Effect of fucoidan on B16 murine melanoma cell melanin formation and apoptosis. Afr J Tradit Complement Altern Med 14(4):149–155

Wells ML, Potin P, Craigie JS, Raven JA, Merchant SS, Helliwell KE, Smith AG, Camire ME, Brawley SH (2017) Algae as nutritional and functional food sources: revisiting our understanding. J Appl Phycol 29:949–982

Westlake JR, Tran MW, Jiang Y, Zhang X, Burrows AD, Xie M (2023) Biodegradable biopolymers for active packaging: demand, development and directions. Sustain Food Technol 1(1):50–72

White LN, White WL (2020) Seaweed utilisation in New Zealand. Bot Mar 63(4):303–313

Wijesekara I, Karunarathna WKDS (2017) Usage of seaweed polysaccharides as nutraceuticals. In: Seaweed polysaccharides. Elsevier, Amsterdam, pp 341–348

Wijesinghe WAJP, Jeon YJ (2012) Enzyme-assistant extraction (EAE) of bioactive components: a useful approach for recovery of industrially important metabolites from seaweeds: a review. Fitoterapia 83(1):6–12

Wu Q, Liu L, Miron A, Klímová B, Wan D, Kuča K (2016) The antioxidant, immunomodulatory, and anti-inflammatory activities of Spirulina: an overview. Arch Toxicol 90:1817–1840

Yang JH (2012) Topical application of fucoidan improves atopic dermatitis symptoms in NC/Nga mice. Phytother Res 26(12):1898–1903

Yang Y, Liu D, Wu J, Chen Y, Wang S (2011) In vitro antioxidant activities of sulfated polysaccharide fractions extracted from *Corallina officinalis*. Int J Biol Macromol 49(5):1031–1037

Yang H, Liu DQ, Liang TJ, Li J, Liu AH, Yang P, Lin K, Yu XQ, Guo YW, Mao SC, Wang B (2014) Racemosin C, a novel minor bisindole alkaloid with protein tyrosine phosphatase-1B inhibitory activity from the green alga *Caulerpa racemosa*. J Asian Nat Prod Res 16(12):1158–1165

Yarosh D (2008) The new science of perfect skin: understanding skin care myths and miracles for radiant skin at any age. Harmony

Yesuraj D, Deepika C, Ravishankar GA, Ranga Rao A (2022) Seaweed-based recipes for food, health-food applications, and innovative products including meat and meat analogs. In: Sustainable global resources of seaweeds. Volume 2: Food, pharmaceutical and health applications. Springer International Publishing, Cham, pp 267–292

Yong WTL, Thien VY, Rupert R, Rodrigues KF (2022) Seaweed: a potential climate change solution. Renew Sust Energ Rev 159:112222

Yuan YV, Westcott ND, Hu C, Kitts DD (2009) Mycosporine-like amino acid composition of the edible red alga, *Palmaria palmata* (dulse) harvested from the west and east coasts of Grand Manan Island, New Brunswick. Food Chem 112(2):321–328

Yusoff FM, Nagao N, Imaizumi Y, Toda T (2019) Bioreactor for microalgal cultivation systems: strategy and development. In: Prospects of renewable bioprocessing in future energy systems. Springer, Cham, pp 117–159

Zhang Q, Li N, Zhou G, Lu X, Xu Z, Li Z (2003) In vivo antioxidant activity of polysaccharide fraction from *Porphyra haitanesis* (Rhodophyta) in aging mice. Pharmacol Res 48(2):151–155

Zhang Q, Li N, Liu X, Zhao Z, Li Z, Xu Z (2004) The structure of a sulfated galactan from *Porphyra haitanensis* and its in vivo antioxidant activity. Carbohydr Res 339(1):105–111

Zhang L, Liao W, Huang Y, Wen Y, Chu Y, Zhao C (2022) Global seaweed farming and processing in the past 20 years. Food Prod Process Nutr 4(1):23

Zhao YS, Eweys AS, Zhang JY, Zhu Y, Bai J, Darwesh OM, Zhang HB, Xiao X (2021) Fermentation affects the antioxidant activity of plant-based food material through the release and production of bioactive components. Antioxidants 10(12):2004

Zia S, Khan MR, Shabbir MA, Aslam Maan A, Khan MKI, Nadeem M, Khalil AA, Din A, Aadil RM (2022) An inclusive overview of advanced thermal and nonthermal extraction techniques for bioactive compounds in food and food-related matrices. Food Rev Intl 38(6):1166–1196

Chapter 15
Present Scenario of Seaweed Cultivation, Existing Challenges, and the Way Forward for Sustainable Culture in Bangladesh

Md. Simul Bhuyan, Mrityunjoy Kunda, Monika Das, Istiak Ahamed Mojumder, Upasana Pattanayak, Sumi Akter, Md. Mohidul Islam, Md. Tarikul Islam, and Sobnom Mustary

Abstract Seaweeds, an important component of the Blue Economy of Bangladesh, have the potential to significantly increase both the national GDP and the standard of living in coastal communities. Seaweed with huge medical value and an excellent cuisine has exporting prospects as seafood to earn huge amounts of foreign cash. Moreover, the biochemical, medicinal, and cosmetics industries use seaweed as a component of different products. Many Southeast Asian nations have a highly developed seaweed industry (e.g., China, Indonesia, and the Philippines). An area of 47,201 km^2, or 32% of the country, is made up of 19 districts in Bangladesh's coastal region. The coastal environment, which includes muddy and sandy coastlines, estuaries, and mangrove areas, can provide an appropriate habitat for seaweed growth. Eight out of the 133 kinds of seaweed that are abundantly available in Bangladesh are commercially significant. Nowadays, seaweed production is being tried on Bangladesh's coastline, and many government and non-government organizations are carrying out various seaweed culture/research programs. Some of the findings are

M. S. Bhuyan (✉)
Bangladesh Oceanographic Research Institute, Cox's Bazar, Bangladesh

Department of Aquatic Resource Management, Faculty of Fisheries, Sylhet Agricultural University, Sylhet, Bangladesh

M. Kunda
Department of Aquatic Resource Management, Faculty of Fisheries, Sylhet Agricultural University, Sylhet, Bangladesh

M. Das
Department of Fisheries, Matshya Bhaban, Dhaka, Bangladesh

I. A. Mojumder
Department of Zoology, University of Chittagong, Chittagong, Bangladesh

U. Pattanayak
Department of Zoology, K. K. S. Women's College, Balasore, Odisha, India

N. Trivedi et al. (eds.), *Recent Advances in Seaweed Biotechnology*,
https://doi.org/10.1007/978-981-96-0519-4_15

good but not up to the mark. They are attempting to lessen the issues they are now encountering. Several strategies are being used to address the risk issues. For Bangladesh's seaweed culture to be effective, ongoing study is required to improve the technologies for the production of various commercially significant seed stocks and their advancements. To find ideal locations for widespread seaweed cultivation, thorough surveys must also be carried out for the expansion of seaweed cultivation.

Keywords Seaweed culture · Present status · Risks and challenges · Solutions · Coastal area

15.1 Introduction

Seaweed is a marine alga and is typically categorized as a plant that thrives on rocky beaches, estuaries, and deep-sea regions up to 180 m below the surface (Bhuyan et al. 2024a, b; Bhuyan 2023; Bhuyan et al. 2023a, b; Islam et al. 2020). Seaweed is a form of primitive plant that doesn't have proper roots, stems, or leaves and is an essential component of marine ecology. In addition to serving as the principal food supply and home for a wide variety of other marine creatures, seaweeds can dominate the whole marine benthic ecosystem (Theuerkauf et al. 2022; Siddiqui et al. 2019). Mollusks, baby fish, and crustaceans depend on the habitat that seaweeds offer for them to survive. Huge forests may be created by species such as kelps, which give a place for numerous species to spawn and serve as nurseries for young animals (Morrison et al. 2014). Additionally, seaweed produces oxygen (O_2) using atmospheric carbon dioxide (CO_2) through the photosynthesis process and thus manufactures its energy. It is believed that more than half of the O_2 on Earth is produced by photosynthetic bacteria and algae (macro and micro) (Mapelli-Brahm et al. 2023; Shlosberg et al. 2022).

S. Akter
Department of Marine Bioresource Science, Chattogram Veterinary and Animal Sciences University, Chattogram, Bangladesh

School of Ocean Science and Engineering, Division of Coastal Sciences, The University of Southern Mississippi, Hattiesburg, MS, USA

M. M. Islam
Marine Fisheries & Technology Station, Bangladesh Fisheries Research Institute, Cox's Bazar, Bangladesh

College of Marine Ecology and Environment, Shanghai Ocean University, Shanghai, China

M. T. Islam
Bangladesh Oceanographic Research Institute, Cox's Bazar-4730, Bangladesh

S. Mustary
Department of Biological Sciences, Birkbeck, University of London, London, UK

Seaweed species may be found in large numbers in all coastal zones across the world, including tropical, temperate, and polar locations (Bhuyan et al. 2024a, b). Seaweed is considered the plant of the millennium because of its positive effects on the environment, society, and economy (Pessarrodona et al. 2022; Wernberg et al. 2019). They are commercially important due to their numerous uses in the food, pharmaceuticals, textiles, cosmetics, and agriculture industries (Sultana et al. 2022; Sharma and Sharma 2017). Seaweed extract is used as a medicinal product that acts as an antioxidant, antifungal, antibacterial, and anticancer. Different types of secondary metabolites are produced for a variety of biological purposes (Bhuyan et al. 2020; Smale et al. 2013; Harley et al. 2012).

Kılınç et al. (2013) reported that seaweed is now a highly adaptable product that is frequently utilized for direct human consumption. Seaweeds have been used as major food items in the diets of the Japanese, Chinese, Filipinos, and Hawaiians for generations and are regarded as a delicacy by all of these cultures (Mouritsen et al. 2018). Different species of green seaweeds (*Enteromorpha*, *Ulva*, *Caulerpa*, and *Codium*) are mainly used as food sources and frequently consumed with rice as cooked veggies or as raw salads. *Porphyra*, *Laminaria*, and *Undaria* are utilized for cooking fish, meat curries, and soups (Siddiqui et al. 2019; Ahmed and Taparhudee 2005).

Seaweeds have been primarily grown and consumed in Asia as traditional food sources for decades, and thus Asian cuisine is starting to gain recognition and popularity in Western nations as well (Sultana et al. 2022). In addition, seaweed's popularity as a vegetable is increasing in the West, and dietary literature claims about its nutritional and/or therapeutic benefits. These seaweeds are eaten directly by residents of the Asia-Pacific islands. These are mostly snacks, soups, salads, noodles, stews, desserts, garnishes, drinks, and side dishes (Feldner 2012; Ling 2012). Products made from seaweed that are edible occur in many forms, including fresh, dried, powdered, and flaked (Debbarma et al. 2022; Ferdouse et al. 2018).

About 221 species of seaweed are commercially valuable, and ten species of seaweed are grown extensively (Mohamed 2015; Nayar and Bott 2014). *Eucheuma* sp. is in second place constituting 17 percent of the overall production of farmed seaweeds, followed by *Saccharina japonica*, *Kappaphycus alvarezii*, *Undaria pinnatifida*, and *Porphyra* sp. (Veeragurunathan et al. 2022; Pereira 2011). Red seaweeds, including *Gelidium*, *Gracilaria*, and *Pterocladis*, are essential for human sustenance as well as other purposes, namely as a bacterial substrate in labs and a binder in food items (Ferdouse et al. 2018). Carrageenan, a substance extracted from *Eucheuma* and *Kappaphycus*, is utilized in cosmetics, food processing, and industrial applications (Rupert et al. 2022). Brown seaweed and green seaweed are significant food sources in Asia, and they are most frequently consumed in different forms (raw, dried, or cooked) as a source of dietary fiber, protein, and minerals (Pérez-Lloréns et al. 2023; Seth and Shanmugam 2016).

Carrageenan, agar, and alginates generated from seaweeds, account for about 40% of the hydrocolloid market for food products worldwide; the other products are obtained from various animals, microorganisms, and land plants (Farghali et al. 2023; Seisun and Zalesny 2021). Carrageenan, the most significant extract, comes from the red seaweed found in Carraigin, Ireland, which has been used for more than a millennium as gelatin and for conventional healing (Tuvikene 2021). Pet food, dairy, meat, and, to a lesser extent, the pharmaceutical sectors extensively use carrageenan. According to Pereira et al. (2009), three primary forms of carrageenan are mostly derived from *E. cotonii*, *E. spinosum*, and *Chondrus crispus*, respectively. Iota carrageenan is more elastic and flexible, while lambda gives dairy products their creamy texture, and kappa carrageenan generates strong, stiff gels that are frequently employed as food additives. Carrageenan concentrations in food items typically range from 0.005% to 2% by weight (Dhingra 2021). Agar has been in high demand for more than 300 years ago, as a culinary ingredient in most Asian nations, while alginates have been employed in the textile printing business since around 1020 (Singh et al. 2023; Ferdouse et al. 2018). The primary source of agar is *Gracilaria* sp., of which 2.7 million tons (tons) were reportedly harvested in China in 2015 (Banik et al. 2023). China's contribution to the global trade in alginates, particularly agar, is substantial despite the country's high domestic consumption (Porse and Rudolph 2017). Agar is now mostly manufactured and exported by Chile and China. In response to the worldwide growing demand, supplies to international trade have also risen from Indonesia, Morocco, and Thailand (Mantri et al. 2023).

15.1.1 Global Status of Seaweed Production

According to the statistics of the Food and Agriculture Organization (FAO), more than 35 million tons of seaweed were produced globally in 2019. According to Behera et al. (2022), the most widely grown seaweeds are from the red and brown category. *Gracilaria* is grown for agar, *Eucheuma* and *K. alvarezii* are grown for carrageenan (a gelling agent), and the remainder is consumed with little processing (Rimmer et al. 2021). China, Korea (North and South), Indonesia, Japan, Malaysia, the Philippines, and Zanzibar are the seaweed producers, among them China, Indonesia, and the Philippines are the leading producers (Kaleeswaran and Betsy 2020). Figure 15.1 depicts the global culture and capture production of seaweed in 2018.

15.1.2 Worldwide Capture and Culture Production

Seaweed simply grows and supports healthy ocean habitats. Rising seaweed is an effective strategy to supply a rising population with food that is incredibly nutritious (Duarte et al. 2022; Yong et al. 2022). In contrast to terrestrial crops, seaweed grows quickly and doesn't require land, freshwater, fertilizer, or pesticides (Punitha et al. 2022; Ghadiryanfar et al. 2016). Statistics of global seaweed capture and culture production are presented in Figs. 15.2 and 15.3 (FAO 2018).

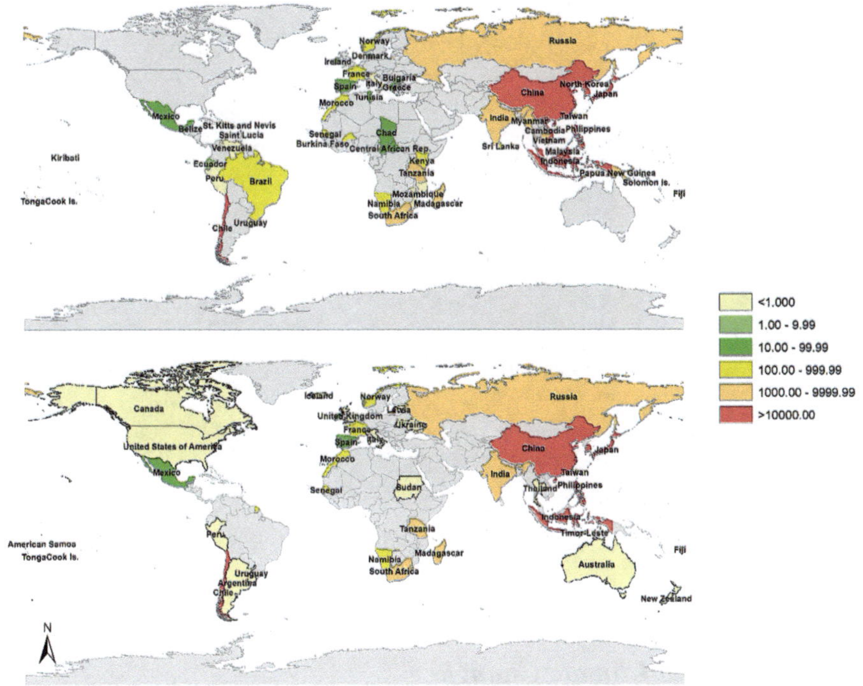

Fig. 15.1 State of global seaweed production by different countries. (**a**) culture and (**b**) capture. Color scale in wet metric tons. (Data source: FAO (2018))

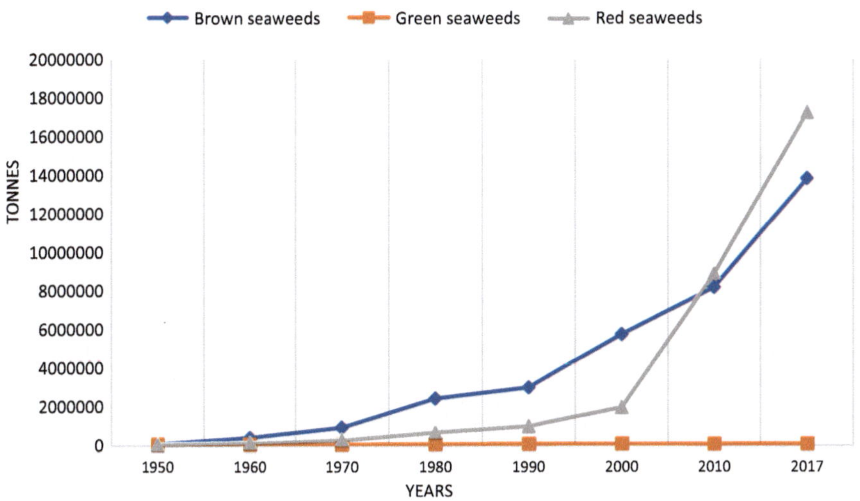

Fig. 15.2 Global scenario of culture production of different types of seaweed (tons). (Data source: FAO (2018))

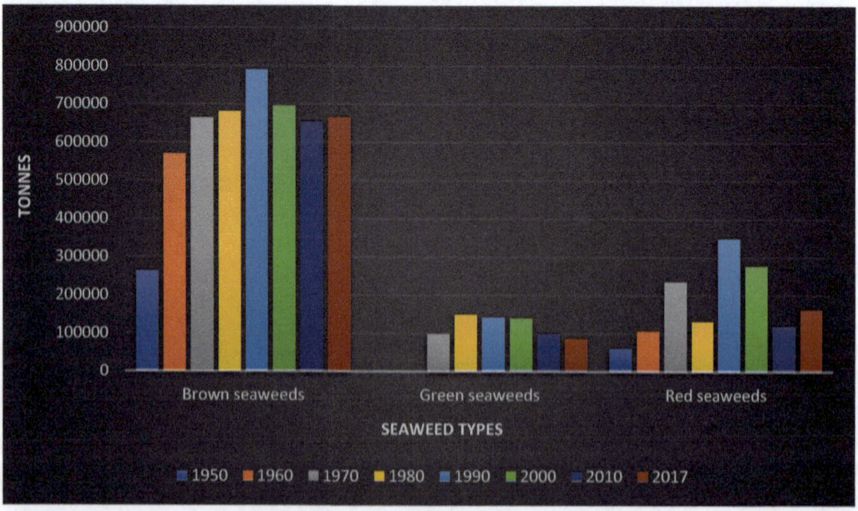

Fig. 15.3 Global scenario of capture production of different types of seaweed (tons). (Data source: FAO (2018))

15.1.3 Role in Climate Change

Seaweeds provide various benefits that are helpful in fighting against climate change (Fig. 15.4). Seaweed farming aids in the carbon (C) stored in sediments and offers prospects for climate change mitigation (Ross et al. 2023; Ricart et al. 2022; Bhuyan et al. 2021, 2022, 2023a). Seaweed from nature approximately absorbs 173 million metric tons of CO_2 yearly (Fujita et al. 2023; Burns 2023; Colombo et al. 2023). Over a thousand metric tons of CO_2 may be absorbed by a square kilometer of seaweed (on average). Seaweed may assist in decarbonizing by substituting emission-intensive items with low-C foods, animal feed, fossil-based plastic substitutes, fertilizers, textiles, and biofuels (DeAngelo et al. 2023; Lubchenco and Haugan 2023).

15.1.4 Ecosystem Services of Seaweed

Seaweed provides enormous ecosystem services that are essential for human and aquatic life (Fig. 15.5). Coastal protection, nitrogen absorption, and C storage are all vital functions of seaweeds (Cotas et al. 2023; Troell et al. 2023; Zheng et al. 2019).

Fig. 15.4 Different types of positive aspects of seaweed cultivation in terms of climate change mitigation and adaptation

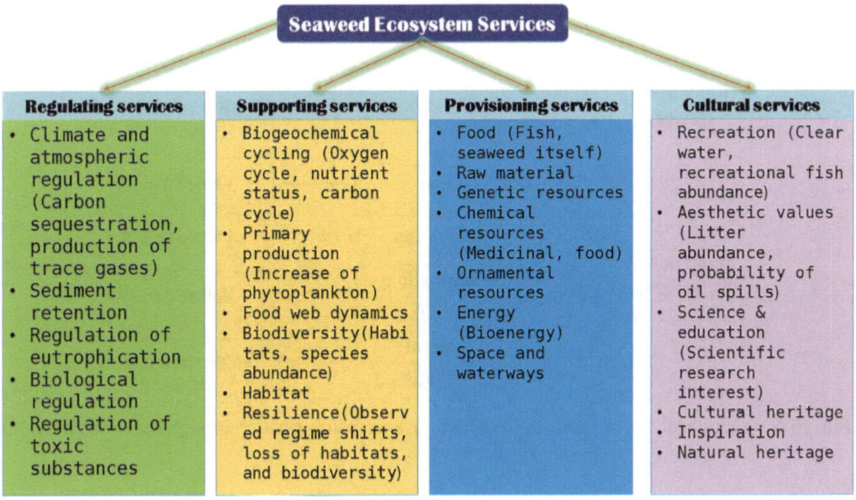

Fig. 15.5 Various ecosystem services provided by seaweed

15.1.5 Controlling Excess N and P in the Marine Environment

According to Malone and Newton (2020), the main factor contributing to coastal eutrophication is nutrient intake (Fig. 15.6), and highly productive coastal lagoon ecosystems are one example of a semi-enclosed system where effects might be amplified (Aires et al. 2019). Dead zones developed in the Gulf of Mexico and the Chesapeake Bay are a result of excessive nitrogen and phosphorus accumulation from point sources and stormwater runoff (Bailey et al. 2020), and thus ecology may be disturbed as a result of eutrophication. Through the removal of excess nutrients from water bodies and the subsequent release of O_2 as a byproduct, seaweed can lessen eutrophication (Ferreira et al. 2011; Dubinsky and Stambler 1996). To support their growth, seaweed draws up nitrogen, phosphorous, and carbon from the water. Inorganic nutrients are successfully removed from the systems when seaweed is collected from nature (Racine et al. 2021; Reid et al. 2013).

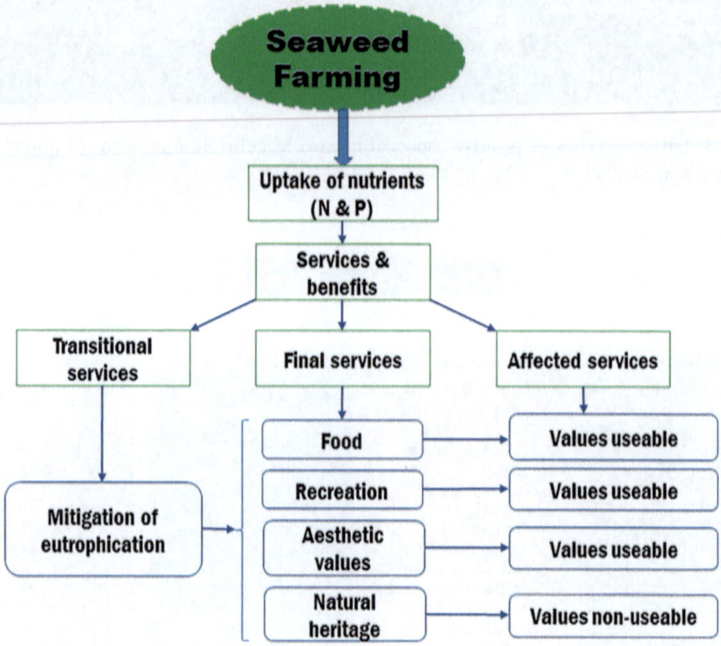

Fig. 15.6 Positive effects of removing excess nutrients from seawater through seaweed cultivation

15.1.6 Market Value of Seaweed

The Food and Agriculture Organization of the United Nations stated that the total worldwide algae output was 36 million tons of wet weight in 2020 in its most recent State of World Fisheries and Aquaculture (SOFIA) report (FAO 2022). This comprises wild and farmed seaweed and microalgae; however, the latter amounts are less significant because they total less than 100,000 tons of wet weight. The worldwide algae business is currently valued at USD 16.5 billion (Bn), and approximately, mostly consists of food goods for human use (FAO 2022). According to a World Bank assessment from 2023, the development potential of 11 growing seaweed markets might reach US$11.8 Bn by 2030. The worldwide market for seaweed farming is anticipated to grow from US$ 20.57 Bn in 2022 to US$ 43.66 Bn by 2032 with a CAGR of 7.8% A significant portion of the world's seaweed is grown in East and South Asia. About 5–10% of the world's aquaculture output comes from the cultivation of seaweed. In addition, according to Persistence Market Research's (2022) analysis of the worldwide aquaculture industry, red seaweed and brow seaweed are the second and third-largest species, respectively. Commercially important seaweed market size, by region, 2018–2030 was presented by Polaris Market Research Analysis (Fig. 15.7).

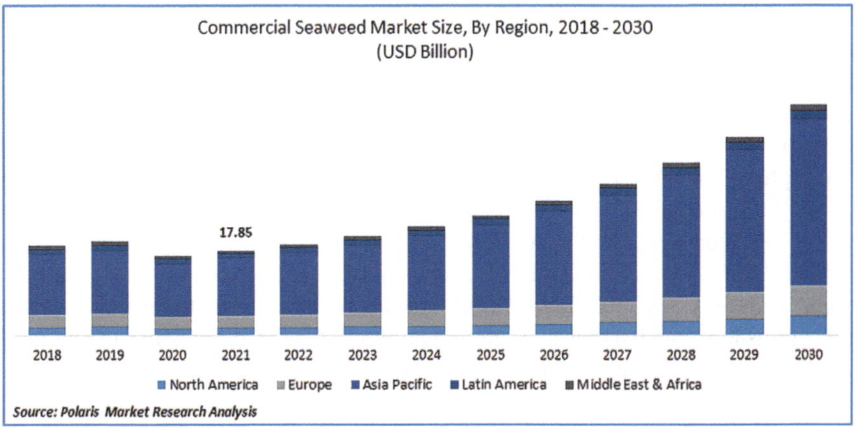

Fig. 15.7 Commercial seaweed market size, by region, 2018–2030 (USD billion). (Figure used with the legal permission from Polaris Market Research, https://www.polarismarketresearch.com/industry-analysis/commercial-seaweeds-market#:~:text=Report%20Summary,10.5%25%20during%20the%20forecast%20period)

15.2 Seaweed Research History in Bangladesh

Seaweeds are essential marine ecosystem engineers who may change the abiotic and biotic components of the environment and also stimulate primary producers (Hossain et al. 2020). Islam (1976) found 165 species in Bangladesh. Islam and Aziz (1982, 1987), and others updated the seaweed recordings between the 1970s and the 2000s with the addition of fresh data from St. Martin's Island (SMI). Aziz and Rahman (2011), Aziz et al. (2010), Islam et al. (2004, 2010), Aziz and Islam (2009), and Chowdhury and Ahmed (2007) have all made significant contributions to seaweed research. Ahmed et al. (2008, 2009) identified 197 different types of seaweed along the Bangladeshi coast. DoF (2014) reported five species from the Bakhkhali-Moheshkhali ecosystem of Cox's Bazar, ten species from the planted mangrove forest, and 140 species of seaweed from SMI. Nevertheless, 39 seaweed species were found in the coral reef environment of SMI, according to Aziz et al. (2015). Aftab Uddin (2019) listed 150 taxa of seaweed in Bangladesh. BFRI (2020) recorded 132 species of seaweed, of which 28 species are Chlorophyta, 35 species are Ochrophyta, and 69 species are Rhodophyta. On the west shore of Saint Martin's Island, 34 different kinds of seaweed were also discovered by Mamun et al. (2020).

According to a review of all the literature, Bangladesh's coastal and nearshore seas are home to more than 77 genera and 250 species of seaweed, of which 45% are rhodophytes, 31% are ochrophytes, and 24% are chlorophytes. Over the past 45 years, there has been a rise of about 85 species, and 32 of those species, including eight species of Chlorophyta, 12 species of Ochrophyta, and 12 species of Rhodophyta, are common in Bangladesh's coastal water. Four species of Chlorophyta and ten species of Rhodophyta might be cultivated in the coastal water of Bangladesh. Chlorophyta species predominate in the upper intertidal zone of SMI, while Phaeophyta and Rhodophyta species flourish in the lower intertidal and subtidal zones. SMI, the Teknaf coast, Inany Beach, Rezukhal, and the Nuniarchara coast are major seaweed ecosystems. Additionally, seaweed farming has potential in the Chowfoldondi coast, Sonadia Khal, Moheshkhali, and Kutubdia islands. Although some seaweeds, such as *Catenella*, are known to thrive in mangrove ecosystems, their prevalence and socioeconomic potential are less well-known (Hossain et al. 2020). Seaweed farming on the coast of Bangladesh over the last 45 years is depicted in Fig. 15.8.

15.3 Prospects of Seaweed in Bangladesh

Seaweeds are recognized as a highly valued marine product that has huge industrial applications worldwide (Bhuyan 2023; Bhuyan et al. 2020; Islam et al. 2019). Due to the ideal climatic conditions and accessibility of commercially significant species, Bangladesh's southeast coast offers good prospects for seaweed farming (Shaika et al. 2022). A 25% increase in food output, 5 million tons of plant protein,

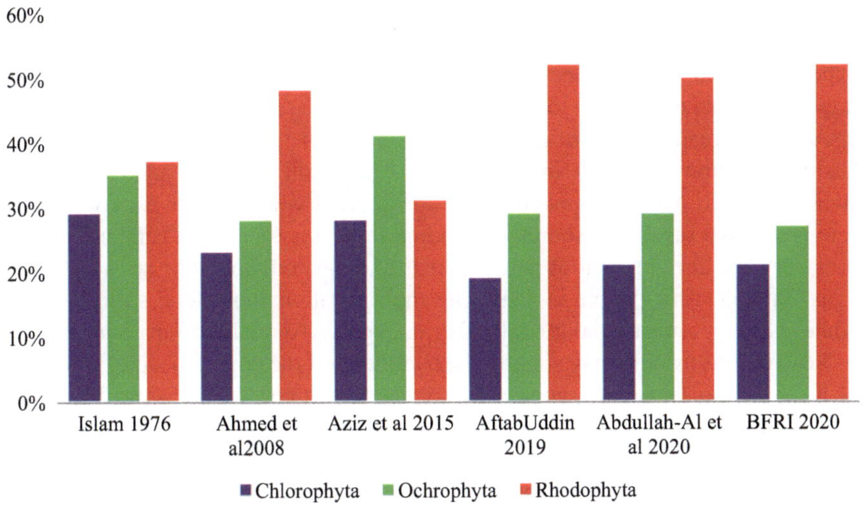

Fig. 15.8 Seaweed farming on the coast of Bangladesh over the last 45 years. (Data source: Hossain et al. (2020))

1.5 million tons of oil, and 125 million megawatt-hours of bioenergy might be contributed by the increased production. Additionally, 15 million tons of C might be sequestered from the environment, and 1 million tons each of nitrogen and phosphorus could be removed from the environment. There might be a total of 100,000 direct and indirect jobs that might be added to the seaweed industry. According to Hossain et al. (2020), the advantages of seaweed cultivation might fairly contribute to Bangladesh's growth of its blue economy and reflect on 26 objectives of the eight sustainable growth goals.

Although there is a large market for seaweed in Bangladesh, there is currently very little output from natural sources. A survey found that the coastal regions of Bangladesh had plentiful seaweed, and the intertidal and subtidal zones are rich in seaweed resources (Sarker et al. 2021). These resources hold enormous promise for the growth of Bangladesh's seaweed-based industries. The major benefit of seaweed cultivation in Bangladesh is the preservation of natural seaweed beds although seaweed collection is prohibited from natural sources. Seaweed culture might be a profitable sector for Bangladesh's coastal towns since it uses little resources, yields high returns, and employs many coastal people. From Cox's Bazar to the Sundarbans, there are many appropriate locations for seaweed cultivation (Ahmed and Taparhudee 2005). The SMI is a seaweed hotspot and is ideal for the growth of seaweed. The common species found on SMI include *Sargassum*, *Hypnea*, *Halymenia*, *Hydroclathrus*, and *Amphiora*. The species with the greatest availability of iodine is *Hypnea*. Nuniarchara and Moheshkhali Channel in Cox's Bazar are places where seaweed is grown (Bhuyan et al. 2020).

Seaweed may readily be added to the nation's small export portfolio, which would significantly aid in alleviating poverty and the ongoing unemployment issue.

Seaweed culture might be introduced in the nation's coastal regions, which could be another option for the people's source of income in addition to the country's strong export potential. On the shore, there are a variety of edible seaweed species. It is important to make an effort to create items that are suited for Bangladeshi consumers and to spread awareness of them. Seaweed is one of the essential sources of healthy nutrition for people in Bangladesh, where millions of people live near the shore. Seaweed may be supplied to the inland areas where people experience thyroid issues as a result of a shortage of iodine since it is easily dried and maintained properly. The development of new products and methods for using seaweed might assist Bangladesh's people in satisfying their nutritional and food security needs while also increasing the value of all fisheries exports (Ahmed and Taparhudee 2005).

Seaweed cultivation may support the livelihood of the coastal people. To be sure of it, 35% of respondents suggested designating farming zones, while 30% highlighted improving the infrastructure for harvesting and processing. The same number of respondents supported raising knowledge of food and medicinal properties while focusing on use for human consumption. Some respondents also mentioned that further investigation is needed and the development of storage capacity and potential might be equally essential to the recipient (Hossain et al. 2020).

Seaweed harvesting is possible every 15 days during its cultivation period. A farmer may make between BDT 12,000 to 14,000 (109.11–127.29 USD) in each season by investing just BDT 1200 (10.91 USD) with a yield of 12–14 kg/m^2 of seaweed. The cost of dried seaweed is between 150 and 170 BDT/kg compared to 50 BDT/kg for raw seaweed (Bhuyan et al. 2020). However, the price of seaweed is now increasing day by day. The price of dried seaweed ranged between 250 and 300 BDT/kg, while the raw seaweed costs 120–140 BDT/kg.

To increase their resilience and decrease their susceptibility, women should be empowered in the coastal villages (Alam and Rahman 2018). It may be done by offering practical training in the making of value-added products and seaweed culture, as well as input supports and soft credit provisions, which were requested by 49%, 41%, 8%, and 2% of respondents, respectively (Hossain et al. 2020). Seaweed farming improved the capability of buying cattle, upscale apparel, and sturdy goods for everyday use. Some households were able to resolve family disputes and send their kids to school because of the cash received from seaweed farming (Farhaduzzaman et al. 2023). Seaweed cultivation has become a new sustainable income source for coastal communities. Additionally, the seaweed producers showed a greater understanding of how seaweeds may help with gastrointestinal, cardiovascular, hormonal, diabetic, and other health issues. Vigorously advertising the advantages of seaweed will help this prospective industry expand faster (Farhaduzzaman et al. 2023; Akhtar et al. 2022).

In Bangladesh, more than a million individuals are at risk for developing goiter and associated illnesses. The iodine shortage alone causes significant thyroid and mental health issues in humans. When it comes to controlling the synthesis of thyroid hormone, most seaweed is a far better option than iodized salt or medications since it contains more iodine than seawater (Triggiani et al. 2009). Seaweed is an excellent source of I, as well as Ca, P, Fe, Na, K, Mg, S, Cu, Zn, and Co. In addition to being up to 10 times richer in Ca than cow's milk, expectant and breast-feeding women, as well as undernourished children should take some seaweed regularly as a part of their diet. From the aforementioned perspective, seaweed should be cultivated in Bangladesh to provide nourishment to the people.

15.4 Present Status of Seaweed Cultivation in Bangladesh

Although seaweed production is rapidly increasing in the world, Bangladesh currently only has a rudimentary stage of seaweed production (Hossain et al. 2020). Red, green, and brown algae have been observed in Bangladesh's coastal and estuary regions in recent years (Aziz et al. 2015). A variety of efforts have been made in Bangladesh, especially during the past 20 years, to support the socioeconomic progress of neglected coastal people (Zafar 2007; Islam et al. 2017). Seaweeds may be found in Bangladesh in the littoral and sublittoral zones from the Sundarbans Mangrove forest to SMI. SMI, the Teknaf coast, Inani Beach, the Rezukhal and Nuniarchara coast, the Chowfoldondi coast, the Sonadia, Moheshkhali, Kutubdia islands, and the mangrove forest might be the other potential seaweed farming locations. The majority of seaweed growing locations now exist in Bangladesh's southeast coastal region. Major seaweed cultivating locations in the coastal water of Bangladesh are shown in Fig. 15.9.

Seaweed cultivation can last anywhere between 4 and 6 months (i.e., November–April) depending on the influence of the southeast monsoon. Seaweed cultivation is carried out by 300 households in the Nuniarchara, Inany Beach, and Rezu Khal

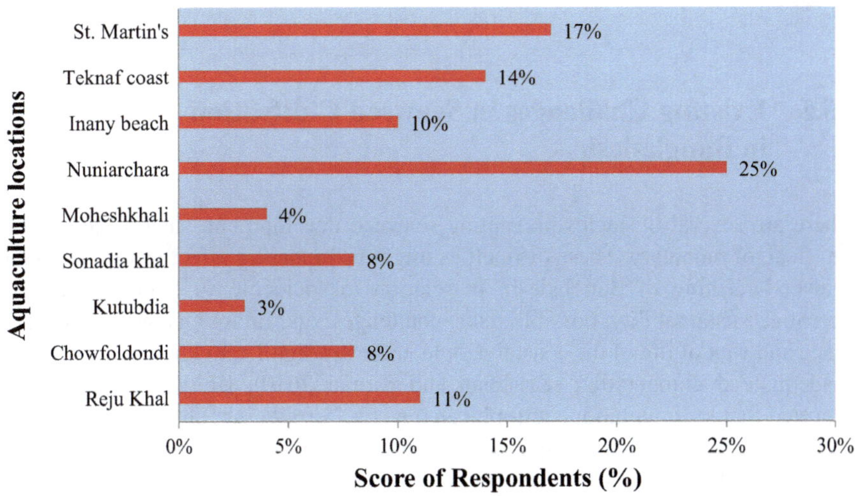

Fig. 15.9 Major seaweed farming locations in the coastal water of Bangladesh. (Data source: Hossain et al. (2020))

Table 15.1 Livelihood assets and socioeconomic impacts (Farhaduzzaman et al. 2023)

Assets for living	Social and economic effects
Financial	NGOs promote joining community groups for financial and/or credit benefits
Physical	Purchase livestock, fine clothing, and long-lasting items Buy a new house or remodel their present one Purchase or leasing land for agricultural reasons
Human	Ability to confidently utilize a variety of technology gadgets such as Android phones In terms of health, seaweeds provide health benefits that are becoming increasingly well-known
Social	The capability of resolving marital issues Reduce family difficulties Preventing kid school desertion
Natural	A big coastal area where growing seaweed is a profitable industry

regions. From such cultivation plots, a total of 390 MT of wet seaweeds are taken out each year, which is equal to 59 MT of dry weight (Hossain et al. 2020).

With the correct legislative, financial, technological, and institutional backing, seaweed culture will reduce the strain on and stop the abuse of other resources. This graph shows how seaweed culture has changed the farmers' multiple income sources. Table 15.1 outlines livelihood resources and socioeconomic effects.

Name of Various Organizations Working on Seaweed in Bangladesh
Seaweed culture is becoming popular in Bangladesh. Different universities, research organizations, financial institutions, NGOs, private organizations, and international organizations are working on the cultivation and product development of seaweed (Table 15.2).

15.5 Existing Challenges in Seaweed Cultivation in Bangladesh

There are several obstacles preventing seaweed development, such as ecological, physical, or monetary. These difficulties might be hindering effective or sustainable seaweed farming in Bangladesh. In relation to sociocultural features, financial potential, sustainability, possible risks, challenges, and future changes in the resilience and way of life of the coastal people, there are significant knowledge gaps and inadequate documentation (Krishnan and Kumar 2010). Bhuyan et al. (2023b) reported that sedimentation, eutrophicated water, hermit crab, current velocity, garbage, exotic/unwanted sp., kayaking, and engine boats were the main challenges in seaweed cultivation which were noticed in Rezu Khal, Cox's Bazar, Bangladesh. Some challenges in seaweed cultivation are outlined in Table 15.3.

Table 15.2 Different organizations that are working on seaweed in Bangladesh

Types	Name of organizations
Universities	Institute of Marine Sciences, University of Chittagong Chattogram Veterinary and Animal Sciences University (CVASU) Department of Oceanography, University of Dhaka Department of Fisheries Technology, Bangladesh Agricultural University Department of Fishing and Post Harvest Technology, Sher-e-Bangla Agricultural University Department of Oceanography, Shahjalal University of Science and Technology Department of Fisheries Technology, Bangabandhu Seikh Mujibur Rahman Agricultural University Bangabandhu Seikh Mujibur Rahman Maritime University Sher-e-Bangla Agricultural University Department of Aquatic Resource Management, Sylhet Agricultural University, Sylhet
Research organizations	Bangladesh Oceanographic Research Institute (BORI) Bangladesh Council for Scientific and Industrial Research (BCSIR) Bangladesh Fisheries Research Forum (BFRF) Bangladesh Agricultural Research Council (BARC) Bangladesh Fisheries Research Institute (BFRI) Bangladesh Agriculture Research Institute (BARI) National Institute of Biotechnology (NIB)
Financial institutions	Palli Karma-Sahayak Foundation (PKSF)
NGOs	Integrated Development Foundation (IDF) Coast Trust
Private organizations	Bangladesh Marine Fisheries Association Falcon International Limited Jahanara Agro Ltd Mahi Agro Nourish
International organizations	ECOFISH II, WorldFish Food & Agriculture Organization (FAO) Japan International Cooperation Agency (JICA)
Implementing authority	Department of Fisheries

15.5.1 Possible Approaches to Overcome These Challenges for Successful Seaweed Cultivation in Bangladesh

Approaches or ways to minimize/mitigate/reduce these risks/challenges for sustainable seaweed cultivation are listed below.

- Marine spatial planning needed
- Culture commercially important seaweed species
- Identify suitable seaweed culture area
- Develop advanced seaweed aquaculture technique

Table 15.3 Existing challenges in seaweed cultivation in Bangladesh

Types of challenges	Details of challenges
Social	Human food has a low acceptance rate Conflict over land ownership and use Illegal poaching in the seaweed production area
Technical	Selecting the right commercial species Proper culture site selection Appropriate cultural methods are lacking Limited knowledge about the factors (temperature, salinity, turbidity, DO, BOD, COD, nitrate, phosphate, etc.) that affect seaweed growth Availability of commercial seed is a challenge, we cannot depend on nature Disease outbreak Sedimentation Pollution The absence of proper post-harvest technologies and infrastructure Lack of knowledge for developing value-added products Lack of expert personnel There is no value chain analysis of seaweed
Economic	Lack of funds for long-term research Proper marketing channel Limited financial assistance value-added product development Low budget for seaweed training Very limited investment from private sectors

- Tissue culture of seaweed to ensure seed for culture
- Develop seaweed value-added product
- Motivate and increase local consumption
- Generation of seaweed market both locally and internationally
- Generating skilled manpower
- Increase the funding by local financial institutions
- Emphasizing the long-term research
- Public-private partnerships
- Consultation among different stakeholders
- Development of national policies and frameworks
- Integrations among different research organizations and universities

15.6 Conclusions

Seaweeds have recently gained significant importance due to their global distribution, renewable nature, and variety of uses. Bangladesh has a lot of potential for seaweed cultivation. Techniques for growing seaweed are very simple and farmers can easily adopt it. As a result, government agencies have focused more on developing seaweed farming under rural employment initiatives to flourish the coastal economy in the wake of the maritime triumph. Due to the lack of agricultural inputs, the necessity for land, and the requirement for fresh water, seaweed farming is gaining popularity as it requires no agricultural inputs, and the necessity of land and freshwater is also not required. The livelihood of coastal populations is greatly impacted by the cultivation of seaweed. Women can work in seaweed farming, ensuring women's empowerment in the process. There are a few obstacles in seaweed farming that must be resolved for its further effective promotion.

Acknowledgments The authors are grateful to the Bangladesh Oceanographic Research Institute for this research opportunity.

References

Aftab Uddin S (2019) Seaweeds of Bangladesh. Institute of Marine Sciences, University of Chittagong, Bangladesh, 174 pp

Ahmed N, Taparhudee W (2005) Seaweed cultivation in Bangladesh: problems and potentials. J Fish Environ 28:13–21

Ahmed ZU, Begum ZNT, Hassan MA, Khondker M, Kabir SMH, Ahmad M, Ahmed ATA, Rahman AKA, Haque EU (eds) (2008) In: Ahmed ZU (ed) Encyclopedia of flora and fauna of Bangladesh. Algae, Chlorophyta (Aphanochaetaceae- Zygnemataceae), vol 3. Asiatic Society Bangladesh Press, Dhaka, 812 pp

Ahmed ZU, Khondker M, Begum ZNT, Hassan MA, Kabir SMH, Ahmad M, Ahmed ATA, Rahman AKA (eds) (2009) In: Ahmed, ZU (ed) Encyclopedia of Flora and Fauna of Bangladesh. Algae, Charophyta-Rhodophyta, (Achnanthaceae-Vaucheriaceae), vol 4. Asiatic Society Bangladesh Press, Dhaka, 543 pp

Aires T, Muyzer G, Serrão EA, Engelen AH (2019) Seaweed loads cause stronger bacterial community shifts in coastal lagoon sediments than nutrient loads. Front Microbiol 9:3283

Akhtar A, Khan MS, Hasan M, Farhaduzzaman AM, Osman MH, Shovon MNH (2022) Baseline survey report on seaweed cultivation, processing, and marketing for employment generation in Bangladesh's coastal poor communities. Int J Innov Sci Res Technol 7:1679–1687

Alam K, Rahman MH (2018) The role of women in disaster resilience. In: Handbook of disaster risk reduction & management. World Scientific Publishing Co. Pte. Ltd, USA, pp 697–719

Aziz A, Islam S (2009) Marine algae of St. Martin's Island, Bangladesh. VII. *Acrochaetium nurulislamii* sp. nov. and new records of Acrochaetium (Rhodophyceae). Bangladesh J Bot 38:145–151

Aziz A, Rahman MT (2011) Marine algae of St. Martin's Island, Bangladesh. XII. New records of red and green algae. Bangladesh J Bot 40(1):41–45

Aziz A, Islam S, Chowdhury AH (2010) Marine algae of St. Martin's Island, Bangladesh. IX. New records of green algae (Chlorophyceae). Bangladesh J Plant Taxon 17(2):193

Aziz A, Towhidy S, Alfasane MA (2015) Sublittoral seaweed flora of the St. Martin's Island, Bangladesh. Bangladesh J Bot 44(2):223–236

Bailey A, Meyer L, Pettingell N, Macie M, Korstad J (2020) Agricultural practices contributing to aquatic dead zones. In: Ecological and practical applications for sustainable agriculture. Springer, Singapore, pp 373–393

Banik U, Mohiuddin M, Wahab MA, Rahman MM, Nahiduzzaman M, Sarker S et al (2023) Comparative performances of different farming systems and associated influence of ecological factors on *Gracilaria* sp. seaweed at the south-east coast of the Bay of Bengal, Bangladesh. Aquaculture 574:739675

Behera DP, Vadodariya V, Veeragurunathan V, Sigamani S, Moovendhan M, Srinivasan R et al (2022) Seaweeds cultivation methods and their role in climate mitigation and environmental cleanup. Total Environ Res Themes 3:100016

BFRI (2020) Seaweeds of Bangladesh coast (Abundance, distribution & taxonomic list). Bangladesh Fisheries Research Institute, Ministry of Fisheries & Livestock, Government of Bangladesh, Dhaka

Bhuyan MS (2023) Ecological risks associated with seaweed cultivation and identifying risk minimization approaches. Algal Res 26:102967

Bhuyan MS, Sharif ASM, Islam MM, Mojumder IA, Das M, Islam MS (2020) Blue economy and the prospect of seaweed in Bangladesh. J Mar Sci Res Oceanogr 3:1–2

Bhuyan MS, Sharif ASM, Islam MM, Islam MS, Hoq ME (2021) Seaweed a powerful tool for climate change mitigation that provides various ecological services. In: Modelling earth systems and environment. Springer Publication, Switzerland

Bhuyan MS, Husain SKA, Chowdhury E, Bat L (2022) Assessment of carbon sequestration capacity of seaweed in climate change mitigation. J Clim Change 8:1–8

Bhuyan, M.S., Mahmood, S., Haider, B., Kunda, M., Husain, S.A., Chowdhury, E., Senapathi, V., Sivakumar, K., & Elangovan, M. (2023a). Experimental Cultivation of Seaweed on the Coast of Cox's Bazar, Bangladesh: Identifying the Effects of Environmental Parameters on Seaweed Growth. Journal of Applied Life Sciences and Environment, 56(3):413-436.

Bhuyan MS, Haider SMB, Kundu M, Husain SKA, Chowdhury E (2023b) Challenges in cultivation of seaweed: a recent experience from Rezu Khal, Cox's Bazar Coast, Bangladesh. J Oceanogr Mar Environ Syst 7:01–09

Bhuyan MS, Haider SMB, Kunda M, Islam MT, Husain SA, Chowdhury E, Mojumder IA (2024a) Is seaweed culture a sustainable approach for climate change adaptation? J NOAMI 38(1–2):75–88

Bhuyan, M. S., Kunda, M., Bakar, M. A., Senapathi, V., Husain, S. A., Chowdhury, E., ... & Pandit, D. (2024b). Heavy metal and mineral analysis of cultivated seaweeds from Cox's Bazar Coast, Bay of Bengal, Bangladesh: a human health risk implication. Discover Oceans, 1(1), 11.

Burns W (2023) Governance of ocean-based carbon dioxide removal research under the United Nations Convention on the Law of the Sea. Me L Rev 75:37

Chowdhury AH, Ahmed R (2007) *Neurymenia fraxinifolia* (Mert.) J. Ag.-a new record of a marine red alga for Bangladesh. Bangladesh J Bot 36(1):81–83

Colombo SM, Roy K, Mraz J, Wan AH, Davies SJ, Tibbetts SM et al (2023) Towards achieving circularity and sustainability in feeds for farmed blue foods. Rev Aquac 15(3):1115–1141

Cotas J, Gomes L, Pacheco D, Pereira L (2023) Ecosystem services provided by seaweeds. Hydrobiology 2(1):75–96

DeAngelo J, Saenz BT, Arzeno-Soltero IB, Frieder CA, Long MC, Hamman J et al (2023) Economic and biophysical limits to seaweed farming for climate change mitigation. Nat Plants 9(1):45–57

Debbarma J, Viji P, Rao BM, Ravishankar CN (2022) Seaweeds: potential applications of the aquatic vegetables to augment nutritional composition, texture, and health benefits of food and food products. In: Sustainable global resources of seaweeds volume 2: food, pharmaceutical and health applications. Springer, Cham, pp 3–54

Dhingra D (2021) Seaweeds–a potential source of food, feed and fertiliser. Agric Eng Today 45(3):19–25

DoF 2014. National Fish Week Compendium (In Bengali). Department of Fisheries, Ministry of Fisheries and Livestock, Bangladesh, 144p

Duarte CM, Bruhn A, Krause-Jensen D (2022) A seaweed aquaculture imperative to meet global sustainability targets. Nat Sustain 5(3):185–193

Dubinsky ZVY, Stambler N (1996) Marine pollution and coral reefs. Glob Chang Biol 2(6):511–526

FAO (2018) Fishery statistical collections. Food and Agriculture Organization of the United Nations, Rome. http://www.fao.org/lshery/statistics/

FAO (2022) The state of world fisheries and aquaculture 2022: towards blue transformation. Food and Agriculture Organization (FAO), Rome. https://www.fao.org/documents/card/en/c/cc0461en

Farghali M, Mohamed IM, Osman AI, Rooney DW (2023) Seaweed for climate mitigation, wastewater treatment, bioenergy, bioplastic, biochar, food, pharmaceuticals, and cosmetics: a review. Environ Chem Lett 21(1):97–152

Farhaduzzaman AM, Khan MS, Hasan M, Islam MR, Osman MH, Shovon MNH, Haider SMB, Kunda M, Islam MT, Bhuyan MS (2023) Seaweed culture, post-harvest processing, and market generation for employment of coastal poor communities in Cox's Bazar. J Appl Life Sci Environ 56:231–244

Feldner SM (2012) A Cook's journey to Japan: 100 homestyle recipes from Japanese kitchens. Tuttle Publishing, USA

Ferdouse F, Holdt SL, Smith R, Murúa P, Yang Z (2018) The global status of seaweed production, trade and utilization. Globefish Res Programme 124:I

Ferreira JG, Andersen JH, Borja A, Bricker SB, Camp J, Da Silva MC et al (2011) Overview of eutrophication indicators to assess environmental status within the European Marine Strategy Framework Directive. Estuar Coast Shelf Sci 93(2):117–131

Fujita R, Augyte S, Bender J, Brittingham P, Buschmann AH, Chalfin M et al (2023) Seaweed blue carbon: ready? or not? Mar Policy 155:105747

Ghadiryanfar M, Rosentrater KA, Keyhani A, Omid M (2016) A review of macroalgae production, with potential applications in biofuels and bioenergy. Renew Sust Energ Rev 54:473–481

Harley CD, Anderson KM, Demes KW, Jorve JP, Kordas RL, Coyle TA, Graham MH (2012) Effects of climate change on global seaweed communities. J Phycol 48(5):1064–1078

Hossain MS, Chowdhury MSN, Alamgir M (2020) Seaweeds for blue economy in Bangladesh. Food and Agriculture Organization of the United Nations, p 87

Islam AKMN (1976) Contribution to the study of the marine algae of Bangladesh. Bibl Phycol 19:1–253

Islam AKMN, Aziz A (1982) Addition to the list of marine algae of St. Martins Island, Bangladesh II. Brown, red and blue green algae. Nova Hedwigia 36:643–657

Islam AKMN, Aziz A (1987) Addition to the list of the marine algae of St. Martin's Island, Bangladesh. III. Red algae. Nova Hedwig 45:211–221

Islam AKMN, Aziz A, Parvin R (2004) Marine algae of St. Martin's Island, Bangladesh. II. Brown Algae. Bangladesh J Plant Taxon 11(1):1–7

Islam S, Aziz A, Chowdhury AH (2010) Marine algae of St. Martin's Island, Bangladesh. VIII. New records of red algae (Rhodophyceae). Bangladesh J Bot 39(1):87–96

Islam MM, Khan MSK, Hasan J, Mallick D, Hoq ME (2017) Seaweed *Hypnea* sp. culture in Cox's Bazar coast, Bangladesh. Bangladesh J Zool 45(1):37–46

Islam MM, Hoq ME, Haque MA, Khan MSK, Hasan J, Ali MZ, Mahmud Y (2019) Seaweeds of Bangladesh coast. Bangladesh Fisheries Research Institute, Mymensingh

Islam MM, Hoq ME, Ali MZ, Hasan SJ, Sharif ASM, Bhuyan MS (2020) Diversity and occurrence of seaweeds from the south-eastern coast of Bangladesh. Indian J Geo Mar Sci 49:1–10

Kaleeswaran V, Betsy CJ (2020) Critical analysis of seaweed production potential in India in comparison to world major producers. J Aquac Tropics 35(1–4):117–133

Kılınç B, Cirik S, Turan G, Tekogul H, Koru E (2013) Seaweeds for food and industrial applications. In: Food industry. IntechOpen, UK

Krishnan M, Kumar RN (2010) Socio economic dimensions of seaweed farming in India. CMFRI Special Publication No. 104. Central Marine Fisheries Research Institute, Kochi. ISSN 0972 – 2351

Ling KF (2012) The food of Asia: featuring authentic recipes from master chefs. Tuttle Publishing, USA

Lubchenco J, Haugan PM (2023) The ocean as a solution to climate change: five opportunities for action. In: The Blue Compendium: from knowledge to action for a sustainable ocean economy. Springer International Publishing, Cham, pp 619–680

Malone TC, Newton A (2020) The globalization of cultural eutrophication in the coastal ocean: causes and consequences. Front Mar Sci 7:670

Mamun AA, Akhtar A, Rahman MF, Kamal AHM, Karim NU, Hassan ML (2020) Habitat structure and diversity patterns of seaweeds in the coastal waters of Saint Martin's Island, Bay of Bengal, Bangladesh. Reg Stud Mar Sci 33:100959

Mantri VA, Kambey CS, Cottier-Cook EJ, Usandizaga S, Buschmann AH, Chung IK et al (2023) Overview of global *Gracilaria* production, the role of biosecurity policies and regulations in the sustainable development of this industry. Rev Aquac 15(2):801–819

Mapelli-Brahm P, Gómez-Villegas P, Gonda ML, León-Vaz A, León R, Mildenberger J et al (2023) Microalgae, seaweeds and aquatic bacteria, archaea, and yeasts: sources of carotenoids with potential antioxidant and anti-inflammatory health-promoting actions in the sustainability era. Mar Drugs 21(6):340

Mohamed G (2015) Current trends and prospects of seaweed farming in India. Rome, Food and Agriculture Organization of the United Nations

Morrison M, Jones EG, Consalvey M, Berkenbusch K (2014) Linking marine fisheries species to biogenic habitats in New Zealand: a review and synthesis of knowledge. Ministry for Primary Industries, Wellington

Mouritsen OG, Rhatigan P, Pérez-Lloréns JL (2018) World cuisine of seaweeds: science meets gastronomy. Int J Gastron Food Sci 14:55–65

Nayar S, Bott K (2014) Current status of global cultivated seaweed production and markets. World Aquac 45(2):32–37

Pereira L (2011) A review of the nutrient composition of selected edible seaweeds. In: Seaweed: ecology, nutrient composition and medicinal uses, vol 7, no 4. Nova Science Publishers, Inc., Coimbra, pp 15–47

Pereira L, Critchley AT, Amado AM, Ribeiro-Claro PJ (2009) A comparative analysis of phycocolloids produced by underutilized versus industrially utilized carrageenophytes (Gigartinales, Rhodophyta). J Appl Phycol 21:599–605

Pérez-Lloréns JL, Critchley AT, Cornish ML, Mouritsen OG (2023) Saved by seaweeds (II): traditional knowledge, home remedies, medicine, surgery, and pharmacopoeia. J Appl Phycol 35:1–20

Persistence Market Research (2022) 10 years of historical data, analysis and forecast, 346p. https://www.persistencemarketresearch.com/market-research/seaweed-cultivation-market.asp

Pessarrodona A, Assis J, Filbee-Dexter K, Burrows MT, Gattuso JP, Duarte CM et al (2022) Global seaweed productivity. Sci Adv 8(37):eabn2465

Porse H, Rudolph B (2017) The seaweed hydrocolloid industry: 2016 updates, requirements, and outlook. J Appl Phycol 29(5):2187–2200

Punitha P, Priyadharshini P, Nanthini Devi K, Dinesh Kumar S, Roopavathy J, Begum A et al (2022) Effect of seaweed liquid extract as an organic biostimulant on the growth, fatty acids and high-value pigment production of Vigna radiata. Biomass Convers Biorefinery 14:1–13

Racine P, Marley A, Froehlich HE, Gaines SD, Ladner I, MacAdam-Somer I, Bradley D (2021) A case for seaweed aquaculture inclusion in US nutrient pollution management. Mar Policy 129:104506

Reid GK, Chopin T, Robinson SMC, Azevedo P, Quinton M, Belyea E (2013) Weight ratios of the kelps, *Alaria esculenta* and *Saccharina latissima*, required to sequester dissolved inorganic nutrients and supply oxygen for Atlantic salmon, *Salmo salar*, in integrated multi-trophic aquaculture systems. Aquaculture 408:34–46

Ricart AM, Krause-Jensen D, Hancke K, Price NN, Masqué P, Duarte CM (2022) Sinking seaweed in the deep ocean for carbon neutrality is ahead of science and beyond the ethics. Environ Res Lett 17(8):081003

Rimmer MA, Larson S, Lapong I, Purnomo AH, Pong-Masak PR, Swanepoel L, Paul NA (2021) Seaweed aquaculture in Indonesia contributes to social and economic aspects of livelihoods and community wellbeing. Sustain For 13(19):10946

Ross FW, Boyd PW, Filbee-Dexter K, Watanabe K, Ortega A, Krause-Jensen D et al (2023) Potential role of seaweeds in climate change mitigation. Sci Total Environ 885:163699

Rupert R, Rodrigues KF, Thien VY, Yong WTL (2022) Carrageenan from *Kappaphycus alvarezii* (Rhodophyta, Solieriaceae): metabolism, structure, production, and application. Front Plant Sci 13:859635

Sarker S, Akter M, Rahman MS, Islam MM, Hasan O, Kabir MA, Rahman MM (2021) Spatial prediction of seaweed habitat for mariculture in the coastal area of Bangladesh using a Generalized Additive Model. Algal Res 60:102490

Seisun D, Zalesny N (2021) Strides in food texture and hydrocolloids. Food Hydrocoll 117:106575

Seth A, Shanmugam M (2016) Seaweeds as agricultural crops in India: new vistas. In: Innovative saline agriculture. Springer, India, pp 441–473

Shaika NA, Ritu JR, Khan S, Rao AR (2022) Prospects and challenges in commercialization of seaweeds in Bangladesh. In: Sustainable global resources of seaweeds volume 1: bioresources, cultivation, trade and multifarious applications. Springer, Cham, pp 225–247

Sharma P, Sharma N (2017) Industrial and biotechnological applications of algae: a review. J Adv Plant Biol 1(1):01–25

Shlosberg Y, Schuster G, Adir N (2022) Harnessing photosynthesis to produce electricity using cyanobacteria, green algae, seaweeds and plants. Front Plant Sci 13:955843

Siddiqui AAM, Kashem MA, Mondal MAI, Shafiuddin M (2019) Commercially important seaweed cultivation and its potentials for the coastal areas of Cox's Bazar, Bangladesh. Int J Fish Aquat Stud 7(5):463–470

Singh M, Kapoor S, Bagchi T, Paul S, Kar S (2023) Seaweed farming: an environmental and societal perspective. In: Industrial microbiology and biotechnology: emerging concepts in microbial technology. Springer, Singapore, pp 605–641

Smale DA, Burrows MT, Moore P, O'Connor N, Hawkins SJ (2013) Threats and knowledge gaps for ecosystem services provided by kelp forests: a northeast Atlantic perspective. Ecol Evol 3(11):4016–4038

Sultana F, Wahab MA, Nahiduzzaman M, Mohiuddin M, Iqbal MZ, Shakil A et al (2022) Seaweed farming for food and nutritional security, climate change mitigation and adaptation, and women empowerment: a review. Aquac Fish 8:463

Theuerkauf SJ, Barrett LT, Alleway HK, Costa-Pierce BA, St. Gelais A, Jones RC (2022) Habitat value of bivalve shellfish and seaweed aquaculture for fish and invertebrates: pathways, synthesis and next steps. Rev Aquac 14(1):54–72

Triggiani V, Tafaro E, Giagulli VA, Sabbà C, Resta F, Licchelli B, Guastamacchia E (2009) Role of iodine, selenium and other micronutrients in thyroid function and disorders. Endocr Metab Immune Disord Drug Targets (Formerly Current Drug Targets-Immune, Endocrine & Metabolic Disorders) 9(3):277–294

Troell M, Henriksson PJ, Buschmann AH, Chopin T, Quahe S (2023) Farming the ocean–seaweeds as a quick fix for the climate? Rev Fish Sci Aquac 31(3):285–295

Tuvikene R (2021) Carrageenans. In: Handbook of hydrocolloids. Woodhead Publishing, Duxford, pp 767–804

Veeragurunathan V, Anand KGV, Grace PG., Gurumoorthya U, Krishnan SG (2022) Marine macroalgal products, Scieng Publications, Villupuram, Tamilnadu

Wernberg T, Krumhansl K, Filbee-Dexter K, Pedersen MF (2019) Status and trends for the world's kelp forests. In: World seas: an environmental evaluation. Academic Press, London, pp 57–78

World Bank (2023) Global seaweed: new and emerging markets report, 2023. https://openknowledge.worldbank.org/handle/10986/40187

Yong WTL, Thien VY, Rupert R, Rodrigues KF (2022) Seaweed: a potential climate change solution. Renew Sust Energ Rev 159:112222

Zafar M (2007) Seaweed culture (*Hypnea* sp.) in Bangladesh- seaweed culture manual-1. Institute of Marine Science and Fisheries, University of Chittagong, Chittagong, 14p

Zheng Y, Jin R, Zhang X, Wang Q, Wu J (2019) The considerable environmental benefits of seaweed aquaculture in China. Stoch Env Res Risk A 33:1203–1221

Chapter 16
Bioplastic from Seaweeds: Current Status and Future Perspectives

Ritu Sharma, Arijit Sankar Mondal, and Nitin Trivedi

Abstract The increasing demand for plastic and its products worldwide has resulted in a serious pollution issue with plastic trash, or "white pollution," which persists in the environment for centuries before degrading. The demand for synthetic plastics derived from petrochemicals can be replaced by bioplastics, which can be used as an alternate source of plastic. Seaweeds or macroalgae have a huge acceptance among all other renewable resources as a source of bioplastics as they have high polysaccharide content, no freshwater and arable land requirements, limited nutrient requirements, and higher growth rates compared to terrestrial plants. Although there are several conventional methods for extracting seaweed hydrocolloids, increased demand has highlighted some of these techniques' shortcomings, most notably in terms of consistency and efficiency of the final product. "Green extraction" techniques have demonstrated great potentiality (reproducible, faster, and environmentally friendly) for increasing production in an economical manner. Nonetheless, this chapter covers the present status, progress, limitations, and future prospects of seaweeds to better understand their potential as a resource for bioplastic synthesis in the near future.

Keywords Bioplastics · Seaweed bioplastics · Seaweed extraction methods · Bioplastic challenges · Macroalgae

---

R. Sharma · A. S. Mondal
Department of Microbiology, Guru Nanak Institute of Pharmaceutical Science and Technology (Life Science), Kolkata, West Bengal, India

N. Trivedi (✉)
Marine Biotechnology, Gujarat Biotechnology University, Gandhinagar, Gujarat, India
e-mail: nitin.trivedi@gbu.edu.in

N. Trivedi et al. (eds.), *Recent Advances in Seaweed Biotechnology*,
https://doi.org/10.1007/978-981-96-0519-4_16

16.1 Introduction

Global plastic production and consumption are rising in response to the growing global demand for plastic materials. Plastic production in the world rose to roughly 368 million metric tonnes in 2020, up from 359 million metric tonnes in 2018, and is projected to triple by 2050 as of 2018 (Rajpoot et al. 2022; Chia et al. 2020). It has led to a serious environmental problem with plastic garbage that has impacted both marine and terrestrial life. Large-scale plastic waste causes "white pollution" that persists in the environment for eras before degrading. The depletion of fossil fuel resources, climate change, and greenhouse gas emissions may result from the use of these traditional synthetic polymers derived from petrochemical sources. With all of these serious consequences, methods for reducing plastic waste generation and accumulation, such as chemical treatments, recycling, landfills, and incineration, while still available, appear to be suboptimal for reducing plastic pollution in the future (Chia et al. 2020; Lim et al. 2021). Currently, plastic industries have been accountable for a large amount of plastic waste, particularly through packaging production; therefore, we must identify innovative, sustainable resources for manufacturing biodegradable packaging. To reduce plastic waste, researchers are looking at the development of "bioplastics" as a sustainable option (Lomartire et al. 2022). Bioplastics, also known as biodegradable plastics, are quickly degraded and have characteristics similar to typical synthetic plastics (polyethylene terephthalate, polystyrene, high-density polyethylene, low-density polyethylene, polyvinyl chloride, etc.) (Chia et al. 2020; Lim et al. 2021). The most promising strategy in this field is the use of biomaterials based on polysaccharides (Aleksanyan 2023). Studies have previously demonstrated that a variety of materials, including vegetable oils, corn, potatoes, food waste, wood, and crops, can be used to make bioplastic. Nevertheless, because it not only competes with agriculture for food and feed but also uses a lot of water, land, and nutrients; this type of bioplastic production is not long-term sustainable (Lim et al. 2021). Thus, to create novel biodegradable plastics, it is essential to synthesize new bio-based biodegradable plastic polymers by gaining a thorough understanding of biomass structure. Seaweed/macroalgae have a huge scope because they can produce films, demonstrating the most recent concept for creating bioplastics (Lim et al. 2021). Seaweeds are divided into three classes based on their pigments: brown seaweeds (Ochrophyta), green seaweeds (Chlorophyta), and red seaweeds (Rhodophyta), which are rich in polysaccharide composition. Today, global macroalgae farming and production have expanded to 35,762,504 tonnes (wet weight) in the last 20 years, with artificial farming accounting for 97% of global aquaculture output (Sharma et al. 2023; Zhang et al. 2022; Trivedi et al. 2022a, b). The high biomass, quick reproduction, affordability, and ease of maintenance in a variety of conditions have made macroalgae a focus of interest for the industrial sector. Furthermore, seaweed bioplastics, in contrast to synthetic plastics, break down readily in soil in 4–6 weeks without running the risk of producing microplastics and releasing oxygen as a consequence of photosynthesis (Lim et al. 2021). There are numerous extraction methods (physical, chemical, and biological) to make seaweed bioplastics, and the process is primarily dependent on the species, kind of solvent, effects on the environment, cost, quantity, and difficulties with scaling up. Furthermore, seaweed-derived polysaccharides have unique optical, thermal, mechanical, antibacterial, and antioxidant properties, and their biodegradability makes them great candidates for bioplastic production (Lim et al. 2021). Nonetheless, this chapter covers the current status, progress, limitations, and future prospects of seaweeds in order to provide a better understanding of their potential as a resource for bioplastic synthesis in the near future.

16.2 Potential Bioactive Components and Physicochemical Characteristics of Seaweed Hydrocolloids for Bioplastic Production

All three types of seaweeds have the presence of different types of polysaccharides (structural and storage) which are a promising source of bioplastics. These include agar, carrageenan, alginate, ulvan, fucoidan, cellulose, etc. (Lomartire et al. 2022). The physicochemical characteristics of seaweed phycocolloids are based on seaweed species, environmental parameters, extraction processes, and treatment techniques. Agar consists of agarose and agaropectin and is obtained from red algae (*Gracilaria* sp. and *Gelidium* sp.). The primary polysaccharide in agar, agarose, is a linear polymer composed of repeating units of 3,6-anhydro-L-galactose and D-galactose (Alba and Kontogiorgos 2019). Agarose is responsible for the gelling and filming properties of agar (Rhim 2011). With reduced tensile strength and water vapor permeability, agar produces thermo-reversible gels that are insoluble in cold water. The degree of methoxylation increases the gelation temperature of agar (Alba and Kontogiorgos 2019; Lomartire et al. 2022). Rheological properties such as lucidity and tension of agar gels could be enhanced by the incorporation of sugars (Bixler and Porse 2011). Agar films, on the other hand, are biologically inert, making them appropriate for elastic and soft gels. Due to their minimal hydroscopic characteristic, they are an excellent choice for packing materials (Lomartire et al. 2022). Carrageenans are mostly obtained from red algae, which are composed of galactose units that alternate between being sulfated and non-sulfated, connected by α-1,3-glycosidic and β-1,4-galactose linkages. The position of the sulfate group in relation to the galactose unit determines the classification of carrageenans into distinct types. All types of carrageenan including the three structural commercial carrageenans (κ-carrageenan, λ-carrageenan, and ι-carrageenan) are soluble in water. Carrageenan is soluble in water in the following way: λ-carrageenan is most soluble in comparison to ι-carrageenan, and κ-carrageenan is least soluble (λ- > ι- > κ-). The only carrageenan that dissolves in cold water is λ-carrageenan. *Kappaphycus alvarezii* and *Eucheuma denticulatum* are primary sources of κ- and ι-carrageenan, while *Gigartina skottsbergi* and *Sarcothalia crispata* provide λ-carrageenan (Alba and Kontogiorgos 2019). κ-carrageenan demonstrated higher tensile strength, elasticity, and moisture permeability in contrast to ι-carrageenan but it also displayed reduced opacity (Paula et al. 2015). When it comes to gelation, λ-carrageenan can serve as a thickening agent while not producing gel; nevertheless, at a concentration of 0.5%, κ-and ι-carrageenan can produce thermo-reversible gels. However, the gelling temperature and gel strength increase with cation concentration. While ι-carrageenan needs Ca^{2+} ions to generate soft and elastic gels, κ-carrageenan can form stiff and brittle gels in the presence of K^{+} ions (Alba and Kontogiorgos 2019). Despite their good gelling, thickening, and stabilizing properties, they are hydrophobic. It can be resolved by combining with hydrophobic compounds to reinforce the matrix (Sedayu et al. 2019). Alginates account for up to 40% of the dry weight in brown seaweeds. They are alginic acid polymers containing β-D-mannuronic acid (M) and α-L-guluronic acid (G) monomeric units connected by 1,4 linkages. For commercial alginate production, the main brown seaweed species are *Laminaria hyperborea*, *Laminaria digitata*, *Macrocystis pyrifera*, *Ascophyllum nodosum*, and *Laminaria japonica*. In addition to being great stabilizers and thickeners, alginates have the ability to produce edible films. The preferential binding of divalent and multivalent cations in alginate is a key physical characteristic that determines alginate's ability to form gels. Based on the M/G ratio and length of the chemical structure, the physicochemical and mechanical properties of gels differ. A higher M/G ratio results in stronger and more elastic gels, while heat-stable, strong, and brittle gels are the results of a low M/G ratio. Because of the hydrophilic nature of alginate, calcium is added to the alginate matrix for increased stability and membrane resilience (Lomartire et al. 2022; Alba and Kontogiorgos 2019). Green seaweeds (*Ulva lactuca*, *Ulva ohnoi*, *Ulva compressa*, *Cladophora pellucida*, etc.) produce ulvan (sulfated polysaccharide) comprising sulfate, xylose, rhamnose, and glucuronic acid. The inclusion of both hydrophobic and hydrophilic groups in ulvan contributes to its capacity to produce thermo-reversible films and is polyanionic in nature (Kidgell et al. 2019; Lim et al. 2021). Ulvan is water soluble and hygroscopic with low intrinsic viscosity and ion binding ability (Wang et al. 2023). The presence of rare sugars such as iduronic acid and sulfate rhamnose, which resemble mammalian glycosaminoglycans, is the most notable characteristic of ulvan. The majority of ulvan was harvested using traditional techniques. Nonetheless, it can be combined with other polymers to enhance the films' mechanical qualities (Kidgell et al. 2019; Lim et al. 2021) (Fig. 16.1).

16.3 Extraction Methods for Seaweed Hydrocolloids

Seaweed hydrocolloids are often extracted and processed into films to create seaweed bioplastics. There are several conventional extraction methods such as washing, hot water, drying, formaldehyde pretreatment, acid treatment, alkali treatment, and precipitation for the extraction of seaweed polysaccharides. However, the extraction procedures vary with the seaweed and polysaccharide application. The

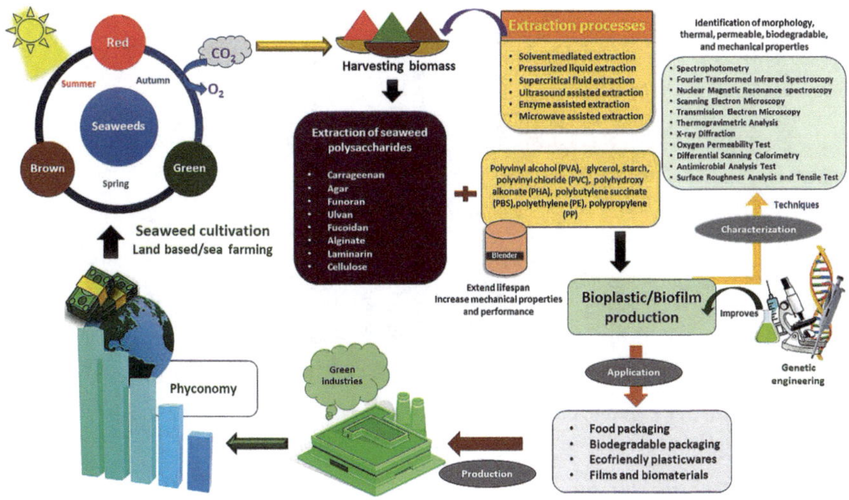

Fig. 16.1 Schematic representation of the bioplastic production from seaweeds. (Cinar et al. 2020; Lim et al. 2021; Lomartire et al. 2022)

extraction methods of different seaweed hydrocolloids are provided in Table 16.1. Among all these techniques, hot water is the most commonly used solvent for hydrocolloid (polysaccharide) extraction (Lim et al. 2021). Maceration is the next but extracted with water at high temperatures. The choice between alkali and acid pretreatment depends on the species and polysaccharide extraction (Lomartire and Gonçalves 2022). The agar and carrageenan are mainly extracted using hot water extraction followed by alkali treatment to improve gel-forming abilities since they show weak rheological properties on water extraction (Lomartire et al. 2022). However, the alginate is pretreated with formaldehyde and acid (hydrochloric acid) prior to alkali treatment. The formaldehyde pretreatment increases the alginate yield alongside removing pigments. Simultaneously, the acid pretreatment not only removes the formaldehyde residues and phenolic content but also converts the insoluble salt to a soluble form. Later, the alginate is neutralized and extracted with alkali treatment (Abdul Khalil et al. 2018). Similarly, ulvan is primarily pretreated with enzymes or acid followed by final extraction using hot water treatment. All the extracted hydrocolloids are finally precipitated, filtered, and dried, leaving the crude product. However, these extensively used conventional extraction methods have several drawbacks, including significant energy consumption, time, and expense. Besides, the chemical solvent utilized in conventional extraction methods has negative effects on the film made from seaweed hydrocolloids (Lim et al. 2021). Furthermore, in order to achieve the best yield, a huge quantity of chemical solvents is employed, which are not sustainable and pose significant risks to human health and the environment when discharged. Therefore, to overcome these limitations, "green extraction" methods using innovative technologies have been introduced that are cost-effective and eco-friendly efficient in replacing petrochemical-based

Table 16.1 Current status of bioplastic/active-film production from seaweeds

Seaweed species	Type of seaweeds	Component of seaweed used	Method of extraction	Plasticizer	Additional components added	Seaweed collection site	References
Kappaphycus alvarezii	Red seaweed	κ-carrageenan, refined carrageenan, semi-refined carrageenan	Alkali treatment	Glycerol (0.9% v/v)	Cellulose nanofibers from palm waste	TACARA Sdn.	Yahaya et al. (2023)
Ulva rigida	Green seaweed	Polysaccharide	Centrifugation	Glycerol, triethyl citrate	–	Phetchaburi, Thailand	Sonchaeng et al. (2023)
Rugulopteryx okamurae	Brown seaweed	–	–	Glycerol	–	IFAPA, Puerto Real, Spain	Santana et al. (2022)
Kappaphycus alvarezii	Red seaweed	Carrageenan	–	Polyethylene glycol (PEG-3000)	–	Mandapam, Rameshwaram, India	Sudhakar et al. (2021)
Ulva lactuca	Green seaweed	Polysaccharide	Acid extraction or enzymatic chemical extraction	Glycerol or sorbitol	–	Taboulba and Sayada (Monastir-Tuunisia)	Guidara et al. (2020)
Himanthalia elongata and *Palmaria palmata*	Brown seaweed and red seaweed	Polysaccharide	Centrifugation	Glycerol (0.5 g/g biopolymer)	Chitosan	Porto Muiños (Cerceda, A Coruña, Spain)	Albertos et al. (2019)
Ulva sp.	Green seaweed	Carbohydrates	Acid hydrolysis	Production of polyhydroxyalkanoate (PHA) by *Haloferax mediterranei*	–	Macroalgae photo-bioreactors (MPBR) (Polytiv, Israel)	Ghosh et al. (2019)

Eucheuma cottonii	Red seaweed	Polysaccharide	Cold distilled water extraction	–	Extracted polysaccharides from seeds of the avocado, jackfruit, and durian, and commercially available corn starch was also used	Traditional village cultivation in Wainyapu, Indonesia (South West Sumba)	Yusmaniar et al. (2019)
Sargassum latifolium	Brown seaweed	Alginate and fucoidan	Acidic extraction followed by alkali treatment	Glycerol (0.3 g/g alginate	$CaCl_2$	Intertidal zone of Hurghada, Egypt	Gomaa et al. (2018)
Acanthophora spicifera	Red seaweed	Sulfated polysaccharide	Centrifugation	20% and 30% glycerol	–	Gulf of Mannar, Tamil Nadu, India	Ganesan et al. (2018a)
Padina tetrastromatica	Brown seaweed	Fucoidan	Centrifugation	Sorbitol (30% of protein)	3% gelatin extracted from the swim bladder of rohu (*Labeo rohita*)	Hare Island, Gulf of Mannar, India	Govindaswamy et al. (2018)
Kappaphycus alvarezii and *Ulva fasciata*	Red seaweed and green seaweed	Semi-refined carrageenan	Alkali (KOH) extraction	Glycerol 30% (w/w)	–	Red seaweed from Gulf of Mannar and green seaweed from Mandapam, India	Ganesan et al. (2018b)

"–" Not available

synthetic plastics (Lomartire and Gonçalves 2022). Alternative extraction and processing methods include supercritical fluid extraction (SFE), reactive extrusion, pressurized solvent extraction (PSE), photobleaching, ultrasound-assisted extraction (UAE), microwave-assisted extraction (MAE), and enzyme-assisted extraction (EAE). Unlike conventional methods, these advanced green extraction approaches use sustainable solvents such as water, eutectic solvents, ionic liquids, and surfactants which minimize the use and discharge of hazardous chemicals (Abdul Khalil et al. 2018). Besides, other advantages include high yield and short extraction time, whereas they possess few limitations such as high equipment expenses and skills. However, there are very limited studies related to the application of the extraction of hydrocolloids from seaweed (Lim et al. 2021).

16.4 Seaweed Bioplastic Challenges and Scaling Up Strategies

Despite the promising results of algae-based bioplastics in the laboratory, commercialization and large-scale production of algae-based bioplastics face a few challenges. The main issue with seaweed-derived bioplastics is that they have poor water, alkali, or acid resistance when used alone (Lim et al. 2021). Second, although having a faster growth rate than terrestrial plants, the increasing demand for plastics does not meet the production requirement for seaweed-derived films, causing concerns because seaweed polysaccharides are needed for various other applications. Additionally, the seaweed cultivation system is the other issue regarding bioplastic production since the seaweed composition is highly influenced by several cultivation parameters (nutrients, temperature, light, season, and contamination), all of which have an impact on the properties of the film (Chia et al. 2020). Last but not least, the total cost of the films made from seaweed products is not sustainable economically. Unlike traditional extraction methods, green extraction methods have shown a significant potentiality (reproducible, quicker, and eco-friendly) for cutting manufacturing and final product costs. However, the most significant benefits of these revolutionary green technologies are possible only by overcoming obstacles such as high capital expenses and process stability in order to make them commercially feasible at the industrial scale (Lomartire and Gonçalves 2022; Abdul Khalil et al. 2018). Therefore, to overcome these aforementioned challenges, researchers are trying to find some alternatives (blenders and genetic engineering) to make seaweed-derived bioplastics industrially scalable. To extend the life of bioplastics, increase their mechanical performance, and improve their characteristics, different seaweed species can be blended or mixed with other materials. A variety of biodegradable materials (poly-butylene adipate-co-terephthalate (PBAT), polybutylene succinate (PBS), polylactides (PLA), poly-hydroxyalkonate (PHA), and starch blends) and nonbiodegradable materials (polyethylene (PE), polypropylene (PP), polyethylene terephthalate (PET), polyvinyl chloride (PVC), and polytrimethylene terephthalate (PTT)) can be implemented to increase the mechanical, surface, and dimensional stability of packaging products. In addition, a number of compatibilizers and plasticizers, such as glycerol, octanoic acid, 1,4-butanediol, carboxymethylcellulose (CMC), maleic anhydride (MA), potassium peroxide sulfate (KPS), and dimethyl sulfoxide (DMSO), can be added to enhance the bioplastics' processibility, water resistance, and flexibility (Cinar et al. 2020). Unlike plants, eukaryotic cells, and other complex organisms, algae have a simpler genetic makeup, which makes genetic engineering easier on them. Comparably, genetic engineering can also be an effective technique to overcome the issues with seaweed-derived bioplastics and boost the production of the required components by modifying genes. The modification of algal strains to synthesize components (such as polyhydroxy butyrate (PHB)) generated by bacteria for the manufacturing of bioplastics can be made possible by genetic engineering. However, there are certain constraints, including screening and isolation of desirable strains, a costly process, gene migration, stability, and applicability for industrial scale. Additionally, moral concerns about how foreign gene transfer from genetically modified organisms (GMOs) affects native species. Despite the risks of genetic manipulation, genetic engineering has immense promise for enhancing the production of polysaccharides or other biomolecules required for the production of bioplastics by modifying genes within a given algae strain (Chia et al. 2020). Furthermore, the employing of an "omics" (genomics, transcriptomics, and metabolomics) approach can be a significant tool for revealing new insights into the evolutionary and ecological patterns of seaweeds. Although there are few research publications available on optimizing the purification of high-quality nucleic acids, it appears to be a vital aspect for future seaweed genome studies. Currently, seaweed omics research is primarily restricted to pharmaceutical and nutraceutical applications; however, if breakthroughs are made in an equally balanced manner, researchers may be able to use this study to bridge the knowledge gap surrounding the production of bioplastics (Patwary et al. 2021).

16.5 Conclusion and Future Perspectives

Plastics made of petroleum might release harmful compounds even when they are disposed of properly. Moreover, most recyclable plastics are not currently ending up in recycling plants. According to statistics, 79% of all plastic trash ends up in landfills, which has exacerbated issues such as microplastic pollution and marine garbage. Furthermore, dioxins produced by burning plastic garbage have a negative effect on both the environment and human health (Geyer et al. 2017; Chia et al. 2020). Industries are required to maintain the quality, cost, and environmental friendliness of manufacturing processes into account in order to have sustainable production. Due to specific constraints, bioplastics are currently the only viable method for addressing this issue although they are still the subject of controversy and debate. The primary problem was the utilization of renewable resources, which are not sustainable for long-term use and compete with agriculture for food and feed. Second, only 1% of bioplastic is produced annually worldwide currently;

nevertheless, if the market grows to its full potential, bioplastic could be sold for $970 a tonne and simultaneously minimize greenhouse gas emissions by 116%. As a result, algae are a viable option for the biorefinery concept of sustainable bioplastics production since they can lower the long-term use of nonrenewable resources without competing with existing sources (Rajpoot et al. 2022; Chia et al. 2020; Trivedi et al. 2022b). Plastic problems may be resolved by using raw seaweed and environmentally friendly production techniques to create bioplastic (Lomartire et al. 2022). Several companies such as Marine Innovation, Evoware, B'Zeos, Notpla, C-Combinator, FlexSea, SWAY, Oceanium, SoluBlue, Loliware, Kelpi, and ZeroCircle are currently commercializing seaweed-based bioplastics (Tennakoon et al. 2023). In addition to exploring the potential of seaweed-derived bioactive compounds for the development of novel and natural drugs or healthy food products and supplements, the marine pharmacology and food industries are interested in using seaweed for biodegradable, safe, and hygienic packaging. Seaweed packaging preserves the products without changing their properties in an environmentally responsible manner. While seaweeds generate antimicrobial compounds, such as phenols, fatty acids, carbohydrates, proteins, and minor compounds, seaweed-derived biofilms may also display antibacterial capabilities in addition to their biodegradability trait. Furthermore, enhancing food shelf life and preventing the growth of foodborne pathogens could be achieved by including entire seaweed or seaweed extracts with antimicrobial activity in food packaging manufacturing (Lomartire et al. 2022). Despite all of these advantages, the majority of the ingredients (plasticizers, compatibilizers, and blenders) used for producing seaweed bioplastics are bio-based rather than fully biodegradable, which has a detrimental effect on the ecosystem. Consequently, scientists are considering using genetic engineering, which is still in its early stages, to solve this problem. Nevertheless, the development of technology and ongoing research and development in the field of bioplastics is important because the study of algae-based bioplastics is still in its experimental or early stages and is not possible to be commercialized at an industrial scale.

References

Abdul Khalil HPS, Lai TK, Tye YY et al (2018) A review of extractions of seaweed hydrocolloids: properties and applications. Express Polym Lett 12:296–317

Alba K, Kontogiorgos V (2019) Seaweed polysaccharides (agar, alginate, carrageenan). In: Encyclopedia of food chemistry. Elsevier, Amsterdam, pp 240–250

Albertos I, Martin-Diana AB, Burón M et al (2019) Development of functional bio-based seaweed (*Himanthalia elongata* and *Palmaria palmata*) edible films for extending the shelflife of fresh fish burgers. Food Packag Shelf Life 22:100382

Aleksanyan KV (2023) Polysaccharides for biodegradable packaging materials: past, present, and future (Brief Review). Polymers 15(2):451

Bixler HJ, Porse H (2011) A decade of change in the seaweed hydrocolloids industry. J Appl Phycol 23:321–335

Chia WY, Ying Tang DY, Khoo KS et al (2020) Nature's fight against plastic pollution: algae for plastic biodegradation and bioplastics production. Environ Sci Ecotechnology 4:100065

Cinar SO, Chong ZK, Kucuker MA et al (2020) Bioplastic production from microalgae: a review. Int J Environ Res Public Health 17(11):3842

Ganesan AR, Shanmugam M, Palaniappan S et al (2018a) Development of edible film from *Acanthophora spicifera*: structural, rheological and functional properties. Food Biosci 23:121–128

Ganesan AR, Shanmugam M, Bhat R (2018b) Producing novel edible films from semi refined carrageenan (SRC) and ulvan polysaccharides for potential food applications. Int J Biol Macromol 112:1164–1170

Geyer R, Jambeck JR, Law KL (2017) Production, use, and fate of all plastics ever made. Sci Adv 3(7):e1700782

Ghosh S, Gnaim R, Greiserman S et al (2019) Macroalgal biomass subcritical hydrolysates for the production of polyhydroxyalkanoate (PHA) by *Haloferax mediterranei*. Bioresour Technol 271:166–173

Gomaa M, Fawzy MA, Hifney AF et al (2018) Use of the brown seaweed *Sargassum latifolium* in the design of alginate-fucoidan based films with natural antioxidant properties and kinetic modeling of moisture sorption and polyphenolic release. Food Hydrocoll 82:64–72

Govindaswamy R, Robinson JS, Geevaretnam J et al (2018) Physico-functional and anti-oxidative properties of carp swim bladder gelatin and brown seaweed fucoidan based edible films. J Packag Technol Res 2:77–89

Guidara M, Yaich H, Benelhadj S et al (2020) Smart ulvan films responsive to stimuli of plasticizer and extraction condition in physico-chemical, optical, barrier and mechanical properties. Int J Biol Macromol 150:714–726

Kidgell JT, Magnusson M, de Nys R et al (2019) Ulvan: a systematic review of extraction, composition and function. Algal Res 39:101422

Lim C, Yusoff S, Ng CG et al (2021) Bioplastic made from seaweed polysaccharides with green production methods. J Environ Chem Eng 9:105895

Lomartire S, Gonçalves AMM (2022) Novel technologies for seaweed polysaccharides extraction and their use in food with therapeutically applications—a review. Food Secur 11(17):2654

Lomartire S, Marques JC, Gonçalves AMM (2022) An overview of the alternative use of seaweeds to produce safe and sustainable bio-packaging. Appl Sci 12:3123

Patwary ZP, Paul NA, Nishitsuji K et al (2021) Application of omics research in seaweeds with a focus on red seaweeds. Brief Funct Genomics 20(3):148–161

Paula GA, Benevides NMB, Cunha AP et al (2015) Development and characterization of edible films from mixtures of κ-carrageenan, ι-carrageenan, and alginate. Food Hydrocoll 47:140–145

Rajpoot AS, Choudhary T, Chelladurai H et al (2022) A comprehensive review on bioplastic production from microalgae. Mater Today Proc 56:171–178

Rhim J-W (2011) Effect of clay contents on mechanical and water vapor barrier properties of agar-based nanocomposite films. Carbohydr Polym 86(2):691–699

Santana I, Félix M, Guerrero A et al (2022) Processing and characterization of bioplastics from the invasive seaweed *Rugulopteryx okamurae*. Polymers 14(2):355

Sedayu BB, Cran MJ, Bigger SW (2019) A review of property enhancement techniques for carrageenan-based films and coatings. Carbohydr Polym 216:287–302

Sharma R, Mondal AS, Trivedi N (2023) Anticancer potential of algae-derived metabolites: recent updates and breakthroughs. Futur J Pharm Sci 9:44

Sonchaeng U, Wongphan P, Pan-utai W et al (2023) Preparation and characterization of novel green seaweed films from *Ulva rigida*. Polymers 15(16):3342

Sudhakar MP, Peter DM, Dharani G (2021) Studies on the development and characterization of bioplastic film from the red seaweed (*Kappaphycus alvarezii*). Environ Sci Pollut Res 2021(28):33899–33913

Tennakoon P, Chandika P, Yi M et al (2023) Marine-derived biopolymers as potential bioplastics, an eco-friendly alternative. iScience 26(4):106404

Trivedi N, Sharma R, Mondal AS et al (2022a) Recent advances in biotechnology of seaweeds: an overview. In: Ranga Rao A, Ravishankar GA (eds) Sustainable global resources of seaweeds, vol 1. Springer, Cham, pp 627–644

Trivedi N, Mondal AS, Sharma R et al (2022b) Marine macroalgal biorefinery: recent developments and future perspectives. In: Bhatia SK, Mehariya S, Karthikeyan OP (eds) Algal biorefineries and the circular bioeconomy, vol 1. CRC Press, Boca Raton, pp 1–36

Wang H, Cao Z, Yao L et al (2023) Insights into the edible and biodegradable ulvan-based films and coatings for food packaging. Food Secur 12(8):1622

Yahaya WAW, Azman NAM, Adam F et al (2023) Exploring the potential of seaweed derivatives for the development of biodegradable plastics: a comparative study. Polymers 15(13):2884

Yusmaniar Y, Syafei DI, Arum M et al (2019) Preparation and characterization of seaweed based bioplastic blended with polysaccharides derived from various seeds of Avocado, Jackfruit and Durian. J Phys: Conf Ser 1402:055097

Zhang L, Liao W, Huang Y et al (2022) Global seaweed farming and processing in the past 20 years. Food Prod Process Nutr 4:23

GPSR Compliance

The European Union's (EU) General Product Safety Regulation (GPSR) is a set of rules that requires consumer products to be safe and our obligations to ensure this.

If you have any concerns about our products, you can contact us on ProductSafety@springernature.com

In case Publisher is established outside the EU, the EU authorized representative is:

Springer Nature Customer Service Center GmbH
Europaplatz 3
69115 Heidelberg, Germany

Batch number: 08158389

Printed by Printforce, the Netherlands